T0186415

Biology and Ecology of Carp

Biology and Ecology of Carp

Editors

Constanze Pietsch

Zurich University of Applied Sciences (ZHAW)
Institute of Natural Resource Sciences (IUNR)
Waedenswil
Switzerland

Philipp E. Hirsch

Department of Environmental Sciences
University Basel
Basel
Switzerland

CRC Press
Taylor & Francis Group
Boca Raton London New York

CRC Press is an imprint of the
Taylor & Francis Group, an **informa** business

A SCIENCE PUBLISHERS BOOK

Cover photograph reproduced by kind courtesy of Mr. Michel Roggo.

CRC Press
Taylor & Francis Group
6000 Broken Sound Parkway NW, Suite 300
Boca Raton, FL 33487-2742

First issued in paperback 2019

© 2015 by Taylor & Francis Group, LLC
CRC Press is an imprint of Taylor & Francis Group, an Informa business

No claim to original U.S. Government works

ISBN-13: 978-1-4822-0664-7 (hbk)
ISBN-13: 978-0-367-37756-4 (pbk)

Library of Congress Cataloging-in-Publication Data

Biology and ecology of carp / Constanze Pietsch, Philipp E. Hirsch, editors.
    pages cm
    "A CRC title."
    Includes bibliographical references and index.
    ISBN 978-1-4822-0664-7 (hardcover : alk. paper) 1. Carp. 2. Carp--Ecology. 3.
    Carp--Physiology. I. Pietsch, Constanze, editor. II. Hirsch, Philipp (Philipp E.) editor.

QL638.C94B53 2015
597'.482--dc23
                                                    2014044906

Visit the Taylor & Francis Web site at
http://www.taylorandfrancis.com

and the CRC Press Web site at
http://www.crcpress.com

Preface

Carp are unlike other group of fish species. In fact, they are special to such an extent, that we feel a whole book dedicated to all aspects of the biology and ecology of carp is warranted.

While justifying the many aspects of carp in only a single volume may seem overambitious to some others might wonder whether there is really so much to say about a single fish. Despite this being a daunting task, we are confident that this book offers something to learn for the expert and the novice alike.

The introduction of the book first provides a detailed account on the natural history of carp and the genetics of carp (Chapter 1). Here we learn how humans from earlier on shaped the genetic identity of carp through aquaculture activities and how today we can differentiate several races or strains of carp. Artificial selection for desirable traits is not only common in aquaculture: Japanese ornamental koi with their different color varieties are also subject to selective breeding and such elaborate techniques as scale-transplantations are performed to perfect special color types (Chapter 2). Exactly how carp are bred and crossed artificially is discussed in great detail in the aquaculture and nutrition part of the book. This part constitutes the backbone of the volume providing in-depth knowledge on both state-of-the art techniques and the latest scientific advances in the field of carp aquaculture. Proceeding from a presentation of the history of carp aquaculture in Europe and Asia the reader learns about classical and high-end forms of aquaculture today (Chapter 3). Chapter 4 discusses the earliest life stages of carp—larvae —and how they are most efficiently produced and raised following artificial or natural reproduction in aquaculture . Interestingly, whether carp offspring becomes male or female is all but predetermined and in aquaculture today manipulation of the sex is a common tool presented in Chapter 5. Regardless of sex, carp in aquaculture will have to be fed optimally and sustainably therefore Chapter 6 presents both traditional ways of feeding carp and topical alternatives at hand to replace fish meal.

This volume also takes a close look at carp in the part parasitology and immunology. Especially relevant for the aquaculture industry are parasites and disease agents of carp that are described in Chapter 7. Chapter 8 follows the parasites and disease agents by taking a more detailed look into molecular mechanisms behind a carp's immune responses to pathogens.

The part on ecology describes how carp interact with their natural environment. In Chapter 9 the feeding ecology of carp, details are given on how carp feed in the wild and how this relates to the effects that carp can have on the ecosystem. Intriguingly, carp can have profound influence of their natural environments and when introduced can proliferate in the wild outside their native range. The associated management implications are also discussed in a comprehensive chapter on carp as invasive species (Chapter 10). While considered a nuisance by some, carp are a highly desired by others as game fish in recreational fishing. Chapter 11 introduces carp fishing and presents detailed data on growth and management of natural carp populations. In the last part of the book, toxicology combines both, the inside-view on carp from the immunology and parasites part, and the outside-view on carp interactions with their environment. Chapter 12 presents the effects of pesticides on carp and Chapter 12 reviews the effects of natural toxins from, e.g., fungi on carp, highlighting the importance of carp as a model organism.

The volume in all its breadth and detail lives up to a comprehensive treatise of one of the most fascinating creatures that man has lived with for thousands of years. We hope the reader will find the content and its presentation as inspiring as the authors' commitment to an outstanding book.

Constanze Pietsch
Philipp E. Hirsch

Contents

Part IV: Ecology

Part V: Toxicology

The Editors

Dr. Constanze Pietsch enthusiasm for fishing started very early in her childhood due to recreational fishing activities of her father and brothers. Due to her deep interest natural sciences it was not astounding that she chose to study biology at the Humboldt-University in Berlin in Germany. However, her studies focused on plant physiology, freshwater ecology and aquatic toxicology before she deepened her knowledge in fish physiology, endocrinology and immunology during her PhD using common carp as the model species at the Institute of Freshwater Ecology and Inland Fisheries in Berlin. After obtaining her PhD in 2008 she moved to the Program Man-Society-Environment as a post-doc at the University of Basel in Switzerland, where she further worked with carp to elucidate possible effects of natural toxins including mycotoxins and anthropogenic toxic chemicals on cell functions, fish nutrition, immunology and fitness. Constanze Pietsch is currently working as a scientific researcher in the Aquaculture and Aquaponics Group at the Zurich University of Applied Sciences in Waedenswil, Switzerland, where her work focuses on fish nutrition. She also leads the educational course for future aquaculturists and teaches in environmental analytics. Her broad knowledge in many aspects of fish and aquaculture alleviated her participation as an editor for the present book in which she mostly coordinated and edited the contributions to the biological aspects of carp.

Dr. Philipp E. Hirsch is an evolutionary ecologist with a background in fish ecology. His interest in carp was first raised culinarily when preparing an annual christmas carp with his father when he was a child. He followed his fascination for fish ecology which led him through three universities in Germany and after graduating as a master in biology at the University of Constance he continued his education in Sweden at the University of Uppsala. His work expanded from classical fish ecological questions to invasive species and phenotypic processed triggered in native and invading fish populations. After he obtained his PhD in 2011 he continued to work on fish ecology and invasive species from both basic and applied aspects within the research centre for sustainable energy and water supply (FoNEW) and the Program Man-Society-Environment as a post-doc at the University of Basel in Switzerland. His strong interest and experience in fish ecology and invasive species complements Constanze Pietsch's competence on the more molecular side of carp biology. He assembled the chapters for the ecology part of the book. Not least because he is a passionate angler and educated in the ecology of species invasions he oversaw the inclusion of both the aspect of recreational fishing and invasive carp into the volume.

List of Contributors

Dr. Pia Bartels is an ecologist with a broad interest in community, ecosystem and evolutionary ecology. Her current research focuses on the impacts of rapid ecological and evolutionary changes on community and ecosystem dynamics. She is also interested in the coupling of spatially separated food webs on different scales and in the effects of different environmental variables on food web coupling.

Dr. Jasminca Behrmann-Godel is an evolutionary ecologist. Combining the scientific fields of ecology, animal behavior, parasitology, population genetics and molecular evolution, she aims to study interactions between individuals and their natural environment. Thereby she uses aquatic animals, mainly regional freshwater fish species as model organisms.

Servaas de Kock is an expert koi culturist who, for many years, has supplied show-quality fish to koi-keepers, dealers and breeders around the world. He has a grasp of the many facets of this subject and contributed to numerous publications in the field.

Dr. vet. med. Radka Dobsikova graduated from the Faculty of Veterinary Hygiene and Ecology of the University of Veterinary and Pharmaceutical Sciences Brno, Czech Republic in 1998 and then participated in a doctor study program in ecology, ecotoxicology and radiobiology resulting in a PhD in 2004. Since then, she works as an assistant professor at the Department of Veterinary Public Health and Animal Welfare of the FVHE UVPS Brno. Her work focuses on aquatic toxicology, where she evaluates the effects of chemical substances, preparations and pharmaceuticals on the organisms of aquatic environment. She is a member of research projects in the field of aquatic toxicology and ecotoxicology.

Dr. Boris Gomelsky has 35 years of fish genetics research experience in the United States, Israel and Russia. The major areas of his specialization are chromosome set manipulation methods, sex control, inheritance of morphological traits, distant hybridization and cytogenetics of fish reproduction. Currently Dr. Gomelsky is a Professor in the Aquaculture Division at Kentucky State University where, in the last 15 years, he has taught classroom and online courses in fish genetics and fish reproduction.

Dr. Gert Füllner was born in 1958 in Weißenberg, a little town in Saxony (Germany). He studied Aquaculture and Fisheries Sciences at Humboldt-University in Berlin from 1978 to 1983. In 1983 he started his work in fish health service in a commercial fish farm in Königswartha (Germany). From 1984 he had changed as scientist to the Institute for Inland Fisheries in Berlin-Friedrichshagen. Since 1991 he is the Chief of the Fisheries Department in Saxon State Institute for Agriculture, today's State Agency for Environment, Agriculture and Geology in Königswartha. Gert Füllner works amongst others, on the fields of carp pond farming, restoring of Atlantic salmon in Elbe River and its tributaries, and fisheries management of new lakes that arose from former brown coal mines in Saxony. For many years he is assistant professor for the lecture "Management of Warmwater Ponds and Fish Breeding" at Humboldt-University in Berlin, which includes the field of carp farming too.

Dr. Pavel Hartman graduated from the Mendel University in Brno (former Czechoslovakia) in 1967 at Agriculture Faculty with specialization on fishery and aquaculture. Since 1971 Dr. Hartman worked in the Central Laboratory for Chemistry of the formerly state-owned Fishery and Aquaculture Enterprise. His dissertation was devoted to chalking requirements of carp ponds in relation to hydrochemical and environmental conditions. Since 1992 to 2010 he overtook directorship of carp farming company owned by city of Ceske Budejovice. These days he works at University of South Bohemia as an expert for water quality parameters in pond aquaculture.

Prof. Dr. Brendan J. Hicks is a Professor in the School of Science at the University of Waikato in Hamilton, New Zealand. In his 35-year career as a freshwater ecologist his research has covered the influence of forestry on fish in streams, stable isotopes and aquatic food webs, and otolith microchemistry. Most recently he has focused on the ecology and management of invasive fish, which in New Zealand include common carp. Brendan has pioneered the use of boat electrofishing to quantify freshwater fish communities in New Zealand.

Dr. Brian Huser was a water resources scientist at Barr Engineering before moving to the Department of Aquatic Sciences and Assessment at the Swedish University of Agricultural Sciences. He is a limnologist who is interested in the effects of climate, pollution, and biota on the cycling of nutrients and metals in surface waters and sediment. Brian was a Fulbright Scholar in 2001/2002.

Dr. Natalia Ivonne Vera-Jimenez is a fish immunologist with emphasis on innate immunity. She has worked with Toll-like receptors, immunomodulation, DAMP and PAMP immune recognition, cell production of reactive oxygen species and wound healing in fish. She studied Marine Biology at the Jorge Tadeo Lozano University (Colombia). Later did a Master in Animal Sciences at Wageningen Universisty (The Netherlands), followed by a PhD in Fish Immunology and Wound Healing at The Technical University of Denmark

(Denmark). Currently she is working on the establishment of new *in vivo* and in vitro fish models for the evaluation of immunotoxicity as a postdoc at the Danish National Food Institute (DTU Food).

Dr. Klaus Kohlmann is scientist at the Department of Ecophysiology and Aquaculture at the Leibniz-Institute of Freshwater Ecology and Inland Fisheries, Berlin, Germany (www.igb-berlin.de). His research is focused on genetic aspects of aquaculture and the characterization and evaluation of aquatic genetic resources. The studies on common carp dealt with the genetic variability and differentiation of wild populations and cultured stocks as well as the phylogeography of the species applying various marker types (allozymes, microsatellite loci, mtDNA).

Dr. Michael Engelbrecht Nielsen has through the last decade been working with the teleost immunssystem. The focus has been in relation to the interplay between production challenges and the response from the immune system. Involving research on both innate and adaptive immune responses as well as modulation of these. Present attention relates to tissue regeneration and immunotoxicology.

Dr. Nicholas Ling is an Associate Professor in the School of Science at the University of Waikato. His research includes fish physiology, specifically stress and exercise physiology, comparative haematology and the effects of pollutants on fish health. He also contributes to the University of Waikato's research into the management of invasive fish, and has published on the use of rotenone to control invasive fish populations.

Dr. Henrik Ragnarsson-Stabo is a researcher at the Institute of Freshwater Research at the Swedish University of Agricultural Sciences. His research interests include ecosystem analysis, fishery management and fish ecology. He has a PhD in Limnology, focused on the biodiversity of fish communities, at Uppsala University.

Gregor Schmidt studied Agriculture (B.Sc.) and Aquaculture and Fishery Sciences (M.Sc.) at the Humboldt-University in Berlin (Germany) from 1998 to 2005. From 2005 to 2011 he worked as a scientific assistant at the Bavarian State Research Center for Agriculture, Institute for Fisheries, in Starnberg (Germany). There he worked on different projects for reproduction, aquaculture, product quality and welfare of many cultivated European freshwater fish species (cyprinids, salmonids, and percids). Since 2012 he has been a scientist at the Institute for Fisheries of the State Research Center for Agriculture and Fisheries in Mecklenburg–Vorpommern, Germany where he is managing a pilot farm for pikeperch. The aim of this project is the development of a sustainable pikeperch aquaculture in recirculation systems.

Dr. Josef Velisek is an Associate Professor in the field of aquatic toxicology at the Faculty of Fisheries and Protection of Waters, University of South Bohemia in Ceske Budejovice. His research activities during the past 10 years included evaluation of lethal and sublethal hematological, biochemical, and physiological responses of fish and crayfish to environmental pollutants, mainly pesticides. A second research activity is fish anesthesia.

Dr. Geert Frits Wiegertjes heads the fish health and immunology unit of the Cell Biology and Immunology group at Wageningen University, The Netherlands. He combines state-of-the-art techniques developed at technological zebrafish platforms with immunological research on fish species of relevance to aquaculture such as common carp, leading to fundamental but strategic research on immune responses to infectious agents such as parasites and viruses, to improve fish health.

Part I

Genetics of Common Carp and Koi Carp

Genetics is the underlying framework of life. Recent research has revealed that the genetic identity of common carp (*Cyprinus carpio* L.) is essential for future breeding and conservation purposes, and it is necessary to investigate and conserve genetically determined species properties. This is especially important for valuable properties of genetically pristine less domesticated carp strains which can be used for cross-breeding with aquaculture carp strains to improve their growth performance or phenotypic appearance. Recent research allowed mapping the special genetic properties of carp by using advanced high-throughput research tools. The results obtained revealed that carp are outstanding cyprinids with respect to their karyotype. Genetic analyses also allowed unraveling the genetic distance and relationships between different carp strains. Understanding the basic genetic basis of certain phenotypic characteristics of carp allows fish breeders to more actively influence the outcome of crossing carp strains with different properties. The latter is certainly important in aquaculture where improvements in growth performance without an increase in feed conversion ratio are desired. Genetics of carp is the central theme of the first chapter of this book while fish nutrition and manipulation of chromosomes to yield mono-sex populations are reviewed in the second part of this book focusing on aquaculture of common carp. During the last few decades hobbyists have been fascinated by koi carp, an ornamental form of common carp. Koi carp has also attracted the attention of professionals on account of their genetic traits. The second chapter of this book accounts for this growing interest in koi carp by providing detailed information on the genetic background of koi carp.

1

The Natural History of Common Carp and Common Carp Genetics

Klaus Kohlmann

Introduction

The common carp, *Cyprinus carpio* L., belongs to the family Cyprinidae, carps and minnows, which have over 2,400 species in 220 genera (Nelson 2006) and represents the largest family of freshwater fishes in the world.

The aim of this chapter is to provide a review of information on the natural history of the species and its genetics. The parts on natural history cover basic aspects such as the DNA content of cells and their karyotype. Hypotheses on the evolution of common carp including the time and place of its origin, the division into subspecies and its natural distribution range are discussed. Special attention is paid to the controversial debate on the origin and spread of European common carp. The parts on common carp genetics describe the genetic bases of its scale cover and body colouration. Different types of genetic markers and their application in studies on the status of subspecies, population structure, phylogeny and phylogeography are reviewed. Finally, a perspective of common carp genetic research in the light of recent developments in next generation sequencing technologies is given.

Leibniz-Institute of Freshwater Ecology and Inland Fisheries, Department of Ecophysiology and Aquaculture, Müggelseedamm 310, 12587 Berlin, Germany.
Email: kohlmann@igb-berlin.de

Infobox: Terminology in genetic analyses

The inherited **genotype** together with epigenetic factors and non-hereditary environmental variation contributes to the **phenotype** of an individual. Each gene can express a number of alternative forms which are called **alleles**. The combination of alleles of genes at adjacent locations on a chromosome that are inherited together forms a **haplotype**. Diploid individuals have two alleles at a given gene. If these alleles are identical the individual is **homozygous**, if they are not the individual is **heterozygous**. **Homozygosity** refers to the amount of homozygous genes, **heterozygosity** to the amount of heterozygous genes.

Pleiotropy occurs if a gene influences multiple, seemingly unrelated phenotypic traits at the same time. In fish, for example, pleiotropic effects of genes controlling scale patterns or body coloration on morphology, physiology and productivity have been observed.

The number, size and shape of **chromosomes** determine an individual's **karyotype**. Each chromosome consists of two sister chromatids that are linked at the **centromere**. The position of the centromere defines their classification as meta- or submetacentric (= the centromere is located in or near the middle of the chromosome), or telo- or subtelocentric (= the centromere is located at or near the end of the chromosome). Karyotypic information can help to identify **genome duplications** which occurred in several fish families.

Phylogeny aims to reconstruct the history of organismal lineages (e.g., species or groups of species) as they changed through time (evolution) and space (phylogeography). It implies that the different species arose from previous, common ancestral forms via descent, and that all organisms are connected by the passage of genes from one generation to the next. Accordingly, the natural history, geographical origin and population structure of fish can be analyzed by molecular biological methods.

Methods used for genetic analyses:

The prerequisite for genetic analyses is the presence of **polymorphisms** (i.e., different variants) in genes (i.e., the DNA sequences) or in their products (i.e., the proteins). In case of **DNA-based markers** the **polymerase chain reaction (PCR)** is the central method that allows to amplify the DNA sequence of interest. PCR products can then be used for a wide range of downstream applications such as fragment lengths determination, digestion by restriction enzymes or direct sequencing.

Currently, **microsatellite loci** and—increasingly—**single nucleotide polymorphisms (SNPs)** are the genetic markers of choice to examine the structure and relationships of populations or the consequences of captive breeding and interactions between wild and farmed populations. In contrast, polymorphisms of the **mitochondrial DNA (mtDNA)** are preferred in phylogenetic studies due to its maternal inheritance, rapid mutation rate and lack of recombination. The importance and use of other genetic markers (**random amplification of polymorphic DNA [RAPD]**, **amplified fragment length polymorphisms [AFLPs]**, and **restriction fragment length polymorphisms [RFLPs]**) is steadily decreasing.

Natural History of Common Carp

The DNA content of haploid common carp cells varies between 1.61 and 2.03 pg depending on the method of measurement (biochemical analysis, bulk fluorometric assay, Feulgen densitometry, flow cytometry), the type of cells examined (liver, red blood cells, sperm), and the species used as standard (*Bufo terrestris, Cricetus cricetus, Gallus domesticus, Homo sapiens, Mus musculus, Strongylocentrotus purpuratus*) (Gregory 2013). A typical common carp karyotype consists of 2n = 100 chromosomes (Fig. 1.1). Based on the position of the centromere about half of them can be classified as meta- or submetacentric (= the centromere is located in or near the middle of the chromosome), the remaining are telo- or subtelocentric (= the centromere is located at or near the end of the chromosome) (Kirpitchnikov 1999). In rare cases 98, 102 or 104 chromosomes are reported (Klinkhardt et al. 1995). Although morphologically differentiated sex chromosomes do not exist (Kirpitchnikov 1999), genetic sex determination generally follows the XY system (Guerrero-Estévez and Moreno-Mendoza 2010) with influence of a recessive autosomal mutation (*mas*-1) that causes female XX individuals homozygous for this mutation to undergo masculinization (Komen et al. 1992).

Since the number of its chromosomes is twice as high as in most other cyprinids (Klinkhardt et al. 1995), common carp has been assumed to be of tetraploid origin. Recent results of a comparative mapping study between common carp and zebrafish (*Danio rerio*, 2n = 50 chromosomes) based on a common carp consensus linkage map with 732 markers (627 microsatellite loci and 105 Single Nucleotide Polymorphisms [SNPs]) constructed by Zhang et al. (2013a) using their own linkage map and two maps previously constructed by Zheng et al. (2011) and Wang et al. (2012a) confirmed this hypothesis. Additional evidence for common carp tetraploidy is provided by the fact that many loci are expressed in duplicate. Ferris and Whitt (1977) estimated the proportion of still expressed duplicated enzyme loci in common carp to be about 52%. Similar values (48 and 60%, respectively) were found by Zhang et al. (2008a) and David et al. (2003) for duplicated common carp microsatellite loci.

Two karyotypic observations suggest that the common carp tetraploidy is a result of allotetraploidization (= species hybridization) rather than of autotetraploidization (= genome doubling): no chromosomes seem to have been lost in the tetraploidization event, and no chromosome quadrivalents had been observed in meiotic nuclei (Ohno et al. 1967), indicating that the two original diploid chromosome sets were already different from each other at the time of tetraploidization. In addition, the phylogeny of Cyprinids fits allotetraploidy because the clades of diploid and tetraploid species coalesce before the divergence of the diploid parents of the common carp (David et al. 2003).

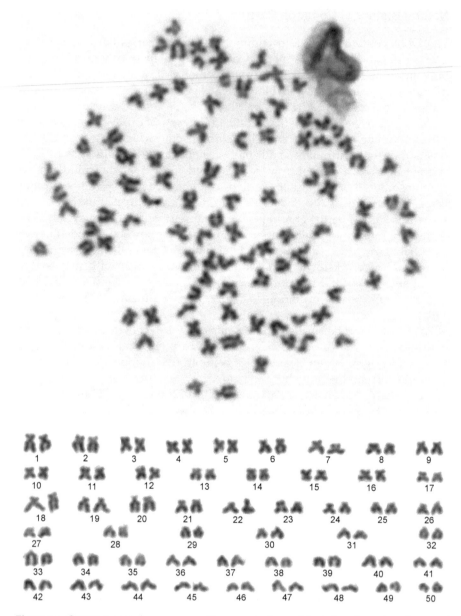

Figure 1.1. Common carp chromosomes. Above: metaphase plate, below: karyotype (arranged by M. Debowska, Dept. of Ichthyology, UWM Olsztyn, Poland, according to Anjum et al. 1997, Anjum and Jankun 1998).

As the catostomid fishes (suckers) of North America represent an entire family derived from a tetraploidization event 50 million years ago (Uyeno and Smith 1972) and due to their similar levels of still expressed duplicate enzyme loci (about 50%; Ferris and Whitt 1979), the common carp tetraploidy was assumed to be of a similar age (Kirpichnikov 1981). Indeed, this estimate was supported by a study of Zhang et al. (1995) who compared the two c-*myc* genes of common carp and found a 94.2% amino acid identity of the two deduced peptides which suggests that the tetraploidization event occurred 58 million years ago. In contrast, sequence analyses of other genes performed by Larhammar and Risinger (1994) indicated that duplicate common carp loci have been diverging for some 16 million years only. Assuming that the genome doubling arose through allotetraploidization, this means that the two diploid donor species started to diverge from each other 16 million years ago. The fusion of their genomes that gave rise to the tetraploid common carp thus occurred more recently than this, but the exact time of the tetraploidization event could not be deduced from these sequence comparisons (Larhammar and Risinger 1994). Studying the sequences of microsatellite flanking regions David et al. (2003) later on also estimated a more recent common carp genome duplication that had occurred about 12 million years ago. The differences between all of these estimates might be caused by the small data sets they were based on. The development of high-throughput sequencing (or next generation sequencing) technologies made—among others—the transcriptome more easily accessible. Transcriptome analysis is a powerful tool that has been used to study various genome features including genome duplication in a plant species (Srivastava et al. 2011). Using this tool Wang et al. (2012b) were able to identify one round of genome duplication in common carp and estimated that it had occurred between 5.6 and 11.3 million years ago. So far the lowest estimate of 5.6 million years fits very well with the lower time limit for the tetraploidization event set by the fossil record which describes the genus *Cyprinus* in late Miocene approximately 5 million years ago (Cavender 1991).

Both the divergence time and the proportion of duplicates in common carp place it among the few vertebrates in which a recent genome duplication occurred (David et al. 2003). This relatively young duplication of the genome is additional to the duplications that are proposed for all vertebrates and is also additional to that preceding the teleost fish radiation (Amores et al. 1998, Taylor et al. 2001, Wolfe 2001).

The natural distribution range of common carp is vast. Its western part includes Eastern Europe, Turkey, some fluvial basins of the Transcaucasus, the Aral Sea, the Syr-Darya and Amu-Darya Rivers and their tributaries, Lakes Zaisan and Issik Kul, together with some small lakes in Central Asia and Kazakhstan. The eastern part of its range comprises the Amur River, numerous rivers and lakes in China and Vietnam, together with the adjacent countries of South Asia (Berg 1964, Kirpitchnikov 1967). On the other hand, it is unlikely that wild common carp ever existed in Indonesia; the numerous varieties of Indonesian carp are apparently domesticated forms of the common carp introduced from China or Vietnam in the 17th century or earlier and from Europe in the 20th century (Steffens 1980).

Early studies of morphological differences (mainly in the number of vertebrae and scales on the lateral line; Kirpitchnikov 1943, 1967) between

European and Amur River common carp led to the establishment of two subspecies: *Cyprinus carpio carpio*, found throughout Europe, the Caucasus and Central Asia, and *Cyprinus carpio haematopterus*, the Far Eastern subspecies inhabiting the Amur basin and lakes and rivers of South-Eastern China (Berg 1964, Svetovidov 1933, Nikolsky 1956, both cited in Kirpitchnikov 1999). This taxonomic classification was later on strongly supported by biochemical and molecular-genetic studies on the evolution, phylogeography and population structure of wild and domesticated common carp that have been carried out both in Europe (e.g., Czech Republic, France, Germany, Hungary, Poland, and Russia) and Asia (e.g., China, Japan, and Vietnam) within the last two decades (for reviews see Chistiakov and Voronova 2009, Gui and Zhu 2012, Vilizzi 2012). Additionally, the existence of a third subspecies *Cyprinus carpio rubrofuscus* (synonymous to *Cyprinus carpio viridiviolaceus*, http://fishbase. org) that might have diverged from *Cyprinus carpio haematopterus* in China has been suggested based on polymerase chain reaction amplified restriction fragment length polymorphisms (PCR-RFLPs) of the mitochondrial *ND5/6* region (Zhou et al. 2003a) and sequence analysis of the mitochondrial cytochrome *b* and control regions (Zhou et al. 2004a) using identical common carp strains. The distribution of this proposed subspecies is mainly restricted to the south of the Nanling Mountains in China and Vietnam (Wu 1977). However, in their phylogenetic study using mitochondrial *COII* and D-loop sequences Wang et al. (2010a) found that the Nanling Mountains did not form a border separating *Cyprinus carpio rubrofuscus* from *Cyprinus carpio haematopterus*. They concluded that human-mediated translocations have confounded their ability to identify subspecies of common carp in China. Thus, the validity of this third subspecies remains questionable.

The geographical origin of the species *Cyprinus carpio* and in particular its European subspecies *Cyprinus carpio carpio* has also been a subject of long lasting and controversial debates. For example, Balon (1995) hypothesized that the ancestral form with a greater number of pharyngeal teeth probably evolved in the area of the Caspian Sea at the end of the Pliocene, i.e., no longer than 2.6 to 3.6 million years ago. However, more recent studies of mitochondrial DNA sequences identified the wild population inhabiting the ancient Lake Biwa in Japan as the most basal lineage of present-day common carp examined so far (Mabuchi et al. 2005, 2006). The estimated main split between Lake Biwa and all other Eurasian common carp appeared to be approximately 1.7 to 2.5 million years ago (Mabuchi et al. 2005). However, this estimated coalescent time is much older than the oldest fossil record of the species (0.5 million years ago from Paleo-Lake Biwa). This indicates that the fossil population could not be the direct ancestor of common carp: although the fossil population might be the direct ancestor of the current wild population of Lake Biwa, coalescence of the mitochondrial DNA sequences and the origin of common carp may have occurred elsewhere (Mabuchi et al. 2005). Additional fossil data demonstrate that several extinct *Cyprinus* species inhabited the Lake Biwa ca. 3.3 million years ago, and these species were also widely distributed in present-day China and Japan (Nakajima 1986).

On the basis of such fossil data, Nakajima (1994, 2003, cited in Mabuchi et al. 2005) hypothesized that one of these extinct species might have been the direct ancestor of living common carp. Although the sampling of common carp mitochondrial DNA sequences throughout its distribution range is still too sparse to conclusively demonstrate the place of origin of the species, the occurrence of an ancient common carp lineage in Lake Biwa supports Nakajima's hypothesis, indicating that East Asia, including the Japanese area, is the most probable place of origin of present-day wild common carp (Mabuchi et al. 2005, 2006).

Conflicting hypotheses were also published on the origin and spread of European common carp. The studies of Kohlmann et al. (2003, 2005) based on allozymes, PCR-RFLP of the mitochondrial *ND-3/4* and *ND-5/6* gene regions, and microsatellite loci strongly supported a common ancestor of European and Central Asian common carp, and simultaneously demonstrated a deep divergence of these two groups from East Asian common carp which included Amur River wild and Japanese ornamental Koi carp. In contrast, Froufe et al. (2002) suggested an Asian ancestry and single introduction of common carp into the Danube River Basin based on the identity of a partial, 565 bp long D-loop sequence in 21 European individuals from the upper Danube River in Austria and Hungary with four Japanese ornamental Koi carp, although all five Asian Amur River wild carp displayed unique haplotypes differing from one to 12 base substitutions from the European haplotype. Finally, Zhou et al. (2003b) found indications for different ancestors of European domesticated common carp: PCR-RFLP analysis of the mitochondrial *ND-5/6* gene region (approximately 2.4 kb) and D-loop sequences (928 bp) clustered German mirror carp with the European subspecies but Russian scattered scaled mirror carp with the East Asian subspecies. When evaluating the conclusions of Froufe et al. (2002) and Zhou et al. (2003b) it has to be considered, however, that both studies did not include any common carp from Central Asia. Moreover, these conflicting results might have been caused by confounding effects of natural range expansions with large-scale human-mediated translocations/introductions and/or breeding activities including hybridization of subspecies. For example, Russian common carp breeds could be divided with only rare exceptions into two main groups corresponding to their breeding history by Random Amplification of Polymorphic DNA (RAPD) and microsatellite markers: one group originated from European common carp, whilst the second one showed substantial admixture with Asian Amur River wild carp (Ludanny et al. 2006, 2010).

Based on the results of an extensive D-loop sequence polymorphism study including 24 wild and domesticated populations representing a geographical range from Western Europe (Spain) to Central Asia (Uzbekistan) and considering available literature data Kohlmann and Kersten (2013) suggested the following scenario on the origin and spread of European common carp as the most probable one.

After its emergence as a species in East Asia a first wave of expansion led to a continuous distribution of common carp in Eurasia from the Don and Danube Rivers in the West to the Amur River drainage basin and China in the Far East which was broken up into eastern and western populations during multiple Pleistocene glaciations (Berg 1964). Their geographical and reproductive isolation finally resulted in the formation of the two subspecies *Cyprinus carpio carpio* in the West and *Cyprinus carpio haematopterus* in the East. A separation of both subspecies in the Middle or Lower Pleistocene is genetically supported by two estimates of divergence times based on the same mutation calibration rate of 0.76% divergence per one million years for cyprinid mitochondrial DNA (Zardoya and Doadrio 1999): about 500,000 years have been estimated based on 4,705 bp long sequences encompassing the *ND-3/4* and *ND-5/6* gene regions (unpubl. data, cited in Kohlmann et al. 2003), and about 900,000 years were deduced from combined, approximately 2,070 bp long cytochrome *b* and D-loop region sequences (Zhou et al. 2004a).

The current European wild common carp then evolved from western populations, i.e., the *Cyprinus carpio carpio* subspecies. The centre of its origin could be located in or close to Central Asia since the mitochondrial DNA variability and the number of endemic haplotypes are highest there [D-loop sequences from Kohlmann and Kersten (2013) as well as PCR-RFLP based *ND-3/4* and *ND-5/6* composite haplotypes from Kohlmann et al. (2003)]. On the other hand, the lack of endemic haplotypes in combination with a generally lower total number of haplotypes in Europe suggests a relatively recent bottleneck event that would be compatible with a postglacial, i.e., 8,000 to 10,000 years ago, spread into Europe (Borzenko 1926, Hankó 1932, Bănărescu 1960; all cited by Balon 1995). According to Balon (1995) the westernmost boundary of the natural distribution range was the Danube River's piedmont zone located in the Middle Danube between the present-day cities of Bratislava (Slovak Republic) and Budapest (Hungary). The further distribution of common carp west and north of this area was then clearly caused by humans. Balon (1995) also provided historical evidence that the Romans were the first who cultured common carp collected from the Danube River, and that the tradition of the 'piscinae' was continued after the collapse of the Roman Empire in monastery ponds throughout the Middle Ages. The remarkable prevalence shift of two D-loop haplotypes being equally frequent in wild and wild/feral to only one haplotype dominating or being fixed in domesticated common carp populations observed by Kohlmann and Kersten (2013) could be explained either by a bottleneck/ founder effect at early stages of cultivation and domestication in Europe or by selection acting on neighbouring coding mitochondrial DNA genes. Since this prevalence shift was accompanied by a general reduction in variability of the D-loop (Kohlmann and Kersten 2013), the *ND-3/4* and *ND-5/6* gene sequences (Kohlmann et al. 2003) as well as presumably neutral nuclear microsatellite markers (Kohlmann et al. 2005), a bottleneck hypothesis seems to be more reasonable. Moreover, since only a single haplotype is dominating or fixed, such a bottleneck event must have occurred early in the domestication process of European common carp.

In contrast, the eastern populations, i.e., the *Cyprinus carpio haematopterus* subspecies, did not contribute to the natural distribution or domestication of common carp in Europe. The occasional occurrence of single individuals genetically related to East Asian common carp might be due to recent human-mediated introductions and subsequent admixture/hybridization with native European common carp (Kohlmann et al. 2003). Moreover, a simultaneous postglacial spread of *Cyprinus carpio carpio* from Central Asia eastward up to the Irtse and Amur Rivers creating a secondary contact or hybrid zone with the East Asian subspecies (Wang et al. 2010a, Zhu et al. 2011, Torgunakova et al. 2012) might have contributed to the confusion about the origin of European common carp.

The common carp is not only considered to be the oldest cultured and most domesticated freshwater fish species in the world (Balon 2004) but also the first fish species to be subject to large-scale transfers and introductions (Welcome 1988). As mentioned above, wild common carp were taken from the Danube River for culture by the Romans and subsequently distributed widely throughout Europe. Independently, the species was spread within China for similar purposes. During a period beginning at about the middle of the 19th century, the common carp was spread in a series of introductions until it achieved a near global distribution and is present now in most areas where climatic conditions permit its survival and reproduction (for details on continents, countries and years of introduction see Welcome 1988). As a result, numerous local varieties of common carp have been developed through a combination of forces including geographic isolation, adaptation, accumulation of mutations and natural as well as human selection, crossbreeding and introgression (Hulata 1995).

In contrast to the still increasing production of domesticated common carp, wild populations are declining and threatened in many areas of their natural distribution range. In addition to negative anthropogenic effects such as habitat destruction, genetic interactions with domesticated individuals (either escaped from farms or stocked into open waters) might be contributing factors (Flajšhans and Hulata 2006). Special attention should be paid to avoid introgressive hybridization among subspecies. Introgression of the East Asian into the European/Central Asian subspecies has already been observed in a German wild/feral population (Kohlmann et al. 2003) and an Uzbek wild population (Murakaeva et al. 2003) both indicating hybridization with Amur River wild carp and the domesticated Ropsha carp, a derivative of it, respectively. Similarly, cryptic invasions of European common carp genotypes had been detected in Japan by Mabuchi et al. (2008).

Common Carp Genetics

The earliest genetic investigations on common carp concerned the inheritance of the scale cover and body colours.

Rudzinsky (1928a,b) was the first to discover that the 'complete' scale cover typical for wild common carp is dominant over the 'mirror' form found in domesticated populations. Subsequent studies (Kirpitchnikov and Balkashina 1935, 1936, Kirpitchnikov 1937, 1945, 1948, Golovinskaya 1940, 1946; all cited in Kirpitchnikov 1999) identified two unlinked autosomal loci (*S* and *N*), each with two alleles, that control the scale cover of four phenotypes (Fig. 1.2):

a) Scaled carp, homozygous *SSnn* or heterozygous *Ssnn*,
b) Mirror carp, recessive homozygous *ssnn*,
c) Linear carp, heterozygous *SSNn* or *SsNn*,
d) Leather or naked carp, heterozygous *ssNn*.

The remaining genotypes *SSNN*, *SsNN* and *ssNN* are not viable and die at hatching or shortly thereafter (Probst 1949a,b, 1950, 1953, Wohlfarth et al. 1963; all cited in Kirpitchnikov 1999).

Recently, Casas et al. (2013) performed a large number of crosses between different scale cover phenotypes and observed a high degree of variation within the mirror carp phenotype due to which they divided it into two sub-types: classical mirror and irregular. They also analyzed the survival rates of offspring groups and found significant differences between Asian and European crosses. Whereas leather x leather crosses involving at least one parent of Asian origin or hybrid with Asian parent(s) showed the expected 25% early lethality (due to the lethality of the *NN* genotype), those with two Hungarian leather parents did not. As an explanation, they suggested that the tested Hungarian leather carp might possess a new mutant *s* allele with

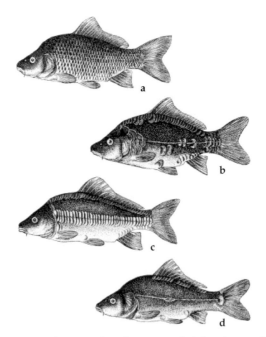

Figure 1.2. Common carp scale cover phenotypes. a. scaled, b. mirror, c. linear, d. leather or naked (from Kirpitchnikov 1999).

stronger effects and/or a new mutant N allele with milder effects on the scale pattern. Another possibility would have been the presence of a weak N allele that causes the loss of scales but not the lethality in homozygotes.

In addition to controlling the scale cover both S and N genes also show a wide range of pleiotropic effects on morphology, physiology and productivity (Table 1.1).

The common carp also comprises a wide variety of pigmented forms. The following genes affecting colouration have been identified (Kirpitchnikov 1999):

- German blue (bl_d): The skin of these individuals contains a smaller number of guanine crystals and fewer guanophores (alampia). This mutation has been observed in Germany, Israel and other countries. The viability and growth rate of the blue common carp are somewhat reduced, but not consistently lower than those of the typical yellowish-

Table 1.1. Pleiotropic effects of scale genes in common carp (from Kirpitchnikov 1999).

Parameter	Scaled carp	Mirror carp	Linear carp	Leather carp	Reference
Weight of 1 year old fishes, favourable conditions*	100	93–96	85–88	70–80	1, 2, 3
Weight of 2 year old fishes*	100	94–96	86–91	83–84	2, 4
Mean number and variation of soft rays in the dorsal fin	18.8 (17–22)	18.7 (17–22)	16.4 (12–19)	15.4 (5–18)	1, 5
Mean number of soft rays in the anal fin	4.96	5.00	3.82	3.56	1
Mean number of soft rays in the pelvic fin	14.7	14.3	14.3	13.1	4
Variation of mean number of gill rakers	24.6–25.1	24.3–24.8	19.4–21.6	18.5–20.5	1, 4, 5
Mean number of gill lamellae	88.6	83.5	82.3	83.2	4
Mean number of pharyngeal teeth	9.22	9.58	7.63	7.44	1
Ability to fin regeneration*	100	76	39	19	2
Erythrocyte count (10^6/ml)	1.93	1.99	1.76	1.69	6
Haemoglobin (g/%)	9.02	8.87	8.18	8.28	6
Critical temperature (°C)	37.6	37.5	36.8	36.6	6
Survival time (min) under oxygen deficit	210	210	132	132	6
Immunological reactivity	fast	fast	slow	slow	7
Intensity of fat metabolism	low	low	high	very high	8, 9
Total survival of 1 year old fish, optimal conditions*	100	91–98	87–93	80–92	1, 2

* expressed as percentage of the value in scaled carp.
References: 1: Kirpitchnikov (1945, 1948), 2: Probst (1953), 3: Tsvetkova (1974), 4: Steffens (1966), 5: Golovinskaya (1940), 6: Chan (1969), 7: Lukjanenko and Sukacheva (1975), 8: Golovinskaya et al. (1974), 9: Tsvetkova (1969).

grey form (Probst 1949b). These blue common carp are sometimes referred to as 'transparent' (Komen 1990).

- Polish blue (bl_p): Phenotypically, these fish resemble the German blue common carp. Most probably, they represent an independent mutation at the same locus. The Polish blue common carp observed in a fish farm at Ochaby grew fast during their first year of life, then more slowly, so that three-year old fish were half the size of the normally pigmented ones (Wlodek 1963).
- Red or 'gold' (g): Red common carp are found in many countries. Their pigmentation is the result of a reduction of both melanin synthesis and melanophores. The growth rate and viability of the European red common carp are somewhat reduced (Moav and Wohlfarth 1968). Whilst the red pigmentation is a recessive mutation in European common carp, it segregates as a dominant character in crosses between Israeli red common carp and the Chinese big-belly variety.

The most striking and famous colour patterns, however, are expressed by Japanese ornamental koi carp which are described in a separate chapter of this book.

In order to study the genetic diversity and relationships of wild and domesticated common carp populations, several types of polymorphic markers have been analyzed. These include proteins (i.e., allozymes) and DNA-based markers such as microsatellites, Amplified Fragment Length Polymorphisms (AFLPs), RFLPs, RAPD, and mitochondrial DNA variability (Chistiakov and Voronova 2009). Advantages and disadvantages of each marker type have been reviewed in an aquaculture research context by Liu and Cordes (2004).

Kirpitchnikov (1999) listed 18 polymorphic proteins coded by 33 loci that have been described in common carp. Summarizing the available data on the degree of genetic variation he estimated the proportion of polymorphic loci to be 28%, the average heterozygosity to be 0.060, and the average number of alleles per locus to be 1.4–1.5. An exceptional high number of alleles were observed at the glucose-6-phosphate isomerase loci of wild common carp from northern Vietnam (Kohlmann et al. 1997). In general, the common carp values for the proportion of polymorphic loci (P) and average heterozygosity (H) are slightly higher than those reported for fishes as a whole (Paaver 1983: 109 species, P = 19.1%, H = 0.063; Nevo et al. 1984: 120 species, P = 20.9%, H = 0.051; Kirpitchnikov 1987: 131 species, P = 19.6%, H = 0.053).

Protein markers have mainly been used to study the genetic variability and relationships of common carp at regional levels. For example, information is available for domesticated and wild populations cultured or sampled in Japan (Macaranas et al. 1986), Italy (Cataudella et al. 1987), Estonia (Paaver and Gross 1991), Hungary (Csizmadia et al. 1995), Poland (Anjum 1995), France (Desvignes et al. 2001), and Czech Republic (Šlechtova et al. 2002) as well as for populations introduced into Indonesia (Sumantadinata and Taniguchi 1990), Australia (Davis et al. 1999), Israel (Ben-Dom et al. 2000) and Tunisia (Louati et al. 2011). In rare cases, larger geographical regions were covered and both subspecies *Cyprinus carpio carpio* and *Cyprinus carpio haematopterus* were examined and could directly be compared (Brody et al. 1979, Kohlmann and Kersten 1999, Murakaeva et al. 2003).

Microsatellites or Simple Sequence Repeats (SSRs) consist of stretches of tandemly repeated nucleotide motifs of the DNA that are usually one to six base pairs long (Litt and Luty 1989, Tautz 1989). They often express high levels of polymorphism based on size differences due to varying numbers of these repeat units contained by alleles at a given locus. DeWoody and Avise (2000), considering 524 microsatellite loci in nearly 40,000 individuals of 78 fish species, found that freshwater fish display levels of population genetic variation (mean heterozygosity, H = 0.460, and mean numbers of alleles per locus, a = 7.5) roughly similar to those of non-piscine animals (H = 0.580 and a = 7.1). Most microsatellite loci are relatively short, ranging from a few to a few hundred repeats. In common carp, the overall mutation rates of microsatellite loci have been estimated to be 2.53×10^{-4} (Zhang et al. 2010) and 5.56×10^{-4} (Yue et al. 2007) per locus per generation, and are believed to be caused by polymerase slippage during DNA replication, resulting in differences in the number of repeat units (Levinson and Gutman 1987, Tautz 1989). Moreover, microsatellites are inherited in a Mendelian fashion as codominant markers and appear to be frequent and randomly distributed throughout eukaryotic genomes. In fishes, they have been estimated to occur as often as once every 10 kb (Wright 1993). Hundreds of microsatellite loci have already been isolated from the common carp genome and their numbers are still increasing. For example, Zhang et al. (2013a) finally used 787 microsatellite markers for the construction of their common carp linkage map. Because of these favourable characteristics (high polymorphism, small locus size, fast evolution, codominant inheritance, high abundance and even genomic distribution) microsatellites became an extremely popular marker type in a wide variety of genetic investigations in fishes, e.g., studies of population and stock structure, genome mapping, parentage and kinship analyses (Liu and Cordes 2004).

Similar to protein markers microsatellite loci of common carp have so far mainly been used to characterize the structures of native and/or introduced populations at regional levels, e.g., in The Netherlands (Tanck et al. 2000), Israel (David et al. 2001), France (Desvignes et al. 2001), Hungary (Bártfai et al. 2003, Lehoczky et al. 2005a), Turkey (Memiş and Kohlmann 2006), Czech Republic (Hulak et al. 2010), Russia (Ludanny et al. 2010), Croatia (Tomljanović et al. 2013), Iran (Yousefian 2011), China (Zhou et al. 2004b, Liao et al. 2006, Li et al. 2007), Vietnam (Thai et al. 2007), Bangladesh (Mondol et al. 2006), and Australia (Haynes et al. 2009, 2010). The study of Kohlmann et al. (2005) is the only one covering 22 wild, feral and domesticated populations spanning a geographical area from Europe (Spain, Germany, Czech Republic, Poland, Hungary) through Central Asia (Uzbekistan) to East and South-East Asia (Russia, China, Japan, Vietnam). A few other applications of microsatellites in common carp were aimed at estimating heritabilities of growth-related, processing and quality traits (Vandeputte et al. 2004, Kocour et al. 2007) as well as investigating sperm competition (Kaspar et al. 2007, 2008) using parentage assignment.

RAPD as well as AFLP markers do not require any prior molecular information and thus are applicable to any species, including fishes. However, RAPD markers are subject to low reproducibility due to the low annealing temperature used in the PCR amplification (Liu and Cordes 2004). Novelo et al. (2010) tested the reliability and inheritance of RAPD markers in two consecutive generations of common carp and found that 20 markers were fully reproducible and demonstrated dominant Mendelian inheritance, 24 markers showed Mendelian inheritance but were non-reproducible in some progenies, and 13 markers displayed ratios in some progenies that did not fit simple Mendelian inheritance patterns. Therefore, they recommend to first identify a set of reproducible RAPD markers that demonstrate Mendelian inheritance before application of the RAPD technique in population studies.

In common carp both RAPD and AFLP markers have been used less extensively and were replaced by more suitable and efficient markers, in particular microsatellites, as their availabilities increased and genotyping efforts (both in time and costs) decreased. For example, RAPD markers have been used exclusively to characterize Russian common carp breeds (Ludanny et al. 2006) and cultured populations in India (Basavaraju et al. 2007), and to study the possibility of predicting heterosis on the basis of genetic distances between populations (Dong and Zhou 1998). In two other studies RAPD markers were combined with microsatellite loci to characterize two Hungarian common carp broodstocks (Bártfai et al. 2003) and with mitochondrial DNA *COII* gene sequences to analyze the phylogenetic relationships of ornamental koi carp, and the Chinese Oujiang colour carp and Long-fin carp (Wang and Li 2004). AFLP markers were used solely in common carp to examine the genetic variability of two cultured stocks and second generation gynogens in Indonesia (Wang et al. 2000), or in combination with microsatellites to study the extent of polymorphism and the genetic relationships between eight common carp populations including five varieties of ornamental koi carp (David et al. 2001, 2007), and to construct genetic linkage maps based on gynogenetic haploids (Cheng et al. 2010) or using F2 populations derived from a cross between Chinese Purse red common carp and Xingguo red common carp (Wang et al. 2012c).

So far, eight complete common carp mitochondrial DNA genome sequences have been deposited in NCBI GenBank (http://www.ncbi.nlm. nih.gov/genbank/; last accession date: 25 July 2013). Only the sequences from a putative Taiwanese strain (Chang et al. 1994, GenBank ID: X61010), a Japanese Lake Biwa wild carp (Mabuchi et al. 2006, GenBank ID: AP009047) and a Chinese Oujiang colour carp (Wang et al. 2013, GenBank ID: JX188253) had been published. The remaining five sequences are unpublished direct submissions by Hu et al. (2011) originating from Chinese Purse red carp (GenBank ID: JN105357), Xingguo red carp (GenBank ID: JN105353), Yangtze River wild carp (GenBank ID: JN105354) and Russian scattered scaled mirror carp (GenBank ID: JN105352) and by Wang et al. (2012) obtained from Chinese Huanghe carp (GenBank ID: JX188254). The total sequence lengths range from 16,575 bp in the Taiwanese carp to 16,582 bp in the Chinese Purse red carp, encode the same set of genes (13 proteins, 2 rRNAs, and 22 tRNAs), and the gene structure and order are identical to those usually found in other vertebrates (Chang et al. 1994, Mabuchi et al. 2006).

Most applications of mitochondrial DNA in common carp genetics, however, are based on partial or complete mitochondrial gene segments using either enzymatic digestion of PCR products (PCR-RFLP) or direct sequencing. For example, PCR-RFLP analyses have been performed on combined *ND-3/4* and *ND-5/6* gene regions (Gross et al. 2002, Kohlmann et al. 2003, Lehoczky et al. 2005b, Memiş and Kohlmann 2006), *ND-5/6* gene regions alone (Zhou et al. 2003a,b), and cytochrome *b* as well as 16S *rDNA* (Tsipas et al. 2009). Partial or complete sequences were obtained for cytochrome *b* (Zhou et al. 2004a, Mabuchi et al. 2005, Imsiridou et al. 2009, Tsipas et al. 2009, Torgunakova et al. 2012), *COII* (Wang and Li 2004, Mabuchi et al. 2006, Wang et al. 2010a,b), *ATPase6/8* (Thai et al. 2005, Mabuchi et al. 2006), 16S *rDNA* (Tsipas et al. 2009) and D-loop (Froufe et al. 2002, Zhou et al. 2003b, 2004a, Thai et al. 2005, 2006, Mabuchi et al. 2005, 2006, 2008, Imsiridou et al. 2009, Tsipas et al. 2009, Wang et al. 2010a,b, Kohlmann and Kersten 2013).

Apart from describing the genetic structures and relationships of investigated populations and answering taxonomic, phylogenetic or phylogeographic questions (see part on natural history of common carp) the major outcome of several independent marker studies was a reduced variability of domesticated/cultured stocks in comparison to wild or feral common carp populations both in Europe (Kohlmann et al. 2003, 2005, Wang et al. 2010b, Tomljanović et al. 2013) and East Asia (Thai et al. 2007, Wang et al. 2010b, Xu et al. 2012a). These observations can either be explained by founder effects at the beginning of cultivation or genetic drift in combination with sometimes strong selection pressures during cultivation due to the high fecundity of common carp females leading to usually small effective population sizes. Considering declining wild populations, these findings underline the high priority that should be given to the protection of common carp genetic resources.

The progress in the development of Next Generation Sequencing (NGS) technologies had also a great impact on the genetic research in common carp. For example, the rapid and cost effective isolation of large numbers of microsatellite (Ji et al. 2012a) and SNP markers (Xu et al. 2012b) enables the construction of high-density linkage maps (Zhang et al. 2013a, Zhao et al. 2013). Such maps provide the basis for Quantitative Trait Locus (QTL) analyses and genome-wide association studies. QTL-associated markers have already been detected in common carp for traits such as cold tolerance (Sun and Liang 2004), standard length and body mass (Zhang et al. 2007, 2008b), feed conversion ratio (Li et al. 2009), head size (Liu et al. 2009), activity of lactate dehydrogenase (Mao et al. 2009) and growth-related (Wang et al. 2012c) as well as complex body shape-related traits (Zhang et al. 2013b).

Once identified, QTLs can be used in Marker Assisted Selection (MAS) to improve the efficiency of breeding programs. In addition to linkage maps, a first Bacterial Artificial Chromosome (BAC) based physical map of the common carp genome has been constructed by Xu et al. (2011). Sequencing the whole common carp genome was completed in China in 2011 (Bejing Institute of Genomics, Chinese Academy of Sciences: http://english.big. cas.cn/ns/es/201105/t20110509_69516.html) and a draft assembly that still contains many fragments has been generated in Europe (Zhang et al. 2011). The characterization of the common carp transcriptome (Ji et al. 2012b, Wang et al. 2012b) and exome (Henkel et al. 2012) will lead to reliable gene predictions for the entire genome and will facilitate future functional studies, for example on the effects of exogenous (environmental) factors, e.g., pathogens and pollutants, on gene expression. As a supplement to the NCBI GenBank, a special database solely dedicated to host the accumulating common carp genome information has been established by the Chinese Academy of Fishery Sciences (CarpBase: A Database of Common Carp Genome Project. http://carpbase.org/index.php).

All of these new and fascinating applications will contribute to a better understanding of common carp evolution, biology and ecology, to a scientifically sound management and conservation of its genetic resources in wild populations, a sustainable use of domesticated populations as well as to the genetic improvement of cultured common carp to meet the demands of changing markets.

Conclusion

The common carp is an evolutionally tetraploid species that arose most probably through species hybridization. Latest transcriptome analyses date the tetraploidization event back to 5.6 to 11.3 million years ago. Genetic data and fossil records indicate East Asia as the place of common carp's origin. Its initially continuous, vast natural distribution range from Eastern Europe in the West throughout Central Asia to China, Japan and Vietnam in the East was broken up into eastern and western populations during multiple Pleistocene glaciations. Their geographical and reproductive isolation finally resulted in the formation of the two subspecies *Cyprinus carpio carpio* in the West and *Cyprinus carpio haematopterus* in the East. The validity of a third subspecies (*Cyprinus carpio rubrofuscus*) which might have diverged from *Cyprinus carpio haematopterus* in southern China remains questionable. Current European wild and domesticated common carp evolved from the western subspecies *Cyprinus carpio carpio*.

The common carp is not only the oldest cultured and most domesticated freshwater fish species in the world but also the first fish species subjected to wide scale transfers and introductions. Today, common carp has achieved a near global distribution and is present in most areas where climatic conditions permit its survival and reproduction. As a result, numerous local varieties have been developed. In contrast to the increasing production of common carp for human consumption wild populations are endangered in many regions of their natural distribution range.

The earliest genetic investigations of common carp concerned its scale cover and body colouration. The two genes *S* and *N* controlling the scale cover also show a wide range of pleiotropic effects on morphology, physiology and productivity. Therefore, scaled and mirror carp became the preferred type for culture. Later on, biochemical and DNA based markers were used to study the status of subspecies, population structure, phylogeny and phylogeography. If wild common carp populations were compared with domesticated stocks a reduced genetic variability of the latter was often observed. The conservation of still existing wild common carp genetic resources has therefore a high priority. The progress in the development of next generation sequencing technologies had a great impact on the genetic research in common carp. High-density linkage maps were constructed and a whole common carp genome was sequenced in China. Quantitative trait loci were identified and could be used in marker assisted selection to improve the performance of cultured common carp.

References

Amores, A., A. Force, Y.L. Yan, L. Joly, C. Amemiya, A. Fritz, R.K. Ho, J. Langeland, V. Prince, Y.L. Wang, M. Westerfield, M. Ekker and J.H. Postlethwait. 1998. Zebrafish hox clusters and vertebrate genome evolution. Science 282: 1711–1714.

Anjum, R. 1995. Biochemical and chromosomal genetic characteristics of several breeding populations of common carp, *Cyprinus carpio* L. Ph.D. Thesis. Olsztyn University of Agriculture and Technology, Faculty of Water Protection and Freshwater Fisheries. Olsztyn, Poland.

Anjum, R., M. Jankun, K. Kohlmann and P. Kersten. 1997. Silver and chromomycin A$_3$ (CMA$_3$) staining of nucleolus organizer regions in the chromosomes of ornamental (Koi) common carp, *Cyprinus carpio*. Cytobios 90: 73–79.

Anjum, R. and M. Jankun. 1998. NOR-bearing chromosomal associations revealed through silver and sequential chromomycin A$_3$ staining in the mirror carp, *Cyprinus carpio* L. Caryologia 51: 167–171.

Balon, E.K. 1995. Origin and domestication of the wild carp, *Cyprinus carpio*: from Roman gourmets to the swimming flowers. Aquaculture 129: 3–48.

Balon, E.K. 2004. About the oldest domesticates among fishes. J. Fish Biol. 65 (Supplement A): 1–27.

Bănărescu, P. 1960. Einige Fragen zur Herkunft und Verbreitung der Süsswasserfischfauna der europäisch-mediterranen Unterregion. Arch. Hydrobiol. 57: 16–134.

Bártfai, R., S. Egedi, G.H. Yue, B. Kovács, B. Urbányi, G. Tamás, L. Horváth and L. Orbán. 2003. Genetic analysis of two common carp broodstocks by RAPD and microsatellite markers. Aquaculture 219: 157–167.

Basavaraju, Y., D.T. Prasad, K. Rani, S.P. Kumar, U.D. Naika, S. Jahageerdar, P.P. Srivastava, D.J. Penman and G.C. Mair. 2007. Genetic diversity in common carp stocks assayed by random-amplified polymorphic DNA markers. Aquac. Res. 38: 147–155.

Ben-Dom, N., N.B. Cherfas, B. Gomelsky and G. Hulata. 2000. Genetic stability of Israeli common carp stocks inferred from electrophoretic analysis of transferrin, phosphoglucomutase and glucose-6-phosphate isomerase. Bamidgeh 52: 30–35.

Berg, L.S. 1964. Freshwater Fishes of the U.S.S.R. and the Adjacent Countries, Vol. 2. Israel Program for Scientific Translations, Jerusalem.

Brody, T., D. Kirsht, G. Parag, G. Wohlfarth, G. Hulata and R. Moav. 1979. Biochemical genetic comparison of the Chinese and European races of the common carp. Anim. Blood Groups Biochem. Genet. 10: 141–149.

Borzenko, M.P. 1926. Materialy po biologii sazana (*Cyprinus carpio* Linne) (Data on the biology of the wild carp). Izv. Bakin. Ikhtiol. Lab. 2: 5–132.

Casas, L., R. Szűcs, S. Vij, C.H. Goh, P. Kathiresan, S. Németh, Z. Jeney, M. Bercsényi and L. Orbán. 2013. Disappearing scales in carps: re-visiting Kirpichnikov's model on the genetics of scale pattern formation. PLoS ONE 8(12): e83327.

Cataudella, S., L. Sola, M. Corti, R. Arcangelli, G. La Rosa, M. Mattoccia, M. Cobelli Sbordoni and V. Sbordoni. 1987. Cytogenetic, genic and morphometric characterization of groups of common carp, *Cyprinus carpio*. pp. 113–129. *In*: K. Tiews (ed.). Proceedings of the World Symposium on Selection, Hybridization, and Genetic Engineering in Aquaculture, 27–30 May 1986, Bordeaux, France. Vol. 1. Heenemann Verlagsgesellschaft, Berlin.

Cavender, T.M. 1991. The fossil record of the Cyprinidae. pp. 34–54. *In*: I.J. Winfield and J.S. Nelson (eds.). Cyprinid Fishes: Systematics, Biology, and Exploitation. Chapman and Hall, London.

Chan, M.T. 1969. Variability of some physiological traits in common carps of different genotype. pp. 117–123. *In*: B.I. Cherfas (ed.). Genetika, Selekcija i Hybridizacija Ryb. Nauka Publ., Moscow.

Chang, Y., F. Huang and T. Lo. 1994. The complete nucleotide sequence and gene organization of carp (*Cyprinus carpio*) mitochondrial genome. J. Mol. Evol. 38: 138–155.

Cheng, L., L. Liu, X. Yu, D. Wang and J. Tong. 2010. A linkage map of common carp (*Cyprinus carpio*) based on AFLP and microsatellite markers. Anim. Genet. 41: 191–198.

Chistiakov, D.A. and N.V. Voronova. 2009. Genetic evolution and diversity of common carp *Cyprinus carpio* L. Cent. Eur. J. Biol. 4: 304–312.

Csizmadia, C., Z. Jeney, I. Szerencses and S. Gorda. 1995. Transferrin polymorphism of some races in a live gene bank of common carp. Aquaculture 129: 193–198.

David, L., P. Rajasekaran, J. Fang, J. Hillel and U. Lavi. 2001. Polymorphism in ornamental and common carp strains (*Cyprinus carpio* L.) as revealed by AFLP analysis and a new set of microsatellite markers. Mol. Genet. Genomics 266: 353–362.

David, L., S. Blum, M.W. Feldman, U. Lavi and J. Hillel. 2003. Recent duplication of the common carp (*Cyprinus carpio* L.) genome as revealed by analyses of microsatellite loci. Mol. Biol. Evol. 20: 1425–1434.

David, L., N.A. Rosenberg, U. Lavi, M.W. Feldman and J. Hillel. 2007. Genetic diversity and population structure inferred from the partially duplicated genome of domesticated carp, *Cyprinus carpio* L. Genet. Sel. Evol. 39: 319–340.

Davis, K.M., P.I. Dixon and J.H. Harris. 1999. Allozyme and mitochondrial DNA analysis of carp, *Cyprinus carpio* L., from south-eastern Australia. Mar. Freshw. Res. 50: 253–260.

Desvignes, J.F., J. Laroche, J.D. Durand and Y. Bouvet. 2001. Genetic variability in reared stocks of common carp (*Cyprinus carpio* L.) based on allozymes and microsatellites. Aquaculture 194: 291–301.

DeWoody, J.A. and J.C. Avise. 2000. Microsatellite variation in marine, freshwater and anadromous fishes compared with other animals. J. Fish Biol. 56: 461–473.

Dong, Z. and E. Zhou. 1998. Application of the random amplified polymorphic DNA technique in a study of heterosis in common carp, *Cyprinus carpio* L. Aquac. Res. 29: 595–600.

Ferris, S.D. and G.S. Whitt. 1977. The evolution of duplicate gene expression in the carp (*Cyprinus carpio*). Experientia 33: 1299–1301.

Ferris, S.D. and G.S. Whitt. 1979. Evolution of the differential regulation of duplicate genes after polyploidization. J. Mol. Evol. 12: 267–317.

Flajšhans, M. and G. Hulata. 2006. Common carp—*Cyprinus carpio*. 7 pp. *In*: D. Crosetti, S. Lapègue, I. Olesen and T. Svaasand (eds.). Genetic effects of domestication, culture and breeding of fish and shellfish, and their impacts on wild populations. GENIMPACT project: Evaluation of genetic impact of aquaculture activities on native populations. A European network. WP1 workshop "Genetics of domestication, breeding and enhancement of performance of fish and shellfish", Viterbo, Italy, 12th–17th June, 2006. http://genimpact.imr.no/.

Froufe, E., I. Magyary, I. Lehoczky and S. Weiss. 2002. mtDNA sequence data supports an Asian ancestry and single introduction of the common carp into the Danube Basin. J. Fish Biol. 61: 301–304.

Golovinskaya, K.A. 1940. The pleiotropic action of scale genes in the common carp. Doklady (Reports) Akademii Nauk SSSR (Moscow) 28: 533–536.

Golovinskaya, K.A. 1946. On the linear form of the cultivated common carp. Doklady (Reports) Akademii Nauk SSSR (Moscow) 54: 637–640.

Golovinskaya, K.A., N.B. Cherfas and L.I. Tsvetkova. 1974. Results of evaluation of the reproductive function in gynogenetic common carp females. Trudy VNIIPRCH (Moscow) 23: 2–26.

Gregory, T.R. 2013. Animal Genome Size Database. http://www.genomesize.com.

Gross, R., K. Kohlmann and P. Kersten. 2002. PCR-RFLP analysis of the mitochondrial *ND-3/4* and *ND-5/6* gene polymorphisms in the European and East Asian subspecies of common carp (*Cyprinus carpio* L.). Aquaculture 204: 507–516.

Guerrero-Estévez, S. and N. Moreno-Mendoza. 2010. Sexual determination and differentiation in teleost fish. Rev. Fish Biol. Fish. 20: 101–121.

Gui, J.F. and Z.Y. Zhu. 2012. Molecular basis and genetic improvement of economically important traits in aquaculture animals. Chinese Sci. Bull. 57: 1751–1760.

Hankó, B. 1932. Ursprung und Verbreitung der Fischfauna Ungarns. Arch. Hydrobiol. 23: 520–556.

Haynes, G.D., D.M. Gilligan, P. Grewe and F.W. Nicholas. 2009. Population genetics and management units of invasive common carp *Cyprinus carpio* L., in the Murray-Darling Basin, Australia. J. Fish Biol. 74: 295–320.

Haynes, G.D., D.M. Gilligan, P. Grewe, C. Moran and F.W. Nicholas. 2010. Population genetics of invasive common carp *Cyprinus carpio* L. in coastal drainages in eastern Australia. J. Fish Biol. 77: 1150–1157.

Henkel, C.V., R.P. Dirks, H.J. Jansen, M. Forlenza, G.F. Wiegertjes, K. Howe, G.E.E.J.M. van den Thillart and H.P. Spaink. 2012. Comparison of the exomes of common carp (*Cyprinus carpio*) and zebrafish (*Danio rerio*). Zebrafish 9: 59–67.

Hulak, M., V. Kaspar, K. Kohlmann, K. Coward, J. Tešitel, M. Rodina, D. Gela, M. Kocour and O. Linhart. 2010. Microsatellite-based genetic diversity and differentiation of foreign common carp (*Cyprinus carpio*) strains farmed in the Czech Republic. Aquaculture 298: 194–201.

Hulata, G. 1995. A review of genetic improvement of the common carp (*Cyprinus carpio* L.) and other cyprinids by crossbreeding, hybridization and selection. Aquaculture 129: 143–155.

Imsiridou, A., A. Triantafyllidis, D.A. Baxevanis and C. Triantaphyllidis. 2009. Genetic characterization of common carp (*Cyprinus carpio*) populations from Greece using mitochondrial DNA sequences. Biologia 64: 781–785.

Ji, P., Y. Zhang, C. Li, Z. Zhao, J. Wang, J. Li, P. Xu and X. Sun. 2012a. High throughput mining and characterization of microsatellites from common carp genome. Int. J. Mol. Sci. 13: 9798–9807.

Ji, P., G. Liu, J. Xu, X. Wang, J. Li, Z. Zhao, X. Zhang, Y. Zhang, P. Xu and X. Sun. 2012b. Characterization of common carp transcriptome: sequencing, *de novo* assembly, annotation and comparative genomics. PLoS ONE 7(4): e35152.

Kaspar, V., K. Kohlmann, M. Vandeputte, M. Rodina, D. Gela, M. Kocour, S.M. Hadi Alavi, M. Hulak and O. Linhart. 2007. Equalizing sperm concentrations in a common carp (*Cyprinus carpio*) sperm pool does not affect variance in proportions of larvae sired in competition. Aquaculture 272S1: S204–S209.

Kaspar, V., M. Hulak, K. Kohlmann, M. Vandeputte, M. Rodina, D. Gela and O. Linhart. 2008. *In vitro* study on sperm competition in common carp (*Cyprinus carpio* L.). Cybium 32 suppl.: 303–306.

Kirpitchnikov, V.S. 1937. A major genes for scale cover in the common carp. Biologichesky Journal (Moscow) 6: 601–632.

Kirpitchnikov, V.S. 1943. Experimental taxonomy of the wild carp *Cyprinus carpio* L. I. Growth and morphological characteristics of the Taparavan, Volga-Caspian and Amur wild carps under conditions of pond rearing. Izvestiya Akademii Nauk SSSR 4: 189–220.

Kirpitchnikov, V.S. 1945. The effect of rearing conditions on viability, growth rate and morphology of the carps with different genotypes. Doklady (Reports) Akademii Nauk SSSR (Moscow) 47: 521–524.

Kirpitchnikov, V.S. 1948. A comparative characteristics of four major varieties of the cultured common carp cultivated in the North of the USSR. Izvestija VNIORCH (Leningrad) 26: 145–170.

Kirpitchnikov, V.S. 1967. Homologous hereditary variation and evolution of wild common carp (*Cyprinus carpio* L.). Genetika 3: 34–47.

Kirpichnikov, V.S. 1981. Genetic Bases of Fish Selection. Springer Verlag, Berlin, New York.

Kirpitchnikov, V.S. 1987. Genetics and selection of fishes. Nauka Publ., Leningrad.

Kirpitchnikov, V.S. 1999. Genetics and Breeding of Common Carp (revised by R. Billard, J. Repérant, J.P. Rio and R. Ward). INRA Editions, Paris.

Kirpitchnikov, V.S and E.I. Balkashina. 1935. Materials on genetics and selection of the common carp. I. Zoologichesky Journal (Moscow) 14: 45–78.

Kirpitchnikov, V.S. and E.I. Balkashina. 1936. Materials on genetics and selection of the common carp. II. Biologichesky Journal (Moscow) 5: 327–376.

Klinkhardt, M., M. Tesche and H. Greven. 1995. Database of fish chromosomes. Westarp Wissenschaften, Magdeburg.

Kocour, M., S. Mauger, M. Rodina, D. Gela, O. Linhart and M. Vandeputte. 2007. Heritability estimates for processing and quality traits in common carp (*Cyprinus carpio* L.) using a molecular pedigree. Aquaculture 270: 43–50.

Kohlmann, K. and P. Kersten. 1999. Genetic variability of German and foreign common carp (*Cyprinus carpio* L.) populations. Aquaculture 173: 435–445.

Kohlmann, K. and P. Kersten. 2013. Deeper insight into the origin and spread of European common carp (*Cyprinus carpio carpio*) based on mitochondrial D-loop sequence polymorphisms. Aquaculture 376-379: 97–104.

Kohlmann, K., P. Kersten and M. Luczynski. 1997. New polymorphism in glucose-phosphate isomerase (GPI) of a wild common carp (*Cyprinus carpio* L.) from Vietnam. Anim. Genet. 28: 313.

Kohlmann, K., R. Gross, A. Murakaeva and P. Kersten. 2003. Genetic variability and structure of common carp (*Cyprinus carpio*) populations throughout the distribution range inferred from allozyme, microsatellite and mitochondrial DNA markers. Aquat. Living Resour. 16: 421–431.

Kohlmann, K., P. Kersten and M. Flajšhans. 2005. Microsatellite-based genetic variability and differentiation of domesticated, wild and feral common carp (*Cyprinus carpio* L.) populations. Aquaculture 247: 253–266.

Komen, J. 1990. Clones of common carp, *Cyprinus carpio*: New perspectives in fish research. Ph.D. thesis. Agricultural University Wageningen, Wageningen.

Komen, J., P. de Boer and C.J.J. Richter. 1992. Male sex reversal in gynogenetic XX females of common carp *Cyprinus carpio* L. by a recessive mutation in a sex-determining gene. J. Hered. 83: 431–434.

Larhammar, D. and C. Risinger. 1994. Molecular genetic aspects of tetraploidy in the common carp *Cyprinus carpio*. Mol. Phylogenet. Evol. 3: 59–68.

Lehoczky, I., I. Magyary, C. Hancz and S. Weiss. 2005a. Preliminary studies on the genetic variability of six Hungarian common carp strains using microsatellite DNA markers. Hydrobiologia 533: 223–228.

Lehoczky, I., Z. Jeney, I. Magyary, C. Hancz and K. Kohlmann. 2005b. Preliminary data on genetic variability and purity of common carp (*Cyprinus carpio* L.) strains kept at the live gene bank at Research Institute for Fisheries, Aquaculture and Irrigation (HAKI) Szarvas, Hungary. Aquaculture 247: 45–49.

Levinson, G. and G.A. Gutman. 1987. High frequency of short frameshifts in poly-CA/GT tandem repeats borne by bacteriophage M13 in *Escherichia coli* K-12. Nucleic Acids Res. 15: 5323–5338.

Li, D., D. Kang, Q. Yin, X. Sun and L. Liang. 2007. Microsatellite DNA marker analysis of genetic diversity in wild common carp (*Cyprinus carpio* L.) populations. J. Genet. Genomics 34: 984–993.

Li, O., D. Cao, Y. Zhang, Y. Gu, X. Zhang, C. Lu and X. Sun. 2009. Studies on feed conversion ratio trait of common carp (*Cyprinus carpio* L.) using EST-SSR marker. J. Fisheries China 33: 624–631.

Liao, X., X. Yu and J. Tong. 2006. Genetic diversity of common carp from two largest Chinese lakes and the Yangtze River revealed by microsatellite markers. Hydrobiologia 568: 445–453.

Litt, M. and J.A. Luty. 1989. A hypervariable microsatellite revealed by *in vitro* amplification of dinucleotide repeat within the cardiac muscle actin gene. Am. J. Hum. Genet. 44: 397–401.

Liu, Z.J. and J.F. Cordes. 2004. DNA marker technologies and their applications in aquaculture genetics. Aquaculture 238: 1–37.

Liu, J.H., Y. Zhang, Y.M. Chang, L.Q. Liang, C.Y. Lu, X.F. Zhang, M.J. Xu and X.W. Sun. 2009. Mapping QTLs related to head length, eye diameter and eye cross of common carp (*Cyprinus carpio* L.). Yi Chuan 31: 508–514 (in Chinese).

Louati, M., L. Bahri-Sfar, K. Kohlmann and O.K. Ben-Hassine. 2011. Current genetic status of common carp (*Cyprinus carpio* L.) introduced into Tunisian reservoirs. Cybium 35: 189–199.

Ludanny, R.I., G.G. Chrisanfova, V.A. Vasilyev, V.K. Prizenko, A.K. Bogeruk, A.P. Ryskov and S.K. Semyenova. 2006. Genetic diversity and differentiation of Russian common carp (*Cyprinus carpio* L.) breeds inferred from RAPD markers. Russ. J. Genet. 42: 928–935.

Ludanny, R.I., G.G. Chrisanfova, V.K. Prizenko, A.K. Bogeruk and S.K. Semyenova. 2010. Polymorphism of microsatellite markers in Russian common carp (*Cyprinus carpio* L.) breeds. Russ. J. Genet. 46: 572–577.

Lukjanenko, V.I. and G.A. Sukacheva. 1975. Peculiarities of the immunological reactivity of four common carp genotypes. pp. 62–76. *In*: O.N. Bauer (ed.). Materials of the 6th All-Union Conference on Fish Diseases. Pitscheprom Publ., Moscow.

Mabuchi, K., H. Senou, T. Suzuki and M. Nishida. 2005. Discovery of an ancient lineage of *Cyprinus carpio* from Lake Biwa, central Japan, based on mtDNA sequence data, with reference to possible multiple origins of koi. J. Fish Biol. 66: 1516–1528.

Mabuchi, K., M. Miya, H. Senou, T. Suzuki and M. Nishida. 2006. Complete mitochondrial DNA sequence of the Lake Biwa wild strain of common carp (*Cyprinus carpio* L.): further evidence for an ancient origin. Aquaculture 257: 68–77.

Mabuchi, K., H. Senou and M. Nishida. 2008. Mitochondrial DNA analysis reveals cryptic large-scale invasion of non-native genotypes of common carp (*Cyprinus carpio*) in Japan. Mol. Ecol. 17: 796–809.

Macaranas, J.M., J. Sato and Y. Fujio. 1986. Genetic characterization of cultured populations of Japanese common carp. Tohoku J. Agric. Res. 37: 21–29.

Mao, R.X., F.J. Liu, X.F. Zhang, Y. Zhang, D.C. Cao, C.Y. Lu, L.Q. Liang and X.W. Sun. 2009. Studies on quantitative trait loci related to activity of lactate dehydrogenase in common carp (*Cyprinus carpio* L.). Yi Chuan 31: 407–411 (in Chinese).

Memiş, D. and K. Kohlmann. 2006. Genetic characterization of wild common carp (*Cyprinus carpio* L.) from Turkey. Aquaculture 258: 257–262.

Moav, R. and G. Wohlfarth. 1968. Genetic improvement of yield in carp. FAO Fish Reports (44) No. 4: 12–29.

Mondol, R.K., S. Islam and S. Alam. 2006. Characterization of different strains of common carp (*Cyprinus carpio* L.) (Cyprinidae, Cypriniformes) in Bangladesh using microsatellite DNA markers. Genet. Mol. Biol. 29: 626–633.

Murakaeva, A., K. Kohlmann, P. Kersten, B. Kamilov and D. Khabibullin. 2003. Genetic characterization of wild and domesticated common carp (*Cyprinus carpio* L.) populations from Uzbekistan. Aquaculture 218: 153–166.

Nakajima, T. 1986. Pliocene cyprinid pharyngeal teeth from Japan and west Asia Neogene cyprinid zoogeography. pp. 502–513. *In*: T. Uyeno, R. Arai, T. Taniuchi and K. Matsuura (eds.). Indo-Pacific Fish Biology. Ichthyological Society of Japan, Tokyo.

Nakajima, T. 1994. Cyprinid fishes. pp. 235–275. *In*: Research Group for Natural History of Lake Biwa (ed.). The Natural History of Lake Biwa. Yasaka shobo, Tokyo (in Japanese).

Nakajima, T. 2003. Common carp as an introduced fish. pp. 86–87. *In*: K. Nakai, T. Nakajima and A. Rossiter (eds.). Alien Species: Their Biology, Impact and Control. Lake Biwa Museum, Kusatsu (in Japanese).

Nelson, J.S. 2006. Fishes of the World. John Wiley & Sons, Hoboken, New Jersey.

Nevo, E., A. Bailes and R. Ben-Shlomo. 1984. The evolutionary significance of genetic diversity: ecological, demographic and life history correlates. pp. 13–213. *In*: G.S. Mani (ed.). Evolutionary Dynamics of Genetic Diversity. Springer-Verlag, Berlin, Heidelberg, New York, Tokyo.

Nikolsky, G.V. 1956. Fishes of the Amur-river Basin. Akademija Nauk SSSR Publications, Moscow.

Novelo, N.D., B. Gomelsky and K.W. Pomper. 2010. Inheritance and reliability of random amplified polymorphic DNA-markers in two consecutive generations of common carp (*Cyprinus carpio* L.) Aquac. Res. 41: 220–226.

Ohno, S., J. Muramoto, L. Christian and N.B. Atkin. 1967. Diploid-tetraploid relationship among old-world members of the fish family Cyprinidae. Chromosoma 23: 1–19.

Paaver, T. 1983. Biochemical genetics of common carp *Cyprinus carpio* L. Valgus Publ., Tallin.

Paaver, T. and R. Gross. 1991. Genetic variability of *Cyprinus carpio* L. stocks reared in Estonia. Soviet Genetics: 839–846 (translated from Genetika [26] 1990: 1269–1278).

Probst, E. 1949a. Vererbungsuntersuchungen beim Karpfen. Allgemeine Fischerei-Zeitung 74: 436–443.

Probst, E. 1949b. Der Blauling-Karpfen. Allgemeine Fischerei-Zeitung 74: 232–238.

Probst, E. 1950. Der Todesfaktor bei der Vererbung des Schuppenkleides des Karpfens. Allgemeine Fischerei-Zeitung 75: 369–370.

Probst, E. 1953. Die Beschuppung des Karpfens. Münchner Beiträge für Fluss- und Abwasserbiologie 1: 150–227.

Rudzinsky, E. 1928a. Über Kreuzungsversuche bei Karpfen. Fischerei-Zeitung 30: 593–597.

Rudzinsky, E. 1928b. Über Kreuzungsversuche bei Karpfen. Fischerei-Zeitung 31: 613–618.

Šlechtova, V., V. Šlechta, M. Flajšhans and D. Gela. 2002. Protein variability in common carp (*Cyprinus carpio*) breeds in the Czech Republic. Aquaculture 204: 241–242.

Srivastava, A., W.L. Rogers, C.M. Breton, L. Cai and R.L. Malmberg. 2011. Transcriptome analysis of *Sarracenia*, an insectivorous plant. DNA Res. 18: 253–261.

Steffens, W. 1966. Die Beziehungen zwischen der Beschuppung und dem Wachstum sowie einigen meristischen Merkmalen beim Karpfen. Biologisches Zentralblatt 85: 273–287.

Steffens, W. 1980. Der Karpfen, *Cyprinus carpio*. Die Neue Brehm-Bücherei, 5. Auflage. A. Ziemsen Verlag, Wittenberg Lutherstadt.

Svetovidov, A.N. 1933. Über den europäischen und ostasiatischen Karpfen (*Cyprinus carpio* L.). Zoologischer Anzeiger 104: 257–268.

Sumantadinata, K. and N. Taniguchi. 1990. Comparison of electrophoretic allele frequencies and genetic variability of common carp stocks from Indonesia and Japan. Aquaculture 88: 263–271.

Sun, X.W. and L.Q. Liang. 2004. A genetic linkage map of common carp (*Cyprinus carpio* L.) and mapping of a locus associated with cold tolerance. Aquaculture 238: 165–172.

Tanck, M.W.T., H.C.A. Baars, K. Kohlmann, J.J. Van der Poel and J. Komen. 2000. Genetic characterization of wild Dutch common carp (*Cyprinus carpio* L.). Aquac. Res. 31: 779–783.

Tautz, D. 1989. Hypervariability of simple sequences as a general source for polymorphic DNA markers. Nucleic Acids Res. 17: 6463–6471.

Taylor, J.S., Y. Van de Peer, I. Braasch and A. Meyer. 2001. Comparative genomics provides evidence for an ancient genome duplication event in fish. Philos. Trans. R. Soc. Lond. B Biol. Sci. 356: 1661–1679.

Thai, B.T., C.P. Burridge, T.A. Pham and C.M. Austin. 2005. Using mitochondrial nucleotide sequences to investigate diversity and genealogical relationships within common carp (*Cyprinus carpio* L.). Anim. Genet. 36: 23–28.

Thai, B.T., T.A. Pham and C.M. Austin. 2006. Genetic diversity of common carp in Vietnam using direct sequencing and SSCP analysis of the mitochondrial DNA control region. Aquaculture 258: 228–240.

Thai, B.T., C.P. Burridge and C.M. Austin. 2007. Genetic diversity of common carp (*Cyprinus carpio* L.) in Vietnam using four microsatellite loci. Aquaculture 269: 174–186.

Tomljanović, T., T. Treer, V.Č. Čubrić, T. Safner, N. Šprem, M. Piria, D. Matulić, R. Safner and I. Aničić. 2013. Microsatellite-based genetic variability and differentiation of hatchery and feral common carp *Cyprinus carpio* L. (Cyprinidae, Cypriniformes) populations in Croatia. Arch. Biol. Sci. 65: 577–584.

Torgunakova, O.A., V.E. Chrisanfov, V.K. Prizenko, A.K. Bogeruk, T.A. Egorova and S.K. Semyenova. 2012. Polymorphism of the cytochrome oxidase b gene (cyt b) in Russian populations of common carp and wild common carp. Russ. J. Genet. 48: 102–109.

Tsipas, G., G. Tsiamis, K. Vidalis and K. Bourtzis. 2009. Genetic differentiation among Greek lake populations of *Carassius gibelio* and *Cyprinus carpio carpio*. Genetica 136: 491–500.

Tsvetkova, L.I. 1969. Certain peculiarities of lipid metabolism in common carp of four different genotypes. pp. 190–202. *In*: K.A. Golovinskaya (ed.). Pond Fish Breeding. VNIIPRCH Publ., Moscow.

Tsvetkova, L.I. 1974. Comparative studies of one-year-old common carps of four different genotypes. I. Growth characteristics of the fishes of different genotypes under the conditions of individual and communal rearing. Trudy VNIIPRCH (Moscow) 23: 36–41.

Uyeno, T. and G.R. Smith. 1972. Tetraploid origin of the karyotype of catostomid fishes. Science 175: 644–646.

Vandeputte, M., M. Kocour, S. Mauger, M. Dupont-Nivet, D. De Guerry, M. Rodina, D. Gela, D. Vallod, B. Chevassus and O. Linhart. 2004. Heritability estimates for growth-related traits using microsatellite parentage assignment in juvenile common carp (*Cyprinus carpio* L.) Aquaculture 235: 223–236.

Vilizzi, L. 2012. The common carp, *Cyprinus carpio*, in the Mediterranean region: origin, distribution, economic benefits, impacts and management. Fisheries Manag. Ecol. 19: 93–110.

Wang, Z., P. Jayasankar, S.K. Khoo, K. Nakamura, K. Sumantadinata, O. Carman and N. Okamoto. 2000. AFLP fingerprinting reveals genetic variability in common carp stocks from Indonesia. Asian Fisheries Sci. 13: 139–147.

Wang, C. and S. Li. 2004. Phylogenetic relationships of ornamental (koi) carp, Oujiang color carp and Long-fin carp revealed by mitochondrial DNA COII gene sequences and RAPD analysis. Aquaculture 231: 83–91.

Wang, C., H. Liu, Z. Liu, J. Wang, J. Zou and X. Li. 2010a. Mitochondrial genetic diversity and gene flow of common carp from main river drainages in China. Freshw. Biol. 55: 1905–1915.

Wang, C., S. Li, Z.T. Nagy, I. Lehoczky, L. Huang, Y. Zhao, X. Song and Z. Jeney. 2010b. Molecular genetic structure and relationship of Chinese and Hungarian common carp (*Cyprinus carpio* L.) strains based on mitochondrial sequence. Aquac. Res. 41: 1339–1347.

Wang, X., X. Zhang, W. Li, T. Zhang, C. Li and X. Sun. 2012a. Mapping and genetic effect analysis on quantitative trait loci related to feed conversion ratio of common carp (*Cyprinus carpio* L.). Acta Hydrobiol. Sinica 36: 177–196.

Wang, J., J. Li, X. Zhang and X. Sun. 2012b. Transcriptome analysis reveals the time of the fourth round of genome duplication in common carp (*Cyprinus carpio*). BMC Genomics 13: 96.

Wang, J., A. He, Y. Ma and C. Wang. 2012c. Genetic map construction and quantitative trait locus (QTL) analysis on growth-related traits in common carp (*Cyprinus carpio* L.) Afr. J. Biotechnol. 11: 7874–7884.

Wang, B., P. Ji, J. Wang, J. Sun, C. Wang, P. Xu and X. Sun. 2013. The complete mitochondrial genome of the Oujiang color carp, *Cyprinus carpio* var. *color* (Cypriniformes, Cyprinidae). Mitochondrial DNA 24: 19–21.

Welcome, R.L. (comp.). 1988. International introductions of inland aquatic species. FAO Fish. Tech. Pap. 294. Rome.

Wlodek, J.M. 1963. Der blaue Karpfen aus der Teichwirtschaft Landek. Acta Hydrobiologica 5: 383–401.

Wohlfarth, G., M. Lahman and R. Moav. 1963. Genetic improvement of carp. IV. Leather and line carp in fish ponds of Israel. Bamidgeh 15: 3–8.

Wolfe, K.H. 2001. Yesterday's polyploids and the mystery of diploidization. Nat. Rev. Genet. 2: 333–341.

Wright, J.M. 1993. DNA fingerprinting in fishes. pp. 58–91. *In*: P.W. Hochachka and T. Mommsen (eds.). Biochemistry and Molecular Biology of Fishes. Elsevier, Amsterdam.

Wu, X.W. 1977. Monographs of Cyprinidae in China. Shanghai People's Publisher, Shanghai.

Xu, P., J. Wang, J. Wang, R. Cui, Y. Li, Z. Zhao, P. Ji, Y. Zhang, J. Li and X. Sun. 2011. Generation of the first BAC-based physical map of the common carp genome. BMC Genomics 12: 537.

Xu, L.H., C.H. Wang, J. Wang, Z.J. Dong, Y.Q. Ma and X.X. Yang. 2012a. Selection pressures have driven population differentiation of domesticated and wild common carp (*Cyprinus carpio* L.). Genet. Mol. Res. 11: 3222–3235.

Xu, J., P. Ji, Z. Zhao, Y. Zhang, J. Feng, J. Wang, J. Li, X. Zhang, L. Zhao, G. Liu, P. Xu and X. Sun. 2012b. Genome-wide SNP discovery from transcriptome of four common carp strains. PLoS ONE 7(10): e48140.

Yousefian, M. 2011. Genetic variations of common carp *(Cyprinus carpio* L.) in south-eastern part of Caspian Sea using five microsatellite loci. World J. Zool. 6: 56–60.

Yue, G.H., L. David and L. Orban. 2007. Mutation rate and pattern of microsatellites in common carp (*Cyprinus carpio* L.) Genetica 129: 329–331.

Zardoya, R. and I. Doadrio. 1999. Molecular evidence on the evolutionary and biogeographical patterns of European cyprinids. J. Mol. Evol. 49: 227–237.

Zhang, H., N. Okamoto and Y. Ikeda. 1995. Two c-*myc* genes from a tetraploid fish, the common carp (*Cyprinus carpio*). Gene 153: 231–236.

Zhang, Y., L.Q. Liang, Y.M. Chang, N. Hou, C.Y. Lu and X.W. Sun. 2007. Mapping and genetic effect analysis of quantitative trait loci related to body size in common carp (*Cyprinus carpio* L). Yi Chuan 29: 1243–1248 (in Chinese).

Zhang, Y., L. Liang, P. Jiang, D. Li, C. Lu and X. Sun. 2008a. Genome evolution trend of common carp (*Cyprinus carpio* L.) as revealed by the analysis of microsatellite loci in a gynogenetic family. J. Genet. Genomics 35: 97–103.

Zhang, Y.F., Y. Zhang, C.Y. Lu, D.C. Cao and X.W. Sun. 2008b. Correlation analysis of microsatellite DNA markers with body weight, length and height of common carp (*Cyprinus carpio* L.). Yi Chuan 30: 613–619 (in Chinese).

Zhang, Y., C.Y. Lu, D.C. Cao, P. Xu, S. Wang, H.D. Li, Z.X. Zhao and X.W. Sun. 2010. Rates and patterns of microsatellite mutations in common carp (*Cyprinus carpio* L.). Zool. Res. 31: 561–564.

Zhang, Y., E. Stupka, C.V. Henkel, H.J. Jansen, H.P. Spaink and F.J. Verbeek. 2011. Identification of common carp innate immune genes with whole-genome sequencing and RNA-seq data. Journal of Integrative Bioinformatics 8: 169.

Zhang, X., Y. Zhang, X. Zheng, Y. Kuang, Z. Zhao, L. Zhao, C. Li, L. Jiang, D. Cao, C. Lu, P. Xu and X. Sun. 2013a. A consensus linkage map provides insights on genome character and evolution in common carp (*Cyprinus carpio* L.) Mar. Biotechnol. 15: 275–312.

Zhang, Y., S. Wang, J. Li, X. Zhang, L. Jiang, P. Xu, C. Lu, Y. Wan and X. Sun. 2013b. Primary genome scan for complex body shape-related traits in the common carp *Cyprinus carpio*. J. Fish Biol. 82: 125–140.

Zhao, L., Y. Zhang, P. Ji, X. Zhang, Z. Zhao, G. Hou, L. Huo, G. Liu, C. Li, P. Xu and X. Sun. 2013. A dense genetic linkage map for common carp and its integration with a BAC-based physical map. PLoS ONE 8(5): e63928.

Zheng, X., Y. Kuang, X. Zhang, C. Lu, D. Cao, C. Li and X. Sun. 2011. A genetic linkage map and comparative genome analysis of common carp (*Cyprinus carpio* L.) using microsatellites and SNPs. Mol. Genet. Genomics 286: 261–277.

Zhou, J.F., Z.W. Wang, Y.Z. Ye and Q.J. Wu. 2003a. PCR-RFLP analysis of mitochondrial DNA ND5/6 region among three subspecies of common carp (*Cyprinus carpio* L.) and its application to genetic discrimination of subspecies. Chinese Sci. Bull. 48: 465–468.

Zhou, J.F., Q.J. Wu, Y.Z. Ye and G.J. Tong. 2003b. Genetic divergence between *Cyprinus carpio carpio* and *Cyprinus carpio haematopterus* as assessed by mitochondrial DNA analysis, with emphasis on origin of European domestic carp. Genetica 119: 93–97.

Zhou, J., Q. Wu, Z. Wang and Y. Ye. 2004a. Molecular phylogeny of three subspecies of common carp *Cyprinus carpio*, based on sequence analysis of cytochrome b and control region of mtDNA. J. Zool. Syst. Evol. Res. 42: 266–269.

Zhou, J., Q. Wu, Z. Wang and Y. Ye. 2004b. Genetic variation analysis within and among six varieties of common carp (*Cyprinus carpio* L.) in China using microsatellite markers. Genetika 40: 1389–1393.

Zhu, J., L. Zhong, C. Zhang, H. Liu and B. Li. 2011. Sequence variation and secondary structure analysis of the first ribosomal internal transcribed spacer (ITS-1) between *Cyprinus carpio carpio* and *C. carpio haematopterus*. Biochem. Genet. 49: 20–24.

2

Japanese Ornamental Koi Carp: Origin, Variation and Genetics

Servaas de Kock[1,*] and Boris Gomelsky[2]

Introduction

The ornamental form of the carp, *Cyprinus carpio* L., provides a rich source of investigation for both science and commercial possibilities . For researchers, however, there is also the challenge of cultural and language barriers that add a dimension of the mystic. On the other hand, many koi hobbyists and professionals are not familiar with published scientific information on koi genetics. This chapter intends to better equip all kinds of readers with an understanding of the origin, variation and genetics of koi.

This chapter was written collectively by a koi professional and writer, who ran a large koi farm and authored several books on koi keeping (De Kock and Watt 2006), and a fish geneticist, who has studied inheritance of different traits in koi for 20 years.

Emergence of Koi

At least three strains of common carp were traditionally reared by farmers in the mountainous Yamakoshi region known as Yamakoshi Nijuusongou (Twenty Villages) between Nagaoka and Ojiya cities in Niigata Prefecture north-west of Tokyo (Japan) during the first year of Tenmei era (1781–1788)

[1] KoiNet, PO Box 1643, Gans Bay, 7220, Cape, RSA.
Email: sdk@koinet.net
[2] Aquaculture Research Center, Kentucky State University, 103 Athletic Drive, Frankfort, KY 40601, USA.
Email: boris.gomelsky@kysu.edu
* Corresponding author

(Koshida 1931). The fry were usually bought from merchants at Uono River and other rivers and brought inland to be reared in wooden house ponds or small terraced paddy fields and reservoir ponds, to supplement their diet when isolated during high snowfalls every year. The farmers eventually started to culture carp themselves, sharing the paucity of resources amongst them. There are three native feral variants namely *asagi* (multicolor), *yamato* (reddish) and *magoi* (blackish) of which the first two found favor in the culture ponds of the Japanese because of good growth and size (Okada 1960, Okubo and Sato 1975, Suzuki and Yamaguchi 1980). The word *magoi* (true carp) being a general term referring to the wild carp in the Japanese rivers. In the Nishikigoi fraternity, however, koi ancestry is traced to the *asagi, tetsu* (iron) and *doro* (rust) variants of the *magoi*. A 1913 catalog of fishes of Japan (Jordan et al. 1913) lists under *Cyprinus carpio* the Sarasa as a form found in Shinano Prefecture, the river running through the Yamakoshi region.

Theories abound why this area in particular is recognized as the birthplace of Nishikigoi bordering on the mystic. The unusual environmental conditions are often cited, but an extended period of drought and famine during 1781–1788 impacted on stock that reduced the gene pool, occurrence of genetic mutations, idle time and the curiosity of man are probably the main reasons. According to local memory, it was shortly after this that colored aberrations appeared. These were called *kawarigoi* (changed carp) and included all types of single and patterned forms including *higoi* (red carp), *shirogoi* (white carp), *magoi* (true carp or black carp), *asagi* (blue carp), *ironashigoi* (colorless carp) and *moyogoi* (patterned carp) (Koshida 1931, Kuroki 1981, Kuroki 1987, Kataoka 1989, Hoshino and Fujita 2009). During the pre-1915 period the development concentrated on the red and white pattern—first the simpler *Sarasa* and later the more florid *Sakura*. As the color drawings used for identification during the 1914 Taisho Expo in Tokyo clearly show, these were not modern koi (Fig. 2.1). Most of the effort went into stabilizing the red-white pattern on the dorsal parts and these fish were known by their different pattern name and collectively as *irogoi* (color carp), *hanagoi* (flower carp) and *moyogoi* (patterned carp). The Shusui originated about the same time from crossing domestic cultured *asagi* with imported leather carp from Europe (see below).

Irogoi improved rapidly in the period after 1915 to about 1942 with the creation of the early Kohaku out of the Sakura and tri-color variety Taisho Sanshoku (a white koi with red pattern and black patches) and other varieties through crossing and selection. Largely still isolated and under harsh environmental and political conditions, limited natural resources and no technological assistance, peasants steadfastly kept up development with a new vigor created by the public exposure of the 1914 Expo. By 1929 the first Yamakoshi Farming Koi Show was held in Higashiyama. The Showa Sanshoku (a black koi with white and red superimposed) tricolor was established by 1935. Thus the popular *gosanke* group that includes the three major color types Kohaku, Sanke and Showa, was formed (Fig. 2.2).

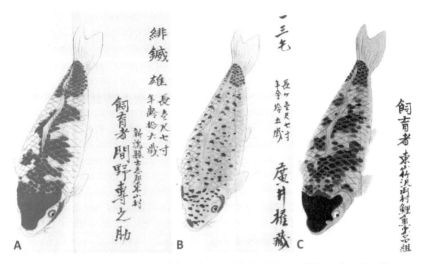

Figure 2.1. Catalog drawings of some fish at the 1914 Taisho Expo in Tokyo showing the state of development. A. Kohaku; B. Gonzo Sanke; C. Kurokihan (Classic Ki Utsuri) (Reproduced with permission from owner Yokio Isa).

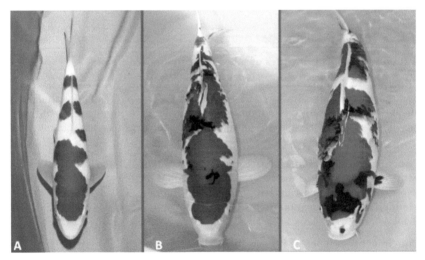

Figure 2.2. *Gosanke*. The three koi varieties most in demand today. A. Kohaku, one year-old (*tosai*), 18 cm TL; B. Sanke, 65 cm TL; C. Showa, 57 cm TL. The Sanke and Showa are two year-old (*nisai*) immature females. Note the different red hues of these 'unfinished' fish. Color not yet fully developed and body conformation incomplete.

Development came to a virtual halt during the World War II, but then resumed though still mainly isolated and practiced as a hobby. Only with improved air travel (around 1950) and packing material and technology (around 1963) did exports become viable which further stimulated production. Improved marketing by organized koi shows (around 1960) further increased demand. As the international marketing machine gained momentum from 1970 to 1990, koi became a global commodity. Masayuki Amano claimed in 1971 that up to 87% out of 908 families in the Yamakoshi region were producers of 'fancy carp'. Development was stepped up and resulted in as many as 120 different popular varieties known today. Production increased in several other centers including Hiroshima, Okayama, Karume, Shizuoka and Yamanashi where the land was more flat and the conditions more favorable and warmer. Breeders like Manabu Ogata, Kentaro Sakai and Michio Maeda added a different approach to production and marketing by producing both quantity and quality to an ever-discerning world market. A new name, Nishikigoi (錦鯉, brocaded carp) emerged when efforts were made to promote the trade in the 1970's. This name was originally expressed in 1918 by Mr. Kei Abe, but never caught on. He was head of the Niigata Prefecture's Fisheries Department and according to Koshida (1931), was earlier responsible for introducing Mendel's principles of inheritance to the farmers of Yamakoshi during lecture tours. Dr. Takeo Kuroki was instrumental in the formation of the *Zen Nippon Airinkai* (ZNA) in May 1968 to promote the hobby. In 1970 the breeders and trade followed with the *Zen Nihon Nishikigoi Shinkokai* (All Japan Nishikigoi Promotion Association) that in 2003 became the International Nishikigoi Promotion Center (INPC).

Infobox: Mendel's principles of inheritance in modern terms:

Principle of dominance: Each gene is represented by two copies located in **homologous chromosomes**. There are different forms of one gene—**alleles**. One allele is **dominant** (*A*) to the other **recessive** allele (*a*). A **homozygous** individual has the same alleles of gene (*AA* or *aa*), while a **heterozygous**—has different alleles (*Aa*).

Principle of segregation: During formation of gametes alleles segregate so that each gamete receives one or the other allele with equal likelihood. Upon fertilization, two gametes combine randomly producing different allele combinations. Basic **Mendelian ratios** for **monohybrid cross** (one gene is involved) are: **1:0** (crosses *AA* x *AA*, *AA* x *Aa*, *AA* x *aa*), **1:1** (cross *Aa* x *aa*) and **3:1** (cross *Aa* x *Aa*).

Principle of independent assortment: During gamete formation segregating alleles of different genes assort independently of each other. Basic **Mendelian ratio** for **dihybrid cross** (two genes are involved) is **9:3:3:1** (*AaBb* x *AaBb*). Different interactions between genes can modify this ratio to 9:7 (complimentary gene action), 9:3:4 (recessive epistasis), 12:3:1 (dominant epistasis), 13:3 (recessive suppression) and 15:1 (duplicate genes).

Stringent selection and culling is necessary in order to deliver koi for sale as well as future breeding material. This is a variety specific operation whereby between 5–40% of fry survive a selection at the age of 5–40 days based on color variation and developmental criteria. It is followed up with 3–5 further selections over the first year of the fish's life reducing the numbers of surviving juveniles to 1–3% of the original spawn.

As the hobby grew internationally, so farms sprung up all over the world driven by the need to meet demand. The respected Peter Waddington (Waddington 2009) estimated that the period of the 'serious koi hobbyist' in Japan was from 1965 to 1990 and peaked in 1978. He wrote: "After 1990, the domestic market for koi continued to decline annually to become as it is today in 2008 when only the very low grade koi produced by the breeders are sold to pet shop suppliers in volume" and "the Japanese Nishikigoi market in and around 1969 formed almost 100% of production in the world whilst in 2007 it only amounted to under 20% and that includes koi of ALL prices and qualities." That may be true with respect to the traditional markets in the West. He need not have worried. New markets in South-East Asia are creating a seemingly insatiable market for the higher quality product and fuel further development. The idiosyncrasies of different cultures also drive a continued range of varieties. The Japanese will remain ahead in the developmental and technological departments of this industry for a long time to come.

Genealogy of Nishikigoi

Various attempts were made to piece together the genealogy of koi based on breeders' notes and other accounts (Koshida 1931, Kuroki 1987, Watt and De Kock 1996, Hoshino and Fujita 2009). There are also numerous accounts from anonymous writers in the Japanese Nichirin and Rinko magazines, but they have no supporting evidence. The resultant tree, which is shown in Fig. 2.3, remains shrouded in myth, and questions need to be asked if the endeavor is worth it, but it remains of interest to the hobby and keen hobbyists. For breeders it may be a tool to plan their breeding, whether of value or not.

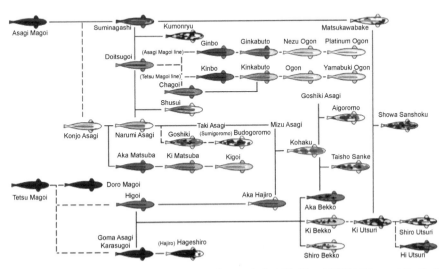

Figure 2.3. Gene tree of Nishikigoi (Adapted after Koshida 1931, Kuroki 1987, Watt and De Kock 1996, Hoshino and Fujita 2009 Redrawn by Gertan Agenbach).

For inexperienced breeders and hobbyists, it may help to put the different varieties in perspective.

The main events leading to the development of Nishikigoi today are summarized in Table 2.1 from data in an anonymous document in the files of the *Japan Nishikigoi Promotion Association*, together with extensive studies by Hoshino and Fujita (2009) and some other authors.

Evaluation Criteria

Because koi is an ornamental fish, its value is affected by perceived qualities and breeders aim to improve those qualities through parentage and selection criteria. While color and color pattern play a part in determining the value and marketability of mass produced, commercial fish, those are not prime factors in determining the value of individual, high quality fish from well-known producers. These breeders created lines that have shown consistent, predictable and sought after qualities. Great value is attached to such a lineage (De Kock and Watt 2006). The essential elements determining the value of fish that breeders try to improve genetically are listed below.

Body conformation. From a human perspective, the koi should have a symmetric, tub-like body with dynamic, broad shoulders. The body gives the producer insight into the potential for growth over the adult life of the koi and, therefore, is one of the most important value determinants. Late maturation and lower fecundity ensure a longer 'show life'. Males with a plumper body form can compete with the females in the show arena for a longer period, but reproduce better in an artificial, one-on-one environment without the competition of slimmer suitors.

Swimming style. A quality of 'graceful' swimming is sought. For older fish words like 'dignity' or 'character' may also be added. Swimming style is the aesthetic value of the biokinetic expression of movement, directly related to skeletal and muscular qualities of the fish. 'Gracefulness' in the swimming style refers not only to powerful but fluid movement but also to the calming effect it has on the viewer. Feeding behavior is the primary cause of this movement, and possibly tameness brought on through domestication and repeated handling may also help in creating this illusion. To swim 'lively' fish must be healthy and in an environment conducive for feeding, i.e., low in ammonia concentration and high in dissolved oxygen. Failing these conditions the koi will exhibit a 'lethargic' or 'listless' swimming style. On the other end of the scale 'frantic' swimming is exhibited when the fish are frightened or tries to escape near-toxic, unfavorable water conditions.

Color quality. The brightness is determined by the number of chromatophores in the skin and their propensity for collecting pigments, normally during their development but also later, from the environment. The purity of color to separate into aesthetically pleasing colors and patterns as viewed through different skin layers. Males tend to be brighter at an early age increasing their initial show value.

Table 2.1. Summary of the key historical events* in the emergence of koi in Japan.

Variety	Date	Place	Originator	Remarks
	1781+			severe drought and famine
Irogoi	1804+	NijuuYamakoshi		emergence from probably Asagi
Asagi Magoi	1868	Ueda River area	Jirohachi Seki	Improving of Asagi line as ornamental
GosukeSarasa	1890	Ojiya	KunizouHiroi	forerunner of Kohaku (Sakura)
Gonzo Sanke	1898			emergence from Asagi magoi
Ki Utsuri	1898	Takezawa	Komasaburo Hoshino	emerged from Tetsu magoi
Doitsugoi	1904	Fukagawa	Shinnosuke Matsubara	introduction of German leather carp
Shusui	1908	Akiyama Fish Farm	Kichigoro Akiyama	German leather carp x Asagi
Akame Kigoi	1914	Akiyama Fish Farm		red eye Kigoi emerged
Sanshoku precursor	1914	Uragara	Heitaro Sato	Shirogoi x Kohaku = tejima-Kohaku
Sanshoku	1917	Yamakoshi	Eisaburo Hoshino	tejima-Kohaku x Bekko
Ki Utsuri	1920	Yamakoshi	Sato Yohei	Ki Bekko x Magoi
Ki Utsuri	1920	Yamakoshi	Eisaburo Hoshino	Ki Utsuri x Ki Utsuri (fixed)
Kigoi	1922	Yamakoshi	Yoshiichi Matsui	improving of Akame Kigoi (red eye)
Kin Kabuto	1924	Yamakoshi	Isematsu Takano	Kinbo x Fijigin
Shiro Utsuri	1925	Yamakoshi	KazouMinemura	Ki Utsuri Sanke x Shiro Bekko
Jintoro Showa	1927	Yamakoshi	Jukichi Hoshino	Ki Utsuri x Matsukawabake x Kohaku
Ginrin Kohaku	1929	Yamakoshi		first exhibition of ginrin
Ossa Showa	1935	Ojiya	Kishichirou Hoshino	Showa x Shiro Utsuri (fixed)
Doitsu Sanke	1941	Yamakoshi	Tahichi Kawakami	
Ogon	1947	Ojiya	Sawata Aoki	Kinkabuto x Ginfushi
DoistuYotsushiro	1950		KunioHiroi	Karasu x Hajiro appear spontaneous
Kumonryu	1953	Yamakoshi	KunioHiroi	DoistuYotsushiro x Karasu lineage
OranjiOgon	1956	Yamakoshi	Masamoto Kataoka	Asagi x Ogon (1963 fixed)
Yamabuki Ogon	1957	Yamakosh	Masamoto Kataoka	
Doitsu Ogon	1958	Yamakoshi	Tomisaku Sakai	Black Doitsu x Ogon

Table 2.1. contd....

Table 2.1. contd.

Variety	Date	Place	Originator	Remarks
Kin Utsuri	1958	Yamakoshi	Kumazo Takahashi	Ogon x Ki Utsuri
Doistu Showa	1959	Yamakoshi	KanekichiFujii	
Hariwake Ogon	1960	Yamakoshi	Tomisaku Sakai	Ogon (4x individuals)
Midorigoi	1960		Kiichiro Suzuki	
Matsuba Ogon	1960	Ojiya	Eizaburo Mano	
Kujaku	1961	Ojiya	Toshio Hirasawa	Matsuba Hariwake x Shusui
ParachinaOgon	1963	Uodu, Toyama.	Tadao Yoshioka	Akame Kigoi x Nezu Ogon
Yamatonishiki	1965		Seikichi Hoshino	
Koshi no Hisoku	1965	Uodu, Toyama	TadaoYoshioha	Shusui x Yamabuki Ogon
Beni Hikari	1990	Ojiya		Doitsu Goshiki x Doitsu Kujaku
Kinkiryu	1990	Yamakoshi	Ichiji Watanabe	Doitsukinryu x Doitsukinshowa Kinryu = Kumonryu x Gin Matsuba
Kikokuryu	1992	Ojiya	Haruo Aoki	Kumonryu x Kikusui
Beni Kikokuryu	1992	Ojiya	Haruo Aoki	Kumonryu x Kikusui
Ginka	1998	Ojiya	Kiyoshi Kase	Ginrin Hajiro x Goshiki x Motoguro Kujaku

*Adapted from data from various anonymous sources, the *Japan Nishikigoi Promotion Association*, *Rinko* and *Nichirin* magazines, and *Nishikigoi Mondo* by Hoshino and Fujita (2009).

Color durability. The ability to maintain or increase the pigment 'loading' of the chromatophores through synthesis and the ability of pigment cells to migrate through different skin layers to 'develop' a pleasing pattern can add to the value of an individual. Intracellular reaction of chromatophores—in particular melanophores—and likely cyanophores to environmental stressors can reduce the value. This could be under both genetic and environmental control.

Color distribution. An aesthetically pleasing pattern to match the particular variety is normally determined by selection. Patterns are not repeatable, but pattern types, pattern edging and shapes do run in families and can be recognized and thus predicted to an extent. Therefore specific bloodlines are more sought-after than others.

These attributes encompass the guidelines set out to nationally accredited judges when judging at koi shows worldwide. As described in Hoshino and Fujita (2009) and De Kock and Watt (2006) judges using a 100-point system award up to 50% to 'body shape' evaluation and 20% to color evaluation. Pattern, 'gracefulness' and 'dignity' carry a 10% contribution each to the overall evaluation. The importance of an aesthetic pleasing body and movement is left in no doubt.

Intentional hybridization and line breeding lead to improvement and proliferation of varieties while stringent selection and culling regime lead to increase in quality and marketability of the variety. This labor-intensive task calls for skilled people of which few are outside of Japan. In Taiwan and perhaps a handful of dedicated producers worldwide, follow the tradition of improving their production and skills. For the rest the resultant breeding is without purpose other than for a multitude of small, colorful fish and thus the world perception of "quality" is compromised (De Kock and Watt 2006).

Nishikigoi do not breed true and always require severe culling during the first culling to remain true due to the variety's constraints. The single colored types like Ogon, Purachina and Chagoi may be exceptions in terms of color, but they do also require rigorous selection for sought-after qualities. For example, at least 60–65% of a Kohaku paring is removed at about 40 days of age. Likewise, 75–80% of a Sanke cross is removed at first culling at 25–30 days of age. With Showa only the black pigmented fry being 5–50% survive the first culling by the 3–5th day after hatching. At least two more culling operations are carried out before the age of six months illustrating how much value is placed on the individual gene make-up.

The Making of Color

While the morphological variation was important in the creation of koi as they are today, the expression of color is certainly central to defining koi as ornamental. Skin pigment cells produce the various colors. Erythrophores and xanthophores contain carotenoid pigments and produce various shades of red or yellow. Melanophores are black and contain melanin. In addition, further hues like green and light blue are formed due to the superposition of pigment cells in the skin.

Leucophores contain granules of guanine resulting in a white appearance. Iridophores contain iridescent plate-like crystals of guanine resulting in a shining, metallic appearance of the skin also known as *hikari*. In combination with yellow, the effect can be golden (e.g., Ogon) and with white it could be silver like Purachina. Platelets of guanine form in the scale structure a radiant, reflecting surface known as *ginrin*. On a red or yellow background there is a golden reflection and when overlaying black, blue or white, silver is reflected.

Cyanophores contain a blue pigment of unknown chemical composition. The melanophores and cyanophores pigment is stored in fibrous organelles that can open or contract under control of the central nervous system. Chromatosomes are derived from the neural crest and migrate to their positions during ontology of the embryo. The melanophores and cyanophores follow the internal route while the other future pigment cells migrate alone to the outer layers of the dermis. Probably this is the reason that red, yellow and white appeared to be overlaying black and blue initially.

Aspects of chromatophores that are exploited during breeding and selection are: (1) pigment cell formation and migration; (2) pigment synthesis; (3) pigment cell translocation; (4) pigment cell interaction between the two main groups of chromatophores (xanthophore-like and melanophore-like cells) and (5) time dependency of all these over the life of the individual.

Inheritance of Color Traits

This part of the chapter has a review of known scientific information on the inheritance of color traits in koi.

The first studies on color inheritance in koi, which became known to the English-speaking scientific community, were performed in Russia (former Soviet Union). In the middle of the 1960s the Japanese government gave a group of koi to the Soviet government as a gift. These fish were delivered to the Research Institute of Freshwater Fish Culture, which is located not far from Moscow, where Dr. V. Katasonov has performed series of studies on color inheritance and described several color-determining genes in koi. Katasonov (1978) showed that melanin formation in wild-type color carp is controlled by dominant alleles of two duplicate (i.e., having similar action) genes designated as B_1/b_1 and B_2/b_2. The presence of one dominant allele at any gene in fish genotype results in the appearance of wild-type color. When wild-type color common carp, homozygous for dominant alleles at both genes (genotype $B_1B_1B_2B_2$), was crossed with koi (genotype $b_1b_1b_2b_2$) the F_1 progeny consisted of wild-type color fish only (genotype $B_1b_1B_2b_2$). When F_1 fish were crossed, the phenotypic ratio 15:1 was observed in F_2 progeny; this ratio is typical for duplicate gene action. The scheme and results of these crosses are given in Fig. 2.4.

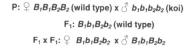

P: ♀ $B_1B_1B_2B_2$ (wild type) x ♂ $b_1b_1b_2b_2$ (koi)

F_1: $B_1b_1B_2b_2$ (wild type)

F_1 x F_1: ♀ $B_1b_1B_2b_2$ x ♂ $B_1b_1B_2b_2$

GAMETES

♀ \ ♂	B_1B_2	B_1b_2	b_1B_2	b_1b_2
B_1B_2	$B_1B_1B_2B_2$ wild type	$B_1B_1B_2b_2$ wild type	$B_1b_1B_2B_2$ wild type	$B_1b_1B_2b_2$ wild type
B_1b_2	$B_1B_1B_2b_2$ wild type	$B_1B_1b_2b_2$ wild type	$B_1b_1B_2b_2$ wild type	$B_1b_1b_2b_2$ wild type
b_1B_2	$B_1b_1B_2B_2$ wild type	$B_1b_1B_2b_2$ wild type	$b_1b_1B_2B_2$ wild type	$b_1b_1B_2b_2$ wild type
b_1b_2	$B_1b_1B_2b_2$ wild type	$B_1b_1b_2b_2$ wild type	$b_1b_1B_2b_2$ wild type	$b_1b_1b_2b_2$ koi

F_2: 15 wild type : 1 koi

Figure 2.4. Inheritance of wild-type color in common carp (duplicate gene action).

The presence of melanin in wild-type color fish is visible already at late embryonic and larval stages while koi embryos and larvae are yellowish and transparent because of lack of black pigment. The control of wild-type color in common carp by two duplicated genes has been confirmed later in several further studies. For example, Cherfas et al. (1992) crossing wild-type color female, heterozygous for two genes (genotype $B_1b_1B_2b_2$), with a koi male (genotype $b_1b_1b_2b_2$) obtained a 3:1 phenotypic ratio in the resulting progeny. This ratio is typical for a test cross in case of duplicate genes.

Later Katasonov et al. (2001) suggested the existence of third alleles at these two duplicate genes (designated as B'_1 and B'_2, respectively), which control the presence of melanin in lower (underlying) skin layer; the dominance relationships between the three alleles was proposed as $B > B' > b$. Katasonov (1974, 1978) also identified that blue body color in koi is controlled by the recessive allele of gene R/r. Blue koi with genotype rr do not have red and yellow pigments. Combination of recessive alleles of the three genes-r, b_1 and b_2 (genotype $rrb_1b_1b_2b_2$) gives the white color of fish. White fish are characterized by the lack of both black and red (yellow) pigments. Katasonov (1973, 1974) also revealed that the inheritance of the trait 'design' which is typical for metallic koi. This trait manifests as a yellow stripe along the dorsal fin combined with a specific ornament on the head and is controlled by the dominant allele of gene D/d. Metallic koi have genotypes DD or Dd while non-metallic koi (genotype dd) are homozygous for recessive allele. When metallic koi are crossed with wild-type color common carp so called 'ghost koi' (or 'ghost carp') appear. Ghost koi (Fig. 2.5) have traits which are controlled by dominant alleles of different genes—clearly visible 'design' from metallic koi and black pigment melanin from wild-type color carp. Although ghost koi is not accepted by koi show standards, this color type is pretty popular in some countries.

Katasonov (1973, 1974, 1976) has also identified trait 'light color' in koi. The light color of fish, which possess this trait, is caused by permanent contraction of melanophores. Fish having this trait are heterozygous for gene L/l (genotype Ll). Homozygotes for dominant alleles (genotype LL) are inviable and perish at larval or fry stage.

Figure 2.5. Ghost koi—the hybrid between metallic koi and wild-type color carp.

Among multicolor traits in koi, the inheritance of the white-red (Kohaku) color complex and Bekko type of black patches, which is observed, for example, in Taisho-Sanke or Shiro-Bekko, have been investigated most frequently. Some information was obtained on the inheritance of the Utsuri type of black patches, which is observed, for example, in Showa-Sanshoku or Shiro-Utsuri. Iwahashi and Tomita (1980) in a study performed at Niigata Prefectural Inland Water Fisheries Experimental Station (Japan) have investigated inheritance of the Kohaku color complex. White-red (Kohaku) fish, as well as solid white (Shiromuji) and solid red (Akamuji) fish originating from Kohaku, were reproduced in three generations and the ratios of color phenotypes in progenies have been investigated. In progenies obtained by crossing of white-red (Kohaku) fish all three color types (solid red, white-red and solid white) were observed. The largest class was white-red fish (68–76%); percentages of solid red and solid white fish varied from 18 to 21% and from 6 to 11%, respectively. Percentages of Kohaku have not been increased in generations as a result of selection. In progenies obtained by crosses of solid-white fish, proportions of solid red fish were low (0–6%); while the percentage of solid white fish increased from 30% in first generation to 80 and 62% in second and third generations, respectively; correspondingly percentage of white-red fish decreased in the last two generations. Crossing of solid red fish gave only solid red fish in progenies in all three generations. Based on their obtained results, Iwahashi and Tomita (1980) concluded that solid red fish originated from Kohaku are pure breed but solid white fish originated of Kohaku are hybrids; Kohaku are also hybrids of solid red and solid white fish. Also, it was suggested that white-red color complex is controlled by numerous genes of white and red colors.

Gomelsky et al. (1996) investigated color variability in normal (amphimictic) and meiotic gynogenetic progenies produced from seven females of several multicolor traits (white-red–Kohaku, white-black–Shiro-Bekko, and white-red-black–Taisho-Sanke and Tancho-Sanke) in a study which was performed in Israel. Normal progenies obtained from Kohaku parents consisted of three color types—solid white, white-red (Kohaku) and solid red; the white-red class was most numerous with proportions varying from 45.7 to 57.2% (see below for data on color variability in gynogenetic progenies). Six color types were observed in progenies obtained from white-red-black (Taisho-Sanke and Tancho-Sanke) and white-black (Shiro-Bekko) parents: solid white, white-red, solid red, white-black, white-red-black and red-black. The white-red-black tricolor type was most numerous in all progenies, including those from white-black bicolor parents. In all normal and gynogenetic progenies obtained from white-red-black and white-black females, the ratios of white : white-red : red and white-black : white-red-black: red-black did not differ significantly. On this basis it was concluded that the occurrence or absence of black patches on the body and the white-red color complex are inherited independently and these traits were discussed and further studied separately (Gomelsky et al. 1996).

Based on large variability in ratios of white : white-red : red fish among progenies and in rate of development of red coverage in white-red fish (from a single tiny spot to nearly complete coverage) it was suggested that the white-red color complex is controlled by many genes (Gomelsky et al. 1996). In a further study on inheritance of white-red complex in koi performed in Israel (Gomelsky et al. 2003), color variability was investigated in three progenies obtained by crossing of white-red (Kohaku) parents with different rates of red-area body coverage. In this study the percentage of coverage by red patches was measured in a sample of white-red fish from each progeny. For measurement of red-area coverage, outlines of the entire body and red patches for both sides of the fish were traced on the plastic film with a permanent marker. The relative red-area coverage was determined as ratio of red patches outlines area to the area of total body outlines. The distribution of white-red fish in sample with regard to red-area coverage, together with the white : white-red : red ratio in progeny was used to estimate the color class distribution in the entire progeny (including the solid white and solid red individuals). In two from three investigated progenies obtained by Kohaku x Kohaku crosses, the frequency of solid red fish was much higher than the frequency of adjacent classes of white-red fish. Thus, in these progenies fish were clearly divided into two groups: fish with white background color (solid white and white-red) and solid red fish. As an example, distribution of fish in one progeny from this study (Gomelsky et al. 2003), which was obtained by crossing of Kohaku female and Kohaku male with 42.0 and 30.4% of body coverage with red patches, respectively, is presented in Fig. 2.6. From 215 fish analyzed in this progeny, 11 fish (5.1%) were solid white, 143 fish (66.5%) were

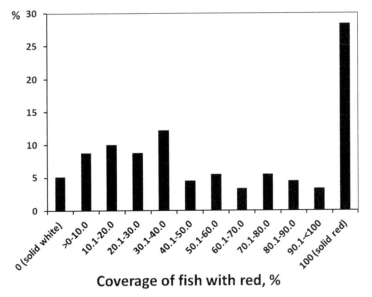

Figure 2.6. Distribution of fish in progeny obtained by Kohaku x Kohaku cross with regard to development of red color (after Gomelsky et al. 2003, with modifications).

white-red and 61 fish (28.4%) were solid red; the coverage of red patches was measured in 60 white-red fish from the progeny. The division of fish into two groups—with white background color (solid white and white-red) and solid red—is clearly visible in the class distribution presented in Fig. 2.6. It was suggested that the appearance of these groups might be explained by the existence of some major color-determining gene(s), which determine(s) the background color (either white or red) of the individual fish (Gomelsky et al. 2003). The development of red patches in fish with a white body background is polygenic and controlled by many genes with alleles that either maintain the white color or induce the appearance of red patches.

Later Novelo and Gomelsky (2009) used image analysis software for determination of red-area coverage in white-red koi in a study performed in the United States. The color variability distribution was described for one progeny obtained by crossing Kohaku x Kohaku; which was similar to the one in the previous study (Gomelsky et al. 2003), where fish were clearly divided into two groups: fish with white body color background and solid red fish. These studies (Gomelsky et al. 2003, Novelo and Gomelsky 2009) have shown that measuring the relative red body coverage provides an opportunity to investigate the development of red color in progenies obtained from Kohaku parents as a trait with continuous variability and provides further information for the better understanding of inheritance of this trait.

As mentioned above, the black pigmentation in wild-type color common carp appears during embryonic development and hatched larvae already have well-developed melanophores. Gomelsky et al. (1996) have described a different mode of development for Bekko type black pigmentation in koi. All hatched larvae obtained in crosses of parents with black pigmentation of Bekko type (Taisho-Sanke, Shiro-Bekko, Tancho-Sanke) expressed no black pigmentation, and developed it later, about one week after transition of larvae to active feeding. Only fish having black pigment at the fry stage develop black patches later. In two progenies obtained by crosses of white-red-black (Taisho-Sanke x Taisho-Sanke) or white-black (Shiro-Bekko x Shiro-Bekko) parents the ratios of pigmented and unpigmented fry did not differ significantly from Mendelian 3:1 ratio. On this basis it was suggested that presence of Bekko black patches is controlled by dominant allele of one gene (*Bl/bl*). Also it was shown (Gomelsky et al. 1996) that the ratio of pigmented: unpigmented fry more correctly reflected original segregation since during further rearing in ponds the proportion of fish with black patches sometimes increased apparently due to their better survival compared with fish without black patches. Gomelsky et al. (1998) have reported the results of a further study (performed in Israel) on inheritance of the Bekko type of black pigmentation in koi. Parents with Bekko pigmentation (Taisho-Sanke and Shiro-Bekko) and without black pigmentation (Kohaku) were used in crosses; a total of 22 progenies were obtained and analyzed. In this study the segregation of fish with and without black pigmentation was usually recorded at the fry stage, at the end of the nursing period. In five of six progenies obtained by Bekko x Bekko crosses typical Mendelian 3:1 segregation was observed. In 15 of 16 progenies obtained by Bekko x Kohaku crosses Mendelian segregation 1:1 was observed; in one progeny obtained by Bekko x Kohaku cross only black pigmented fry were observed. The data obtained in this study confirmed a suggestion made earlier (Gomelsky et al. 1996) that the Bekko type of black pigmentation in koi is controlled by the dominant allele (*Bl*) of one gene *Bl/bl*. Fish with Bekko pigmentation have genotype *Blbl* (most of fish used in crosses) or *BlBl* (one identified fish). It is known that the rate of Bekko pigmentation in koi is highly variable—from a few dispersed black spots to many black patches covering large areas of the body. Gomelsky et al. (1998) noted that the major gene *Bl/bl* determines only the presence or absence of black patches while the rate of their development is apparently under control of multiple genes.

David et al. (2004) reported results of studies (performed in Israel) on inheritance of red color development (Kohaku color complex) as well as Bekko and Utsuri types of black pigmentation in koi. About 40 crosses of white-red (Kohaku), solid white (the authors called this group 'transparent') and solid red fish parents in different combinations (including 20 crosses Kohaku x Kohaku) were obtained and analyzed. The 13 crosses of randomly chosen Kohaku parents gave three phenotypes in offspring: solid white, white-red and solid red; proportions of these three phenotypes were highly variable among crosses; for example, the proportion of white-red offspring ranged from 10.6 to 67.6%. The same three phenotypes were also found in crosses solid white x solid white, solid white x Kohaku, solid red x Kohaku, and solid white x solid red while crosses solid red x solid red resulted in red progeny only. The proportions of phenotypes were studied also in progeny of Kohaku parents with various levels of red coverage and an effect of individual parents was detected. Based on obtained results David et al. (2004) concluded that at least three genes, having intra- and inter-locus interaction, control development of red color in koi. The authors proposed that *Tp* (for transparent) is a locus that dominantly controls the absence of red color (David et al. 2004). When transparent parents are heterozygous at the *Tp* locus, they can have colored offspring. It was also suggested that two other genes (R^a and R^b for red) control the extent of red color, ranging from none to complete red coverage. The completely red parents are probably homozygous to one or the two *R* genes, and thus have only red progeny. It was also suggested that some transparent (solid white) fish that are determined by the R^a and R^b loci rather than by the *Tp* dominant allele should exist.

David et al. (2004) showed that Utsuri and Bekko patterns of black pigmentation appear at different development stages—around hatching and 14-day post-hatching, respectively; and this implies that these patterns are controlled by different genes. Nine Bekko x Bekko crosses and 14 Bekko x no black pigmentation crosses were performed and segregations in progenies (Bekko : no black pigmentation) were recorded and analyzed. An average about 75% of the offspring had the Bekko pattern in nine crosses between Bekko parents, fitting a Mendelian 3:1 ratio. When segregations in progenies were separately analyzed, two of the nine crosses deviated from a 3:1 but fitted to a 2:1 ratio, four deviated from a 2:1 but fitted to a 3:1 ratio and three fitted both ratios. From 14 crosses of Bekko x no black pigmentation, segregations in 13 crosses fitted a Mendelian 1:1 ratio while one cross deviated from this ratio. David et al. (2004) noted that the obtained results are in agreement with those of Gomelsky et al. (1998) who suggested the *Bl* gene having a dominant black-pattern allele. The authors (David et al. 2004) also suggested that the inheritance of the Bekko pattern may be somewhat more complicated; namely some selection against fish with *BlBl* genotype might be operating (probably during maturation). This suggestion was based on several observations. First, none of 17 Bekko parents used in crosses had *BlBl* genotype (all parents were heterozygotes *Blbl*). The author also referred to the study of Gomelsky et al. (1998) where out of 17 fish parents only one was shown to be a *BlBl* homozygote. Second, the variation in the proportion of Bekko offspring was higher in crosses between two Bekko parents than in crosses in which only one parent was a Bekko. Third, some Bekko x Bekko crosses gave segregations close to a 2:1 ratio (suggesting absence of homozygotes *BlBl*).

David et al. (2004) analyzed segregations in nine progenies obtained by Utsuri x Utsuri crosses. Progenies were separated into dark and light pigmented larvae, which were grown separately to the juvenile stage. The light pigmented larvae developed into transparent (solid white), Kohaku and solid red while dark-pigmented larvae developed into several color phenotypes including Utsuri but also wild-type carp color, gray, brown and combination of these colors. Among offspring of nine crosses between Utsuri parents, the proportion of Utsuri juveniles ranged from 0 to 16%, with an average of 4%. Also David et al. (2004) presented data on pigmentation of the hatched-out larvae in crosses where Utsuri parents were used. The average proportion of dark-pigmented larvae in 30 crosses of Utsuri x Utsuri was 24.5%, ranging from 1 to 54.7%. In 14 crosses where one parent was Utsuri and the other had no black pigmentation the average proportion of pigmented larvae was 15.3%, ranging from 0 to 40.9%. Based on the obtained data, David et al. (2004) suggested a complex genetic control of the Utsuri type of black pigmentation with possible environmental effects.

In a further study performed in Israel, Bar et al. (2013) have investigated the development of two types of black pigmentation (Utsuri and Bekko) in koi fry. In two progenies obtained by crossing between Utsuri and Bekko parents, two waves of black pigmentation were observed. Larvae pigmented at hatching became fry with Utsuri type of pigmentation. In groups of larvae which were unpigmented at hatching, the second wave of black pigmentation was observed later and late-pigmented fry developed Bekko type of pigmentation. In three progenies obtained by crossing Bekko x Bekko only the late wave of pigmentation was observed. In all three progenies the segregations of pigmented and unpigmented fry did not differ significantly from the 3:1 Mendelian ratio at some time point but later in two of the three progenies segregations shifted to 2:1. The authors noted that missing fry might be homozygotes for the Bekko pigmentation allele. In this study (Bar et al. 2013) the possible association between black pigmentation in koi and mutations in *Melanocortin receptor 1* (*mc1r*) gene, which cause variability in dark pigmentation in animals, has been investigated.

Recently Gomelsky and Schneider obtained data on timing of black pigmentation appearance and segregation of pigmented and unpigmented larvae in progenies obtained by crossing of Utsuri parents in a study performed in the United States (unpubl. data). One Hi Utsuri (black-red) male was crossed individually with two Hi Utsuri females; also one Hi Utsuri female was crossed with Kohaku male (no black pigmentation). Larvae at hatching had already segregated to pigmented and unpigmented groups as was described by David et al. (2004). Appearance of larvae at 3 day after hatching from Hi Utsuri x Hi Utsuri progeny is shown in Fig. 2.7. Observed segregations of pigmented and unpigmented larvae have also been in the range indicated by David et al. (2004) for corresponding crosses. Proportion of pigmented larvae in two crosses Hi Utsuri x Hi Utsuri were 19.2 and 29.5%, while only 5.6% of pigmented larvae were observed in Hi Utsuri x Kohaku progeny (Gomelsky and Schneider, unpubl. data).

The information presented above shows that the number of studies, which were devoted to investigation of color inheritance in koi, is not large. The scientific information on genetics of color traits in koi is scarce if to compare, for example, with goldfish (Smartt 2001) or some aquarium fish. Apparently, the main reasons of paucity of genetic studies in koi result from two peculiarities of koi culture and evaluation. First, koi culture to a large extent is based on strict and continuous culling. Most of koi color types are not true breeds, i.e., if fish of the same type are crossed; the progeny will be variable and will contain different color types. Second, multicolor koi are characterized by individual variability; the value of multicolor individual fish depends on color pattern, which is determined by size and location of color patches on the body. It is obvious that variability of these traits cannot have strict genetic determination.

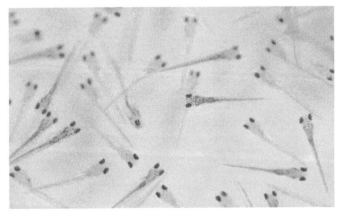

Figure 2.7. Pigmented and unpigmented larvae from Utsuri x Utsuri progeny at 3 day after hatching.

The obtained data on color inheritance show that during development of koi the accumulation of mutations of different genes, which impacted synthesis of the same pigments, has occurred. These mutations have been of both loss-of-function (stopping production of corresponding pigment) and gain-of-function (initiating production of pigment) types. It is very demonstrative in case of genetic control of melanin production in koi. Synthesis of melanin typical for wild-type color common carp is suppressed by mutations of corresponding genes. However, accumulated mutations of other genes resulted in appearance of different types of black pigmentation (Bekko and Utsuri), which are typical for koi. The same regulating mechanisms can be suggested, for example, for red pigment. Loss-of-function mutation of gene controlling synthesis of red pigment could result in appearance of fish with white background color. Further mutations of other genes causing appearance of red patches on white background color could result in Kohaku color type. The combination of mutations of genes controlling synthesis of different pigments could give known color variability of koi.

Variation and Inheritance of Scale Cover Types

There are four main types of scale cover in koi (and common carp). In **scaled** fish (*wagoi*) even rows of scales cover the whole body. **Mirror** (or scattered, *kagamigoi*) fish have large ('mirror') scales, which are scattered on the body and do not find favor with the collector. The mirror fish demonstrate high variability with regard to the reduction of scale cover. Some mirror fish have the body almost completely covered with big scales. In **linear** fish (also called *kagamigoi*) large scales form an even row along the lateral line. In general, there are no scales on the bodies of **leather** fish (also called nude, *kawagoi*); several scales may be found near the base of fins.

The type of scale cover in koi (and common carp) is determined by an interaction of two genes having two alleles each (*S/s* and *N/n*) (Kirpichnikov 1981, 1999). Fish of different scale cover types have the following genotypes: scaled-*SSnn* or *Ssnn*, mirror-*ssnn*, linear-*SSNn* or *SsNn*, and leather-*ssNn*. Fish having the dominant allele *N* in the homozygote state (genotype *NN*) are unviable and perish at the time of hatching. The expected phenotypic ratios in all possible crosses are presented in Table 2.2.

In koi, variants with reduced scale cover are known since 1908 when Kichigoro Akiyama crossed German leather (nude) carp of the Aishgrunder strain with the endemic Asagi strain. As mentioned above, leather carps have *ssNn* genotype, that is they are mutants for both (*S/s* and *N/n*) genes. Therefore, it is not surprising that soon the full range of scale cover types appeared and Japanese names for the various phenotype subclasses like *Yoroi-goi* (armour plated scales), *Aragoi* (jumbled scales), *Ishigaki-rin* (stone wall scales), etc. came in use. Variants with reduced scale cover can be found in all color koi types. This shows that scale cover types and color traits are inherited independently. Variants with reduced scale cover in koi are called *doitsu* (which is the literal translation of 'Germany') that reflect the history of their appearance in Japan (Fig. 2.8).

Table 2.2. Expected phenotypic ratios in crosses of koi with different scale cover types.

Parents and type of crossing	Phenotypes (%)			
	Scaled $SSnn$ and/or $Ssnn$	Mirror $ssnn$	Linear $SSNn$ and/or $SsNn$	Leather $ssNn$
1. Scaled x Scaled:				
$SSnn$ x $SSnn$, $SSnn$ x $Ssnn$	100	0	0	0
$Ssnn$ x $Ssnn$	75	25	0	0
2. Scaled x Mirror:				
$SSnn$ x $ssnn$	100	0	0	0
$Ssnn$ x $ssnn$	50	50	0	0
3. Scaled x Linear:				
$SSnn$ x $SSNn$, $SSnn$ x $SsNn$	50	0	50	0
$Ssnn$ x $SSNn$				
$Ssnn$ x $SsNn$	37.5	12.5	37.5	12.5
4. Scaled x Leather:				
$SSnn$ x $ssNn$	50	0	50	0
$Ssnn$ x $ssNn$	25	25	25	25
5. Mirror x Mirror:				
$ssnn$ x $ssnn$	0	100	0	0
6. Mirror x Linear:				
$ssnn$ x $SSNn$	50	0	50	0
$ssnn$ x $SsNn$	25	25	25	25
7. Mirror x Leather:				
$ssnn$ x $ssNn$	0	50	0	50
8. Linear x Linear:*				
$SSNn$ x $SSNn$, $SSNn$ x $SsNn$	33.3	0	66.7	0
$SsNn$ x $SsNn$	25	8.3	50	16.7
9. Linear x leather:*				
$SSNn$ x $ssNn$	33.3	0	66.7	0
$SsNn$ x $ssNn$	16.7	16.7	33.3	33.3
10. Leather x Leather:*				
$ssNn$ x $ssNn$	0	33.3	0	66.7

* In these crosses 25% of offspring (NN) die; the ratios among viable fish are shown.

It is known (Kirpichnikov 1981, 1999) that the genes for scale cover have wide pleiotropic effect, i.e., they influence many traits. The effect of allele N is especially strong. Linear and leather carps (genotype Nn) have retarded growth rate and decreased survival as compared with scaled and mirror carps (genotype nn); also, allele N influences many physiological, morphological and meristic traits in fish. The pleiotropic effect of allele s is much weaker although some differences between scaled (SS or Ss) and mirror (ss) fish are observed. More detailed information on pleiotropic effect of genes for scale cover can be found in Chapter 1. The Doitsu varieties are less appreciated by the Japanese because of body shape considerations, but in most other countries the bright color qualities are sought after. Growth depression and poor health is common among *doitsu* and a higher probability of individuals suffering from a range of morphological defects can be found. These traits include reduction of fins, barbels and lateral line, and spinal deformities.

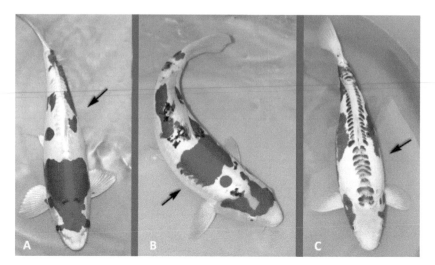

Figure 2.8. Some *doitsu* koi indicating scale phenotypic feature mostly used to identify the genotype. A. Kikusui, leather (nude, *kawagoi*); B. DoitsuSanke, linear (*kagamigoi*); C. Shusui, mirror scattered, *kagamigoi*).

Most *doitsu* varieties like Kikusui, Kumoryu, *doitsu* Purachina, etc. are of the leather class. Shusui is from the mirror class, and it may account for its superior morphological appearance. There is no particular rule since breeders mostly are oblivious of the genetic makeup of their brood stock. Development continues to improve these qualities in *doitsu* koi mainly in terms of growth and body conformation. The Shusui is selected from one of the sub types of the resultant mirror types. In Table 2.2, cross 5, mirror x mirror, will yield the highest proportion Shusui to select from. For the best possible improvement in body conformity and growth, scaled or linear females can be used that are tested to be heterozygous for the S/s gene. The linear phenotype is not generally accepted and culled out. While stringent culling is needed with every mating, many less attractive phenotypes slip through, especially when breeding is intended for export to countries with different preferences.

Cross leather x leather (cross 10 in Table 2.2) produces the highest proportion (66.7%) of leather offspring. However, for non-Shusui type *doitsu* the common breeders practice is crossing females of the mirror type with males of the leather type (cross 7 in Table 2.2); this cross gives 50% mirror fish and 50% leather fish. The 50% mirror component is culled out except a few future mirror females selected for best body conformation and color aspects, but ignoring scale cover. The 50% leather component should produce the marketable product as well as future male parents, while selecting for all the qualities demanded by the variety. Because of possible resemblance between some mirror and linear fish, the test crosses of presumably mirror females are sometimes performed for confirmation of their genotypes. Crossing of scaled females, tested to be heterozygous for the S/s gene, with leather males (cross 4 in Table 2.2) is also used in breeders practice for production of leather fish.

Variation and Inheritance of Scale Reflection

The *ginrin* is phenotype with iridophores in the scales containing reflecting platelets composed primarily of guanine (De Kock and Watt 2006, Gur et al. 2013). *Ginrin* varieties are known in all color types and independent of scale cover type. For exhibiting purposes, *ginrin* varieties are collectively known as *Kinginrin* since it emulates silver when the reflective scale is against a black or white background and gold when viewed against a yellow or red background. The term literally describes the golden and silver visual effect on the appearance of the fish.

Based on appearance, *ginrin* is classified in two main groups (De Kock and Watt 2006). In the first group the guanine forms a lumpy pearl-like deposit in the center of the scale, giving it an additional dimension referred to as *pearl ginrin*. In the second group the guanine platelets are flat, forming mirror-like structures which are judged like a diamond on its reflectivity from multiple angles. *Dia-gin* and *Beta-gin* are the most well-known of the latter group. *Beta-gin* first appeared in Niigata after the turn of the century, and it is speculated that it was introduced via the German Aischgrunder imports of 1904. *Dia-gin* appeared much later in the Hiroshima region and has more popular appeal due to multiple reflective surfaces imitating diamonds. The *ginrin* also has quantitative properties ranging from just having one reflective scale, to a few rows, to down to the lateral line, and more rarely completely wrapping the body. It also seems that the development of *ginrin* is later and the culling regime needs to be adjusted accordingly. *Ginrin*, like the phenotypes with reduced scale cover, also finds expression in other morphological changes noticeably in body conformation, growth and health.

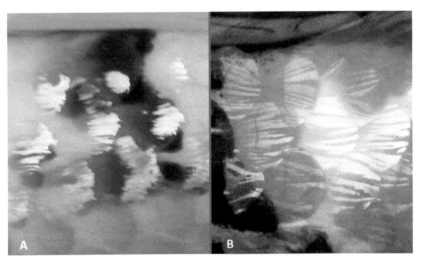

Figure 2.9. *Ginrin*. A. *Beta-gin*, also called Niigata *ginrin* is more lumpy and dull; B. *Dia-gin* or Hiroshima *ginrin* has larger crystal platelets reflective in many angles and is very popular.

Recently, Gomelsky et al. (2015) reported the results of studies on inheritance of Dia-gin or Hiroshima type of ginrin in koi. The fish segregations 'ginrin : non-ginrin' were recorded in three F1 progenies obtained by crosses 'ginrin x non-ginrin' and 16 F2 progenies obtained by crosses 'ginrin x ginrin' and 'non-ginrin x non-ginrin'. The segregations 'ginrin : non-ginrin' in most F1 progenies did not differ significantly from the 1:1 Mendelian ratio while crossing non-ginrin fish from F1 progenies resulted in the appearance of only non-ginrin fish in F2. Based on these data, it was suggested that ginrin trait is controlled by a dominant mutation of one gene (Gr/gr): fish with genotypes GrGr and Grgr have ginrin phenotype while fish with genotype grgr do not possess this trait. In most analyzed F2 progenies obtained by crosses 'ginrin x ginrin' segregations 'ginrin : non-ginrin' were shifted from the 3:1 and 2:1 theoretical ratios due to the deficit of ginrin fish. It was suggested that this shift was caused by a possible negative pleiotropic effect of ginrin mutation on survival of fish (both with genotypes GrGr and Grgr). Survival of ginrin fish was especially impacted in cases of inbreeding. It was recorded that non-ginrin fish were substantially larger than ginrin fish from the same progenies.

Variation and Inheritance of Fin Length

The long-fin variety of koi, sometimes called butterfly koi, is a relatively new morph that has been developed in the last several decades. In the United States, long-fin koi have been developed at Blue Ridge Fish Hatchery (North Carolina, USA) from the middle 1980s by crossing of normal short-fin koi with the long-fin common carp of Asian origin (LeFever 1991, 2010). This long-fin common carp apparently originated from Indonesia, where a local long-fin strain of common carp (also called 'kumpay') has been described. There is information that the long-fin morph of koi was obtained in Japan in the early 1980s by crossing koi with Indonesian long-fin carp. In Japan, long-fin koi are not allowed to be demonstrated at koi competitive shows, yet some breeders do produce a range of popular varieties for the export market. These include all the scale types including *wagoi*, *doitsu* and *ginrin*. Long-fin koi are popular in the United States and most other countries.

Inheritance of long fins in koi was investigated by Gomelsky et al. (2011) in a study performed in the United States. Fish segregations with regard to the presence or absence of long fins in two progenies were recorded and analyzed. In the first progeny, produced by crossing a long-fin koi female with a regular short-fin koi male, the observed segregation of long-fin fish: short-fin fish did not differ significantly from the 1:1 Mendelian ratio. In the second progeny, produced by crossing the same long-fin female with a long-fin male, the observed segregation of long-fin fish: short-fin fish did not differ significantly from the 3:1 Mendelian ratio. Based on these data, it was concluded that the appearance of long fins in koi is under the control of a dominant mutation of one gene (*Lf/lf*). Fish with genotypes *LfLf* and *Lflf* have long fins, while fish with genotype *lflf* do not have this trait. Analysis of fish variability in the second progeny has shown that fish with genotypes *LfLf* and

Lflf did not differ with regard to relative length of the tail; this indicated that the *Lf* allele is characterized by complete dominance over the allele *lf*. The authors noted that since appearance of long fins is controlled by a dominant mutation, it would be relatively difficult to develop a true-breed, long-fin line since the crossing of two heterozygotes (*Lflf*) gives the mixed progeny (Gomelsky et al. 2011). Development of a true-breed, long-fin line could be achieved only by identifying heterozygotes *Lflf* by means of test crosses and removing them from the stock.

Color Variability in Gynogenetic and Triploid Progenies and Production of Clones in Koi

Similar to other fish, it is possible to produce in koi gynogenetic and triploid progenies. Gynogenesis is embryo development under control of only maternal heredity. In order to induce gynogenetic development in koi (as well as in any other fish reproducing by normal sexual mode) eggs should be inseminated by genetically inactivated (usually by irradiation) sperm. For production of viable gynogenetic diploids (2n), haploid chromosome set (n) in eggs should be doubled. This can be achieved by suppression of either second meiotic division (meiotic gynogenesis) or first mitotic (cleavage) division in haploid embryos (mitotic gynogenesis) by strong physical treatments (usually by cold or heat shocks). It needs to be noted that in spite of the fact that fish obtained by induced gynogenesis are originated from female parent only, their genotypes do not copy maternal genotype. During egg development, the recombination of genetic material by chromosome crossing over occurs; therefore gynogenetic fish resulting from suppression of the second meiotic division in eggs (meiotic gynogenesis) do not have maternal genotype. Fish obtained by mitotic gynogenesis are homozygous for all genes because in this case homologous chromosomes are completely identical as products of mitotic reduplication in haploid embryos. Normally, many genes in fish genotypes are at a heterozygous state, therefore completely homozygous mitotic gynogenetic fish cannot have maternal genotype. By suppression of 2nd meiotic division in eggs after insemination of them with intact (non-irradiated) sperm triploid (3n) fish can be obtained. More detailed information on production of gynogenetic and triploid progenies in common carp and koi can be found in Chapter 5 of this book. The data on color variability in gynogenetic and triploid koi are presented below.

Taniguchi et al. (1986) have investigated variability of areas of red and black spots in fish from normal (amphimictic), triploid and meiotic gynogenetic koi progenies in a study performed at Kochi University (Japan). All three progenies were obtained using eggs from white-red Kohaku female and sperm from white-red-black Taisho-Sanke male. For production of triploid progeny cold shock was applied for suppression of 2nd meiotic division in eggs while for production of diploid meiotic gynogenetic progeny both cold shock for suppression of 2nd meiotic division in eggs and genetic inactivation of sperm by UV irradiation were applied. Area of red and black spots on the body was determined by measuring free-hand drawing with an image analyzer. There was no difference between fish from gynogenetic and normal progeny in the frequency distribution of red color area; the authors noted that this suggests that genes for red (R) and white (W) have no allelic relationship. Black spots were not observed in fish of gynogenetic origin; it proved genetic inactivation of the male's chromosomes. The area of black spots was lower in triploids compared with the diploids. The authors noted that this difference may be due to male parent's allele (B) being suppressed as a result of the additive effect of the female parent's allele (b) which negatively affects black color development (Taniguchi et al. 1986).

Gomelsky et al. (1996, 2003) have observed that proportions of white-red fish in meiotic gynogenetic progenies produced from Kohaku females varied from 47.8 to 88.9% (with mean value for six progenies 77.8%) and were usually higher than proportion of white-red fish observed in crosses of Kohaku parents. Gomelsky et al. (1996, 1998) have recorded that proportion of pigmented fish in meiotic gynogenetic progenies obtained from females with Bekko type of pigmentation (Shiro-Bekko and Taisho-Sanke) have varied from 63 to 78% with a mean value for six progenies of 71.7%. It is known that the frequency of heterozygotes in meiotic gynogenetic progenies depends on the crossing-over frequency between the gene and the centromere (see Chapter 5 of the book). Based on the observed proportion of fish with black pigmentation in meiotic gynogenetic progenies (71.7%) obtained from heterozygous ($Blbl$) females, the recombination frequency between gene Bl/bl and centromere was calculated to be about 0.4 (Gomelsky et al. 1996, 1998). In contrast to meiotic gynogenesis, mitotic gynogenesis results to homozygosity for all genes. Therefore mitotic gynogenetic progeny produced from heterozygous females $Blbl$ should, theoretically, consist of two types of homozygotes ($BlBl$ and $blbl$). In one from the two mitotic gynogenetic progenies obtained from heterozygous females $Blbl$ the ratio of pigmented and unpigmented fish was very close to the expected ratio 1:1; however, in the second mitotic gynogenetic progeny the strong prevalence of pigmented fish was observed. It was suggested that low viability of unpigmented gynogens in this progeny was due to the action of some lethal genes, which were expressed only in the absence of dominant allele Bl (Gomelsky et al. 1998).

Meiotic and mitotic gynogenetic progenies in koi have been obtained and investigated at Niigata Prefectural Inland Water Fisheries Experimental Station (Sato and Amita 2001, Sato 2013). In meiotic gynogenetic progenies obtained from Kohaku females, proportions of white-red fish varied significantly and sometimes were higher than typical proportion of white-red fish in Kohaku x Kohaku crosses (60–70%). On the contrary, proportions of white-red fish in mitotic gynogenetic progenies obtained from Kohaku females were very low (0–1%) while proportions of other two color types, solid red and solid white fish, were variable; proportion of solid red fish in these progenies varied from approximately 50 to 95%.

Clones (or groups of genetically identical individuals) in fish can be produced by obtaining the second consecutive gynogenetic generation from females of mitotic gynogenetic origin (for details see Chapter 5). A clone in koi has been produced for the first time at Niigata Prefectural Inland Water Fisheries Experimental Station in 1998 by inducing meiotic gynogenesis from a solid red female obtained earlier by mitotic gynogenesis from Kohaku female. All fish of the obtained clone have been solid red, the same as female of mitotic gynogenetic origin, from which this clone was originated (Sato 2013, S. Sato, pers. comm.).

As mentioned above, offspring obtained by induced gynogenesis do not have maternal genotype. Therefore the method for production of clones in fish by consecutive gynogenetic generations does not allow copying of valuable show-quality koi. The following question arises: is there any other biotechnological method which does give an opportunity to clone grand champions? Theoretically it is possible using so-called Somatic Cell Nuclear Transfer (or SCNT) technology. In 1996 the first cloned mammalian animal (sheep Dolly) was obtained by this method (Wilmut et al. 1997). In this case the nucleus from cultured specialized (mammary) somatic tissue derived from adult sheep was transferred into an egg with preliminary removed original nucleus. The production of Dolly became a widely known scientific achievement and drew attention by public media. Ironically, similar pioneer study in fish was reported a decade before the first cloning of mammal from adult somatic cell. Chinese scientists (Chen et al. 1986) succeeded in producing adult crucian carp by transferring the nucleus from kidney tissue (derived from adult fish) to an egg after removal of the original nucleus. In the last 10–15 years the technology of somatic cell nuclear transfer in fish has been developed further (mostly in model species, zebrafish and medaka) by applying some advanced biotechnological methods (Lee et al. 2002, Wakamatsu 2008, Siripattarapravat et al. 2009). Recently Tanaka et al. (2012) reported the results of a study performed in Japan on attempt of cloning high-quality goldfish breed 'Ranchu' by fin-cultured cell nuclear transplantation. As was reported, several embryos with transferred nuclei have reached the hatching stage. In future, somatic cell nuclear transfer technology could be applied for obtaining genetic copies of high-quality koi. However, research and development in Nishikigoi is a slow process because females require several years to mature. Also, application of different experimental treatments to eggs and early embryos could seriously affect other evaluation criteria (besides color pattern) such as body conformation. Therefore, it is doubtful that koi cloning can be effectively done commercially in the near future.

Acknowledgements

The authors thank Katsushi Takeda, Ronnie Watt, Pierre Jordaan, Kiyoko Fujita, Gertan Agenbach and Toru Inouye for valuable and essential assistance, Shoh Sato for providing unpublished data and some articles, Yuka

Kobayashi for interpreting articles in Japanese, Alexander Recoubratsky and Viktor Dementyev for providing some articles and Charles Weibel for help in formatting of figures.

References

Bar, I., E. Kaddar, A. Velan and L. David. 2013. *Melanocortin receptor 1* and black pigmentation in the Japanese ornamental carp (*Cyprinus carpio* var. Koi). Front. Genet. 4:6. doi: 10.3389/fgene.2013.00006.

Chen, H., Y. Yi, M. Chen and X. Yang. 1986. Studies on the development potentiality of cultured cell nuclei of fish. Acta Hydr. Sin. 10: 1–7 (in Chinese); Republished in English: 2010. Int. J. Biol. Sci. 6: 192–198.

Cherfas, N.B., Y. Peretz and N. Ben-Dom. 1992. Inheritance of the orange type pigmentation in Japanese carp (Koi) in the Israeli stock. Isr. J. Aquac.-Bamidgeh 44: 32–34.

David, L., S. Rothbard, I. Rubinstein, H. Katzman, G. Hulata, J. Hillel and U. Lavi. 2004. Aspects of red and black color inheritance in the Japanese ornamental (Koi) carp (*Cyprinus carpio* L.). Aquaculture 233: 129–147.

De Kock, S. and R. Watt. 2006. Koi: A Handbook on Keeping Nishikigoi. Firefly Books Inc., New York.

Gomelsky, B., N.B. Cherfas, N. Ben-Dom and G. Hulata. 1996. Color inheritance in ornamental (koi) carp (*Cyprinus carpio* L.) inferred from color variability in normal and gynogenetic progenies. Isr. J. Aquac. - Bamidgeh 48: 219–230.

Gomelsky, B., N.B. Cherfas and G. Hulata.1998. Studies on the inheritance of black patches in ornamental (Koi) carp. Isr. J. Aquac.-Bamidgeh 50: 134–139.

Gomelsky, B., N. Cherfas, G. Hulata and S. Dasgupta. 2003. Inheritance of the white-red (Kohaku) color complex in ornamental (Koi) carp (*Cyprinus carpio* L.). Isr. J. Aquac.-Bamidgeh 55: 147–153.

Gomelsky, B., K.J. Schneider and A.S. Alsaqufi. 2011. Inheritance of long fins in ornamental koi carp. North Amer. J. Aquac. 73: 49–52.

Gomelsky, B., T.A. Delomas, K.J. Schneider, A. Anil and J.L. Warner. 2015. Inheritance of sparking scales (ginrin) trait in ornamental koi carp. North Amer. J. Aquac. (in press).

Gur, D., Y. Politi, B. Sivan, P. Fratzl, S. Weiner and L. Addadi. 2013. Guanine-based photonic crystals in fish scales form from an amorphous precursor. Angew. Chem. Int. Ed. 52: 388–391.

Hoshino, S. and S. Fujita. 2009. Nishikigoi Mondo (English version). NABA Corporation, Japan.

Iwahashi, M. and M. Tomita. 1980. Genetical studies of nishikigoi (*Cyprinus carpio*). II. On the heredity of Kohaku. Rep. Niigata Inland Waters Fish. Exp. Stn. 8: 74–79 (in Japanese with English summary).

Jordan, D.S., S. Tanaka and J.O. Snyder. 1913. A catalogue of the fishes of Japan. J. Coll. Sci. Imp. Univ. Tokyo 33 (article 1): 1–497.

Kataoka, M. 1989. Nishikigoi dangi (Nishikigoi discussion). Takayoshi KataokaKumagaya-shi, Saitama, Japan (in Japanese).

Katasonov, V.Y. 1973. Investigations of color in hybrids of common and ornamental (Japanese) carp: I. Transmission of dominant color types. Genetika 9: 59–69 (in Russian with English summary).

Katasonov, V.Y. 1974. Investigations of color in hybrids of common and ornamental (Japanese) carp: II. Pleiotropic effect of dominant color genes. Genetika 10: 56–66 (in Russian with English summary).

Katasonov, V.Y. 1976. The lethal action of the gene of light pigmentation in the common carp *Cyprinus carpio* L. Genetika 12: 152–161 (in Russian with English summary).

Katasonov, V.Y. 1978. A study of pigmentation in hybrids between the common carp and decorative Japanese carp: III. The inheritance of blue and orange patterns of pigmentation. Genetika 14: 2184–2192 (in Russian with English summary).

Katasonov, V.Y., V.N. Dementyev and A.V. Klimov. 2001. Genetics of color in the ornamental carp: the inheritance of underlying dark coloration. Russian Journal of Genetics 37: 1210–1211.

Kirpichnikov, V.S. 1981. Genetic Bases of Fish Selection. Springer-Verlag, Berlin.

Kirpichnikov, V.S. 1999. Genetics and Breeding of Common Carp. INRA, Paris.

Koshida, S. 1931. Fish Farming as Side Business for Farmers. Niigata Farmers Association, Niigata, Japan (in Japanese).

Kuroki, T. 1981. Manual to Nishikigoi. Shin Nippon Kyoiku Tosho Co., Shimonoseki-city, Japan.

Kuroki, T. 1987. Modern Nishikigoi, 2nd Ed., Shin Nippon Kyoiku Tosho Co., Shimonoseki-city, Japan.

Lee, K.Y., H. Huang, B. Ju, Z. Yang and S. Lin. 2002. Cloned zebrafish by nuclear transfer from long-term-cultured cells. Nat. Biotechnol. 20: 795–9.

LeFever, W. 1991. The new butterfly koi. Tropical Fish Hobbyist (November): 78–83.

LeFever, R. 2010. The origin of butterfly koi. Pond Trade Magazine (March/April): 12–14.

Novelo, N.D. and B. Gomelsky. 2009. Comparison of two methods for measurement of red-area coverage in white-red fish for analysis of color variability and inheritance in ornamental (koi) carp *Cyprinus carpio*. Aquatic Living Resources 22: 113–116.

Okada, Y. 1960. Studies on the freshwater fishes of Japan II. Special Part. J. Faculty of Fish. Prefec. Univ. of Mie 4: 1–860.

Okubo, E. and R. Sato. 1975. Racial study on "Asagi" carp from the Matsushiro. Bull. Freshwater Fish. Res. Lab. 25: 73–81 (in Japanese with English summary).

Sato, S. 2013. Breeding in Nishikigoi, *Cyprinus carpio* of Niigata prefecture. Fish Genetics and Breeding Science 42: 81–84 (in Japanese with English summary).

Sato, S. and K. Amita. 2001. Induction of mitotic-gynogenetic diploid Nishikigoi, *Cyprinus carpio*. Rep. Niigata Inland Waters Fish. Exp. Stn. 25: 1–5 (in Japanese with English summary).

Smartt, J. 2001. Goldfish Varieties and Genetics. A Handbook for Breeders. Fishing News Books—Blackwell Science, Oxford, UK.

Siripattarapravat, K., B. Pinmee, P.J. Venta, C. Chang and J.B. Cibelli. 2009. Somatic cell transfer in zebrafish. Nature Methods 10: 733–735.

Suzuki, R. and M. Yamaguchi. 1980. Meristic and morphometric characteristics of five races of *Cyprinus carpio*. Jap. J. Ichth. 27: 199–207.

Tanaka, D., A. Takahashi, A. Takai, H. Ohta and K. Ueno. 2012. Attempt at cloning high-quality goldfish breed 'Ranchu' by fin-cultured cell nuclear transplantation. Zygote 20: 79–85.

Taniguchi, N., A. Kijima, T. Tamura, K. Takegami and I. Yamasaki. 1986. Color, growth and maturation in ploidy-manipulated fancy carp. Aquaculture 57: 321–328.

Waddington, P. 2009. Article in Koi-Bito. http://www.koi-bito.com/forum/main-forum/9409-peter-waddington-article.html.

Wakamatsu, Y. 2008. Novel methods for the nucleus transfer of adult somatic cells in medaka fish (*Oryziaslatipes*): use of diploidized eggs as recipients. Dev. Growth Diff. 50: 427–436.

Watt, R. and S. De Kock. 1996. Living Jewels: Koi Keeping in South Africa. Delta Books, Johannesburg, RSA.

Wilmut, I., A.E. Schnieke, J. McWhir, A.J. Kind and K.H.S. Campbell. 1997. Viable offspring derived from fetal and adult mammalian cells. Nature 385: 810–813.

Part II

Aquaculture and Nutrition of Carp

Fish production in aquaculture is increasingly important for human consumption and is a promising supplier of valuable protein sources throughout the world. This trend is exemplified by the continuous growth of the mostly semi-intensive production of cyprinids, including common carp, in China and other Asian countries where it is widely distributed. More traditional forms of carp culture can be found in Eastern Europe using extensive production forms which use ponds that need adequate pond preparation and management strategies during the production season. Different traditions and forms of aquaculture of carp are summarized in Chapter 3.

Throughout the centuries humans have selected carp for traits desirable in aquaculture. A fundamental optimization of the gene pool has been achieved by applying traditional breeding and crossing strategies which are reviewed in Chapter 4. This chapter also demonstrates how traditional and modern techniques for natural and artificial reproduction are applied. Selection in aquaculture today extends towards direct manipulation of the genetic composition of individuals and populations. During the last few decades chromosome set and sex manipulations have been applied to improve the performance of carp. Recently, gene transfer has also successfully been used to improve the weight gain of certain carp strains. These modern molecular biological techniques and their implications are presented in Chapter 5.

Regardless of their genetic origin all cultured and wild carp have to feed. The genetic make-up may determine the maximal weight gain that can be achieved, however only with optimal food supply. Carp are omnivores yet their diet has to supply a certain amount and composition of proteins, lipids and carbohydrates to allow for efficient growth and development of the organism. Dietary nutrients have not only to be supplied in adequate amounts but also of high quality. This is especially important when small fry are fed with artificial feeds. A comprehensive treatise of all matters of carp nutrition is given in Chapter 6. This chapter completes the picture of carp aquaculture in Part 2 of this volume by combining the effect of different management of aquaculture facilities (e.g., pond-drying schemes) with a detailed account of nutritional requirements (e.g., natural food in ponds vs. supplementary feeding). The reader learns about how different diets result in diverging qualities of fillet both in taste and in color.

3

Carp Aquaculture in Europe and Asia

Pavel Hartman,[1] Gregor Schmidt[2,*] and Constanze Pietsch[3]

Introduction

Aquaculture in Europe and Asia has a long history and many developments in aquaculture have taken place recently, especially within the last 200 years. Several historical scenarios have been developed that might have contributed to the early developments of aquaculture. These include the possible use of natural or artificial water bodies for early fisheries or husbandry attempts (Rabanal 1988). The earliest developments in aquaculture are assumed to have taken place at least 2000 years ago (Rabanal 1988). Nevertheless, the precise time as well as the process of emergence of aquaculture still remains unclear, although it almost certainly had been developed independently under different conditions at different locations.

This chapter gives an overview on carp aquaculture with respect to its historical development. Regional differences in traditional fish breeding are also reviewed. Despite the fact that the knowledge on rearing fish is centuries old, the development of different techniques continues up to the present day.

[1] University of South Bohemia in České Budějovice, Faculty of Fisheries and Protection of Waters, South Bohemian Research Center of Aquaculture and Biodiversity of Hydrocenoses, Institute of Aquaculture and Protection of Waters, Husova tř. 458/102, 370 05 České Budějovice, Czech Republic.
 Email: phartman@frov.jcu.cz
[2] Institute for Fisheries, State Research Center Mecklenburg–Vorpommern for Agriculture and Fishery, Branch Office Hohen Wangelin, Malchower Chaussee 1, D-17194 Hohen Wangelin, Germany.
 Email: g.schmidt@lfa.mvnet.de
[3] Zurich University of Applied Sciences, Institute of Natural Resource Sciences (IUNR), Research Group Ecotechnology, Campus Gruental, PO Box, CH8820 Wädenswil, Switzerland.
 Email: constanze.pietsch@zhaw.ch
* Corresponding author

Global Overview on Carp Aquaculture Production

Aquaculture production of common carp contributes significantly to global finfish production. Figure 3.1 shows that the global production of carp steadily increased from 1950 to 1980 after which the growth accelerated. This recent and pronounced increase results indirectly from economic and political changes, and directly from the application of more efficient techniques for carp production such as application of fast growing varieties of carp.

Similar to most other important cyprinids in aquaculture (such as silver carp–*Hypophthalmichthys molitrix*, bighead carp–*H. nobilis*, grass carp–*Ctenopharyngodon idella*, and Indian carp–*Catla catla*) the production of common carp increased over the last decade (Fig. 3.2). During this time period the global production of common carp increased by 28%.

Today, Asia unquestionably dominates the global production of common carp in aquaculture producing more than 90% of the biomass of this species (Table 3.1). Asia has also consistently been the leading continent in carp production over the past decade(s) (Fig. 3.3). The global differences are due to the fact that some regions strongly depend more than others on fish as an important protein source for human consumption. In parts of the world where other protein sources are considered to be more important common carp production in aquaculture has not been established to such a great degree. The different characteristics of carp production in Central and Eastern Europe, as well as in Asia, will be reviewed in the following sections. Since carp are mostly raised in ponds the following section will focus on the characteristics and differences of ponds designed for carp aquaculture.

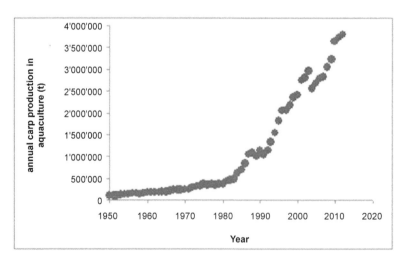

Figure 3.1. Global trend for carp produced in aquaculture (in metric tons). Source: FAO 2011. Fishery statistics: Aquaculture production.

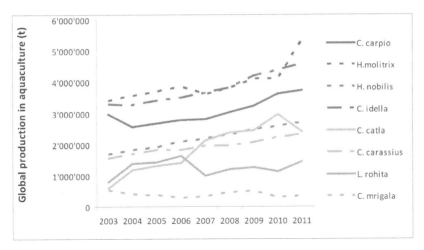

Figure 3.2. Global trend for production of different cyprinid species (in metric tonnes) in aquaculture. Source: FAO 2012. Fishery statistics: Aquaculture production; common carp–*Cyprinus carpio*, silver carp–*Hypophthalmichthys molitrix*, bighead carp–*H. nobilis*, grass carp–*Ctenopharyngodon idella*, Indian carp–*Catla catla*. Goldfish–*Carassius carassius*, Indian major carp–*Labeo rohita*, and Mrigal carp (also called Indian carp)–*Cirrhinus mrigala*.

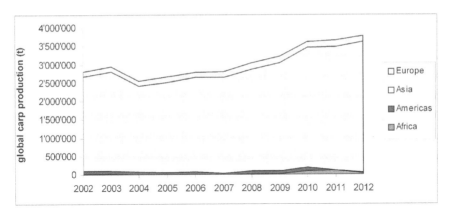

Figure 3.3. Global production of common carp in aquaculture (metric tonnes). Source: FAO 2012. Fishery statistics: Aquaculture production.

Table 3.1. Common carp production in 2011 (FAO 2013).

Continent	Production (metric tonnes)
Africa	111,531*
Americas	39,905
Asia	3,494,922
Europe	179,323
Oceania	457
Grand total	3,826,138

* FAO estimate

Pond Aquaculture

There has been a long tradition of fish culture in ponds. Pond aquaculture today is the main production form for carp due to its typical sustainable agriculture-type production technique. Although adaptations to improve culture methods will also be important in the future, pond aquaculture rests on a strong and historically grown fundament of social, economic and ecological pillars. This is especially true for countries in Central and Eastern Europe. In this section, this historically relevant form of pond aquaculture in water bodies that can be used for carp culturing will be described briefly.

In countries of Central and Eastern Europe, the first pond constructions were recorded in monastic homesteads. The ponds purpose was to shelter and later to rear fish for the Lenten days of the monks. Ponds were also constructed for utilization of water power, as drinking and non-potable water sources, and pond dams provided road network in inaccessible, swampy terrains. In 795 King Charlemagne (known as Karl der Grosse in German, or *Carolus Magnus* in Latin) issued instructions for "administration —keeping and new development of ponds wherever it is beneficial". In Bohemian countries (their territory is at present located in the modern-day Czech Republic and includes Moravia and Silesia) the first ponds were established as early as the 10th century. The establishment of the first ponds coincided with the expansion of monasteries and Christianity. Of particular importance for the development of pond culture was the propagation of the monastic order of Cistercians. The Cistercians were known to be brilliant architects and constructers—medieval hydro-engineers. During their zenith both raceways and ponds were constructed for use of water power. Pond development was also supported by the King of Bohemia Charles IV. Later on pond development as well as carp culture was adopted by the aristocracy. Before the beginning of 17th century, the total surface area of ponds in Bohemia reached 180,000 ha, which became more than twice as much by the end of 14th century.

At that time a number of historical figures were already engaged in the development of carp culture in ponds. As educated scholars they aimed to improve the quality standards of pond farming management and—important for the progress of pond culture—they provided written accounts of their pursuits. They passed on their experience through handbooks treating, for example, reproduction and rearing of carp. Jan Dubravius from Bohemia and Olbrycht Strumiensky from Poland can be mentioned among the earliest innovators. Jan Dubravius was a Moravian bishop, diplomat and savant. He wrote the first book dealing with fish farming in ponds in Europe. His own experiences and those of his predecessors were collected in a technical publication "About Ponds and Fish Living in Them", published in Latin in 1547 in Wrocław (Poland). A second book was released in 1573 wherein Olbrycht Strumiensky described the development of pond culture in Silesia (Czech Republic). The aim, already then, was to advance the practice of pond culture. This included descriptions of the construction of large ponds, and best practice descriptions in pond management with respect to fertilization via overwintering and fish stocking.

Regrettably, the Thirty Years' War (1618–1648) interrupted the growth of pond culture and decreased the total acreage of ponds by 50%. The decline of the pond acreage continued until the middle of the 19th century, when in several European countries a renaissance of pond aquaculture started.

The list of important innovators and establishers of modern-age pond aquaculture includes Tomasch Dubisch (1813–1888). He worked briefly in Hungary and then in Silesia at the biggest pond aquaculture farm in Golyš. As a fish breeder Dubisch deployed a new system of fry rearing to prepare the fish for the large grow-out ponds using the so called green manuring (which will be reviewed later in this chapter). In addition, he conducted selective breeding programs which contributed to the emergence of the Galician (Galician-Polish) carp stock. The Galician carp stock showed enhanced growth rates of 30% compared to other strains. Around the same time, in Poland, Josef Šusta's (1835–1914) book on "Nutrition of carp and its pond retinue" was published and had a great impact. Šusta was among the first to apply a systematic approach to carp rearing. He categorized age groups of fish, established observation-based pond fertilization schemes, feeding guidelines, and standardized methods for the monitoring of growth rates.

The first half of the 20th century including the two World Wars led to a decline of pond aquaculture. The decline had consequences, especially for the diversity of the gene pool of the common carp. Due to losses during fry and fingerling transport, and later on because of frequent outbreaks of SVC (spring viraemia of carp) in the 1950s and beginning of 60s, a number of carp stocks valuable for further breeding refinement ceased to exist.

Nevertheless, from the 1950s onwards pond aquaculture underwent relatively dynamic developments, especially with regard to the amount of produced common carp and by-catch species. The sevenfold increase in pond production in Czech Republic (from 3,100 metric tons in 1949 to 20,200 metric tons in 2010) can be attributed to intensification of natural pond production by effective supplementary feeding, improvement of stock breeding, new methods in reproduction, and finally many capital investments for aquaculture. In Central and Eastern Europe, and similarly in Western Germany, the 1960s to 1980s were characterized by intensification of carp production in ponds. In Eastern European countries carp aquaculture was combined with farming of poultry. Intensification focused mainly on enhancement of feeding efficiency and increasing stocking densities. The transformation of the political systems and a changeover in proprietary rights in Eastern European countries together with the growing awareness for environmental conservation in the recent decades led to the definition of a common objective: adequate utilization of natural pond production with rational supplementary feeding to maintain current production volumes.

Characteristics of Ponds used for Carp Culture

Pond aquaculture is a fish farming method focusing on utilization of surface waters retained in fish ponds or other fish rearing facilities (concrete storage reservoirs, hatcheries, etc.). The primary purpose is to produce fish for food, but also for angling and, alternatively, to rear other water organisms.

A pond, in that context, can be considered an agricultural area used for production of highly valuable products of animal origin. Ponds are characterized by a high abundance of aquatic organisms. This results from a high primary and secondary production which is facilitated by the retention of nutrients in the shallow (i.e., 1–2 m depth) earthen pond reservoirs. Because ponds are so productive and shallow and because light typically enters the largest part of the pond bottom, they can be considered as a single large littoral zone.

One of the basic prerequisites for proper fish farming is the possibility to regularly empty the ponds to ensure complete harvest and completely new fish stocking. Regular emptying is also a way of managing algal and zooplankton blooms to provide natural fish food. The controlled filling and emptying of the ponds is the main difference from fish stocking and angling in "free waters", i.e., rivers and lakes. A fish pond is therefore often considered to be a hydraulic structure with a natural bottom and with the possibility to regulate the water volume which allows draining and harvesting. Fish pond construction entails construction of a dam and other technical features. Depending on the type of construction and water supply, ponds can be classified differently (Table 3.2).

Fish Pond Water Supply and Their Flow Rate

Ponds are usually supplied with surface run-off water. Its sources are rain or snow precipitations in the catchments of fast flowing waters draining directly into the fish pond. An important requirement of fish pond is an adequate water supply that allows controlling the pond filling rate even in the event of water shortages or floods. For mitigating the effects of floods a retention area near the pond is important. This retention allows diversion of an over-supply of water. Losses of pond water occur during:

1. Transpiration processes by water plants
2. Evaporation of the water from the surface
3. Seepage of water through the bottom, dam permeability, and leakages of constructed outlet devices such as weirs and other constructions for water outlet

As a consequence of these losses it is necessary to allow for sufficient water input ensuring water retention times of approx. 90–120 days.

Table **3.2.** Description and differentiation of pond according to their water supply, type of construction.

Classification of ponds according to:	Type of pond	Characterization
Water supply	Drainage	Water supply by atmospheric water, without connection to other water bodies
	Groundwater	Water supply by springs
	Surface waters	Water supply by open waters
Type of construction	Impoundment pond	Flowing waters are impounded by a dam
	Diversion pond	A ditch provides water for the pond from a watercourse
	Spur dike pond	Separated area behind a spur dike
	Lagoon pond	Separated area in a lagoon of an estuary
Type of application	Spawning pond	Shallow pond that is only used for approximately 4 weeks for reproduction in spring
	Nursery ponds	Small pond (max. 2 hectare) for the first feed uptake by larvae
	Stock pond	Medium-sized pond for breeding within the first year
	Rearing pond	Medium-sized pond for breeding within the second year
	Production pond	Big pond for breeding within the third year
	Sobering pond	Small pond with good water supply to store the fish before selling
	Wintering pond	Deep pond with reliable water supply for overwintering of fish

Principles of Fish Pond Aquaculture

The principle of fish rearing in ponds can be explained as a metabolic nutrient conversion. Fish are secondary consumers and utilize food energy that was fixed via photosynthetic assimilation by primary producers (e.g., green algae in the pond), and transferred to primary consumers (zooplankton and benthos which are feeding on algae). These primary consumers represent natural food items for fish. Further details on the trophic role of carp in aquatic ecosystems can be found in Chapter 9 on the feeding ecology of carp. The individual production steps (primary production of algae and plants, secondary production of zooplankton) take place at the same time and in the same environment as the actual fish production.

The level of natural production of ponds or, alternatively, the fertility of fish ponds (trophic status) is defined as the ability of a fish pond to sustain the production of fish biomass within one vegetation period from its own sources of natural food. Natural production can be expressed as weight gain per surface or volume unit (Kostomarov 1958). In the jargon of pond fish culturist the natural production is defined as a weight gain of fish as growth during a certain time period per pond area or, alternatively, per volume of retained water. Natural production of pond aquaculture can be increased by input of nutrients which enhance the growth of fish. See Chapter 6 for details on nutrition of carp in aquaculture.

In Central Europe the production of carp weighing 1.5 to 2 kg designated for human consumption takes three years. If summer temperatures remain lower than average the production period can be up to a year longer. During production the rearing conditions must regularly be adapted to the needs of growing carps. In order to achieve this, fish farmers apply several ponds among which the carp are transferred to meet the requirements of fish at different life stages. Reproduction generally takes place in small ponds in which the hatching of the larvae also occurs. Although a fish farm should possess several of those, spawning ponds only represent 0.1 to 0.3% of the entire area in fish farms (Martyshev 1983). After hatching, the larvae are transferred to nursery ponds where they stay for up to six weeks. These ponds account for approximately 3% of the entire area. When natural food is becoming scarce, the carp are put into the first rearing ponds which cover 10% of the fish farm area. Juvenile carp stay in these ponds until they are transferred to overwintering ponds in autumn. In spring, the one-year-old carp are transported to the second rearing ponds, corresponding to 25% of the entire pond area, where they stay until the following autumn. After the next overwintering the two-year-old carp are bred in production ponds until they reach their marketable size of 1.5 to 2 kg. For production ponds approximately 55% of the entire pond area should be used. Stock ponds usually account for 1% of the entire area.

Natural Factors Affecting Trophic Status of Fish Ponds

Climatic (especially thermal), geographic and hydrologic conditions determine the natural trophic status of ponds. Examples of these conditions are the location of ponds, e.g., on geological bedrock, and factors influencing the nutrient input, e.g., via feeds (which is reviewed in Chapter 6 by G. Füllner), and precipitation which affects run-off into the ponds. Furthermore, pond morphology—defined as the slope and the extent of the littoral zone relative to the volume of the pond is important. In general, ponds with a water surface area exceeding 5 ha with an average depth of 2.5 m–3.8 m will have a higher productivity than deeper ponds. Last but not least the productivity of fish pond is influenced by the amount of sedimentary deposits. The accumulation of too much fine sediments can decrease the concentration of dissolved oxygen in the water and hence the conditions for aerobic metabolism. Moreover, the accessibility of nutrients determines the development rate of aquatic vegetation which further affects other pond organisms. The following section will focus on the chemical conditions that underlie the biological functions of ponds.

Chemical Processes in Fish Ponds

As mentioned above natural production and fish rearing takes place simultaneously in the pond environment. The most essential chemical factor for pond aquaculture is the concentration of oxygen and carbon dioxide (CO_2) together with some minerals in the water.

Gas solubility in water follows several rules: a) solubility of particular gases in water is different (if the solubility index for nitrogen = 1, then solubility for oxygen is 2.3, and 100.4 for carbon dioxide); b) gases dissolve in water according to their partial pressure on the water surface; c) the gas amount in water is constant at a certain temperature and atmospheric pressure on the water surface.

The amount of dissolved oxygen in water has an overruling importance for fish production. Oxygen deficit is defined as difference between the state of equilibrium (100% saturation of water with oxygen at given temperature and pressure) and the lower oxygen saturation value that is currently noted. A high oxygen deficit can induce hypoxia and asphyxiation in fish. However, common carp are able survive at oxygen concentrations as low as 0.4 mg/l water (Dunham et al. 2002).

The main source of oxygen in pond water is photosynthetic assimilation of plants, and subsequently its diffusion from atmosphere. Not merely low oxygen saturation, but also its over saturation can be dangerous for different life stages of fish and other aquatic organisms. Pond waters are characterized by diurnal and seasonal cycles of oxygen levels (Fig. 3.4). Variation in pH and CO_2 takes place simultaneously with the diurnal and seasonal fluctuations of oxygen (Fig. 3.5).

Eutrophic and hypertrophic ponds which are characterized by high levels of nutrients and high algae abundance, typically have a low water transparency (less than 1 m measured as Secchi depth). As a consequence such ponds show considerable differences in oxygen concentrations between surface and bottom-near water layers. The sun-irradiated photic layer at the pond surface express ordinarily high oxygen saturation (frequently

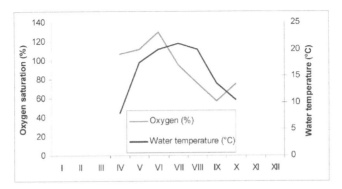

Figure 3.4. Average changes of oxygen levels in carp ponds in Europe during production season, the months are depicted as Roman numerals (1972–1976, according to Hartman et al. 2005).

exceeding normal saturation), while a significant non-irradiated part of the water volume suffers from anoxia from the bottom.

CO_2 is a product of the respiration of aquatic organism and decomposition of organic matter, but it is also an essential component of the photosynthetic assimilation of phytoplankton and plants. CO_2 is a part of the carbonate-hydrocarbon equilibration in water systems (Stegman 1973) whereby the formation of H_2CO_3 from CO_2 is in a stable balance. In addition H_2CO_3 dissociates to HCO_3^-. In pond waters the weight ratio of carbon in CO_2 and H_2CO_3 (free carbon) to carbon bound in HCO_3^- will be 1:5. This ensures a stabile pH in water in the narrow range from 7.2 to 7.8. Depletion of free CO_2 via photosynthetic assimilation results in a pH increase to 8.2–8.3, then CO_2 is gained from the Ca $(HCO_3)_2$ (calcium bicarbonate, also called calcium hydrogen carbonate).

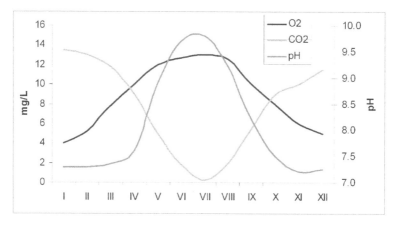

Figure 3.5. Seasonal O_2, CO_2 and pH dynamics in a pond (Hartman 1992).

Relatively high CO_2 amounts are bound to metal ions and are subsequently deposited in pond soils, especially in those with a high proportion of mineralized organic compounds. One of the ways to release CO_2 from the pond soil is the mechanical cultivation of the ground. For ponds with a high water level a current of oxygen-saturated water can be established (Stegman 1973, Hartman 1979).

Husbandry Techniques Influencing Production

The requirements of carp with respect to water quality are shown in Table 3.3. The basis of production is the replenishment of nutrients or equilibration of the carbon amount in relation to nutrients, such as biogenic elements including nitrogen (N), phosphorus (P), and calcium (Ca). The main focus of this measure is to support the natural pond productivity. It was earlier known as organic and mineral fertilization (manuring) of ponds. For photosynthetic assimilation of CO_2 algae in water require a nutrient ratio 106 C : 16 N : 1 P

Table 3.3. Physiological water quality demands of common carp in pond culture (after Füllner et al. (2000) and Adámek et al. (2010) with modifications).

Parameter	Unit	Transiently (short-term) tolerable lowest levels	Optimum range	Transiently (short-term) tolerable highest levels
Water temperature	°C	≥0.5	summer 20–28 winter 1–3	≤38
O_2	mgl^{-1}	≥1.5 at summer water temperatures (without food intake) ≥0.5 at winter water temperatures	5–30	≤40
pH		≥5.5	7–8.3	≤11
Alkalinity	mmol	≥0.2	1.0–6.0	≤8
Ammonia (NH_3)	mgl^{-1}	–	≤0.02	≤0.2
Nitrite	mgl^{-1}		≤0.0004	≤0.02
Hydrogen sulphide (H_2S)	mgl^{-1}	–	≤0.0002	≤0.002
Iron (Fe^{2+})	mgl^{-1}		≤0.05	≤0.1
Carbon dioxide (CO_2)	mgl^{-1}		20	>75
Chemical oxygen demand (COD)*	mgl^{-1}		summer 15–25 over-wintering ponds ≤15	
Suspended solids	mgl^{-1}		25–80	>400

*oxygen consumed from permanganate

(Füllner 2000). From this point of view the carbon acts as the limiting factor in carp ponds. The accessible carbon for photosynthetic assimilation occurs in form of free CO_2 or HCO_3^-. As explained previously a deficit of CO_2 in pond water results in pH level fluctuations.

With respect to supply of pond water with carbon and corresponding nutrients, green manuring is the most natural, cheapest, and therefore ideal method of organic fertilization (Table 3.4). Total application doses over the vegetation time are in ranges of 0.5 t (defined as preparatory dosage used especially in water protection areas) to 5 t ha^{-1} (Schäperclaus and Lukowicz 1998). Beef cattle dung works as carbonaceous fertilizer (Fig. 3.6), and its microbial components support development of organisms consumed by fish. The highest one-time dose of cattle dung is 400 kg ha^{-1}, and represents an increase in five-day biochemical oxygen demand (BOD_5) by 3–4.5 mg l^{-1}. Application of organic manures begins in winter time on the bottom of empty and closed ponds and ends in Europe in the middle of June.

Furthermore liming of ponds is aimed to create bonds of Ca^{2+} ions to CO_2 occurring in HCO_3^- (bicarbonate) and CO_3^{2-} (carbonate). This ensures sufficient buffering capacity of water to changes in the pH (positive acid-neutralizing capacity (or $ANC_{4.5}$ using methyl orange as pH indicator).

Table 3.4. Sources and ways of application of carbonaceous substrates (fertilizers, manures) (Hartman 2011).

Organic substrate (manure)	Proportional composition of raw matter in%	Application doses		Deadline and way of application	
		Standard, semi-intensive culture	(%) Before vegetation period Intensive culture	Before vegetation period (%)	During vegetation period (%)
Plant fast-composts and pond plant matter	Org. comp. 10% N 0.2% P₂O₅ 0.11% CaO 1.5% Dry matter ≤20%	≤10 t ha⁻¹	≤15 t ha⁻¹	100% from previous vegetation period	Mowing from 15. July to 15. August including composting
Green manure * progressive driving of grains and other plant materials	Org. comp. ≤12% N ≤0.35% P ≤0.06% Dry matter ≤15%	1.5–2.5 t ha⁻¹ green manure	2.5–5 tha⁻¹ green manure	0%	100% due to filling time (until July 15th)
Beef cattle dung	Org. comp. 8–15% N 0.5% P₂O₅ 0.25% Dry matter ≤25%	≤0.5 t preparatory dose up to 1.5 tha⁻¹	1.5–3 t ha⁻¹	60% applied to the bottom in heap form	40% applied to the water surface

*cereals, sprouts of rye, wheat or oats, seed amount ≤90 kg ha⁻¹, mustard seed amount of 10 kg ha⁻¹, mix of spring crops+ legumes

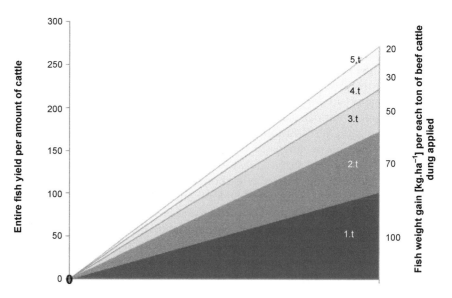

Figure 3.6. Relation of the amount of deployed beef cattle dung (1 to 5 tonnes) and fish weight gain. The efficiency of organic manure fertilization decreases with increasing deployment of the dung (Hartman 1992, Hartman 2012).

Pond liming stabilizes fish production results by increasing alkalinity of the water and hence precluding the negative effects of acidification on natural production. Dejdar (1956) regarded the pond alkalinity level as a criterion for pond yield capacity. The expression alkalinity was later replaced by acid neutralizing capacity (Pitter 2009). Calcium serves a number of important functions. Firstly, Ca^{2+} as a nutrient is part of the body structure of aquatic organisms, especially vertebrates including fish. Secondly, Ca^{2+} is a cation of the carbonate dynamic equilibrium in compounds containing CO_2; carbonate dynamic equilibrium comprises besides CO_2 and H_2CO_3 mainly from water-soluble Ca $(HCO_3)_2$ (calcium bicarbonate) and water-insoluble limestone $CaCO_3$, which is normally deposited in the muddy sediment of the pond (Stegman 1973).

Liming in order to replenish Ca^{2+} is usually carried out outside of the vegetation period on the bottom of empty ponds, but also either on ice or water surface of water-filled ponds. Usually either finely ground limestone at 100–600 kg ha^{-1} or a mixture of burnt lime with limestone (CaO + $CaCO_3$) is used as shown in Table 3.5.

Table 3.5. Used amounts of $CaCO_3$ according to the acid neutralizing capacity ($ANC_{4.5}$) level of the inflowing water (Janeček and Přikryl 1982).

$ANC_{4.5}$ (mmol)	$CaCO_3$ full dose kg ha^{-1}	$CaCO_3$ lowered dose kg ha^{-1}
0.5–1.0	600	400
1.0–1.5	300	200
1.6–2.0	100	0

Ca^{2+} applied in form of CaO (burnt lime) or $Ca(OH)_2$ (water-slaked lime) binds surplus of CO_2 and H_2CO_3 (as result of predominance of dissimilation -decomposition) into Ca$(HCO_3)_2$, and works as disinfectant. Disinfection by liming is usually applied from the second half to the end of the vegetative time (end of August, September) using milled burnt lime. The used amount of CaO are is equivalent to that of free CO_2 and ranges between 40 and 100 kg ha^{-1}.

Sufficient resources of calcium are usually present in pond soil. The calcium is released into the aquatic environment by dissolving of its compounds (mostly $CaCO_3$) with CO_2, i.e., weak carbonic acid (Hartman 2001). This process occurs most intensively during vegetation time, in which CO_2 consumption and production through respiration and assimilation take place. Dissolving of Ca^{2+} in water is enhanced by bioturbational activity of fish, which makes more sediment-deposited calcium compounds reactive.

The management of aquatic vegetation is important for pond productivity. It is realized by biological procedures, and mechanical procedures, i.e., mowing, composting and using herbicides, usually selectively on vegetation of the bond bowl and canals. Management of aquatic vegetation is nowadays provided by bio-procedure (or bio-control) using stocking of grass carp together with common carp and other fish species in mixed stocks. A proportion of grass carp in those kinds of stocks makes 2.5–5% of the stocked amount of the common carp. In temperate climate zones a three-year-old grass carp weighing 1 kg consumes 100–150 kg of raw plant matter during vegetation time. Mechanical removal of 'soft' submerged and 'firm' littoral aquatic vegetation is usually provided by mowing boats. Mowed vegetation is composted in the offshore area of ponds. Mowing of aquatic vegetation in Europe usually starts around July 15th and must be finished by September 15th. The ultimate effect of reducing aquatic vegetation can be understood as a means to increase natural retention and circulation of autochthone nutrients in the pond biotope.

The success of fish production in ponds is also strongly dependent on measures that improve animal hygiene conditions for fish rearing. These include:

- Sludge removal of the pits after draining of the pond with utilization of sediments for agriculture which improved the oxygen level of the water and hence the oxygen that is available for aerobic processes.
- Wintering and summering of ponds with supporting measures such as draining, partial cultivation of the pond bowl, sowing, etc., since these measures are also advantageous for aerobic processes whereas draining allows killing of potential pathogens.

Sludge removal from the fish harvesting ground and pond bowl is needed if high layers of sediments hamper proper pond cultivation. Regular muddy sediments of 10–15 cm in height are regarded as adequate. Pond mud usually accumulates in the deepest part of pond, which often is the fish harvesting ground. This hampers not only appropriate fish harvest, but also results in output of muddy suspension to lower parts of the basin, which causes deterioration of water quality. Lowering of the total pond water level is a result of the so-called pond ageing. This is a justifiable reason to extract the pond sediment load and use it on crop land situated above the pond. Before the sediment load is extracted the pond bowl must be drained in order to partly dry out the pond bottom. Drying-out of the pond sediment load is in favor of aerobic processes, which in turn make nutrients for fish production accessible, and also allows the killing of disease agents in the sediment layers. Pond wintering and its accompanying measures like pond bottom drainage is meant to expose the partly dried-out pond sediments to natural disinfection by frost. The sediment is dehydrated through freezing. This enables aeration of the otherwise muddy and hence less permeable sediment layers. Complete or at least partial wintering can be useful, e.g., in two-years production ponds either biannually or in intervals not exceeding four years.

Partial and complete pond summering means to put the pond out off operation either for a certain part or the complete duration of the vegetative season. This will dry out the pond sediments and allow oxygen enrichment. The summering process should be combined with pond bottom drainage and pond bowl cultivation with sowed plants. This implies not only to sow the field crops but also to reap them and re-fill the pond. The main target of the summering process is the mineralization of organic compounds, and the enhancement of the availability of nutrients after re-filling. It is recommended to carry out pond summering every 10–12 years. This low frequency of summering processes makes it feasible to combine a summering with necessary maintenance of other pond constructions such as, e.g., weirs and dams.

Several measures that indirectly improve production of fish pond aquaculture include the adjustment of stocking density and species diversity in the fish stock. The use of supplementary feeding can influence pond productivity by leaching of nutrients from fish excrements and undesirable leaching of nutrients from un-consumed feed into the pond water. Furthermore, stocking densities and species diversity in fish stocks determines predation activity and introduction of nutrients of pond ecosystem into nutrient circulation (Steffens 1985). Accordingly, fish stocks in pond aquaculture can be classified by diversity of fish species, age classes and stocking density. Properly estimated stocking densities enhance productivity by approaching an optimal conversion of nutrients into fish biomass (Knösche et al. 1998).

The omnivore common carp features a rather low digestive ability for raw fiber and vegetable protein. Therefore, in comparison to phosphorus bound in natural food, phosphorus bound in crop proteins is digestible to only 25–28% for common carp. The rest (72–75%) is in turn released back into the environment (Steffens 1985). The US National Research Council for Fish Nutrition (1993) gives total phosphorus conversion levels of merely 25%. A slightly higher phosphorus retention from crops of 32% was found by Jirásek et al. (2005). The efficiency of total nitrogen digestibility from vegetable proteins is almost analogous (Fig. 3.7). By contrast retention of total phosphorus and total nitrogen originating from natural food of carp ranges between 80–90% or even more.

Finally, measures stabilizing and dampening pond fertility are important for pond productivity and comprise actions that lead to reduced nutrient

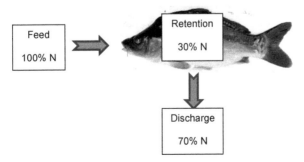

Figure 3.7. Approximate total nitrogen efficiency in common carp using crops as supplemental feed (Jirásek et al. 2005).

input and organic pollution from the pond basin, reduced stocking density, and orientation towards well-balanced and mixed multi-species stocks.

Measures stabilizing and dampening pond fertility include the reduction of nutrient input into the pond basin. This can be realized by proper cultivation of surrounding agricultural land (i.e., the restrictive use of fertilizer) and other ponds located in the pond basin. Furthermore, control of organic and mineral manure deposition, removal of aquatic poultry farming in the catchment, and reduction of nutrients from municipal pollution entering as run-off are important measures to reduce nutrient input. In addition, application of oxidizing disinfectants such as chlorinated lime [$Ca(ClO)_2 + CaO$] or potassium permanganate ($KMnO_4$) are used for short-term reduction of metabolic processes connected to nutrient turn-over in ponds. Chlorinated lime and potassium permanganate hamper in a short-term manner development of primary pond production and the microbial metabolism.

Common Carp Rearing

Three- and Four-year Production Cycles

Climatic conditions of Central European regions together with market requirements predestine the common carp to be reared to market size at least within three vegetation periods commonly known as 'summer'. A three-year production cycle is mostly performed in Central Europe. Accordingly, an individual size of K_3 intended for processing ranges between 1.5–2.5 kg. In colder regions a four-year production cycle is preferred using so called two-heat ponds. The individual market size of common carp originating from this type of production cycle ranges between 2.5–3 kg or even more (Table 3.6), and is intended to be sold live for the Christmas market.

Advantages and disadvantages of four-year production cycles include the fact that both sexes of common carp reach maturity and higher individual weights within four years. This, in contrast to, e.g., salmonids, improves market value by higher proportion of edible parts relative to gonads. It furthermore permits reduction of seasonality of carp production—a longer production cycle allows carrying out a summer harvest of marketable fish. In addition, lower stocking densities of fry and fingerlings lead to a relatively high stability of production spread-out to four years. However, there is still a risk of higher losses of larger K_3 during overwintering. In order to counteract this, lower stocking densities can be used that lead to better access to the feed, an increased growth performance of the fish and better fish condition which results in less mortality. In contrast, higher stocking densities require higher maintenance efforts and feeding ratios, and two-heat farming requires a high water retention capacity in riverine basins in order to provide enough well-tempered water.

Table 3.6. Time schedule and weight distribution of a four-year production cycle of common carp. Pond fish farmer's year consists of winter and "heat". The "heat" must be understood as vegetative period during which the natural as well as supplemental food intake takes place that can in turn be converted to growth.

Fry and fingerling on-growing ponds		Main or chief production ponds	
Fry	Fingerlings used for stocking	K_3 used for human consumption or for re-stocking	Market-sized carp
K_0....... K_1	K_1 .... K_2	K_2 .... K_3	K_3........ K_4
heat + winter	heat + winter	heat + winter	heat (over-wintering just in case of spring harvest)
10% of water surface	25–30% of water surface	60–65% of water surface	
$K_r < 2.5$ g, K_1 15–35 g	K_2 150–400 g	$K_3 > 1.2$ kg for human consumption $K_3 = 0.6$–1.2 kg for re-stocking	$K_4 > 2$ kg

Rearing of Market-sized Carp

Rearing of market-sized carp is realized in main production ponds, which correspond to 65% of the total water surface of a particular pond homestead focused for market-sized fish production. These ponds are usually large, 1–2.5 m in depth below the dam next to the outlet device, and are normally constructed at the end of the pond systems with the possibility for wintering.

Even in the first half of the vegetative season (usually in May) depending on the amount of natural food sources (decreasing numbers of coarse and middle-sized zooplankton) a supplementary feeding can be meaningful. Schlott et al. (2011) suggested requirements for supplementary feeding based on the developmental rate of the zooplankton ≥500 µm in size. Developmental rate expressed as volume of zooplankton (ml) per liter pond water was assessed after conservation and sedimentation, and three levels were determined. At level one, an insufficient zooplankton volume is characterized by less than 0.2 ml/l and supplementary feeding with a protein feed mixture at 2–4% of fish biomass daily is recommended. At level two, a sufficient volume of 0.2–0.55 ml/l zooplankton requires supplementary feeding with field cereals at 2% of fish biomass. At level three, a better-than-average volume of 0.55–0.8 ml/l zooplankton is found and supplementary feeding with field cereals should be occurring at 0.5% of fish biomass.

Janeček and Přikryl (1982) graded the daily requirements of feed based on water temperature and oxygen amount (Table 3.7). It should be noted that especially in highly trophic ponds, the daily feed amount must also be estimated with consideration of pH level since respiration of animals and plants influences the pH levels via CO_2 excretion, and the trophic status in turn depends on the amount of feed that is used.

Table 3.7. Daily feed amounts (field crops or feed mixtures based on field crops) according to water temperature and dissolved oxygen amount (Janeček and Přikryl 1982).

Minimal amount of oxygen (mg/l)	Daily feed ratio (% fish biomass) at particular water temperatures						
	10–11°C	12–13°C	14–15°C	16–17°C	18–19°C	20–21°C	22–26°C
7 mg/l O_2	0.6	0.9	1.4	2.0	3.0	4.0	5.0
6 mg/l O_2	0.4	0.6	0.9	1.4	2.0	3.0	4.0
5 mg/l O_2	0.2	0.4	0.6	0.9	1.4	2.0	3.0

Stocking Density and Polyculture

Individual weight gain of different age groups of carp is related to production capacity of a pond (Walterem 1934, Füllner et al. 2000). With increasing stocking density weight gain per surface unit increases to the upper limit. After the upper limit has been exceeded, increasing stocking density results in a decrease of the individual weight gain as well as weight gain per surface unit. Weight gain per surface unit can be boosted by replenishment of nutrients consumed by the pond biocenosis by manuring and supplementary feeding. The production optimum per surface unit depends on the applied age groups of carp and on an optimal (i.e., not too high) stocking density.

Pond Harvest

Pond harvest is the terminal stage of the productive phase of rearing. Pond harvesting usually takes place in autumn. It allows an inventory of how well fish are growing in a particular pond over a certain period. It also ensures accessibility of fish for the autumn and, especially, the Christmas market. During the season before Christmas common carp show a reduced feeding activity over several weeks. This results in a fine, not too fatty muscle texture, and not least, an outstanding taste. The reduced digestive activity of carp also makes the guts tastier and provides a good edibility of gonads. From the legislative point of view, pond harvest is classified as lowering of water volume in the pond in order to concentrate and harvest fish stock. The harvest is conducted using methods of large-scale fish catch such as seines or tow nets (see below). During fish catch, transient oxygen deficits may occur. While oxygen is stable in inlet water, oxygen levels in tow nets and fish harvesting grounds significantly decrease during draught because concentrated fish get stressed leading to a higher oxygen requirement. Oxygen content of inflowing water and exchange from air are often not sufficient to satisfy the increased oxygen demand of the stressed fish.

For an active method of pond harvest, tow (draught) nets are used and can be differentiated into the following types: Tow nets for fry and fingerlings—working with a width of 3–12 m (from scepter to scepter), a depth of the 'core' of 2.5–7 m, and a mesh size of 6–30 mm, whereby usually the 'core' depth is a half of the working width. For tow nets for larger fish with a working width of usually more than 12 m (16–30 m), and a mesh size of 30 mm or more, the core depth is regularly more than one half of the working width, and sometimes as deep as the length of the working width. For passive methods of fish catch, square or rectangular horizontal bottom nets are used—especially for fry and fingerling harvest.

Harvested fish are graded according to size and species (if polyculture is used) in tubs near the ponds. Most consumers desire carp with a maximum table size of 2 kg. Graded fish are weighted, counted, loaded on trucks and transported to fish storage reservoirs where they are kept without feeding until distribution. Fry and fingerlings are usually counted using volumetric methods, i.e., liter of concentrated fry' and transported to be stocked in the over-wintering ponds.

Fish transport

Performance and safety of fish transports depend on biological factors and technical equipment. Cyprinids intended for transport have to be starved at least 24 hours before loading and transport. Starved fish have a reduced metabolism consuming less oxygen and the occurrence of faeces within the transport vessels is reduced with starved fish. Fish must be in good health and fitness condition, i.e., sufficient storage of glycogen in hepatopancreas, and stress and injuries must be avoided. Transport devices must be equipped with a supply source of oxygen (oxygen cylinders) or aerators that can aerate the transport vessels in order to allow sufficient oxygen concentrations depending on the temperature and the age class of the fish (Table 3.8.).

Table 3.8. Hourly oxygen consumption (mg) per kg carp (K) at different temperatures (according to Szymansky 1973, Guziur et al. 2003).

Age class and species	15°C	20°C	24°C
$K_{advanced\ fry}$ to K_1	376 mg	576 mg	799 mg
K_1 to K_2	184 mg	288 mg	400 mg
K_2 to K_3	94 mg	144 mg	202 mg

Over-wintering

Over-wintering means stocking of fish in an appropriate environment (ponds), which is safe for their survival over winter. Over-wintering ponds must meet the requirements of fish welfare in a period of total starvation. Therefore, over-wintering ponds have to be well prepared during autumn, i.e., mowing and composting of offshore vegetation and liming for alkalinity reserve ($ANC_{4.5} = 2$ mM). In addition vegetation in the output canal has to be mowed and equipped with devices for deflection of additional run-off stemming from snow melt. An overall sufficient flow-through needs to be secured to stabilize pH and oxygen supply (pH 6.5–8, >3 l/ha/s).

Common carp at winter temperatures of 1–4°C rest at places or in nests on the bottom with removed sediment located mostly near the outflow. At low water temperature carp lower their metabolism. This physiological state is referred to as 'winter rest' and can last over several months. For well-prepared over-wintering ponds with an average depth of 1 m, a stocking density 10 tons per hectare (1 kg per square meter) is recommended, i.e., 30 K_1 or 4 K_2 individuals (Schäperclaus and Lukowicz 1998). The same authors present the most important stressor types that need to be avoided for an optimal over-wintering:

a) too high stocking density
b) bad water quality
c) injuries caused by handling, parasites or diseases
d) too much vegetation or muddy sediments
e) disturbance (predators or winter sports)
f) strong water currents

Photosynthetic assimilation is the main source of oxygen during over-wintering. Oxygen deficits are solved by means of 'window opening'. This term refers to the cutting of holes into the ice cover or using special mechanic defrosters. The goal of this 'window opening' is to allow for gas exchange between the water and the atmosphere.

Storage of Living Fish

Rearing and storage of living fish before consumption is considered to be a phase of sobering. This sobering is a state of reduced metabolism triggered by food-deprivation. Its aim is to improve the taste and nutritional characteristics of fish muscle. The most likely factor realizing this improvement is the depletion of excess fat in the fish. This leads to a more meagre and tender meat. The reduction of fat reserves probably also clears the fish from at least a part of the fat-solvable aromatic substances that can cause a 'muddy' taste (Vejsada 2008). The storage is terminated by distribution to the consumer. Proper welfare of fish has to be ensured over this period. Ponds for sobering are usually artificial concrete reservoirs or cages supplied with oxygen saturated water. Oxygen requirements in this kind of ponds are given in Table 3.9.

In case of long storage of living fish the stocking density of K_4 ranges between 60–120 kg/m^3 with regard to individual weight. Stocking densities during short-term life fish storage (<4 weeks) can be up to 200 kg/m^3 (Mareš and Nováček 1983).

Table 3.9. Oxygen consumption (g) per ton of market-sized K_4 per hour (Kujal 1980).

Temperature (°C)	2	4	6	8	10	12	14	16
O_2 consumption (g)	8	12	18	26	36	48	63	79

Aquaculture in Europe

Statistical data indicate that common carp production in Europe will not increase as rapidly as the global trend might suggest. However, common carp will remain an important species in aquaculture in those areas where there is a long-standing tradition of producing carp. Carp are mostly consumed domestically and live or freshly dressed fish are required by the consumers. Thus it is not expected that more elaborate post-catch processing of common carp will gain any further importance in Europe.

The main carp production in Europe is located in the Eastern part. In these regions carp pond aquaculture is a fundamental element of people's rural traditions. After the collapse of the Soviet regime, carp production was cut into half. But since that time production is stable and in the last decade it increased by 24%. In contrast, production in Western Europe was much lower in the past and decreased by more than 40% in the last decade (Fig. 3.8). The reason for this are manifested changes in consumers' behaviors. In European Soviet countries carp production was supported as an important protein source for human consumption. In Western Europe carp was mostly substituted by sea fish products and other highly-desired salmonid species (e.g., rainbow trout, Atlantic salmon). Nevertheless, over the last few years carp has gone through a renaissance as an environment-friendly, sustainable and healthy product. The production of organic carp together with a quality label has been established locally all over Central Europe, but accounts for only small percentages in each country. The rationale behind such eco-labels is an increased consumer acceptance by ensuring a sustainable resource use in extensive or semi-intensive production systems.

However, a strong influence on carp production was also seen from recreational fisheries in Europe. Carp are no longer produced only for human consumption, but also for stocking of natural waters and water reservoirs for angling purposes. The importance of carp for recreational fisheries, also commonly known as sport fishing, is reviewed in Chapter 11 by Ragnarsson-Stabo. Generally angling of carp has led to the demand for more active specimen on the hook than the domesticated carp, so that wild forms or hybrids of wild carp, and domesticated strains are used to satisfy these requirements. In addition, wild carp are also required for re-stocking natural waters, where a re-estabishment of the natural fauna is desired. In addition, the importance of by-catch species increases. Many farmers are focused on production of by-catch in polyculture with carp (see below). Finally, it should be mentioned that apart from table size production, the generation of fingerlings for stocking programs is of high economical relevance due to a high revenue of fingerlings.

Modern Aquaculture in Western Europe

In Central Europe numerous fish farms concentrate on breeding and rearing of carp. For example, more than 100 farms with altogether 133 carp strains have been identified in Germany in 2007 (Muller-Belecke et al. 2007). These

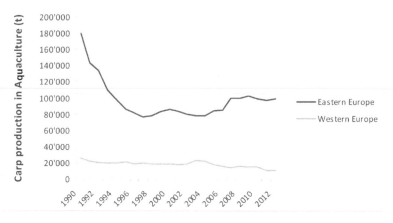

Figure 3.8. Carp production in Eastern Europe (without Russia) and Western Europe. Source: FAO 2012. Fishery statistics: Aquaculture production.

comprised mostly mirror carp strains and, to a much lesser extent, scaled carp. In total 11 regionally distinct common carp stocks are being kept in Germany in different rearing places. These separated populations are defined by local nomenclature. Interestingly, scaling as a characteristic of carp has increased in importance relative to traditional characteristics such as growth performance and vitality. Survival rates in extensive production systems with low stocking densities are generally higher than in intensive production. Farmers aim to select carp specimen for reproduction with respect to characteristics desired by the consumers (which includes scaling patterns). With respect to scaling, attention is especially paid to a homogenous size and pattern of scaling. Scaling types are differentiated between the presence and absence of a continuous row of dorsal scales and scales at the beginning of the tail fin and the pectoral fins. Scaling types also show regional differences. Depending on the region, also different body shapes are preferred, e.g., slim forms or compact high-backed forms. Which consumer preferences apply is related to centuries-old regional breeding tradition that made customers acquainted with certain carp characteristics.

Modern Aquaculture in Eastern Europe

An important issue in aquaculture in Eastern Europe is the conservation of traditional aquaculture in addition to follow-up activities that aim to ensure sustainability and development of economic and social structure. Furthermore, implementation of procedures minimizing unfavorable effects on the environment is desired.

Today, in the Czech Republic, the most common genetic sources of common carp have become the strains of South Bohemian scaled carp (C 73) and South Bohemian scale-less carp (BV). These stocks are conserved by pedigree breeding and valued for their good growth and health under standard rearing conditions. In addition, there are seven other stocks regarded as genetic sources in the Czech Republic.

The Amur wild carp (AS) belongs to the most important of the imported strains classified as breeding reserve. This stock is based on the original wild riverine population of the common carp (*Cyprinus carpio haematopterus*, most recently *Cyprinus rubrofuscus*) from the Amur area in the Far East. The import was carried out in 1983 for scientific purposes and for breeding. The desired effect was a heterosis in the F1 generation in case of hybridization with the Czech stocks. The Amur wild carp strain is characterized by a broad high back and ridge. It shows a high resistance against diseases (i.e., SVC and KHV (koi herpes-virosis)) and high fry survival (from K_0 to K_1).

The Ropsha carp was selected in 1930s in the former USSR using the Amur strain and Galician carp. This stock is reared in Northern regions of Russia, e.g., southward of Sankt Petersburg. It was imported to the former Czechoslovakia as an important stock for hybridization. The desired traits were high non-specific resistance to bacterial and viral diseases resulting in high survival of fry and fingerlings. In the third on-growing year however, its growth rate decreases, probably due to early maturation.

The Hungarian scale-less carp strain was imported in 1972. This strain performed especially well across standard rearing conditions and became the most common scale-less stock in the Czech Republic, it is used for pedigree breeding as well as a hybridization component for the generation of F1 generations with other strains.

In Hungary 17 strains are kept as life gene bank at the Research Institute for Fishery (HAKI) in Szarvas. The Polish life gene bank contains 18 stocks kept at the Research Institute for Ichthyology and Aquaculture in Golysz kept at the Fishery Station in Zator which belongs to the Institute for Inland Fishery in Olsztyn (Flajšhans et al. 2008).

Carp as a Dish in Europe

Carp is a popular product in many European countries. While the market in Northern and Southern Europe is rather small, carp is an important food in Central and Eastern Europe. The latter are therefore considered to be the European centers of carp pond aquaculture. The ways of traditional preparation of carp products depend on the region. Carps can be baked, boiled or fried. In some regions even parts of viscera (mainly gonads) are prepared in diverse ways. During last years the consumers' preferences have changed. Traditional dish from carp is not consistent with consumers' request for quick and easy prepared meals. This is why new carp products were developed over the last years. In the meantime carp is offered fried as part of fish and chips and is processed and marinated for fish salads. Another breakthrough in marketing is machining the fillets with bone-cutters. Bone-cutters trench bones into small pieces. Thus consumption is not negatively affected by fish bones. Thereby new groups of consumers could be reached and new markets could be opened up. However, carp remains a seasonal product and will continue to be consumed at religious festivals (Christmas, Easter) and in Lenten season.

Carp Aquaculture in Asia

Aquaculture of carp species in Asia is strongly dominated by China (Fig. 3.9). In addition, production of carp species clearly increased within the last decade, probably due to the fact that fish provide a considerable part of the total protein that is consumed (Prein and Ahmed 2000). Carp are particularly sought-after because of their meat quality, high market value and good growth performance (Dey et al. 2005).

In China the increase of aquaculture within the last decade accounts for more than 300%, whereby production of carp species presents more than 80% of the total production of freshwater fish in aquaculture. For the other selected Asian countries the increase of carp species production is quite similar, although the total amount of fish is clearly lower compared to China.

Figure 3.10 shows that the proportion of common carp in the production of selected Asian countries mostly accounts for 20%, or less. An exception is Indonesia where approximately 80% of the cyprinid species that are produced in freshwater aquaculture are common carp.

Carp production in Asian countries is clearly dominated by China. Similar to most Asian countries China's fish production systems are extensive or semi-intensive and polyculture is often applied (Dey et al. 2005).

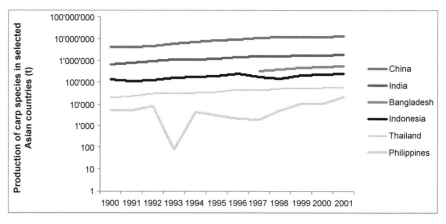

Figure 3.9. Production of carp species (including common carp, silver carp, grass carp, bighead carp, Indian carp species, and crucian carp) in selected Asian countries from 1990 until 2001. Source: FAO 2003. Fishery statistics: Aquaculture production.

Polyculture Systems

Carp rearing together with complementary (side) fish species is very common (Woynarovich et al. 2010). It is realized especially in large main production ponds in both, three-year and four-year production cycles. The main fish for production in pond aquaculture is the common carp whose stocks are often supplemented with other cyprinids, catfish or pikeperch. The success of fish polyculture is mainly based on optimal resources utilization. This utilization

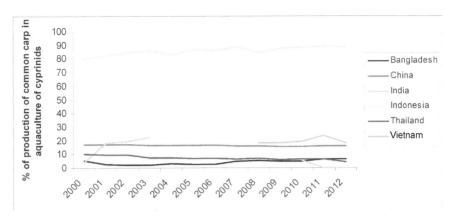

Figure 3.10. Percentage of production of common carp in relation to the entire annual production of cyprinid species in selected Asian countries. Source: FAO 2011. Fishery statistics: Aquaculture production.

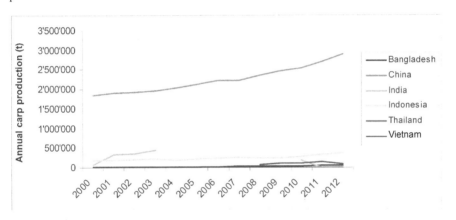

Figure 3.11. Carp production (metric tons) in selected Asian countries. Source: FAO 2011. Fishery statistics: Aquaculture production.

is achieved by the fact that species with entirely or partly different feeding habits or diet preferences are reared together consuming nutrients across a wider spectrum of the food chain. Although carp is the main fish species in polyculture, the economic importance of the other species is increasing. This is especially true for predatory fish such as catfish and pikeperch. During the last years even aquaculture of sturgeon for caviar production in carp ponds has been established. Besides their importance for human consumption, fish are also used as stocking material. The animal species that can be reared together with carp are selected according to differences they show compared to common carp. The complementary species to carp should be focused on the selected based on differences in the supplementary species' preferred diet or feeding habit, their behavior and their preferred habitat in a pond, but the selected species need to be similar to carp concerning their temperature tolerance. In addition, the selection of animal species for polyculture together with common carp is influenced by their intended final use and economic value. The most popular animal species that are chosen for carp polyculture are reviewed here. Although the combined production of different species seems to be a simple concept, polyculture in carp ponds requires exact planning which has recently been summarized by Woynarovich et al. (2010).

A common cyprinid that is used in polyculture with carp is the tench (*Tinca tinca*) (Fig. 3.12A). This fish species is native to almost all of Europe and Asia. Although its meat is considered to be of rather high value for human consumption, its economic importance is low. Mainly recreational fishing leads to a demand for stocking of tench. The growth performance of tench is lower than for carp, and it requires at least four years to reach a marketable size of >300 g. In contrast to carp, no intentional breeding is used for tench. In lakes, the tench functions as a benthic-feeding cyprinid with an autecology similar to that of carp. However, the robustness of this species within fish production systems is considered inferior to that of carp. Nevertheless, the tolerance of tench to water parameters is similar to that of other cyprinids.

The production of tench is achieved in separated age groups which are reared with carp of the same age. The percentage of tench in fish stocks commonly constitutes 10–15%. If tench are raised in monoculture in small ponds, feeding with formulated diets has to be done at established feeding tables. This can lead to production of marketable tench within three years. Reproduction of tench is realized naturally in ponds or artificially in hatcheries. Although the latter method has many advantages, mostly natural reproduction is performed. For natural reproduction a sun-exposed pond of 0.5 ha is filled with water and 10–20 spawning pairs are introduced. Spawning occurs at temperatures of approximately 20°C. Since these temperatures are reached relatively late in the annual course in mid-European countries and also due to its low growth, juvenile tench stay in the ponds until autumn and are then transferred to over-wintering ponds. Breeding in hatcheries is conducted similar to that of carp although the use of hormones for stimulation is indispensable. The approved method is the use of carp pituitaries for hormonal stimulation of breeding, but the dose has to be increased for tench. The injection is performed between the ventral fins and the eggs can be stripped within 24 hours at 20°C. The swollen eggs are moderately sticky so that removal of adhesive compounds is not necessary prior to incubation in Zuger glasses. The larvae emerge after three days and should be transferred to rearing tanks by using the out-flowing water. In the tanks the larvae adhere to substrate. The yolk sack will be depleted after four days, after which the larvae start external feeding. For this, the smallest available zooplankton is selected.

In Asia polyculture of common carp often includes grass carp (*Ctenopharyngodon idella*), silver carp (*Hypophthalmichthys molitrix*), and bighead carp (*Hypothalmichthys nobilis*). These three species originated from East Asia where they already have been cultured for several thousand years. In the 1960s these species were transported to Central Europe via Russia. While these three carp species are used as the main stocking species in China

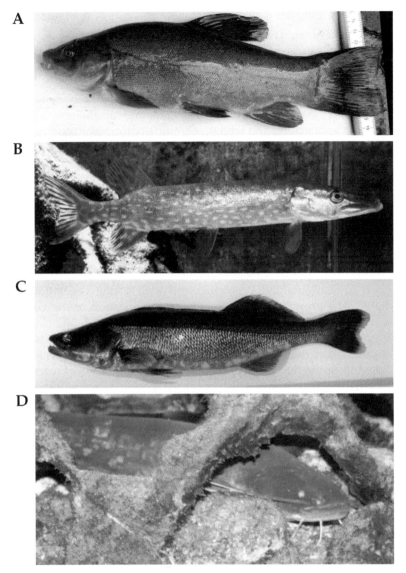

Figure 3.12. Species for polyculture with common carp; A = tench; B = pike; C = pikeperch; D = European catfish, pictures taken by G. Schmidt.

together with other fish species under intensive production conditions in ponds, their importance in Central Europe is less pronounced. In Europe polyculture wants to ensure that growth of common carp will not to be negatively influenced, as common carp is still the main target species. Fortunately, plant-feeding fish from East Asia at low stocking densities positively influence the productivity of ponds. Benthic organisms and zooplankton profit from their way of feeding since nutrients from fish faeces can be used in their diet and more benthic organisms and zooplankton are available to carp. Still, the feeding of larvae of different carp species does not differ from each other. Their feeding habits do change, however, after the juvenile fish reach a size of 4–6 cm. This is accompanied by an elongation of the intestine two-fold (grass carp) up to seven-fold (silver carp) (Schäperclaus and Lukovicz 1998). In addition, the gill apparatus changes in silver carp and bighead carp, so that these species are able to filter plankton (Schäperclaus and Lukovicz 1998). Bighead carp also consume up-floating algae and filamentous cyanobacteria and feed on benthic organisms. In contrast, silver carp with their fine gill rakers are able to feed on phytoplankton (diatoms and green algae) and small zooplankton. Only at water temperatures above 16°C grass carp start feeding on plants. Below that they provide competition to common carp. Juvenile grass carp prefer filamentous algae and soft parts of submerged plants while older fish consume nearly all water plants including stems. In addition terrestrial plants are used. However the food is not digested completely and the fine faeces fertilize the pond. This in turn can provide growth for microorganisms and plankton which can also be consumed by filter-feeding fish. For the bringing-up of juveniles only ponds rich in nutrients and vegetation and with a long vegetation period are suitable. The growth of fish depends on the duration of the vegetation period, water temperature and food supply. Thus, in moderate climate the rearing up to a marketable size takes at least four years whereas in warmer climate production can be cut into half. Herbivorous cyprinids can only be reproduced in farms that are able to provide large amounts of warm water. Usually only a few spawning fish are reared with twice as many males as females in the groups. These spawners are reared in separate nutrient-rich ponds all around the year. In spring these fish are transferred to small but deep ponds which are supplied with warm water leading to water temperatures of 18–20°C. After 1,300 to 1,400°d (day degrees) the eggs of silver and grass carps are ripe which is achieved in bighead carp at 1,350 to 1,450°d. At this time the fish should be transferred to the hatchery and one female together with two males should be allowed to adapt to a tank. Then the water temperature is increased to 25°C and ovulation is induced by injection of pituitary extract as described for common carp in Chapter 4 by G. Schmidt. The dose depends on the volume of the female (4–6 mg per kg fresh weight). Males receive 2 mg per kg fresh weight. The injection is conducted twice within 12 hours whereby the first injection only contains 10% of the entire amount that has been calculated. After that the fish are transferred back to tanks while separate tanks are used for males and females. At 25°C the eggs are fully mature after 8 hours (grass carp), 9 hours (silver carp), and 9.5 hours (bighead carp), respectively. The sperm of several males is obtained by pipetting and mixed with the eggs. The addition of a sodium chloride solution (0.4%) may improve the fertilization rate. After two minutes of careful agitation the eggs are washed by addition of fresh water. Due to the washing step a removal of adhesive compounds can be omitted. The incubation of eggs is performed in jars at 25°C. The inflow should be kept as low as possible to keep the eggs floating. After 12 hours the inflow of water is increased because the sensitivity of the eggs to external disturbances has decreased and their oxygen demand further increases. After approximately 30 hours the larvae are hatching and they are flushed into tanks where they stay for three to four days until external food uptake occurs. After that the fry is transferred to properly prepared, warm weaning ponds (in most cases 20,000 to 100,000 larvae/ha) and first feeding occurs. Supplementary feeding is achieved with grain meal or legumes meal (<100 μm) that is mixed with the inflowing water of the ponds. Formulated diets can also be used. At least four weeks of rearing of fish are recommended before further transferring the fish. After that the juvenile fish can be raised in polyculture with common carp. The ratio of silver carp to bighead carp depends on the supply of water plants and plankton. Grass carp can also be fed with grass cut in summer. Ideally, this should be applied before feed application for common carp. Removal of the fish from the pond should preferably occur at temperatures below 10°C. Since non-native carp species are used in this type of polyculture, the construction of the pond should be adapted to polyculture so that the escape of the fish is prevented. If not serious negative consequences for the surrounding environment can be expected (see, e.g., Chapter 10 on carp as an invasive species).

Polyculture of common carp also includes pike (*Esox lucuis*) in some cases which is the only member of the family of *Esocidae* in Europe (Fig. 3.12B). The pike is a predator that was raised in ponds together in the past to decimate competitors for food for carp. Although its fillet is very tasty, the high number of bones makes it less desirable for human consumption. Nevertheless the interest in stocking for recreational fisheries is considerable. Males can reach maturity already after two years with less than 30 cm of length, whereas females follow the year after that at approximately 45 cm length. Depending on their age females generate up to 20,000 to 40,000 eggs per kg body weight. Spawning of pike already occurs at water temperatures of 8°C in flooded meadows or in small bays and estuaries with dense macrophyte populations. The maintenance of a brood stock is time-consuming since big pikes cannot be reared together due to high cannibalism. Breeding is therefore often achieved with wild specimen. Artificial and natural spawning is possible in fish farms. For the latter, several males are put together with one female in a shallow pond overgrown with grass. Bigger ponds can also be stocked with several spawning groups. At water temperatures between 8 and 12°C the pikes spawn in portions and the fish are removed from the pond immediately thereafter. Pike larvae hatch at 100–120 d°. At that time they have a size of approximately 800 μm and can be found attached to substrate. After further 80–120 d° the larvae have developed to free swimming. The swim bladder is filled at the water surface and external feeding starts. At this time point the pond is drained and pike fry is collected. Alternatively artificial breeding in hatcheries can be performed. Therefore, fish ready for spawning are transferred to indoor tanks where the use of pituitary extract is generally not necessary. In rare cases a low dose of pituitary extract (1 mg per kg body weight) can be used at 24 hours prior to extraction of sperm.

Sperm can be stored at 4°C for several days. For artificial reproduction the animals are anaesthetized and dried to prevent water contact and swelling of eggs which prevents fertilization. The eggs are striped into bowls and after addition of sperm the mixture is incubated for approximately one hour. After that water is added or a fertilization solution (1.5% urea, 0.7% salt) so that all eggs are covered. The eggs are mixed and transferred to Zuger or MacDonald jars. The inflow is adjusted so that the eggs are slowly agitated but do not agglutinate. Dead eggs must be removed. The larvae should not hatch in the jar because they would adhere to its wall. Thus the eggs are spread on nets which are placed in tanks or raceways. After hatching the nets are removed and a substrate is put into the tanks for adherence of the larvae until up-swimming. Pike can receive first feed in good fertilized ponds or also in tanks. Stocking density in nursery ponds should range between 20,000–24,000 specimen per ha (Calas 1934). The nutritional base is zooplankton which needs to be supplied in high density to reduce cannibalism that occurs very early. The fish need to be removed and transferred to other ponds when they reach a size of 3 cm at latest. When a sufficient amount of fish for predation is supplied pike can reach individual length of more than 25 cm until the beginning of the first winter.

Polyculture of common carp is also possible by using pikeperch (*Sander lucioperca*) (Fig. 3.12C). Pikeperch actually inhabit the entire Eurasian freshwater and brackish water systems. They prefer turbid and summer-warm water bodies. This species is thus, a classical by-fish on carp pond farming. Pikeperch thereby use the natural food that is supplied by the pond. Within the first summer zooplankton and benthic organisms are consumed, and at the end of the first year of living they start predation. Supplying enough fish for feeding is challenging for pond farming and limits the production of big marketable pikeperch. Thus, only a small number of fish reaches this size of >1 kg usually within four years. Nevertheless the economic importance of pikeperch for pond farming has increased during the last years. Males reach maturity usually within the second or third year, whereas the females often require an additional year for that. For reproduction one female is put together with one or two males in a freshly filled shallow pond. Reproduction starts in spring at water temperatures of 8–10°C. Pikeperch lay eggs of approximately 2 mm on substrate where the sticky eggs attach. Females are able to provide 100,000 to 200,000 eggs per kg of body weight. The development of eggs depends on temperature and hatching occurs at approximately 70–90 d°. For artificial breeding the spawning pairs need to be offered proper substrate for attachment of the eggs. If artificial breeding is intended to include stripping a previous injection of hormonal substances (carp pituitary or synthetic substances) in the peritoneal cavity is necessary. After transferring the eggs into jars the inflow needs to be adjusted so that the eggs are not agitated but still are supplied with enough fresh water. Hatching can then be expected after 70–100 d°. The newly hatched larvae are transported to new tanks with the outflowing water. In most cases these larvae at densities of 500,000 larvae per ha approximately are brought into properly prepared ponds without any additional fish stock but with high zooplankton abundance. The latter will often be too low after six weeks and the young pikeperch (0.5–3 g) will have to be transferred to rearing ponds (to max. 20,000/ha). To these ponds cyprinids (e.g., roach) ready to spawn are put so that after feeding on plankton and benthos enough food will be available for the continuing phase of nutrition. In the second year maximal stocking densities of 300 pikeperch per ha can be achieved.

For polyculture with common carp, European catfish (*Silurus glanis*) can be used as well (Fig. 3.12D). This catfish species originated from Central Europe and Western Asia, but stocking activities have led to its spreading to Western and Southern Europe. In pond farms these catfish can reach a size of 2–5 kg. Only a few farms keep a stock of adult fish for breeding due to laborious rearing efforts. The fish for breeding need to be reared in separate ponds with a lot of food (fish). Water temperatures of >10°C in spring makes separation of male and female specimen necessary. With increasing temperatures (>20°C) pairs are put together in small ponds (<1 ha) for spawning. A cave shelter for spawning consisting of wood is constructed. After spawning the eggs are removed from the pond and raised in net cages in separate ponds.

Natural reproduction is not as safe as reproduction in a hatchery. The latter is achieved by injection of hormones (3–5 mg pituitary/kg fresh weight) and after 24 hours eggs can be obtained from anaesthetized females. Stripping of males is rather difficult, which is the reason for killing males and removing the testis surgically. After removal it is grated and fluid containing sperm cells is added to the eggs followed by addition of physiological salt solution. The eggs are then transferred to jars and treated with alkaline protease (1%) at the following day. Hatching will occur after 60–70 hours at 24°C and the newly hatched larvae are transferred to shaded tanks where the larvae attach to the walls. After four days mouth, intestinal tract and swim bladder have developed and the larvae are attracted by external food. In hatcheries they can be fed *Artemia* naupli. Alternatively they can be stocked in ponds at maximum densities of 100,000 specimen/ha. After six weeks of culturing the young catfish can be reared in monoculture with feeding dry feeds. In most cases, however, these fish are added to ponds with other fish species including carp.

In Asian countries polyculture of carp species with shrimp is also conducted including black tiger shrimp, *Penaeus monodon*, and freshwater prawns such as *Macrobrachium rosenbergi* (Nair and Salin 2007). These types of polyculture may become more important in the future.

In conclusion, polyculture of carp with other fish species is quite common and has proven to be successful. Therefore the main reason for complementary fish species is better utilization of the wide spectrum of natural nourishment, thereby improving natural weight gain per surface unit. After Janeček and Příkryl (1982) the proportion of side fish species in relation to the total production can make up to 5–10% for tench, 2.5–3% for grass carp, <25% silver carp, <10% for bighead carp, 10–15% for coregonids (*C. lavaretus* and

C. peled), and <3% carnivore fish (European catfish, pikeperch, Northern pike). These proportions of side fish species are usually applied without elevation of supplementary feeding or nutrient replenishment. The above mentioned ratios can thus be used to determine the proportion of side fish production in carp ponds that can be achieved by using exclusively natural food production (Guziur 2003, Wojda 1993). By this, polyculture allows to diversify the range of products and achieve better economic efficiency.

Acknowledgement

The authors would like to thank Dr. Viktor Svinger, Ph.D. (Fachberatungfür Fischerei des Bezirks Oberfranken, Ludwigstraße 20, 95444 Bayreuth, viktor. svinger@bezirk-oberfranken.de) for translation of Czech into English. The work was financially supported by the Ministry of Education, Youth and Sports of the Czech Republic - projects "CENAKVA" (No. CZ.1.05/2.1.00/01.0024), "CENAKVA II" (No. LO1205 under the NPU I program).

References

Adámek, Z., J. Helešic, B. Maršálek and M. Rulík. 2010. Aplikovanáhydrobiologie. VydalaJihočeskáuniverzita v ČeskýchBudějovicích. Fakultarybářstvíaochranyvod Vodňany, ISBN 978-80-87437-09-4, 350 pp. (in Czech).
Bayerische Landesanstalt für Landwirtschaft. 50 Jahre Aussenstelle für Karpfenteichwirtschaft Höchstadt des Instituts für Fischerei, LfL–Fischerei, Information, 34 pp. (in German).
Dejdar, E. 1956. Ertragsteigerung der tschechoslowakischen Teiche durch mineralische und organische Düngung., Deutsche Fischerei Zeitung No. 11., 1956 (in German).
Dey, M.M., F.J. Paraguas, R. Bhatta, F. Alam, M. Weimin, S. Piumsombun, S. Koeshandrajana, L.T.C. Dung and N.V. Sang. 2005. Carp production in Asia: Past trends and present status. *In*: D.J. Penman, M.V. Gupta, M.M. Dey (eds.). Carp Genetic Resources for Aquaculture in Asia. WorldFish Center Contribution No. 1727. WorldFish Center, Malaysia.
Dunham, R.A., N. Chatakondi, A. Nichols, T.T. Chen, D.A. Powers and H. Kucuktas. 2002. Survival of F2 transgenic common carp (*Cyprinus carpio*) containing pRSVrtGH1 complementary DNA when subjected to low dissolved oxygen. Marine Biotechnology 4: 323–327.
Dvořák, J., P. Hartman, J. Holický, I. Rusov and M. Siegel. 1981. Odchovkapříhoplůdkun arybnícíchStátníhorybářstvío.p.–METODIKA, ÚtvarchovurybStátníhorybářstvío.p., SeverografiaChomutov, č.135/81, 1–22 (in Czech).
Faina, R. 1983. Využívánípřirozenépotravykaprem v rybnících, Metodika č.8, Výzkumnéhoústavurybářského ahydrobiologického, Vodňany-1983, 16 pp. (in Czech).
Flajšhans, M., M. Kocour, P. Ráb, M. Hulák, V. Šlechta and O. Linhart. 2008. Genetika a šlechtěníryb. Jihočeskáuniverzita v Č. Budějovicích ,Výzkumnýústavrybářský a hyrobiologickýveVodňanech , 232 pp., ISBN 978-80-85887-82-2 (in Czech).
Füllner, G., N. Langner and M. Pfeifer. 2000. Karpfenteichwirtschaft. Bewirtschaftung von Karpfenteichen. Gute fachliche PraxisFreistaat Sachsen, Sächsisches Landesanstalt für Landwirtschaft, Referat Fischerei–Königswarta, Oktober 2000, pp. 130 (in German).
Gela, D., M. Kocour, M. Rodina, M. Flajšhans, P. Beránková and O. Linhart. 2009. Technol otieřízenéreprodukcekapraobecného (*Cyprinus carpio* L.), Metodika VÚRH Vodňany, (technologickářada), FROV JU Vodňany, 2009, 99, pp. 43 (in Czech).
Guziur, J., H. Bialowas and W. Milczarewicz. 2003. Rybactwostawowe, OficynaWydawnicza „HOŽA" Warszawa-2003, ISBN 83-85038-82-5, 384 pp. (in Czech).
Hartman, P. 1979. Využitíkysličníkuuhličitého (CO_2) zednarybníků k optimalizacihodnot pH vody. Živočišnávýroba–vědeckýčasopis, roč 24, UVTIZ Praha-1979, 847–854 pp. (in Czech).
Hartman, P. 2001. Pond aquaculture in Central and Eastern Europe in the 21st. Century, VodňanyCzech Republik, May 2–4, 2001. Disposable calcium and its suplementation in Č.Budějovice Magistrate´s Ponds, 12 pp.

Hartman, P. 2004. Šetrnýzpůsobvápněnírybníků. Metodika VÚRH Vodňany-73/2004. 12 pp. (in Czech).

Hartman, P., H. Chromý, I. Rusov and M. Siegel. 1984. (State Fisheries, ČeskéBudějovice): The Use of Aeration Technology to Adjust the Oxygen Balance of the Černodubský Pond. Živočišnávýroba 29, (12) UVTIZ Praha 1984, 1125–1130.

Hartman, P. and R. Mikl. 2004. Dynamikarůstukapra v rybnícíchměstaČeskéBudějovice, Sborníkzesympozia, 55 let výukyrybářskéspecializacenaMZLU v Brně" 2004, 66–70 (in Czech).

Hartman, P., I. Přikryl and E. Štědronský. 2005. Hydrobiologie, třetípřepracovanévydání, InformatoriumPraha 2005, ISBN 80-7333-046-6 , 359 pp. (in Czech).

Janeček, V. and I. Přikryl. 1982. Chovnásadových a tržníchkaprů v intenzifikačníchrybnících, Metodika č.2/1982, VÚRH Vodňany, 16 pp. (in Czech).

Janeček, V. and I. Přikryl. 1992. Polykulturníobsádkykapra s býložravýmirybami a línem, Edicemetodik č. 38, VÚRH–JihočeskáuniverzitaVodňany, 16 pp. (in Czech).

Jirásek, J., J. Mareš and L. a Zeman. 2005. Potřebaživin a tabulkyvýživnéhodnotykrmiv pro ryby, Mendlovazemědělská a lesnickáuniverzita v Brně, MZe ČR a Komisevýživy a krmeníhospodářskýchzvířat ČAZV Praha, 2005, Projekt NAZV č.QD0211, ISBN 80-7157-646-8, 68 pp. (in Czech).

Knösche, R., K. Schreckenbach, M. Pfeifer and H. Weissenbach. 1998. Phosphor und Stickstoffbilanzen von Karpfenteichen. Zeitschrift für Ökologie und Naturschutz, 7/1998, 181–189 (in German).

Kostomarov, B. 1958. Rybářství, vydala ČSAZV veStátnímzemědělskémnakladatelství Praha, 1958, 353 pp. (in Czech).

Kouřil, J. and J. Hamáčková. 1982. Odkrmranéhoplůdkukapravežlabech–metodika č.3/82. Výzkumnýústavrybářský a hydrobiologický, Vodňany, 15 pp. (in Czech).

Kujal, B. 1980. Parametrysádek pro sádkováníkaprů, projekt 114/01, Hydroprojekt ČeskéBudějovice, normativ pro projektování, závaznýod 1.8.1980, 3 pp. (in Czech).

Nair, C.M. and K.R. Salin. 2007. Carp polyculture in India. Practices, emerging trends. Global Aquaculture Advocat. Jan./Feb. 2007, 54–56.

NRC. 1993. US National Research Council. Nutrient requirements of fish. National Academy Press, 114 p.

Pitter, P. 2009. Hydrochemie, Praha: Vydavatelství VŠCHT, 2009, ISBN 978-80-7080-701-9, 579 pp. (in German).

Pokorný, J., Z. Lucký, S. Lusk, M. Pohunek, Jurák, E. Štědronský and O. Prášil. 2004. Velkýency klopedickýrybářskýslovník. NakladatelstvíFraus Plzeň, ISBN 80-7238-117-2, 1. vydání, 649 pp. (in Czech).

Rabanal, H.R. 1988. History of Aquaculture. ASEAN/UNDP/ FAO Regional Small-Scale Coastal Fisheries Development Project, Manila, Philippines.ASEAN/SF/88/Tech.7. FAO, United Nations.

Schäperclaus, W. and M. Lukowicz. 1998. Lehrbuch der Teichwirtschaft, 4.neubearbeitete Auflage, Parey Buchverlag Berlin-1998, ISBN 3-8263-8248-X, 590 pp. (in German).

Schlott, K., M. Fichtenbauer and C. Bauer. 2011. Bedarfsorientierte Fütterung in der Karpfenteichwirtschaft, Das Absetzvolumen von Zooplankton, Gebharts 33, Ökologische Station Waldviertel, ISBN 3-901605-35-5, 36 pp. (in German).

Spangenberg, R. and K. Schreckenbach. 1984. Syndrom energetickéhodeficitu u kaprů a možnostijehopřekonání. Zeitschrift für Binnenfischerei der DDR, roč. 31 č. 9, 1984, (doslovnýpřeklad pro účelyútvaruchovuryb SR o.p.), 271–278 (in Czech).

Steffens, W. 1985. Grundlage der Fischernährung, VEB Gustav Fischer Verlag Jena 1985, 1. Auflage, Lizenznummer 261 700/136/85, LSV 2924/4614, 226 pp. (in German).

Stegman, K. 1973. Vápněníkaprovýchrybníků, překlad v Čs. rybníkářství,č.4/73, původGospodarkarybna č. 10/1973, 6–12 (in Czech).

Štěpánek, M., V. Bernátová and V. Kiřík. 1979. Hygienickývýznamživotníchdějůvevodách. AvicenumPraha, 1. vyd. 1979, [15] s. barev.obr.příl., No. 735 21-08/25, 587 pp. (in Czech).

Šusta, J. 1938. Výživakapra a jehodružinyrybničné –novézákladyrybochovurybničného" českévydánínákladem ČSAV 1938 uspořádali a poznámkamiopatřili Univ. prof. Dr. Karel Schäfernaa Ing. Dr. Bořivoj Dvorak, 224 pp. (in Czech).

Vejsada, P. 2008. Vlivvýživynavybranévlastnosti masa tržníhokapra (*Cyprinus Carpio* L.), Zeměděls káfakultaJihočeskéuniverzity č. Budějovice 2008, doktorskádisertačnípráce, 128 pp. (in Czech).

Woynarovich, A., T. Moth-Poulsen and A. Péteri. 2010. Carp polyculture in Central and Eastern Europe, the Caucasus and Central Asia: a manual. FAO Fisheries and Aquaculture Technical Paper. No. 554. Rome, FAO, 73 p.

4

Reproduction and First Feeding of Carp

Gregor Schmidt

Introduction

The genetic potential of common carp (*Cyprinus carpio*) and its domestication for the last centuries are the reasons for the continuous success of this species in aquaculture. Carp can be raised in freshwater and brackish water (Garg 1996), show omnivorous feeding behavior and are extremely tolerant to many environmental conditions including oxygen concentrations and temperature.

Reproduction and breeding of common carp have a long tradition and it has been practiced for thousands of years. Carp is produced today using different techniques including semi-intensive and extensive production procedures. Reproduction and breeding are thereby practiced with a fine tradition. In Europe, carp aquaculture mainly uses ponds (Fig. 4.1), and reproduction and first feeding of carp have consequently been adapted to pond aquaculture conditions in many cases.

Numerous fish farms in Central Europe have specialized on breeding of carp. Accordingly, it was shown that in 2007 more than 100 farms used altogether 133 different carp strains in Germany for reproduction (Muller-Belecke et al. 2007). This included mirror carp strains and less frequently, scaled carps. A few farms also produce progenies of wild carps but only for stocking purposes. The farms have often bred the same strain for several decades. To prevent the detrimental effects of inbreeding, depression individual spawners from other farms are used for crossing in particular cases. To obtain a stock with positive characteristics individual animals are

Institute for Fisheries, State Research Center Mecklenburg–Vorpommern for Agriculture and Fishery, Hohen Wangelin, Malchower Chaussee 1, D-17194 Hohen Wangelin, Germany.
Email: g.schmidt@lfa.mvnet.de

Figure 4.1. Pond in Germany, picture taken by G. Schmidt.

selected during the rearing of carp to marketable size and less frequently so during the rearing of fish for stocking. The criteria for selection of a breeding stock in a pond farm include improvement of the general fitness (survival, health and adaptability), increasing growth performance and phenotype. For example, the presence and pattern of scales and the ability for adaptation to fluctuating environmental conditions are important criteria for selecting a breeding stock.

Fish that have been selected for supplying the spawning stock are raised separately from fish designated for consumption. Since carp show a high reproductive potential a spawning stock often consists of few animals. The actual size of a spawning stock contains 50 to 100 fish (Müller-Belecke 2007). Maturity is affected by biotic and abiotic factors, water quality, nutrient supply and climate. The climatic conditions in Central Europe allow male carps to achieve maturity in the 3rd to 4th year, whereas females need 4 to 5 years to spawn. Interestingly, the first spawn is never used for breeding, but the animals are allowed to reproduce in the following year. Spawning fish are commonly used for several years which can lead to their use for up to 15 years in exceptional cases. However, the big-sized very old carp are more difficult to handle so that average males are used three to four times and females are used five to six times for spawning. The reproduction of carps is achieved in pure breeding. Cross-breeding in order to use the heterosis effect (i.e., enhancement of traits in offspring as a result of mixing the genetic contributions of the parents) or other biotechnological breeding methods are never used in practice. The selection of spawning fish is always based on their appearance because the performance of the offspring of several females is usually raised together and the efficiency of individual females can thus not be distinguished.

Reproduction Methods

Proper rearing and feeding of potential spawners is important since storage of nutrients into eggs determines the starting quality of the larvae. Consequently, a relationship between the caloric content of carp eggs and the survival of embryos has been observed (Linhart et al. 1995). Thus, the ponds should be selected and prepared according to the needs of the fish. As it has been shown in the previous chapter on "Aquaculture of Carp in Europe and Asia" fish farmers keep several pond types in reserve for carp rearing to meet the requirements of carp at different life stages. Natural reproduction generally takes place in small ponds in which also hatching of larvae occurs. Although a fish farm should possess several of those, spawning ponds typically only represent less than 0.3% of the entire area of fish farms (Martyshev 1983).

For reproduction of carp several methods can be used. Traditionally, reproduction under natural conditions is conducted. However, modern fish farming often uses hatcheries. This is often the case in regions that experience fluctuating weather conditions in spring. In contrast to the situation in spawning ponds that undergo weather fluctuations, more favorable conditions for the fry can be chosen in hatcheries. Besides the welfare of fish cohorts can be checked more easily in tanks.

Natural Reproduction

In general, the control of gametogenesis and induction of spawning via the endocrine system in carp have already been described (Yaron 1995). Under natural conditions the time point for reproduction strongly depends on water temperature and light irradiation. Spawning occurs from May to July when water temperatures exceed 18°C. In most cases reproduction of carp takes place under natural conditions. Therefore, small ponds of few square meters are filled with water. These ponds are only used for reproduction and are left to dry during the rest of the year. Spawning ponds are often exposed to sunlight to reach temperatures above 18°C at the earliest possible point of spring time. The low depth of approximately 50 cm at the deepest point at the outlet ascends to a shallow marginal zone of 20 cm. Spawning ponds are encircled by a dam of one meter above the water level. This prevents intense exposure to wind and the subsequent cooling of the water. The bottom of the pond should be overgrown with firm and stable grass that should not be cut too low before retaining water since contact to ventral parts of the female stimulates ovulation and grass is also used as a substrate for egg deposition. During reproduction a continuous inflow with pre-conditioned water containing a low nutrient load is recommended. Introduction of egg or fry predators or parasites (i.e., through zooplankton) via inflowing water must be prevented, e.g., by applying filtration through a mesh with the pore size of < 1 mm. After preparation of the spawning pond, spawning fish are transferred to the pond whereby females have to be treated with special care because the ovaries are sensitive to pressure due to squeezing. Usually mating group compositions, one female and two males and less frequently 1:1 or 1.5:1 ratios ($\male$ to $\female$) are used. The amount of mating groups per pond depends on the pond size. If no spawning occurs within 3 to 4 days after introduction, the fish should be removed and the pond should be allowed to dry out. In this case, it is also recommended to prepare another pond for spawning to avoid decreasing water quality, for example due to fouling of grass, or the increasing influence of parasites.

Spawning generally takes place early in the morning where the males court the females, drive them through the vegetation until they releases the eggs accompanied by strong beating with the tailfin. The males immediately inseminate the eggs because their swelling upon contact with water leads to the closing of the micropyle. The mating activity can be repeated several times in a row by one mating group. The sticky eggs are disposed on the grass. After mating the adult fish are removed from the pond and retransported to the ponds for rearing of spawning fish. For their capture it might be necessary to temporarily reduce the water level in the spawning pond. It is essential however to prevent direct exposure to sunlight and heat to save the spawn from damage. Later, the pond can again be filled with water.

Artificial Reproduction

Reproduction in a hatchery is advantageous because fry that are able to swim and feed can be reared independently of weather fluctuations under controlled conditions. This also allows a pre-timing or a delay of the spawning time to account for fluctuating environmental conditions. Before spawning fish are transferred to the hatchery where they will be left unfed for several days. Genders are separately put into tanks which are supplied with cold and oxygen saturated water. Then the water temperature is slowly increased to above 18–20°C. The ripeness of eggs can be determined by biopsies. The obtained eggs are incubated in a fixing solution containing 60% ethanol, 30% formalin and 10% glacial acetic acid (Serra solution) for 10 minutes leading to transparent eggs that can be viewed through binoculars. If the nucleus is centrally orientated the female is not yet ready for spawning. In the following time the nucleus migrates to the cell wall and disintegrates (Fig. 4.2). This shows that the female is ready for spawning.

However, for artificial spawning of carp application of a gonadotropin-releasing hormone (GnRH) is necessary in many cases since there is a lack of certain essential environmental stimuli. Due to national differences legal recommendations must be consulted before using such hormones. Injection of dried pituitary glands dissolved in physiological sodium chloride solution (0.65%, w/v) is an established method. For this, pituitaries derived from the same species are extracted by removing the skullcap and the brain. The pituitary is a white organ that is located below the diencephalon in a small dent in the lower part of the cranium. After removal the pituitary should be incubated in acetone for 24 hours followed by drying for several days (Steffens 1984). Dry storage can be continued for several years in the refrigerator. A dried pituitary of a three-year-old carp weighs approximately 3 to 4 mg. For each kilogram of a spawning female one pituitary is used. Slow

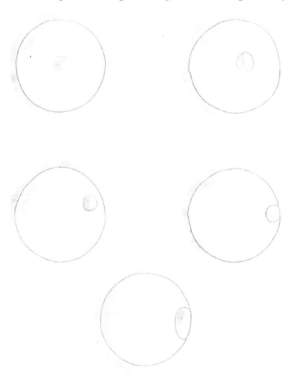

Figure 4.2. Ripeness of carp eggs. In immature eggs the nucleus can be found in a central position, whereas mature eggs show the nucleus in a decentral position; illustrated by Robert Holtmann..

injection of the pituitary suspension in a volume of 1–2 ml generally takes place in the dorsal musculature between the lateral line and the dorsal fin to allow the distribution in muscle tissue (Fig. 4.3). Scaled carps can be injected intra-peritoneally at the ventral fin which must be carried out with care to prevent damage from inner organs. Injection of pituitary suspensions is used for two procedures: One single injection after which the animals spawn after 16 to 18 hours or two injections at an interval of 12 to 24 hours. Note that the first injection contains 10% and the second injection 90% of the suspension. Injection of pituitary extract in males is not mandatory but is often used to facilitate the release and thus increase the amount of sperm. For males a single injection of a low dose into the dorsal musculature is sufficient. It is advisable to do this when the nucleus in the eggs has reached its marginal position in the cell. If the females are treated with two injections, males can already be injected when the nucleus is in a decentral position in the egg.

Alternatively different synthetic GnRHs can be used (Szabo 2002, Mikolajiczyk et al. 2003). Modified non-peptides, such as Buserilin and Gonazon, are promising because they are less degradable and lead to stronger effects. Consequently, lower doses can be used (10 to 20 μg/kg fresh weight).

Handling of spawning fish can lead to a strong stress response in the animals. Thus, it is recommended to anesthetize the animals when working with fish of a greater size or for injections. For the selection and use of anesthesia it is also recommended to refer to the legislation as issued by regional authorities. Generally compounds, such as clove oil, tricaine mesylate (MS 222, Sandoz) or 2-phenoxyethanol are used. With the beginning of ovulation several procedures can be followed: For stripping of eggs, the carp is anaesthetized, dried with a towel to prevent contact of eggs or sperm with

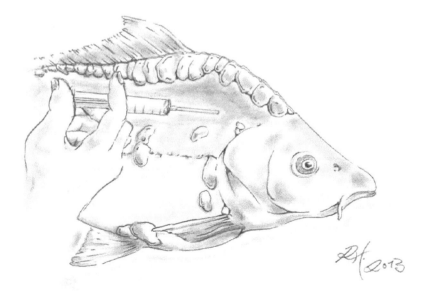

Figure 4.3. Position for injection of the hormonal solution in carp; illustrated by Robert Holtmann.

water. Sperm is obtained mostly from several males together by squeezing the abdomen from both sides. Sperm can be stored in jars in a refrigerator for several days. The stripping of females starts with application of pressure to both sides of the abdomen near the genital papilla which is followed by backwards-directed massaging. The eggs are collected in plastic bowls; the sperm is added and mixed carefully. For this, 100 ml water is added to 1 kg roe. Alternatively, an insemination solution containing 4 g sodium chloride and 3 g urea per liter water can be used which increases the activity of sperm (Woynarrovich 1962). After addition of water or insemination solution the eggs must be agitated continuously with caution to prevent sticking together. The liquid can be decanted after approximately 10 minutes and new liquid can be added. This washing step is repeated several times for one hour until the swelling of eggs has completed.

After swelling the eggs should be transferred to MacDonald or Zuger jars (Fig. 4.4) which makes it necessary to remove adhesive compounds. Therefore, a solution of 0.5 g/L tannin is used from which 50 ml are added to 1 kg eggs. The mixture is firmly agitated and further water is added. The eggs are then washed twice and the entire tannin exposure is repeated by using a lower tannin dose. After thoroughly washing the eggs with water they are placed in MacDonald or Zuger jars (Figs. 4.4 and 4.5). A volume of 1 liter allows the development of 200 ml swollen eggs which equals approximately 20,000 eggs. The inflowing water should be adjusted so that the eggs are sufficiently supplied with oxygen and are agitated slowly. Dead eggs should be removed to prevent fungal contamination. In some cases disinfection is necessary. For this, the use of malachite green oxalate was banned in the EU several years ago, but formalin and peracetic acid may be used. In the jars, hatching takes place within 3 to 4 days. This can be synchronized when the water inflow is reduced for some minutes upon onset of hatching. Newly

Figure 4.4. Zuger Jar for egg incubation. Picture taken by G. Schmidt.

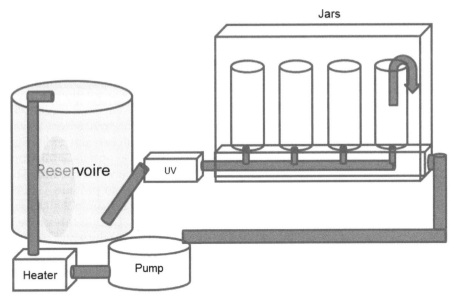

Figure 4.5. Typical setup for a hatchery system.

hatched larvae should be flushed with the outflowing water into a tank. In the tank larvae are collected in a net cage (mesh size: 100 µm) (Tölg et al. 1981).

Spawning in Tanks

For this method a mating group is placed in a tank and shortly after this spawning takes place at the bottom of the tank. After removal of the adult fish, the eggs are skimmed and treated as described above. As an alternative plastic brushes can be spread on the bottom of the tank which simulates the presence of spawning substrates. The eggs will then develop on the brushes. This method often replaces the use of GnRH (Proske 1978). Additional stimulation can be achieved by raising the temperature to 24°C, but apart from this, the fish need rest and further external stressors must be strictly avoided.

Development of Eggs and Hatching

Carp eggs are approximately 1.5 mm in size, but swell to twice this size in water. They are transparent with a slight shade of yellow. Females release 150,000 to 200,000 eggs per kg body weight but the amount also depends on age and condition of the spawning fish. According to Schmeller (1998) the proportion of roe relative to the entire body weight varies between 19 and 48%. A five-year-old female weighting between 2.5 kg and 3.5 kg can yield 500,000 eggs. During their development eggs pass through different stages which lead from the first level of meiosis to eye pigmentation stage and finally to yolk-sac larvae, which means it is ready for hatching. However, a part of the eggs often remains unfertilized which is characterized by a white color followed by fungal contamination (commonly with *Saprolegnia* ssp.).

The incubation time of eggs depends on water temperature. Increasing temperatures decreases the time until hatching which corresponds to 4 days at 20°C and to 5.5 days at 18°C (Yamamoto 1933). During incubation the level of dissolved oxygen should not fall under 6 mg/L (Kaur and Toor 1978). Hatching of an entire batch can take several days. The transparent larvae mostly emerge with the tail first, have a length of 4.8 mm to 6 mm and weigh 6 to 7 mg (Horvath et al. 1984). They have a yolk sac and a continuous fin from the back to the anus. However, they lack mouth, scales and ventral fins. The pectoral fins are present rudimentarily. Glands at the head provide a sticky exudate which is necessary for attachment of larvae to substrate while jerky movements allow the movement to the water surface. During this phase a simple intestinal tract and the mouth gap are developed. At the water surface, the larvae take up air which is subsequently transported via the esophagus and the Ductus pneumaticus to the swim bladder. This allows the fish to position itself better in the water and the exogenous feeding can commence. Although the yolk sac has not yet been depleted carps start feeding on day 3 after hatching. At this time they feed both endogenously and exogenously. Within the next days, dorsal, caudal and anal fins are differentiated and the pectoral fins develop further. Pigmentation intensifies dorsally and proceeds to the ventral side (Fig. 4.6). Approximately two weeks after hatching scales emerge. After completion of scaling by three weeks the larval stages are finalized and only development of sexual organs is missing.

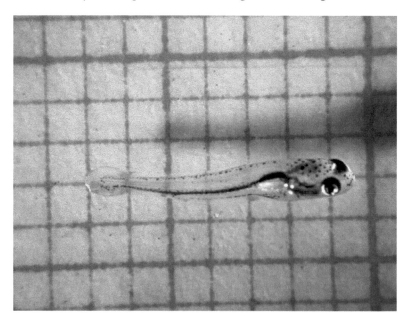

Figure 4.6. Carp larvae before first feeding on the 4th day post hatch, picture taken by G. Schmidt.

After hatching larvae are transferred to nursery ponds where they stay for up to six weeks. These ponds account for approximately 3% of the entire area of a fish farm (Martyshev 1983).

First Feeding

When larvae start swimming, carp are nearly without any exceptions fed under natural conditions. First feeding under controlled conditions in the hatchery is rarely practised.

First Feeding in Ponds

Swimming larvae are transferred from hatchery or spawning ponds to nursery ponds which are small (< 1 hectare) and show low depth (< 1 m). These ponds should be filled with temperate and filtered water only shortly before stocking to prevent extensive development of zooplankton. Generally, 50,000 to 200,000 larvae are used per hectare pond area (Proske 1998). The stocking density depends on the productivity of the pond and the intended growth performance of fry. Depending on the number of larvae a more or less abundant feed supply is present. During the nursery stage, zooplankton has to meet the needs of the fry with respect to size and abundance.

This is achieved by preparing the pond only shortly before stocking with ambient and filtered water. This reduces the rapid development of zooplankton and favors the extensive progress of small zooplankton. When carp larvae shift to exogenous feeding they mostly use rotifers of a size between 200 to 300 μm (Horvath et al. 1984). This is followed by feeding on copepods and cladocereans (Kainz 1998). For details on the feeding ecology and artificial feeding please refer to Chapter 9 of this book by Huser and Bartels. To meet the continuously increasing demand for food it is possible to gradually increase the water volume after stocking. After 5 to 6 weeks the nursery ponds are emptied, since their productivity is exhausted. Transported by the out-flowing water fingerlings are caught which are 4 cm to 5 cm in length and show a body mass of 4 g to 5 g (Steffens 2008). When natural food is becoming scarce carp are put into stock ponds which make up for 10% of the area of a fish farm (Martyshev 1983). Juvenile carp stay in these ponds until they are transferred to wintering ponds in autumn. In some cases rearing in nursery ponds is omitted and larvae that are able to take up food are directly transferred to stock ponds.

First Feeding in a Hatchery

In exceptional cases, e.g., due to bad weather conditions or a lack of zooplankton in the ponds, the first feeding may take place in hatcheries. Therefore, larvae are counted and placed into tanks. Since the small carp larvae cannot be counted automatically, they are dispersed in a tank of known water volume. After that, a defined water volume is taken out and the larvae in this volume are counted which allows estimation of the amount of larvae in the entire tank.

The first feed is given after two days (Fig. 4.7). At this time point larvae are actively searching for food. A successful first feeding strongly depends on adaptation of the feeding to the physiological needs of different life stages with respect to environmental conditions including stocking density, water temperature and quality, light regime and selection of tanks. In addition feeding intensity and frequency, sinking behavior of feed particles, and the way of feed application as well as the characteristics of the feed (form, color, stability) and its quality (texture, taste, anti-nutritive compounds (for these see Chapter 13)) affects weaning success. Besides these technical requirements, the health-status of the larvae also plays a role with respect to their energetic reserves including yolk sac and the development of the intestinal tract (Barrows and Rust 2000). Stocking density is generally limited by water quality. According to Steffens (1981) the optimal stocking density is 50,000 larvae m^{-3}. Although important improvements in processing feed for larvae have been achieved in the last years, first feeding with a formulated diet is often problematical. Carp larvae take up the diet but the digestive capacity of their intestinal tracts, i.e., the enzyme activity are often not able to efficiently digest highly complex feed particles. Thus, feeding only formulated diets during first feeding often leads to deformations, reduced growth and low survival rates (Zrenner 2009). However, first feeding with high quality diets can also lead to adequate growth (Carvalho et al. 1997).

Figure 4.7. Weaning of carp in tanks (on the 21st day post hatch). Picture taken by G. Schmidt.

Table 4.1. Typical composition of starter diets (Schwarz et al. 1995).

	Natural diet	*Artemia salina*[1]	Formulated diet[2]
Particle size (µm)	30–1000	250–400	100–300
Crude protein (%)	54.8–69.8	55 ± 5	45–70
Crude ash (%)	9–21	10 ± 2.5	5–12
Crude fibre (%)	–	-	2–4
Crude fat (%)	5.7–13.2	20 ± 5	5–20

[1]Sorgeloos et al. 2001; [2]manufacturer's declaration.

Artificial Feeding

Similar to other teleosts carp show an essential demand for phosphoglycerides, especially during early (embryonic and larval) developmental stages (Fontagné et al. 1998, Geurden et al. 1995, Tocher et al. 2008). While embryos are supplied with nutrients by the yolk sac prior to first feeding, later developmental stages need to receive an abundance of essential fats via a natural diet. Mainly due to the fish farmers use natural food for first feeding of carp fry. Therefore, often zooplankton from open waters is collected with gauze nets and sorted according to its size. Zooplankton should be washed before use to remove mud and small algae. To avoid damages of the zooplankton, they must be handled with care. During the first days of feeding the smallest zooplankton (diameter < 300 µm) is fed. With the growth of fish the size of the food is increased. As a rule of thumb it can be assumed that the optimal food particle size equals the eye diameter of the fish.

This feeding practice however, has several disadvantages. The development of zooplankton in natural waters depends on biotic and abiotic factors with limited or no possibilities for fish farmers to exert influence on. A continuous availability of suitable natural food cannot be guaranteed. In addition, the collection and the sorting of zooplankton is time-consuming. If zooplankton abundance is sufficiently high, it can be lured by a luminous source and the sorting according to its size can be achieved by pumping across gauzes of a different size. However, the main disadvantage is that the collected zooplankton can be contaminated by bacteria, viruses and parasites and a deficient hygiene may lead to their introduction into the fish rearing tanks.

As an alternative to zooplankton from natural waters, live food organisms can be bred. In practice nauplii of brine shrimp (*Artemia* spp.) are commonly used. For one liter of culture 2 g to 5 g *Artemia* cysts are incubated at 26°C to 28°C in aerated water (pH 7–8) containing 3.5% sodium chloride. This is often done in cylinders or MacDonald jars. After 24 to 26 hours nauplii hatch and 10 minutes before harvest the aeration is stopped. Then, nauplii gather at the bottom of the flask and the shells accumulate at the water surface. Unhatched eggs can be found at the bottom of the jars. It is important to separate the nauplii from the shells and remaining eggs because the latter cannot be digested and excreted by fry although they are ingested. Consumption of egg shells can lead to mortality of carp larvae. Since newly hatched nauplii are attracted by light a luminous source outside at the bottom of a jar leads to their accumulation which can be followed by removal by suction and filtering by using gauzes (100 to 150 μm). To avoid introduction of incubation water to the fish rearing tanks the nauplii should be washed in a sieve with lukewarm water before feeding. The time consuming separation of nauplii from the shells can be omitted when the eggs are de-capsulated by using sodium hypochloric solution before being cultured. Thereby, the color of the cysts changes from brown to gray, white and orange. This is followed by carefully washing the eggs and culturing as described above. The removal of the capsules leads to reduced incubation time of 18 to 20 hours. The enrichment of nauplii with essential fatty acids is only rarely used. Instead, a part of nauplii as a food source for carp larvae is soon substituted by high quality starter feeds.

Feeding of larvae is commonly done at continuous illumination with mean intensity at water surface. For first feeding a permanent supply with natural food is the approved method in aquaculture. Accordingly, a high density of food organisms should be maintained with an optimal value of two to four individuals per ml volume in the tank. However, especially during the first days of feeding not all food is taken up by larvae and this pollutes the rearing system. This of course requires high cleaning efficiency in the mechanical and biological water treatment stages. In addition, the bottom of the tanks should be cleaned daily to prevent the development of a biofilm (mainly fungi) which may lead carp larvae getting entangled in the filaments and hyphae which can lead to larvae mortality.

The transition to dry feed commences on the fourth day after first feeding. For this slowly sinking starter feeds with a diameter of 200 μm are recommended, because the young fish are not yet able to take up food from the surface of the water or the bottom of the tank (For more information on the application of different feeds see Chapter 6).

Feeding Carp Fingerlings

It was shown that carp display a diurnal feeding behavior with higher activity in the mornings compared to the evenings (Yilmaz et al. 2005). Feeding rates also influence growth performance of juvenile carp as has been shown by Shimeno et al. (1997). In this study feeding rates of 1.64% or lower, negatively affected weight gain and also influenced lipid and carbohydrate metabolism after 30 days of experiment.

The study of Yilmaz et al. (2005) also reported that carp fingerlings should not be fed with high energetic diets, e.g., obtained by addition of soy-acid oil to the diet, because diets with lower energy content led to better growth and feed conversion ratios after 60 days of feeding. Although carp are known to digest carbohydrates better compared to salmonids (Yamamoto et al. 2001), the efficiency of carbohydrate digestion clearly varies with carp life stage and microflora in the gut of carp (Krogdahl et al. 2005). The advantageous effect of probiotics and certain compounds of plant origin on growth performance of juvenile carp have been investigated by several research groups (e.g., Singh et al. 2011, Novereirian and Nasrollahzadeh 2012). Nutrition of carp at later developmental stages is reviewed in Chapter 6 of this book by G. Füllner.

Acknowledgement

The authors like to thank Daniel Dagassan for corrections in English and useful comments on the manuscript.

References

Alabaster, J.S. and R. Lloyd. 1980. Water quality criteria for freshwater fish. Butterworth, London.

Albrecht, M.L., S. Fritzsche, B. Rennert and F. Fredrich. 1987. Satzkarpfenproduktion in Warmwasseranlagen. In Technologien, Normen und Richtwerte der Fischproduktion: 65–76 (in German).

Barrows, F.T. and M.B. Rust. 2000. Larval feeding — fish. *In*: R.R. Stickney (ed.). Encyclopedia of Aquaculture. John Wiley & Sons, Inc., New York, pp. 465–469.

Beamish, F.W.H. 1964. Respiration of fishes with special emphasis on standard oxygen consumption. III. Influence of oxygen consumption. Can. J. Zool. 42: 355–366.

Carvalho, A.P., A.M. Escaffre, A. Oliva Teles and P. Bergot. 1997. First feeding of common carp larvae on diets with high levels of protein hydrolysates. Aquaculture international Volume 5 Issue 4, pp. 361–367.

Fontagné, S., I. Geurden, A.-M. Escaffre and P. Bergor. 1998. Histological changes induced by dietary phospholipids in intestine and liver of common carp (*Cyprinus carpio* L.) larvae. Aquaculture 161: 213–223.

Garg, S.K. 1996. Brackish water carp culture in potentially waterlogged areas using animal wastes as pond fertilizers. Aquacult. Int. 4(2): 143–155.

Geurden, I., J. Radünz-Neto and P. Bergot. 1995. Essentiality of dietary phospholipids for carp (*Cyprinus carpio* L.) larvae. Aquaculture 131: 303–314.

Horvath, L., G. Tamás, I. Tölg and J.E. Halver. 1984. Special Methods in Pond Fish Husbandry. Halver Corporation, Budapest.

Kainz, E. 1998. Produktionsverhältnisse und Lebensbedingungen im Fischteich. *In*: W. Schäperclaus and M. Lukowicz (eds.), Lehrbuch der Teichwirtschaft, 157–219. ISBN 3-8263-8248-X (in German).

Kaur, K. and H.S. Toor. 1978. Effect of dissolved oxygen on the survival and hatching of eggs of scale carp. The Progressive Fish-Culturist 40: 35–37.

Krogdahl, A., G.-I. Hemre and T.P. Mommsen. 2005. Carbohydrates in fish nutrition: digestion and absorption in postlarval stages. Aquaculture Nutrition 11(2): 103–122.

Linhart, O., S. Kudo, R. Billard, V. Slechta and E.V. Mikodina. 1995. Morphology, composition and fertilization of carp eggs: A review. Aquaculture 129: 75–93.

Martyshev, F.G. 1983. Pond Fisheries. Translation of Prudovoe Rybovodstvo, Vysshaya Shkola Publishers, Moscow 1973. Publishers: A.A. Balkema/Rotterdam, 451 p.

Mikolajiczyk, T., J. Chyb, M. Sokolowska-Mikolajiczyk, W.J. Enright, P. Epler, M. Filipiak and B. Breton. 2003. Attemps to induce an LH surge and ovulation in common carp (*Cyprinus carpio* L.) by differential application of a potent GnRH analogue, azagly-nafarelin, under laboratory, commercial hatchery, and natural conditions. Aquaculture 223: 141–157.

Muller-Belecke, A., G. Füllner, H. Klinger, R. Rösch, R. Tiedemann, H. Wedekind and U. Brämick. 2009. Aquatische genetische Ressourcen–Laichfischbestände von Wirtschaftsfischarten in Deutschland. Schriften des Instituts für Binnenfischerei e.V. Potsdam-Scrow, Bd. 25: 74 p. (in German).

Novereirian, H.A. and A. Nasrollahzadeh. 2012. The effects of different levels of biogen probiotic additives on growth indices and body composition of juvenile common carp (*Cyprinus carpio* L.). Caspian J. Env. Sci. 10(1): 115–121.

Proske, C. 1978. Über ein praxisgerechtes Verfahren der kontrollierten Vermehrung von Karpfen (*Cyprinus carpio*). Fischer und Teichwirt 29: 25–28 (in German).

Proske, C. 1998. Bewirtschaftung des Karpfenteichs. *In*: W. Schäperclaus and M. Lukowicz (eds.), Lehrbuch der Teichwirtschaft, 247–303. ISBN 3-8263-8248-X (in German).

Singh, P., S. Maqsood, M.H. Samoon, V. Phulia, M. Danish and R.S. Chalal. 2011. Exogenous supplementation of papain as growth promoter in diet of fingerlings of *Cyprinus carpio*. Int. Aquat. Res. 3: 1–9.

Schmeller, H.B. 1998. Vermehren und Vorstrecken von Karpfen. *In*: W. Schäperclaus and M. Lukowicz (eds.), Lehrbuch der Teichwirtschaft, 531–545. ISBN 3-8263-8248-X (in German).

Schreckenbach, K., R. Knösche and K. Ebert. 2001. Nutrient and energy content of freshwater fishes. J. Appl. Ichthyol. 17: 142–144.

Schwarz, F.J., M. Oberle and M. Kirchgessner 1995. Nutrient content of zooplankton in carp ponds. Aquaculture 129: 251–259.

Shimeno, S., T. Shikata, H. Hosokawa, T. Masumoto and D. Kheyyali. 1997. Metabolic response to feeding rates in common carp, *Cyprinus carpio*. Aquaculture 151: 371–377.

Sorgeloos, P., P. Dhert and P. Candreva. 2001. Use of brine shrimp, *Artemia* spp., in marine fish larviculture. Aquaculture 200: 147–159.

Steffens, W. 1981. Industriemäßige Fischproduktion. Jena: VEB Deutscher Landwirtschaftsverlag (in German).

Steffens, 1984. Moderne Fischwirtschaft, Grundlagen und Praxis. Neumann-Neudamm Verlag (in German).

Steffens, W. 2008. Der Karpfen. Westarp-Wissenschaften, ISBN 978-3-89432-649-4 (in German).

Szabo, T., C. Medgyasszay and L. Horvath. 2002. Ovulation induction in nase (Chondrostoma nasus, Cyprinidae) using pituarity extract or GnrH analogue combined with domperidone. Aquaculture 203: 389–395.

Tolg, I., L. Horvath and G. Tamás. 1981. Fortschritte in der Teichwirtschaft. Hamburg and Berlin: Verlag Paul-Pary, ISBN 3-490-10614-8 (in German).

Woynarrovich, E. 1962. Hatching of carp eggs in zuger-glasses and breeding of carp larvae until an age of 10 days. Bamigeh Bull. Fish. Cult. Isr.: 38–46.

Yamamoto, T. 1933. Influence of temperature on the embryonic development of the carp. Bull. Jap. Soc. Fish. 2: 167–174.

Yamamoto, T., T. Shima, H. Furuita, N. Suzuki and M. Shiraishi. 2001. Nutrient digestibility values of a test diet determined by manual feeding and self-feeding in rainbow trout and common carp. Fish. Sci. 67: 355–357.

Yaron, Z. 1995. Endocrine control of gametogenesis and spawning induction in the carp. Aquaculture 129: 49–73.

Yilmaz, E., A. Sahin, M. Duru and I. Akyurt. 2005. The effect of varying dietary energy on growth and feeding behaviour of common carp, *Cyprinus carpio*, under experimental conditions. Applied Animal Behaviour Science 92: 85–92.

Zrenner, M. 2009. Anfütterung von Karpfenbrut (*Cyprinus carpio* L.) unter den Bedingungen der intensiven Aquakultur. Master thesis Technical University Munich, 70 p. (in German).

5

Chromosome Set Manipulation, Sex Control and Gene Transfer in Common Carp

Boris Gomelsky

Introduction

This chapter reviews information on chromosome set manipulation methods, sex control and gene transfer in common carp. Induced gynogenesis, androgenesis and polyploidy are usually considered as chromosome set manipulation methods. Development and application of the hormonal sex reversal are the basis for sex regulation in common carp, as well as in other fish species. The chapter's section on gene transfer includes descriptions of methods for production and properties of transgenic common carp. Common carp is a convenient species for chromosome set manipulation and gene transfer studies because of its high fecundity, established methods of artificial spawning[1] and existence of unique morphological genetic markers. The development of techniques in this chapter is given from a chronological viewpoint. This chapter includes data obtained in koi also, since koi is an ornamental form of common carp. Special attention is given to some important methodological aspects; this information is especially important for scientists and aquaculturists who are new in these areas.

A short description of each technique is provided before giving materials on common carp. More comprehensive general information on each method

---

[1] Demonstration videos on ornamental common carp (koi) artificial spawning can be found at http://www.youtube.com/playlist?list=PLI2Wd3VFTYkNTOW3upWRxKFK3cnq1TBZ5

Aquaculture Research Center, Kentucky State University, 103 Athletic Drive, Frankfort, KY 40601, USA. Email: boris.gomelsky@kysu.edu may be found in primary literature, review articles as well as in books on fish genetics (Cherfas 1981, Hunter and Donaldson 1983, Thorgaard 1983, Purdom 1993, Komen and Thorgaard 2007, Piferrer et al. 2009, Dunham 2011, Gomelsky 2011).

Induced Gynogenesis

In case of gynogenesis embryo development is controlled by maternal heredity only after activation of eggs by insemination. Several all-female forms have been described in fish, which are reproduced by natural gynogenesis (Cherfas 1981, Purdom 1993, Lamatsch and Stöck 2009). Gynogenetic development may be induced in fish, which reproduce by normal sexual mode. In this case genetic inactivation of male chromosomes and restoring of diploidy are attained by artificially applied treatments. Usually male chromosomes are inactivated by sperm irradiation through X-rays, gamma rays or ultraviolet light (UV). Chromosomes and cytoplasmic structures of spermatozoa have different resistance to irradiation; high doses can inactivate chromosomes but spermatozoa can preserve their capability to move and to inseminate eggs.

Insemination of eggs with genetically inactivated spermatozoa results in appearance of haploids, which in fish are morphologically abnormal ('haploid syndrome') and perish at the time of hatching. In hatched larvae the 'haploid syndrome' is characterized by a shortened and curved body, swollen pericardial cavity and undeveloped eyes (Cherfas 1981). For obtaining viable gynogenetic fish, diploidy should be restored. This can be achieved by any of two methods: either by blocking of the second meiotic division in eggs (meiotic gynogenesis) or by blocking of the first mitotic division in haploid embryos (mitotic gynogenesis). For blocking of meiotic or mitotic divisions, strong physical treatments (shocks) are applied at anaphase of the corresponding divisions. Low or high temperatures (cold and heat shocks) or hydrostatic pressure are regularly applied physical treatments. In case of meiotic gynogenesis, applied 'early' shock suppresses the second meiotic division in eggs; as a result, the second polar body is not extruded and fuses with the haploid female pronucleus. In case of mitotic gynogenesis 'late' shock suppresses first mitotic division in haploid embryos; as a result, the diploid nucleus is formed by fusion of two haploid nuclei. Viable gynogenetic diploids may appear without externally applied shocks due to spontaneous blocking of the second meiotic division in eggs (meiotic gynogenesis); the frequency of occurrence of such spontaneous gynogenetic diploids is low (usually less than 0.5%).

Meiotic Gynogenesis

The Russian scientists Romashov et al. (1960) and Golovinskaya et al. (1963) reported the first data on meiotic gynogenesis in common carp. It was shown in these studies that after insemination of eggs by sperm irradiated with high dosages of X-rays (100–200 kR) spontaneous viable gynogenetic diploids appeared with low frequency. Based on segregation of gynogenetic fish by identification of different scale cover types Golovinskaya and Romashov (1966) correctly suggested that gynogenetic diploids appeared due to suppression of the second meiotic division in eggs. They further suggested that heterozygosity in gynogenetic fish resulted from crossing over between gene and centromere.

Further studies on meiotic gynogenesis in common carp were performed by Cherfas (1975, 1977), Cherfas and Truveller (1978), Nagy et al. (1978, 1979, 1983), Nagy and Csanyi (1978, 1982), Gomelsky et al. (1979, 1989, 1992, 1996), Taniguchi et al. (1986), Hollebecq et al. (1986), Linhart et al. (1986, 1987), Wu et al. (1986), Komen et al. (1988), Cherfas et al. (1990, 1993a, 1994a,b), Sato et al. (1990), Sumantadinata et al. (1990), Kim et al. (1993), Shelton and Rothbard (1993), Khan et al. (2000), Aliah and Taniguchi (2000), Sato and Amita (2001), Sato (2013), Alsaqufi et al. (2014) and others. The techniques for production of meiotic gynogenetic progenies and the fundamental data on meiotic gynogenesis in common carp are described below.

Irradiation of Common Carp Sperm for Genetic Inactivation

X-rays or gamma rays were typically used for genetic inactivation of sperm in initial studies on induced gynogenesis in common carp. In later studies, UV-irradiation was predominantly used for this aim largely owing to its availability and applicability. UV-irradiation has low penetration ability as compared with X-rays or gamma rays; therefore, before irradiation with UV, the dense sperm of teleost fish should be diluted and thinly spread. Since the rate of dilution and thickness of diluted sperm vary in different studies the doses of irradiation are difficult to compare (Chourrout 1987). In our studies on induced gynogenesis in koi (Alsaqufi et al. 2014), a FisherBiotech UV microprocessor-controlled Crosslinker (FB-UVXL-1000; Fisher Scientific) was used for sperm irradiation (Fig. 5.1). This model has five 8-W, 254-nm UV bulbs and is equipped with a system for direct metering a total energy per unit area, and delivering of target dosage (3,000 or 4,000 J/m^2) is controlled by device itself. Earlier, this device was successfully used for induction of gynogenesis in black crappie (*Pomoxis nigromaculatus* Lesueur) (Gomelsky et al. 2000). For UV-irradiation, sperm was diluted with saline solution (1 ml of sperm per 9 ml of 0.85% NaCl solution); 2 ml of diluted sperm was placed in 6-cm glass Petri dish with approximate 0.07 cm thickness; six Petri dishes were placed in the Crosslinker simultaneously. Uniform irradiation was achieved by placing the Crosslinker on a shaker table to keep the diluted sperm in motion during treatment hence securing susceptibility of all spermatozoa to the irradiation.

Figure 5.1. Irradiation of common carp sperm with UV Crosslinker.

Suppression of Second Meiotic Division in Common Carp Eggs

Cold or heat 'early' shocks were applied for suppression of the second meiotic division in common carp eggs. The timing of shock application can be standardized for different pre-shock temperatures by expression of the pre-shock time in terms of mitotic interval τ_0 (proposed by Dettlaff and Dettlaff 1961). The values of 1 τ_0 in common carp at different temperatures are presented in Table 5.1. Optimum timing for blocking of the second meiotic division in eggs was determined to be about 0.05–0.15 τ_0 (e.g., 1.4–4.2 min at 20°C) and 0.10–0.20 τ_0 (e.g., 2.8–5.6 min at 20°C) after insemination for cold and heat shock, respectively.

Table 5.1. Values of 1 τ_0 in common carp at different temperatures (after Ignatieva 1979).

Temperature, °C	16.0	17.0	18.0	19.0	20.0	21.0	22.0	23.0	24.0	25.0
Duration of 1 τ_0, min	53	45	38	32	28	24.5	22	20	19	19

Application of Genetic Markers for Verification of Gynogenetic Development

Gynogenetic origin of fish produced in experiments may be confirmed by application of genetic markers. When genetic markers that are morphologically identifiable are used for this purpose, eggs taken from the female with a morphological trait controlled by a recessive allele are inseminated with irradiated sperm taken from male with a trait controlled with a dominant allele. If the chromosomes of the spermatozoa are successfully inactivated by irradiation, the offspring should not have the paternal dominant trait; this will indicate a gynogenetic origin of all obtained fish. For example,

Infobox

Mitosis is a type of cell division typical for somatic (body) cells. During mitotic divisions the diploid level of ploidy (2n) remains unchanged.

Mitotic interval is the duration of one mitotic cycle during synchronous embryonic cleavage divisions of blastomeres. The duration of the mitotic interval depends on temperature; it shortens along with temperature increase. However, the timing of any embryological stage expressed in terms of mitotic intervals for a given temperature is a fixed value.

Meiosis is a type of cell division typical for sex cells. During meiosis the chromosome number from the diploid level of ploidy (2n), typical for somatic cells, is reduced to the haploid level of ploidy (n), typical for gametes (eggs and spermatozoa). The reduction in chromosome number during meiosis results from two consecutive cell divisions during which the chromosomes are only reduplicated once.

Oogenesis is the process of transformation of initial female sex cells (**oogonia**) to **eggs**. After the initial meiotic stages of the first meiotic division primary **oocytes** start to accumulate nutritional substances (mostly yolk) and grow. When growth is completed, oocytes enter to the maturation stage which is triggered by environmental spawning conditions or hormonal stimulation. The first meiotic division completes with extrusion of the **first polar body**. Soon the second meiotic division begins which proceeds until metaphase. At the stage of metaphase II the meiosis is blocked and oocytes are released. **Eggs** are **haploid** (n) but each chromosome consists of two chromatids; the second meiotic division finishes later during the fertilization period.

Nuclear transformations during fertilization period in fish: The head of the spermatozoon penetrates into the egg's cytoplasm, swells and transforms into the **male pronucleus**. Simultaneously the second meiotic division in the egg is completed: the **second polar body** is extruded and the remaining haploid chromosome set transforms into the **female pronucleus**. The male and female pronuclei migrate into the depth of the cytoplasm and fuse forming the metaphase plate of the **first mitotic cleavage division**. After its completion, two **embryonic cells (blastomeres)** are formed.

Genetics of sex determination: The system of sex chromosomes plays two functions: determining the sex of individuals and regulating sex ratio in progeny. One sex, **homogametic**, has two similar sex chromosomes, while another sex, **heterogametic**, has two different sex chromosomes. Heterogametic sex may be either male or female. In the former case (**male heterogamety**, and correspondingly **female homogamety**) the sex chromosomes are usually designated as X and Y (males—**XY**, females—**XX**); in the latter case (**female heterogamety**, and **male homogamety**)—as Z and W (females—**WZ**, males—**ZZ**).

Sex chromosomes determine the sex of individuals and commence the development of reproductive system in the corresponding direction (male or female). This genetic program is realized through a successive chain of hormonal regulations. On this basis, **hormonal sex reversal** is possible; experimental hormonal influence can switch the genetic program to opposite sex but formula of sex chromosomes remains unchanged.

eggs from mirror (large 'mirror' scales scattered on the body) common carp (genotype ss) were inseminated with irradiated sperm from fully scaled male (genotype SS or Ss) (Komen et al. 1988, Cherfas et al. 1993a) or eggs from unpigmented koi without melanin development (genotype $b_1b_1b_2b_2$) were inseminated with irradiated sperm from wild-type fully pigmented common carp (genotype $B_1B_1B_2B_2$) (Gomelsky et al. 1996). In the latter case, dominant alleles (B_1 and B_2) controlling the presence of melanin were used as genetic markers to verify the genetic inactivation of paternal chromosomes. This trait is very suitable for using as a genetic marker as a proof of gynogenesis since the presence of melanin is already easily detectable at the larval stage.

Presently microsatellite DNA markers are widely used for confirmation of gynogenetic origin of fish. Genomes of gynogenetic fish should not have specific microsatellite alleles of the male, whose sperm was irradiated. Microsatellite markers have been used to prove the gynogenetic origin of common carp progenies in studies by Aliah and Taniguchi (2000) and Alsaqufi et al. (2014).

Sex Composition of Meiotic Gynogenetic Progenies

Common carp meiotic gynogenetic progenies consisted of females only (Golovinskaya et al. 1974, Nagy et al. 1978, Gomelsky et al. 1979, Cherfas 1981) proving female homogamety (females—XX, males—XY) in this species. In spite of all-femaleness of meiotic gynogenetic progenies in common carp, induced gynogenesis is not used for direct sex control because of possible inbreeding depression and relative complexity of the procedure since irradiation of sperm and shocks are required for obtaining every all-female progeny. However, all-female progenies of meiotic gynogenetic origin are commonly used in initial experiments on hormonal sex reversal (see below).

Genetic Properties of Meiotic Gynogenic Progenies

Increased homozygosity is the major genetic result of induced gynogenesis. Meiotic and mitotic types of induced gynogenesis differ with regard to the degree of increase in homozygosity. In the case of meiotic gynogenesis, heterozygosity results from crossing over between gene and centromere, and may differ to a great extent for different genes. For genes controlling some morphological traits (scale cover and color mutations) and polymorphic proteins (variability detected by electrophoretic analysis) the frequency of heterozygotes in meiotic gynogenetic progenies in common carp varied from 0.05 to 0.99 (Cherfas and Truveller 1978, Nagy et al. 1979, Cherfas 1981). A similar range in recombination frequency (from 0.01 to 0.96) was observed for 14 microsatellite loci (Aliah and Taniguchi 2000, Alsaqufi et al. 2014). The average heterozygosity for all analyzed loci in common carp is about 0.40. Subsequently, the coefficient of inbreeding (F) for the first meiotic gynogenetic generation in common carp is about 0.60, which is higher than the value of F for self-fertilization (0.50). Therefore, it was suggested to use meiotic gynogenesis for development of inbred lines for application in practical breeding (Golovinskaya 1968, Nagy and Csanyi 1978).

The rate of homozygosity increase is lowered during successive generations of meiotic gynogenesis while the genetic identity (i.e., genetic similarity between individuals) rises. The gradual increase of genetic identity in consecutive generations of meiotic gynogenesis in common carp was studied by transplantation tests. Abramenko and Recoubratsky (1987) used inter-individual transplantations of anal fins in the third consecutive gynogenetic generation. In cases of rejection, transplanted fins were disintegrated because of immunological reactions while in cases of acceptance the degenerative processes have not been observed. The rate of transplant acceptance in the third consecutive gynogenetic generation was 19.4%; based on these data Abramenko and Recoubratsky (1987) calculated the number of histocompatibility genes in common carp. Nagy et al. (1983) showed that 100% of scale transplants were accepted in the fourth consecutive meiotic gynogenetic generation in common carp which indicates that fish of the fourth gynogenetic generation were genetically identical.

Mitotic Gynogenesis

Studies on mitotic gynogenesis in common carp were initiated later than development of meiotic gynogenesis. Investigations on mitotic gynogenesis in this species have been reported by Nagy (1987), Linhart et al. (1987), Gomelsky et al. (1989, 1992, 1998), Sato et al. (1990), Komen et al. (1991), Rothbard (1991), Sumantadinata et al. (1990), Cherfas et al. (1993a,b, 1994b), Yousefian et al. (1996), Sato and Amita (2001), Sato (2013), Alsaqufi et al. (2014) and others.

The 'late' heat shock was usually used for suppression of the first mitotic division in haploid embryos; the optimum timing for shock application was about 1.5–1.9 τ_0 (or 42–53 minutes at pre-shock temperature 20°C) after insemination. Usually mitotic gynogenetic progenies consisted of females only. However, Komen et al. (1992) identified a recessive mutation of sex-determining gene, which resulted in the appearance of males in some gynogenetic progenies.

Mitotic gynogenesis results in homozygosity for all genes (F = 1.0) since in this case homologous chromosomes are the products of simple mitotic reduplication in haploid embryos. Therefore, fish obtained by mitotic gynogenesis are sometimes called 'doubled haploids'.

As mentioned above, meiotic gynogenetic diploids can arise in fish from spontaneous suppression of the second meiotic division in eggs. As a result, late-shocked, presumably mitotic gynogenetic progenies may be 'contaminated' with spontaneous meiotic gynogenic individuals. Genetic markers may be applied for verification of mitotic gynogenesis. Gomelsky et al. (1992) have suggested using the scale cover gene *N* for this purpose, because it has very high recombination frequency. Currently microsatellite DNA markers are widely used for verification of mitotic gynogenesis in fish

(Morishima et al. 2001, Lahrech et al. 2007). For this purpose, microsatellite loci with high meiotic segregation frequencies should be identified based on the proportion of heterozygotes in meiotic (early-shocked) gynogenetic progenies; homozygosity at these loci in late-shocked progenies would indicate their mitotic gynogenetic origin. Recently a study of this type was performed by Alsaqufi et al. (2014) in koi carp. Recombination rates (y) for 10 microsatellite loci were determined; six loci which had y 0.47 and higher were used as markers in two late-shocked gynogenetic progenies; complete homozygosity was revealed at all six loci thus confirming mitotic gynogenetic origin of fish.

Mitotic gynogenesis provides an opportunity to produce clones in fish, i.e., groups of genetically identical individuals. Clones in fish may be produced by two methods:

1. From females which have been obtained by means of mitotic gynogenesis, a second consecutive gynogenetic generation (meiotic or mitotic) is produced. (Usually meiotic gynogenesis is used since embryos are less sensitive to physical treatments at timing of 'early shock' than at timing of 'late shock'.) Resulting individual progenies will be clones and each fish in the progeny will have the maternal genotype and, just as the mother, will be homozygous for all genes. Therefore, clones produced by this method are called 'homozygous'.
2. Females, obtained by mitotic gynogenesis, are crossed with hormonally sex-reversed males (see below) of the same origin. Resulting clones are called 'heterozygous' since heterozygosity for some genes can result from differences in allele composition between parents (i.e., females and sex-reversed males) that are homozygous for all genes because of their mitotic gynogenetic origin.

Komen et al. (1991) reported production of two types (homozygous and heterozygous) of clones in common carp; tissue transplantation tests were used to prove the genetic identity of fish in clones. Later, the creation of clones in common carp using mitotic gynogenesis were reported by Ben-Dom et al. (2001) and Sato (2013) (see Chapter 2 for more detailed information on production of clones in koi).

Induced Androgenesis

In case of androgenesis the embryo development is under control of only paternal chromosomes. In contrast to gynogenesis, there is no natural androgenesis as a mode of reproduction in fish. However, androgenetic development may be induced in species reproducing by means of the normal sexual mode. In this case, inactivation of female chromosomes is achieved by irradiation of eggs. The androgenetic haploids are produced after insemination of eggs with genetically inactivated chromosomes by intact spermatozoa. The paternal haploid chromosome set must be doubled by suppression of the first mitotic division in haploid embryos in order to produce viable diploid androgenetic fish. Just as fish obtained by mitotic gynogenesis, androgenetic diploids are homozygous for all genes. Therefore, like mitotic gynogenesis, androgenesis may be used for production of clones. In species with male heterogamety (males—XY, females—XX) androgenetic progenies consist of females XX and males YY.

Production of diploid androgenetic fish in common carp have been reported by Kondoh et al. (1989), Kondoh and Sato (1990), Grunina et al. (1990) and Bongers et al. (1994). Either UV-irradiation (Kondoh et al. 1989, Kondoh and Sato 1990, Bongers et al. 1994) or X-rays (Grunina et al. 1990) have been used for irradiation of eggs for genetic inactivation of female chromosomes. In all studies, the first mitotic division in androgenetic haploids was suppressed by heat shock. Later, Grunina et al. (1995), Bongers et al. (1999), Yamaha et al. (2003) and Sato (2013) reported identification of viable androgenetic YY males; these males produced all-male progenies after crossing with normal females (XX). Bongers et al. (1995) used androgenesis for obtaining clones of common carp.

Successful interspecies androgenesis between common carp and goldfish was reported by Bercsenyi et al. (1998); in this study genetically inactivated common carp eggs were inseminated with goldfish sperm, and the goldfish haploid chromosome set was doubled by suppression of the first mitotic division.

Spontaneous Gynogenesis and Androgenesis

Gynogenetic or androgenetic development in fish can occur spontaneously without artificial irradiation of sperm or eggs. Gomelsky and Recoubratsky (1990) have demonstrated these possibilities in common carp. It is known that melanin formation in wild-type colored common carp is controlled by dominant alleles of two duplicate genes (B_1/b_1 and B_2/b_2), whereas the genotype of koi, which do not have melanin, is $b_1b_1b_2b_2$. The presence or absence of melanin is already clearly visible in newly hatched larvae (Katasonov 1978). Gomelsky and Recoubratsky (1990) analyzed progenies obtained in two reciprocal crosses. In the first cross, eggs obtained from wild-type colored common carp female (with melanin, i.e., pigmented dark; genotype $B_1B_1B_2B_2$) were fertilized with sperm from the koi male (without melanin; genotype $b_1b_1b_2b_2$); in the second, reciprocal cross, eggs taken from the koi female ($b_1b_1b_2b_2$) were fertilized with sperm from wild-type colored common carp male ($B_1B_1B_2B_2$). It was assumed that both performed crosses should result in appearance of wild-type colored fish, which are heterozygous at both genes (genotype $B_1b_1B_2b_2$). Heterozygous fish appear as fertilization proceeds normally with the fusion of female and male pronuclei. However, in the case of possible inactivation of paternal or maternal chromosomes carrying dominant alleles, larvae without melanin formation would appear in progenies. Thus, the reason of the study by Gomelsky and Recoubratsky (1990) was to reveal unpigmented larvae in the obtained progenies by close observation of all obtained hatched larvae under dissection microscope. In progeny obtained in the first cross ($♀B_1B_1B_2B_2$ x $♂$ $b_1b_1b_2b_2$) 35 unpigmented larvae were found among observed 22.2 thousand larvae (0.158%); all unpigmented larvae in this progeny had severe morphological abnormalities resembling the 'haploid syndrome'. Apparently, these larvae were haploids of androgenetic origin and resulted from spontaneous inactivation of female chromosomes. In progeny obtained in the second cross ($♀$ $b_1b_1b_2b_2$ x $♂$ $B_1B_1B_2B_2$) 31 unpigmented larvae were found among observed 29.9 thousand larvae (0.104%). Nineteen unpigmented larvae were morphologically abnormal; apparently these larvae were haploids of gynogenetic origin and resulted from spontaneous inactivation of male chromosomes. Twelve unpigmeted larvae were morphologically normal; later these larvae swam-up and transferred to active external feeding. Gomelsky and Recoubratsky (1990) have noted that viable unpigmented larvae were undoubtedly diploids and that they could appear by spontaneous diploid gynogenesis, which combined spontaneous inactivation of male chromosomes and spontaneous suppression of 2nd meiotic division in eggs. In the control progeny obtained by crossing of two wild-type colored carp parents ($♀$ $B_1B_1B_2B_2$ x $♂$ $B_1B_1B_2B_2$) no unpigmented larvae were found among observed 30.1 thousand larvae. This study has demonstrated that spontaneous viable diploid fish of gynogenetic origin can appear with low frequency without any treatments to sperm or embryos (Gomelsky and Recoubratsky 1990). The appearance of spontaneous gynogenetic fish becomes more visible in case of distant hybridization (see below).

Induced Polyploidy

Induced polyploidy aims to produce individuals with an increased number of haploid chromosome sets. The primary purpose of this method in aquaculture and fisheries is to obtain triploid fish, i.e., fish whose karyotypes contain three haploid chromosome sets. As a rule, triploid fish are genetically sterile, i.e., they are not capable of producing viable progeny. Triploid fish are also characterized by complete or partial reduction of gonads (Benfey 1999, Piferrer et al. 2009). Disturbances in development of the reproductive systems in triploids are caused by the presence of a third, 'additional' haploid chromosome set, which disrupts the normal process of meiosis in sex cells.

Production of Triploids by Suppression of Second Meiotic Division

The most common method for producing triploid fish is based on suppression of the second meiotic division in eggs after insemination by intact spermatozoa. This method uses the same shocks for doubling of the female chromosome set as a technique for inducing diploid meiotic gynogenesis described above. Application of shock should provide a high frequency of triploids, but at the same time does not reduce embryo survival significantly. In initial studies on induced triploidy in common carp cold shocks were usually used (Ojima and Makino 1978, Gervai et al. 1980, Ueno 1984, Taniguchi et al. 1986, Wu et al. 1986, Cherfas et al. 1990, Linhart et al. 1991). Further studies mostly applied heat shocks (Hollebecq et al. 1988, Recoubratsky et al. 1989, 1992, Gomelsky et al. 1992, 2012a, Cherfas et al. 1993a, 1994c, Basavaraju et al. 2002). Application of heat shocks appears to be more convenient and practical for mass production of triploid progenies since the duration of a typical heat shock (1.5–3.0 minutes) is much shorter than the duration of a typical cold shock (30–60 minutes). Hydrostatic pressure, which is successfully used for production of triploids in many fish species, is not practical for common carp because of the strong adhesiveness of carp eggs (Linhart et al. 1991). Since eggs cannot be stirred in pressure chamber during shock application, eggs stick together; this makes their removal from the chamber and further incubation difficult.

Recoubratsky et al. (1989, 1992) described the method for mass production of triploid common carp using heat shock, which provided 80–100% triploids in progenies with embryo survival of 50–70% of controls. The following parameters of heat shock were recommended as optimal: timing: 0.2 τ_0 after insemination (5–6 minutes at pre-shock temperature 20°C), shock temperature and duration: 40°C and 2 minutes, or 41°C and 1.5 minutes. Effectiveness of this method was proved in further studies (Cherfas et al. 1993a, Gomelsky et al. 2012a). According to this method, the portion of eggs (200–250 g or approximately 140,000–175,000 eggs) is inseminated with sperm in large-volume (10–12 liters) bowls based on standard methodology with adding a water-cow milk mixture (8 L of water per 1 L of milk) approximately 2 minutes after insemination to remove eggs adhesiveness and stirring of eggs with bird feathers. At the time of heat shock, the water-milk mixture is carefully poured off from the bowl containing the eggs and heated water (9–11 L according to the volume of used bowl) is added. The temperature of added water should be 1.5–2.0°C higher than target shock temperature; which is established in the bowl 15–20 seconds after adding of heated water. During shock, the stirring of eggs with feathers should continue to prevent clumping. Stirring should stop 15–20 seconds prior to the end of the shock; the hot water is poured off and the water-milk mixture of room temperature is added.

Methods of Ploidy Determination

Different methods of ploidy determination have been used in experiments on induced triploidy in common carp. These methods included karyological analysis, determination of DNA content by flow cytometric analysis, measurement of erythrocyte nuclei on stained blood smears, determination of number of nucleoli or Nucleolar Organizing Regions (NORs) in nuclei [see, for example, Ojima and Makino (1978), Linhart et al. (1991), Cherfas et al. (1993a)]. From all these methods the determination of ploidy using a flow cytometer can be considered as most advanced, quick and precise (Fig. 5.2).

Another technically advanced device is a Coulter counter which is widely used in aquaculture and fisheries for determination of fish ploidy, in

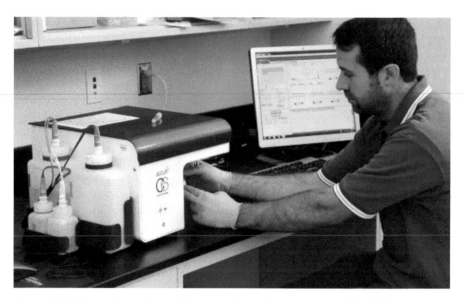

Figure 5.2. Determination of koi ploidy using flow cytometer (Accuri C6, BD Biosciences) at Aquaculture Research Center of Kentucky State University.

particular for verification of ploidy in triploid grass carp in the USA (Allen and Wattendorf 1987, Masser 2002) which has not been used until recently in common carp. A Coulter counter is electronically able to record the size of passing particles. Ploidy determination is based on difference in size of erythrocytes between diploid and triploid fish. Polyploidy in fish (as well as in other animals) increases proportionally the size of cells and nuclei. Therefore, in triploid fish the volume of cells and nuclei is 1.5 times larger than in diploid fish. Gomelsky et al. (2012a) analyzed the ploidy of triploid and diploid koi by two methods, Coulter counter and flow cytometry analysis. For all analyzed fish, the ploidy levels determined using both methods were identical. These results suggest the accuracy of Coulter counter analysis for ploidy determination in common carp.

Gomelsky et al. (1992, 2012a) have suggested using the scale cover gene *N* for identification of ploidy in common carp. Because of high recombination rate and incomplete dominance of this gene it is possible to identify triploids by means of external examination of fish based on the specific scale coverage pattern.

Growth Rate of Triploids

The retardation in somatic growth of normal diploid fish is usually recorded during the period of gonad maturation. As mentioned above, triploid fish are characterized by a reduction in gonad development. It was speculated that rearing of triploids could be commercially more profitable than rearing diploids because the energy not invested in gonad development will be redirected to somatic growth. This advantage should be especially profound when sexual maturation is reached before fish reach a marketable size. Cherfas et al. (1994c) and Basavaraju et al. (2002) assessed triploid common carp for culture in such conditions. In both studies a reduction in gonad development (especially ovaries) in triploid fish was observed, although Cherfas et al. (1994c) recorded several triploid females with well-developed gonads. However, the reduction in gonad development in triploids did not result in an increase of somatic growth rate. Triploid fish grew even slower than diploid fish in almost all comparative trials (Cherfas et al. 1994c, Basavaraju et al. 2002). These results make it doubtful that there is a potential of triploid common carp for commercial rearing.

Recently Gomelsky et al. (2015) reported that triploid koi females had unexpectedly well-developed ovaries and produced mass aneuploid progeny when crossed with normal diploid males. Most of the 110 analyzed juveniles were aneuploid with ploidy ranging from 2.3n to 2.9n with a mean value of 2.6n; two juveniles were diploid. This showed that triploid koi females produced aneuploid eggs with a ploidy range from haploid to diploid level with the modal ploidy level around 1.5n, similar to the production of aneuploid spermatozoa observed earlier for triploid males in fish.

Production of Tetraploids[2]

Triploids in fish can be also produced by crossing of previously obtained tetraploid fish with normal diploid fish. Tetraploid fish can be obtained by suppression of the first mitotic division in normal diploid embryos. Triploid rainbow trout have been successfully obtained by this method (Chourrout et al. 1986, Myers and Hershberger 1991). Successful experiments on mass production of tetraploid common carp larvae by application of heat shocks were reported by Recoubratsky et al. (1989) and Cherfas et al. (1993a). The frequency of tetraploids in progenies was up to 100% in some trials; tetraploids had 200 chromosomes in their karyotypes and had twice higher DNA content than diploids. Common carp tetraploids demonstrated very poor survival: only three tetraploid fingerlings have survived after stocking thousands of tetraploid larvae in ponds (Cherfas et al. 1993a). However, viable tetraploids have been obtained in hybrids of common carp and crucian carp (see below part on distant hybridization).

Sex Control

As mentioned above, development and application of the hormonal sex reversal are the basis for sex regulation in fish. Hormonal sex reversal (or inversion) is a change from the normal process of sex differentiation under the influence of steroid sex hormones so that genotypic females develop

---

[2] As indicated in Chapter 1, common carp has been assumed to be of tetraploid origin since the number of its chromosomes (2n = 100) is twice as high as most other Cyprinids. Here experiments on novel, induced tetraploidization using heat shocks are described.

testes or genotypic males develop ovaries. Sex reversal changes only the fish phenotype, the genotypic formula of sex chromosomes remains unaffected; for example, in case of female homogamety sex-reversed males obtained by hormonal sex reversal of genotypic females XX have the same formula of sex chromosomes (XX). There are two possible ways of using hormonal sex reversal for sex control. The direct method for producing monosex populations (containing only individuals of a single sex) involves hormonal treatment of all reared fish during the period of sex differentiation. The indirect method, or genetic sex regulation, involves crossing normal fish with previously obtained sex-reversed fish. In the case of female homogamety, which has been revealed in common carp, all-female progenies may be produced by crossing normal females XX with sex-reversed XX males (sometimes called 'neomales') obtained by phenotypic hormonal sex reversal of genotypic females. Genetic sex regulation is regarded as the preferred method since there is no need to treat all reared fish with a hormone and hormonally treated fish are not intended for human consumption.

Additionally, in the case of female homogamety (females—XX/males—XY) it is convenient to use gynogenetic progenies consisting of only genotypic females (XX) as material for initial experiments on sex reversal under androgen influence. In further generations, a portion of fish from all-female progenies should be sex-reversed again to renew stock of reversed males (Fig. 5.3). This scheme was realized in common carp, for example, in Russia and Israel (Gomelsky 1985, Gomelsky et al. 1994, Cherfas et al. 1996).

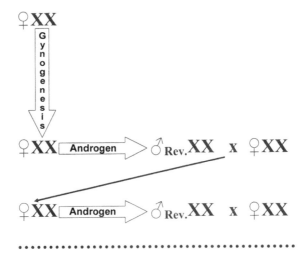

Figure 5.3. Scheme of genetic sex regulation for production of all-female progenies in species with female homogamety (XX/XY).

Technique of Hormonal Treatment for Sex Reversal

For the first time, successful sex reversal of genotypic common carp females by androgen treatment was achieved by Nagy et al. (1981). The androgen

17α-methyltestosterone (MT) was added to the diet at a dose of 100 mg/kg and orally administered to all-female gynogenetic carp at different 36-day periods. The androgen treatment during any 36-day period beginning from 8 to 62 days after hatching resulted in 70–90% males in the experimental groups while the control group of fish receiving a MT-free diet consisted of only females. Mature sex-reversed males (XX) produced all-female progenies when crossed with normal females (XX).

Gomelsky (1985) induced hormonal sex reversal in two consecutive generations of common carp by using the same dosage of MT in a prepared diet (100 mg/kg). In the first experiment with all-female gynogenetic progeny the MT treatment lasted 36 days (62–98 days after transition of fish to external feeding) with fish initially weighting 9 g; the resulting percentage of males was 83%. In the second experiment with all-female progeny from sex-reversed males, fish were treated with MT for 40 days (78–118 days after transition of fish to external feeding) with an initial fish weight of 12 g leading to 51% of males in the MT-treated group. A histological investigation of the sex reversal process in genotypic females revealed that androgen treatment caused cytological differentiation of sex cells towards the male direction; along with spermatogenesis, morphological transformation of primary ovary to normal testes has occurred. Gomelsky (1985) also showed that reversed males did not differ from normal males of common carp with regard to the volume of milt and the fertilizing ability of their spermatozoa.

The effects of 5 weeks of oral administration of MT (doses of 50 and 100 mg/kg) on sex ratio in normal mixed sex progeny of common carp were studied by Komen et al. (1989) whereby the initiation of androgen treatment took place 3, 6 or 10 weeks after hatching. Best results (92.7% of males in experimental group as compared to 64.4% in control) were obtained after administration of 50 mg/kg MT in the diet between 6 and 11 weeks after hatching; earlier or later treatments with MT at concentrations 50 and 100 mg/kg resulted in high percentages of sterile fish. Komen et al. (1989) also noted that for a more precise description of the hormonal treatment procedure the amount of MT per unit of body weight or per unit of body weight gain should be determined.

Wu (1990) noted that 100% functional sex reversal in all-female gynogenetic common carp progenies was obtained by feeding fry a diet containing 30 mg/kg MT from hatching up to 90 days or fingerlings from 30 days after hatching to 120 days.

Gomelsky et al. (1994) compared experimental results on hormonal sex reversal obtained under different climatic conditions. The mean weight of fingerlings at the beginning of androgen treatment was similar in the most successful experiments (up to 82–96% of males in gynogenetic progenies): 2.7–9.2 g and 5.8–9.0 g in Israeli and Russian climatic conditions, respectively. However, the age of the fish at the time these weights were attained was quite different: 27–40 and 63–65 days after hatching, respectively. Based on this comparison Gomelsky et al. (1994) suggested that the weight of the fish (rather than their age) may be used as a practical criterion for determining the appropriate period of androgen treatment.

Gomelsky (1985) and Gomelsky et al. (1994) have described an interesting phenomenon in experiments on induced sex reversal in common carp when fish were kept in closed recirculation systems during hormonal treatment. Feeding of fish with MT-containing diet induced sex reversal in fish fed with hormone-free diet but kept in the same water recirculation system. In some cases, percentages of males in fish, which were influenced only through shared water, was even higher than in MT-fed fish. It was suggested that the MT consumed by fish may be transformed in the liver into soluble, active metabolite(s) and excreted via the bile. This assumption was partially confirmed by results of a special experiment (Yarzhombek and Gomelsky 1993) where bile from common carp, that were fed a diet containing MT at high dosage, was added to the water of an aquarium with guppy (*Poecilia reticulata* Peters) fry. The addition of bile induced the development of secondary male characteristics in all guppies in an aquarium, including genotypic females. Later Hulak (2007) and Hulak et al. (2008) investigated this phenomenon further using closed recirculation systems for experiments on sex reversal in all-female gynogenetic progeny of common carp. Experimental designs in these studies (Hulak 2007, Hulak et al. 2008) were similar to those described in Gomelsky (1985) and Gomelsky et al. (1994): in some tanks of a recirculation system fish were fed with MT-containing diet (100 mg/kg) for 40–50 days from 50th day after fertilization (initial fish weight was 2.1–3.9 g) while in other tanks of the same rearing system fish of the same origin received MT-free diet. Results of this study have confirmed previous observations: from 61.5 to 97.4% and from 85.2 to 100% of males were detected in MT-fed and MT-free-fed groups, respectively. Special analyses detected 0.33 to 2.68 µg/l of MT in water from the recirculation system; these results proved that oral administration of MT to fish resulted in pollution of water in recirculation system with MT, which resulted in sex reversal in MT-free-fed fish kept in the system (Hulak 2007, Hulak et al. 2008). Similar implications of hormone treatment application in closed water systems were also described in experiment with Nile tilapia (*Oreochromis niloticus* L.) (Abucay and Mair 1997, Abucay et al. 1997).

A conceptual model for common carp that considered both age and size in relation to gonad differentiation was proposed by Shelton et al. (1995). The varying results from hormone-induced sex reversal are a result of variations of the fish's growth history. The growth trajectory is controlled by environmental factors, stocking density, temperature and food availability. Selection of fish to be androgen-treated should use growth rate criteria and fish, which deviate widely from the modal size, should be culled for increase of effectiveness of androgen treatment.

Tzchori et al. (2004) demonstrated that fadrozole, an aromatase inhibitor, is able to induce sex reversal in common carp from all-female progenies obtained from sex-reversed males. Thirty-day-old fry with an initial fish weight of 3.5–4.0 g were fed with diet containing fadrozole at doses from 100 to 400 mg/kg feed for 36 or 50 days. From 59 to 97% of males were recorded in fadrozole-treated groups and the efficiency of treatment was dose-dependent. Later, Ogawa et al. (2008) induced 'post-sex differentiational' sex reversal in genotypic females of common carp. One-year-old females of gynogenetic origin were fed with diet containing a fadrozole dose of 200 mg/kg for 24 weeks. At week 8, testicular lobule-like structures and cysts of spermatocytes began to appear. Spermatozoa were initially observed at week 16 accompanied by degradation of oocytes, and they spread to the whole gonad in some individuals by week 18. The authors noted that the obtained results show that sexual bipotentiality is maintained even after sex differentiation in the ovary of the common carp (Ogawa et al. 2008).

Basavaraju et al. (2008) used two androgens (MT and 17α-methyl-dihydroxytestosterone, MDHT) to induce sex reversal of genotypic females in normal mixed sex progenies of common carp. MDHT appeared to be more effective than MT in inducing sex reversal. A maximum of 77.1% of males were obtained in the group of 60-day-old fish fed the diet containing 100 mg/kg MT for 40 days whereby the mean initial fish weight was 0.10 g. When MDHT (50 or 100 mg/kg) was given to 50-day-old fish with a mean initial weight of 0.05 g for 50 days only males were found in the treated groups. When a similar treatment was applied to fish of the same age but with larger initial mean weight (0.24 g) the percentage of males was significantly smaller (approximately 70–80%). Based on the obtained data the authors suggested that not only the age but also the weight of the fish is equally important in determining the appropriate androgen treatment. Sex-reversed XX males producing all-female progenies were identified in androgen-treated groups by test crosses (Basavaraju et al. 2008).

Comparative Rearing of All-female and Mix-sex Progenies

Production of all-female progenies by crossing of sex-reversed and normal fish shows the most promise for increasing yield in practical common carp aquaculture. Sex-reversed males of common carp do not demonstrate abnormalities in development of sperm ducts, which are observed in rainbow trout reversed males (Bye and Lincoln 1986, Geffen and Evans 2000). Normal development of sperm ducts permits the stripping of sperm from sex-reversed common carp males and its use for mass production of all-female progenies.

Cherfas et al. (1996) reported the results of comparative rearing in experiments using all-female and normal mixed sex progenies in Israeli climate whereby most 14-month-old individuals of market-size (about 1 kg) were sexually mature. Rearing of all-female progenies increased production yield by 7–8% which was attributed to sexual dimorphism in body weight since females were 15% heavier than males. An additional advantage of rearing all-female progenies was a prevention of uncontrolled fish reproduction (Cherfas et al. 1996). Kocour et al. (2005) evaluated growth performance of all-female and mixed sex progenies under moderate Central

European climate conditions. In this case, 3-year-old fish (1.5–1.8 kg) from all-female groups showed 6–8% advantage compared to mixed sex groups; at this age all males and half of females were mature. However, the authors noted that the economic benefit from rearing all-female carp populations would be decreased by higher expenses to establish, rear and renew the neomale broodstock (Kocour et al. 2005).

Distant Hybridization

Distant or interspecies hybridization can result in gynogenesis and polyploidy in fish without application of any artificial treatments to gametes or embryos. As mentioned above, appearance of spontaneous gynogenetic diploids was described in common carp. Because of spontaneous suppression of the second meiotic division in eggs, triploids can appear in crossing of normal diploid parents in low frequency. As a rule, rare fish of gynogenetic or triploid origin are hard to identify in pure species (not interspecies hybrids). However, in cases of distant hybridization, when real hybrids having haploid chromosome sets of two different species are inviable, the appearance of spontaneous gynogens or triploids becomes easily detectable. For example, real hybrids from crosses of female silver carp (*Hypophthalmichthys molitrix* Valenciennes) and male common carp are inviable and perish soon after hatching; observed rare survivors in this cross are pure silver carp individuals of gynogenetic origin (Makeeva and Koreshkova 1982, Makeeva 1989). Gomelsky (unpubl. data) observed that rare survivors in crosses of female common carp and male European loach (*Misgurnus fossilis* L., Cobitidae) were gynogenetic individuals of common carp. Vasilyev et al. (1975) showed that rare survivors from crossing common carp females with males of grass carp (*Ctenopharyngodon idella* Valenciennes) are triploid; they appear due to sporadic spontaneous suppression of the second meiotic division in eggs and have two haploid chromosome sets of common carp and one haploid set of grass carp.

Studies on Hybridization of Common Carp with Crucian Carps (Genus Carassius)

Triploidy of fish may also result from backcross hybridization of F_1 hybrid females with males of the parental species. Appearance of triploids in backcross progenies results from the ability of diploid hybrid females to produce diploid eggs (i.e., eggs with unreduced ploidy level relative to ploidy of somatic cells). This phenomenon was described in hybrids between common carp and crucian carps (genus *Carassius*). This genus includes several species which inhabit a wide range of ecosystems throughout Eurasia. One species of this genus (*Carassius auratus* L.) is polytypic and consists of several subspecies; one subspecies of this species (*Carassius auratus auratus*) is widely distributed all over the world as its ornamental form, goldfish. Also, several fish of this genus, for example, Prussian (gibel) crucian carp

Carassius gibelio Bloch and *Carassius auratus langsdorfii* Temminck & Schlegel along with normal bisexual forms (reproducing by normal sexual mode) have all-female forms, which reproduce by means of natural gynogenesis (Cherfas 1981, Lamatsch and Stöck 2009). Ojima et al. (1975) were the first to show that backcross hybrids obtained from crossing F_1 hybrid females (Japanese crucian carp, *Carassius cuvieri* x common carp) with males of common carp were triploids. Later, the ability of hybrid females of common carp with different fish of the genus *Carassius* to produce diploid eggs was described and this phenomenon was used in series of long-term studies on development and investigation of polyploid and gynogenetic hybrid forms. A brief description of these studies is presented below.

Cherfas et al. (1994d) summarized the results of long-term studies on induced gynogenesis and polyploidy in hybrids between Prussian (gibel) crucian carp (*Carassius gibelio* Bloch) (bisexual form) and common carp. The ability of F_1 hybrid females to produce diploid (i.e., with unreduced chromosome number) eggs was responsible for the high yield of diploids (without application of any shocks) in gynogenetic progenies obtained from F_1 females, when eggs were inseminated with genetically inactivated sperm. Cherfas et al. (1994d) reported the reproduction of diploid hybrid females producing diploid eggs in four consecutive gynogenetic generations; Recoubratsky et al. (2012) have already reported obtaining of the seventh consecutive gynogenetic generation of diploid hybrid females. Cytogenetic studies (Emelyanova and Cherfas 1980, Emelyanova 1984) have shown that the ability of diploid hybrid females to produce eggs with unreduced chromosome number was caused by additional chromosome endoreduplication in early oogenesis. Triploid backcross hybrids were obtained when eggs from F_1 females were inseminated with sperm of parental species (Cherfas and Ilyasova 1980, Cherfas et al. 1981); backcross hybrids to common carp had two haploid chromosome sets of common carp and one haploid chromosome set of crucian carp; correspondingly backcross hybrids to crucian carp had two haploid sets of crucian carp and one haploid set of common carp. Usually triploid hybrids were sterile; however some triploid backcross hybrids to crucian carp were partly fertile and produced triploid eggs; triploid gynogenetic progeny has been obtained by insemination of eggs with genetically inactivated sperm (Cherfas et al. 1994d). By crossing of triploid backcrosses to crucian carp hybrid females with males of common carp first all-female (Cherfas et al. 1994d) and later bisexual tetraploid (amphidiploid) (A.V. Recoubratsky, pers. comm.) progenies have been obtained. Recently (in 2012) some fertile females were found also among triploid backcross hybrids to common carp. When crossed with common carp these females produced progeny consisting of both triploids and tetraploids; apparently triploids resulted from spontaneous gynogenesis (A.V. Recoubratsky, pers. comm.). Tetraploids have been obtained by other methods also. Gomelsky et al. (1985, 1988) demonstrated that in contrast to the sterile F_1 hybrid males (XY) sex-reversed males (XX), obtained by hormonal sex reversal of genotypic hybrid females, were fertile and produced unreduced diploid spermatozoa. By crossing diploid hybrid females (XX) and reversed males (XX) all-female (XXXX) allotetraploid (amphidiploid) hybrids were obtained. Tetraploid hybrids also produced diploid gametes; by crossing tetraploid females with tetraploid reversed males, a second generation of all-female tetraploid hybrids has been obtained (Cherfas et al. 1994d).

Liu et al. (2001) summarized the results of long-term studies on formation of tetraploid stocks of red crucian carp (*Carassius auratus* red var.) and common carp hybrids. Both F_1 females and males obtained by crossing of red crucian carp females with common carp males produced haploid eggs and spermatozoa, respectively. However, F_2 hybrids, obtained by crossing of F_1 hybrids, produced diploid eggs and sperm. By crossing F_2 breeders the first tetraploid generation was produced in F_3. This tetraploid progeny consisted of females and males, which produced reduced diploid gametes. It provided the opportunity to reproduce bisexual hybrid tetraploid (amphidiploid) populations *inter se* (between them) in consecutive generations. Liu et al. (2001) reported production of five consecutive generations of tetraploids (or F_3–F_8 counting from initial hybridization); later Liu (2010) reported production of the 15th consecutive tetraploid generation (or F_{18} counting from initial hybridization). Liu et al. (2004) reported the production of gynogenetic progeny by inseminating of eggs from tetraploid females (F_8–F_{10}) with genetically inactivated sperm. The obtained diploid gynogenetic females also produced diploid eggs and were reproduced in consecutive gynogenetic generations. Liu (2010) have reported production of the sixth consecutive gynogenetic generation of diploid hybrids.

Wu et al. (2003) described results of long-term studies on hybridization red variety of common carp and red crucian carp (*Carassius auratus*). The F_1 females between two species produced diploid eggs; by crossing these females with common carp male triploid hybrids having two haploid genomes of common carp and one haploid genome of crucian carp have been obtained. These triploid females produced triploid progeny after using intact (non-irradiated) sperm of parental origin or from other close species for insemination. These results suggest that the penetrated male pronucleus did not fuse with the triploid female pronucleus. Thus artificially obtained triploid hybrid form started reproducing by means of natural gynogenesis. Wu et al. (2003) reported the reproduction of this triploid form in seven gynogenetic generations. Sometimes fusion of the male and female pronuclei did occur and as a result, tetraploid (amphidiploid) fish containing two haploid chromosome sets from each parental species were obtained. These tetraploid females produced unreduced tetraploid eggs and also reproduced by means of natural gynogenesis (Wu et al. 2003, Ye et al. 2009).

Recently Gomelsky et al. (2012b) have shown that hybrid females obtained by hybridization of two ornamental fish, koi and goldfish (*Carassius auratus auratus*), also produce diploid eggs. The backcross progeny (F_1 hybrid females x koi males) obtained without application of any physical treatments to eggs, was triploid. Also, gynogenetic diploid progeny was obtained by insemination of eggs produced by hybrid females with genetically inactivated sperm.

In all studies on hybridization of common carp with crucian carps, which have been described above, the ability of diploid hybrid females to produce unreduced diploid eggs was recorded. This creates the possibility to reproduce hybrids by means of induced gynogenesis and develop different polyploid forms which differ with regard to their mode of reproduction. The variability of outcomes of these studies is remarkable; in this respect the hybridization between common carp and crucian carps appears to be a unique phenomenon among fish.

Gene Transfer

Gene transfer technology provides an opportunity to obtain transgenic organisms which have foreign genes in their genomes. Another name for transgenic organisms is GMO or 'genetically modified organisms'. Introduced genes are called transgenes, i.e., transferred genes. The first production of transgenic mice was reported in 1982. The first article on production of transgenic fish was published in 1985 when Chinese scientists reported creating transgenic goldfish by inserting a gene of human growth hormone (Zhu et al. 1985). In the next several years transgenic fish were produced in several aquaculture species. During a certain period, production and investigation of transgenic fish developed rapidly and significant scientific advances were obtained. However, environmental and food safety concerns, negative public perception towards transgenic organisms, and the lack of permissions and regulations have prevented the large-scale introducing transgenic fish for food production into commercial aquaculture (Myhr and Dalmo 2005, Dunham 2011, Hallerman 2013).

The most common way of introduction of foreign DNA material into a recipient fish is microinjection into early embryos soon after fertilization. There is some probability that during replication of genomic DNA a foreign DNA construct will be inserted into the genomic DNA. Other methods of introduction of foreign DNA material have also been elaborated. Some positive results were obtained using electroporation. According to this technique cells are placed in a buffer solution containing foreign DNA and electric pulses are applied. Under the influence of electric pulses the permeability of membranes increases, allowing foreign DNA to pass into the cell with probability to be inserted into the genome.

Transferred DNA constructs, in addition to structural genes, include regulatory sequences (promoters) that are needed for successful expression of a construct in host genome. In the beginning, DNA constructs that contained promoters of viral or mammalian origin and genes from mammalian genomes were inserted into fish genomes. Later, so called 'all-fish' DNA constructs were used; all components of these constructs originated from different fish species.

Similar to other fish species, experiments on the production of transgenic common carp were mostly aimed to transfer Growth Hormone (GH) genes of mammalian or fish origin. It was supposed that the increased synthesis of growth hormones would enhance fish growth. Similar to the study with goldfish mentioned above, Chen et al. (1989) obtained transgenic common carp by injection of a DNA construct containing a human growth hormone gene driven by a mouse metallothionein gene promoter. As a result transgenic fish grew 11% faster than non-transgenic controls.

Zhang et al. (1990) obtained transgenic common carp by injection of a recombinant plasmid containing the Rous Sarcoma Virus-Long Terminal Repeat (RSV-LTR) promoter linked to rainbow trout growth hormone (GH) cDNA[3]. Expression of the trout GH polypeptide was detected by immunobinding assay in the red blood cells of nine individuals of transgenic fish tested. A part of the progeny derived from crosses between transgenic males and non-transgenic females inherited the foreign DNA. This transgenic progeny grew faster than non-transgenic siblings. Later Chatakondi et al. (1995) showed that F_1 transgenic common carp containing the rainbow trout GH gene had higher protein content and lower fat content in the muscle as compared with the non-transgenic control. Dunham et al. (2002) have investigated survival of the second generation (F_2) transgenic fish when subjected to low dissolved oxygen. It was detected that transgenic common carp in some families had higher percentage and longer times of survival than fish from control groups.

Li et al. (1993) have injected the human growth hormone (hGH) gene capped with mouse metallothionein gene promoter into fertilized eggs of red common carp. The results showed that at the age of 263 days all largest fish weighing 1.4 to 2.9 times more than the average of control were transgenic. Fu et al. (1998) evaluated the growth rate of the fourth generation (F_4) of transgenic red common carp. On average, the F_4 hGH-transgenic red carp showed 19–25% higher specific growth rates than those of controls.

Moav et al. (1995) have injected 'all-fish' DNA constructs consisting of the common carp β-actin promoter fused to the Chinook salmon (*Oncorhynchus tshawytscha* Walbaum) GH cDNA into fertilized eggs of common carp. Twenty-five individual fish, out of 200 males and females raised from injected embryos, contained the foreign construct in a mosaic fashion at sexual maturity, and some of them were crossed with each other as well as with non-transgenic carp. The level of transgene transmittance varied from 32 to 87% when transgenic parents were crossed and from 0 to 50% when transgenic parents were mated with non-transgenic. In a further study growth performance of transgenic fish was evaluated; in some trials transgenic groups showed higher growth rates as compared to the control (Hinits and Moav 1999).

Wang et al. (2001) reported results of injection of other 'all-fish' DNA construct consisting of grass carp (*Ctenopharyngodon idellus*) growth hormone (gcGH) cDNA driven by the β-actin gene promoter of common carp. Transgenic individuals with 'fast growing' effect were detected in growth

---

[3] Complimentary DNA or cDNA is obtained by reverse transcription from mRNA; therefore cDNA represents the gene that was initially transcribed.

rate comparison; among transgenic fish 8.7% had a body weight higher than the heaviest fish in the non-transgenic group. Different characteristics of transgenic common carp having gcGH gene in successive generations (F_2–F_4) were described in further studies (Wang et al. 2006, Guan et al. 2008 and others). Recently, Zhong et al. (2012) presented pedigree of GH-transgenic common carp lines and analyzed their characteristics. The obtained results indicated that GH transgene initiated effects on growth enhancement in the early development stages and played an important role in improving protein synthesis and lipolysis. Also, it was noted that even if fast-growing GH-transgenic homozygous parental fish are generated, the classical selective breeding should be practiced to produce transgenic fish with improved traits that can meet aquaculture demand (Zhong et al. 2012).

One of the environmental concerns for introduction of genetic engineering in commercial aquaculture is that transgenic fish can escape from fish farms and may contaminate local natural populations genetically. Therefore it was suggested to allow raising only of sterile triploid progenies of transgenic fish. Beside this, studies are carried out on application of gene transfer for induction of fish sterility (Wong and Van Eenennaam 2008, Dunham 2011). One possible way for induction of transgenic sterility is to knock-down genes whose products are necessary for development of reproductive systems in fish. Recently, Su et al. (2014) presented results of first experiments on transgenic sterilization in common carp. Four types of constructs were designed to knock-down expression of two primordial germ cell (PGC) proteins, *nanos* and *dead end*, and were electroporated into common carp embryos. At 19 months of age common carp males exposed to these constructs had reduced rates of sexual maturity; however, reduction of female rates of sexual maturity was not conclusively shown. The authors noted that tested constructs appear promising for transgenic sterilization of common carp (Su et al. 2014).

Acknowledgements

The author thanks Kyle Schneider for reading the manuscript and valuable comments, Alexander Recoubratsky for providing unpublished data and some articles, as well as Nina Cherfas and Shoh Sato for providing some articles and Charles Weibel for help in formatting of figures.

References

Abramenko, M.I. and A.V. Recoubratsky. 1987. Determination of number of histocompatibility genes in common carp based on results of transplantation test in third generation of induced gynogenesis. Genetika 23: 1658–1663 (in Russian with English summary).
Abucay, J.S., G.C. Mair, D.O.F. Skibinski and J.A. Beardmore. 1997. The occurrence of incidental sex reversal in *Oreochromis niloticus* L. pp. 729–738. *In*: K. Fitzsimmons (ed.). Tilapia Aquaculture, Proceedings from the 4th International Symposium on Tilapia in Aquaculture held November 1997 in Orlando, Florida.
Abucay, J.S. and G.C. Mair. 1997. Hormonal sex reversal of tilapias: implications of hormone treatment application in closed water systems. Aquaculture Research 28: 841–845.

Aliah, R.S. and N. Taniguchi. 2000. Gene-centromere distances of six microsatellite DNA loci in gynogenetic Nishikigoi (*Cyprinus carpio*). Fish Genet. Breed. Science 29: 113–119.

Allen, S.K. and R.J. Wattendorf. 1987. Triploid grass carp: status and management implications. Fisheries 12: 20–24.

Alsaqufi, A.S., B. Gomelsky, K.J. Schneider and K.W. Pomper. 2014. Verification of mitotic gynogenesis in ornamental (koi) carp (*Cyprinus carpio* L.) using microsatellite DNA markers. Aquaculture Research 45: 410–416.

Basavaraju,Y., G.C. Mair, H.M.M. Kumar, S.P. Kumar, G.Y. Keshavappa and D.J. Penman. 2002. An evaluation of triploidy as a potential solution to the problem of precocious sexual maturation in common carp, *Cyprinus carpio*, in Karnataka, India. Aquaculture 204: 407–418.

Basavaraju, Y., H.M.M. Kumar, S.P. Kumar, D. Umesha, P.P. Srivastava, D. Penman and G.C. Mair. 2008. Production of genetically female common carp, *Cyprinus carpio*, through sex reversal and progeny testing. Asian Fish. Sci. 21: 355–368.

Ben-Dom, N., N. Cherfas, B. Gomelsky, R.R. Avtalion, M. Boaz and G. Hulata. 2001. Production of heterozygous and homozygous clones of common carp (*Cyprinus carpio* L.): evidence from DNA fingerprinting and mixed leukocyte reaction. Isr. J. Aquac.—Bamidgeh 53: 89–100.

Benfey, T.J. 1999. The physiology and behavior of triploid fishes. Rev. Fish. Sci. 7: 39–67.

Bercsenyi, M., I. Magyary, B. Urbanyi, L. Orban, and L. Horvath. 1998. Hatching out goldfish from common carp eggs: interspecific androgenesis between two cyprinid species. Genome 41: 573–579.

Bongers, A.B.J., E.P.C. in't Veld, K. Abo-Hashema, I.M. Bremmer, E.H. Eding, J. Komen and C.J.J. Richter. 1994. Androgenesis in common carp (*Cyprinus carpio* L.) using UV irradiation in a synthetic ovarian fluid and heat shocks. Aquaculture 122: 119–132.

Bongers, A.B.J., J.B. Abarca, B.Z. Doulabi, E.H. Eding, J. Komen and C.J.J. Richter. 1995. Maternal influence on development of androgenetic clones of common carp, *Cyprinus carpio* L. Aquaculture 137: 139–147.

Bongers, A.B.J., B. Zandieh-Doulabi, C.J.J. Richter and J. Komen. 1999. Viable androgenetic YY genotypes of common carp (*Cyprinus carpio* L.). J. Heredity 90: 195–198.

Bye,V.J. and R.F. Lincoln. 1986. Commercial methods for the control of sexual maturation in rainbow trout (*Salmo gairdneri*). Aquaculture 57: 299–309.

Chatakondi, N., R. Lovell, R. Duncan, M. Hayat, T. Chen, D. Powers, T. Weete, K. Cummins and R.A. Dunham. 1995. Body composition of transgenic common carp, *Cyprinus carpio*, containing rainbow trout growth hormone gene. Aquaculture 138: 99–109.

Chen, T.T., Z. Zhu, C.M. Lin, L.I. Gonzalez-Villasenor, R. Dunham and D.A. Power. 1989. Fish Genetic Engineering: a novel approach in aquaculture. *In*: Aquaculture '89 Proceedings. National Shellfish Association, Los Angeles, California.

Cherfas, N.B. 1975. Studies on diploid gynogenesis in common carp. I.Experiments on the mass production of diploid gynogenetic progeny. Genetika 11: 78–86 (in Russian with English summary).

Cherfas, N.B. 1977. Studies on diploid gynogenesis in common carp. II. Segregation with regard to some morphological traits in gynogenetic progenies. Genetika 13: 811–820 (in Russian with English summary).

Cherfas, N.B. 1981. Gynogenesis in fishes. pp. 255–273. *In*: V.S. Kirpichnikov. Genetic Bases of Fish Selection. Springer-Verlag, Berlin.

Cherfas, N.B. and K.A. Truveller. 1978. Studies on diploid gynogenesis in common carp. III. Analysis of gynogenetic progenies with regard to biochemical markers. Genetika 14: 599–604 (in Russian with English summary).

Cherfas, N.B. and V.A. Ilyasova. 1980. Induced gynogenesis in hybrids between crucian carp and common carp. Genetika 16: 1260–1269 (in Russian with English summary).

Cherfas, N.B., B.I. Gomelsky, O.V. Emelyanova and A.V. Recoubratsky. 1981. Triploidy in back-cross hybrids between crucian carp and common carp. Genetika 17: 1136–1138.

Cherfas, N.B., O. Kozinsky, S. Rothbard and G. Hulata. 1990. Induced diploid gynogenesis and triploidy in ornamental (koi) carp, *Cyprinus carpio* L. 1. Experiments on the timing of temperature shock. Isr. J. Aquac.—Bamidgeh 42: 3–9.

Cherfas, N.B., B. Gomelsky, Y. Peretz, N. Ben-Dom, G. Hulata and B. Moaz. 1993a. Induced gynogenesis and polyploidy in the Israeli common carp line Dor-70. Isr. J. Aquac.—Bamidgeh 45: 59–72.

Cherfas, N.B., G. Hulata and O. Kozinsky. 1993b. Induced diploid gynogenesis and polyploidy in ornamental (koi) carp, *Cyprinus carpio* L. 2. Timing of heat shock during the first cleavage. Aquaculture 111: 281–290.

Cherfas, N.B., Y. Perez, N. Ben-Dom, B. Gomelsky and G. Hulata. 1994a. Induced diploid gynogenesis and polyploidy in ornamental (koi) carp, *Cyprinus carpio* L. 4. Comparative study on the effects of high- and low-temperature shocks. Theor. Appl. Genet. 89: 193–197.

Cherfas, N.B., Y. Perez, N. Ben-Dom, B. Gomelsky and G. Hulata. 1994b. Induced diploid gynogenesis and polyploidy in ornamental (koi) carp, *Cyprinus carpio* L. 3. Optimization of heat-shock timing during the 2nd meiotic division and the 1st cleavage. Theor. Appl. Genet. 89: 281–286.

Cherfas, N.B., B. Gomelsky, N. Ben-Dom, Y. Peretz, and G. Hulata. 1994c. Assessment of triploid common carp (*Cyprinus carpio* L.) for culture. Aquaculture 127: 11–18.

Cherfas, N.B., B.I. Gomelsky, O.V. Emelyanova and A.V. Recoubratsky. 1994d. Induced diploid gynogenesis and polyploidy in crucian carp, *Carassius auratus gibelio* (Bloch) × common carp, *Cyprinus carpio* L., hybrids. Aquac. Fish. Manag. 25: 943–954.

Cherfas, N.B., B. Gomelsky, N. Ben-Dom, D. Joseph, S. Cohen, I. Israel, M. Kabessa, G. Zohar, Y. Peretz, D. Mires and G. Hulata. 1996. Assessment of all-female common carp progenies for fish culture. Isr. J. Aquac.—Bamidgeh 48: 149–157.

Chourrout, D. 1987. Genetic manipulations in fish: review of methods. pp. 111–126. *In*: K. Tiews (ed.). Selection, Hybridization and Genetic Engineering in Aquaculture, vol. 2. Heenemann Verlagsgesellschaft mbH, Berlin.

Chourrout, D., B. Chevessus, F. Krieg, A. Happe, G. Burger and P. Renard. 1986. Production of second generation triploid and tetraploid rainbow trout by mating tetraploid males and diploid females - potential of tetraploid fish. Theor. Appl. Genet. 72: 193–206.

Dettlaff, T.A. and A.A. Dettlaff. 1961. On relative dimensionless characteristics of development duration in embryology. Arch. Biol. 72: 1–16.

Dunham, R.A. 2011. Aquaculture and Fisheries Biotechnology: Genetic Approaches, 2nd ed. CABI Publishing, Cambridge, MA, USA.

Dunham, R.A., N. Chatakondi, A. Nichols, T.T. Chen, D.A. Powers and H. Kucaktas. 2002. Survival of F_2 transgenic common carp, *Cyprinus carpio*, containing pRSVrtGH1 cDNA when subjected to low dissolved oxygen. Marine Biotech. 4: 323–327.

Emelyanova, O.V. 1984. A cytological study of maturation and fertilization processes in hybrids between crucian carp and common carp. Cytology (Moscow) 26: 1427–1433 (in Russian with English summary).

Emelyanova, O.V. and N.B. Cherfas. 1980. Results of cytological analysis of unfertilized eggs of F_1 females obtained from crossing female crucian carp × male common carp. Proc. Res. Inst. Pond Fish. 33: 169–184 (in Russian with English summary).

Fu, C., Y. Cui, S.S.O. Hung and Z. Zhu. 1998. Growth and feed utilization by F_4 human growth hormone transgenic carp fed diets with different protein levels. J. Fish Biol. 53: 115–129.

Geffen, A.J. and J.P. Evans. 2000. Sperm traits and fertilization success of male and sex-reversed female rainbow trout (*Oncorhynchus mykiss*). Aquaculture182: 61–72.

Gervai, J., S. Peter, A. Nagy and V. Csanyi. 1980. Induced triploidy in carp, *Cyprinus carpio* L. J. Fish Biol. 17: 667–671.

Golovinskaya, K.A. 1968. Genetics and selection of fish and artificial gynogenesis of the carp (*Cyprinus carpio*). FAO Fish. Rep. 44: 215–222.

Golovinskaya, K.A. and D.D. Romashov. 1966. Segregation for scale cover types in the case of diploid gynogenesis in common carp. Proc. Res. Inst. Pond Fish. 14: 227–235 (in Russian with English summary).

Golovinskaya, K.A., D.D. Romashov and N.B. Cherfas. 1963. Radiation gynogenesis in common carp. Proc. Res. Inst. Pond Fish. 12: 149–167 (in Russian with English summary).

Golovinskaya, K.A., N.B. Cherfas and L.I. Tsvetkova. 1974. Results of evaluation of reproductive function in gynogenetic common carp females. Proc. Res. Inst. Pond Fish. 23: 20–26 (in Russian with English summary).

Gomelsky, B.I. 1985. Hormonal sex inversion in common carp (*Cyprinus carpio* L.). Ontogenez 16: 398–405 (in Russian with English summary).

Gomelsky, B. 2011. Fish Genetics: Theory and Practice. VDM Verlag Dr. Müller, Saarbrücken, Germany.

Gomelsky, B.I., V.A. Ilyasova and N.B. Cherfas. 1979. Studies on diploid gynogenesis in common carp. IV. Gonad state and evaluation of reproductive ability of carp of gynogenetic origin. Genetika 15: 1643–1650 (in Russian with English summary).

Gomelsky, B.I., N.B. Cherfas and O.V. Emelyanova. 1985. On the ability of hybrids between crucian carp and common carp to produce diploid spermatozoa. Proc. Acad. Sci. (Moscow) 282: 1255–1258 (in Russian).

Gomelsky, B.I., O.V. Emelyanova and A.V. Recoubratsky. 1988. Obtaining and some biological peculiarities of amphidiploid hybrids of crucian carp and common carp. Proc. Acad. Sci. (Moscow) 301: 1210–1213 (in Russian).

Gomelsky, B.I., A.V. Recoubratsky, O.V. Emelyanova, E.V. Pankratyeva and T.I. Lekontseva. 1989. Obtaining diploid gynogenesis in carp by thermal shock of developing eggs. J. Ichthyol. 29: 134–137.

Gomelsky, B.I. and A.V. Recoubratsky. 1990. On the possibility of spontaneous andro- and gynogenesis in fishes. J. Ichthyol. 30: 152–155.

Gomelsky, B.I., O.V. Emelyanova and A.V. Recoubratsky. 1992. Application of the scale covergene (N) to identification of type of gynogenesis and determination of ploidy in common carp. Aquaculture 106: 233–237.

Gomelsky, B., N.B. Cherfas, Y. Peretz, N. Ben-Dom and G. Hulata. 1994. Hormonal sex inversion in the common carp (*Cyprinus carpio* L.). Aquaculture 126: 265–270.

Gomelsky, B., N.B. Cherfas, N. Ben-Dom and G. Hulata. 1996. Color inheritance in ornamental (koi) carp (*Cyprinus carpio* L.) inferred from color variability in normal and gynogenetic progenies. Isr. J. Aquac.—Bamidgeh 48: 219–230.

Gomelsky, B., N. Cherfas and G. Hulata. 1998. Studies on the inheritance of black patches in ornamental (koi) carp. Isr. J. Aquac.—Bamidgeh 50: 134–139.

Gomelsky, B., S.D. Mims, R.J. Onders, W.L. Shelton, K. Dabrowski and M.A. Garcia-Abiado. 2000. Induced gynogenesis in black crappie. North Amer. J. Aquac. 62: 33–41.

Gomelsky, B., K.J. Schneider, R.P. Glennon and D.A. Plouffe. 2012a. Effect of ploidy on scale-cover pattern in linear ornamental (koi) common carp *Cyprinus carpio*. J. Fish Biol. 81: 1204–1209.

Gomelsky, B., K.J. Schneider and D.A. Plouffe. 2012b. Koi x goldfish hybrid females produce triploid progeny when backcrossed to koi males. North Amer. J. Aquac. 74: 449–452.

Gomelsky, B., K.J. Schneider, A. Anil and T.A. Delomas. 2015. Gonad development in triploid ornamental koi carp and results of crossing triploid females with diploid males. North Amer. J. Aquac. 77: 96–101.

Grunina, A.S., B.I. Gomelsky and A.A. Neifakh. 1990. Diploid androgenesis in common carp. Genetika 26: 2037–2043 (in Russian with English summary).

Grunina, A.S., A.A. Neifakh and, B.I. Gomelsky. 1995. Investigations on diploid androgenesis in fish. Aquaculture 129: 218.

Guan, B., W. Hu, T. Zhang, Y. Wang and Z. Zhu. 2008. Metabolism traits of 'all-fish' growth hormone transgenic common carp (*Cyprinus carpio* L.). Aquaculture 284: 217–223.

Hallerman, E. 2013. Transgenic fishes: applications, state of the art, and risk concerns. pp. 1698–1713. *In*: P. Christou, R. Savin, B.A. Costa-Pierce, I. Misztal and C.B.A. Whitelaw (eds.). Sustainable Food Production. Springler, New York.

Hinits, Y. and B. Moav. 1999. Growth performance studies in transgenic *Cyprinus carpio*. Aquaculture 173: 285–296.

Hollebecq, M.G., D. Chourrout, G. Wohlfarth and R. Billard. 1986. Diploid gynogenesis induced by heat shocks after activation with UV-irradiated sperm in common carp. Aquaculture 54: 69–76.

Hollebecq, M.G., F. Chambeyron and D. Chourrout. 1988. Triploid common carp produced by heat shock. Colloq. INRA, Versailles 44: 207–212.

Hulak, M. 2007. Sex control of common carp (*Cyprinus carpio* L.). Ph.D. thesis, University of South Bohemia, Research Institute of Fish Culture and Hydrobiology, Vodnany, Czech Republic.

Hulak, M., M. Paroulek, P. Simek, M. Kocour, D. Gela, M. Rodina and O. Linhart. 2008. Water polluted by 17α-methyltestosterone provides successful male sex inversion of common carp (*Cyprinus carpio* L.) from gynogenetic offspring. J. Appl. Ichthyol. 24: 707–710.

Hunter, G.A. and E.M. Donaldson. 1983. Hormonal sex control and its application to fish culture. pp. 223–303. *In*: W.S. Hoar, D.J. Randall and E.M. Donaldson (eds.). Fish Physiology, vol. 9B. Academic Press, New York.

Ignatieva, G.M. 1979. Early Embryogenesis in Fishes and Amphibians. Nauka, Moscow (in Russian).

Katasonov, V.Y. 1978. A study of pigmentation in hybrids between the common carp and decorative Japanese carp: III. The inheritance of blue and orange patterns of pigmentation. Genetika 14: 2184–2192 (in Russian with English summary).

Khan, T.A., M.P. Bhise and W.S. Lakra. 2000. Early heat-shock induced gynogenesis in common carp (*Cyprinus carpio* L). Isr. J. Aquac.—Bamidgeh 52: 11–20.

Kim, E.O., W.G. Jeong and J.S. Hue. 1993. Induction of gynogenetic diploids in common carp, *Cyprinus carpio* and their growth. Bull. Natl. Fish. Res. Devel. Agency (Korea) 47: 83–91.

Kocour, M., O. Linhart, D. Gela and M. Rodina. 2005. Growth performance of all-female and mixed-sex common carp *Cyprinus carpio* L. populations in the Central Europe climatic conditions. J. World Aquac. Soc. 36: 103–113.

Komen, J., J. Duynhouwere, C.J.J. Richter and E.A. Huismann. 1988. Gynogenesis in common carp (*Cyprinus carpio* L.). I. Effects of genetic manipulation of sexual products and incubation condition of eggs. Aquaculture 69: 227–239.

Komen, J., P.A.J. Lodder, F. Huskens, C.J.J. Richter and E.A. Huisman. 1989. Effects of oral administration of 17α-methyltestosterone and 17β-estradiol on gonadal development in common carp, *Cyprinus carpio* L. Aquaculture 78: 349–363.

Komen, J., G. Bongers, C.J.J. Richter, W.B. van Muiswinkel and E.A. Huisman. 1991. Gynogenesis in common carp (*Cyprinus carpio* L.). II. The production of homozygous gynogenetic clones and F_1 hybrids. Aquaculture 92: 127–142.

Komen, J., P. de Boer and C.J.J. Richter. 1992. Male sex reversal in gynogenetic XX females of common carp (*Cyprinus carpio* L.) by a recessive mutation in a sex-determining gene. J. Heredity 83: 431–434.

Komen, H. and G.H. Thorgaard. 2007. Androgenesis, gynogenesis and the production of clones in fishes: a review. Aquaculture 269: 150–173.

Kondoh, S., S. Sato and M. Tomita. 1989. Induction of androgenetic fancy carp. Rep. Niigata Pref. Inland Water Fish. Exp. Stn. 15: 19–23 (in Japanese).

Kondoh, S. and S. Sato. 1990. A method for production of androgenetic fancy carp, *Cyprinus carpio*. Rep. Niigata Pref. Inland Water Fish. Exp. Stn. 16: 107–109 (in Japanese).

Lahrech, Z., C. Kishioka, K. Morishima, T. Mori, S. Saito and K. Arai. 2007. Genetic verification of induced gynogenesis and microsatellite-centromere mapping in the barfin flounder, *Verasper moseri*. Aquaculture 272S: S115–S124.

Lamatsch, D.K. and M. Stöck. 2009. Sperm-dependent parthenogenesis and hybridogenesis in teleost fishes. pp. 399-432. *In*: I. Schön, K. Martens and van P. Dijk (eds.). Lost Sex: The Evolutionary Biology of Parthenogenesis. Springer, Dordrecht.

Li, G., Y. Xie, K. Xu, J. Zou, D. Liu, C. Wu and Z. Zhu. 1993. Growth of growth-hormone-treated transgenic red carp. Aquaculture 111: 320.

Linhart, O., P. Kvasnicka, V. Slechtova and J. Pocorny. 1986. Induced gynogenesis by retention of the second polar body in the common carp *Cyprinus carpio* L., and heterozygosity of gynogenetic progeny in transferrin and LDH-B1 loci. Aquaculture 54: 63–67.

Linhart, O., V. Slechtova, P. Kvasnicka, P. Rab, J. Kouril and J. Hamackova. 1987. Rates of recombination in LDH-B1 and MDH loci phenotypes after 'pb' and 'm' gynogenesis in carp, *Cyprinus carpio* L. pp. 335–345. *In*: K. Tiews (ed.). Selection, Hybridization and Genetic Engineering in Aquaculture, vol. 2. Heenemann Verlagsgesellschaft mbH, Berlin.

Linhart, O., M. Flajshans and P. Kvasnicka. 1991. Induced triploidy in the common carp (*Cyprinus carpio* L.): a comparison of two methods. Aquat. Living Resour. 4: 139–145.

Liu, S., Y. Liu, G. Zhou, X. Zhang, C. Luo, H. Feng, X. He, G. Zhu and H. Yang. 2001. The formation of tetraploid stocks of red crucian carp × common carp hybrids as an effect of interspecies hybridization. Aquaculture 192: 171–186.

Liu, S., Y. Sun, C. Zhang, K. Luo and Y. Liu. 2004. Production of gynogenetic progeny from allotetraploid hybrids red crucian carp x common carp. Aquaculture 236: 193–200.

Liu, S.J. 2010. Distant hybridization leads to different ploidy fishes. Sci. China Life Sci. 53: 416–425.

Makeeva, A.P. 1989. Results of distant hybridization of species from different families of Cypriformes. pp. 153–168. *In*: V.S. Kirpichnikov (ed.). Genetics in Aquaculture. Nauka, Leningrad, Russia (in Russian with English summary).

Makeeva, A.P. and N.D. Koreshkova. 1982. Analysis of gynogenetic progeny of silver carp *Hypophthelmichtix molitrix* (Val.) based on morphology and biochemical markers. Proc. Res. Inst. Pond Fish. 33: 185–210 (in Russian with English summary).

Masser, M.P. 2002. Using grass carp in aquaculture and private impoundments. SRAC Publication No. 3600, Sothern Regional Aquaculture Center, MS, USA.

Moav, B., Y. Hinits, Y. Groll and S. Rothbard. 1995. Inheritance of recombinant carp β-actin/GH cDNA gene in transgenic carp. Aquaculture 137: 179–185.

Morishima, K., I. Nakayama and K. Arai. 2001. Microsatellite–centromere mapping in the loach, *Misgurnus anguillicaudatus*. Genetica 111: 59–69.

Myers, J.M. and W.K. Hershberger. 1991. Early growth and survival of heat-shocked and tetraploid-derived triploid rainbow trout (*Oncorhychus mykiss*). Aquaculture 96: 97–107.

Myhr, A.I. and R.A. Dalmo. 2005. Introduction of genetic engineering in aquaculture: ecological and ethical implications for science and governance. Aquaculture 250: 542–554.

Nagy, A. 1987. Genetic manipulations performed on warm-water fish. pp. 163–173. *In*: K. Tiews (ed.). Selection, Hybridization and Genetic Engineering in Aquaculture, vol. 2. Heenemann Verlagsgesellschaft mbH, Berlin.

Nagy, A. and V. Csanyi. 1978. Utilization of gynogenesis in genetic analysis and practical animal breeding. pp. 16–20. *In*: J. Olah and Z. Krasznai (eds.). Increasing the Productivity of Fishes by Selection and Hybridization. Ferenz Muller Publ., Szarvas, Hungary.

Nagy, A. and V. Csanyi. 1982. Changes in genetic parameters in successive gynogenetic generations and some calculations for carp gynogenesis. Theor. Appl. Genet. 63: 105–110.

Nagy, A., L. Rajki, L. Horvath and V. Csanyi. 1978. Investigation on carp, *Cyprinus carpio* L., gynogenesis. J. Fish Biol. 13: 215–224.

Nagy, A., L. Rajki, J. Bakos and V. Csanyi. 1979. Genetic analysis in carp (*Cyprinus carpio*) using gynogenesis. Heredity 43: 35–40.

Nagy, A., M. Bercsenyi and V. Csanyi. 1981. Sex reversal in carp (*Cyprinus carpio*) by oral administration of methyltestosterone. Can. J. Fish. Aquat. Sci. 38: 725–728.

Nagy, A., Z. Monostory and V. Csanyi. 1983. Rapid development of the clonal state in successive generations of carp (*Cyprinus carpio*). Copeia 3: 745–749.

Ogawa, S., M. Akiyoshi, M. Higuchi, M. Nakamura and T. Hirai. 2008. 'Post-sex differentiational' sex reversal in the female common carp (*Cyprinus carpio*). Cybium 32 (2) suppl.: 102–103.

Ojima, Y. and S. Makino. 1978. Triploidy induced by cold shock in fertilized eggs of the carp. Proc. Jpn. Acad. Sci. (Ser. B) 54: 359–362.

Ojima, Y., M. Hayashi and K. Ueno. 1975. Triploidy appeared in the backcross offspring from funa-carp crossing. Proc. Jpn. Acad. Sci. 51: 702–706.

Piferrer, F., A. Beaumont, J. Falguière, M. Flajšhans, P. Haffray and L. Colombo. 2009. Polyploid fish and shellfish: production, biology and applications to aquaculture for performance improvement and genetic containment. Aquaculture 293: 125–156.

Purdom, C.E. 1993. Genetics and Fish Breeding. Chapman and Hall, London.

Recoubratsky, A.V., B.I. Gomelsky, O.V. Emelyanova and E.V. Pankratyeva. 1989. Obtaining triploid and tetraploid common carp progenies by heat shock. Proc. Res. Inst. Pond Fish. 58: 54–60 (in Russian with English summary).

Recoubratsky, A.V., B.I. Gomelsky, O.V. Emelyanova and E.V. Pankratyeva. 1992. Triploid common carp produced by heat shock with industrial fish-farm technology. Aquaculture 108: 13–19.

Recoubratsky, A.V., E.V. Ivanekha, D.A. Balashov, L.N. Duma, V.V. Duma and K.V. Kovalev. 2012. Triploid crucian carp x common carp hybrids as a new species for aquaculture. Problems of Fisheries (Moscow) 13: 626–642 (in Russian with English summary).

Romashov, D.D., K.A. Golovinskaya, V.N. Belyaeva, E.D. Bakulina, G.L. Pokrovskaya and N.B. Cherfas. 1960. Radiation diploid gynogenesis in fish. Biophysics 5: 461–468 (in Russian with English summary).

Rothbard, S. 1991. Induction of endomitotic gynogenesis in nishiki-goi, Japanese ornamental carp. Isr. J. Aquac.- Bamidgeh 43:145–155.

Sato, S. 2013. Breeding in Nishikigoi, *Cyprinus carpio* of Niigata prefecture. Fish Genetics and Breeding Science 42: 81–84 (in Japanese with English summary).

Sato, S., S. Kondoh and M. Tomita. 1990. Mass production of gynogenetic fancy carp, *Cyprinus carpio*, by heat shock. Rep. Niigata Inland Water Fish. Exp. Stn. 16: 65–71 (in Japanese).

Sato, S. and K. Amita. 2001. Induction of mitotic-gynogenetic diploid Nishikigoi, *Cyprinus carpio*. Rep. Niigata Inland Water Fish. Exp. Stn. 25: 1–5 (in Japanese with English summary).

Shelton,W.L. and S. Rothbard. 1993. Determination of the development duration(τ_0) for ploidy manipulation in carps. Isr. J. Aquac.—Bamidgeh 45:82–88.

Shelton, W.L., V. Wanniasingham and A.E. Hiott. 1995. Ovarian differentiation in common carp (*Cyprinus carpio*) in relation to growth rate. Aquaculture 137: 203–211.

Su, B., E. Peatman, M. Shang, R. Tresher, P. Grewe, J. Patil, C.A. Pinkert, M.H. Irwin, C. Li, D.A. Perera, P.L. Duncan, M. Fobes and R.A. Dunham. 2014. Expression and knockdown of primordial germ cell genes, *vasa, nanos* and *dead end* in common carp (*Cyprinus carpio*) embryos for transgenic sterilization and reduced sexual maturity. Aquaculture 420-421: S72–S84.

Sumantadinata, K., N. Taniguchi and Sugiarto. 1990. Increased variance of quantitative characters in the two types of gynogenetic diploids of Indonesian common carp. Jpn. Soc. Sci. Fish. 56: 1979–1986.

Taniguchi, N., A. Kijima, T. Tamura, K. Takegami and I. Yamasaki. 1986. Color, growth and maturation in ploidy-manipulated fancy carp. Aquaculture 57: 321–328.

Thorgaard, G.H. 1983. Chromosome set manipulation and sex control in fish. pp. 405–434. *In*: W.S. Hoar, D.J. Randall and E.M. Donaldson (eds.). Fish Physiology, vol. 9B. Academic Press, New York.

Tzchori, I., T. Zak and O. Sachs. 2004. Masculinization of genetic females of the common carp (*Cyprinus carpio* L.) by dietary administration of an aromatase inhibitor. Isr. J. Aquac.– Bamidgeh 56: 239–246.

Vasilyev,V.P., A.P. Makeeva and I.N. Raybov. 1975. On the triploidy of distant hybrids of carp (*Cyprinus carpio*) with other representatives of Cyprinidae family. Genetika 11: 49–65 (in Russian with English summary).

Ueno, K. 1984. Induction of triploid carp and their haematological characteristics. Jpn. J. Genet. 59: 585–591.

Wang, W., Y. Wang, W. Hu, A. Li, T. Cai, Z. Zhu and J. Wang. 2006. Effects of the "all-fish" growth hormone transgene expression on non-specific immune functions of common carp, *Cyprinus carpio* L. Aquaculture 259: 81–87.

Wang, Y., W. Hu, G. Wu, Y. Sun, S. Chen, F. Zhang, Z. Zhu, J. Feng and X. Zhang. 2001. Genetic analysis of "all-fish" growth hormone gene transferred carp (*Cyprinus carpio* L.) and its F_1 generation. Chinese Sci. Bull. 46: 1174–1177.

Wong, A.C. and A.L. Van Eenennaam. 2008. Transgenic approaches for the reproductive containment of genetically engineered fish. Aquaculture 275: 1–12.

Wu, C., Y. Ye and R. Cheng. 1986. Genome manipulation in carp (*Cyprinus carpio* L.). Aquaculture 54: 57–61.

Wu, C. 1990. Retrospects and prospects of fish genetics and breeding in China. Aquaculture 85: 61–68.

Wu, Q., Y. Ye and X. Dong. 2003. Two unisexual artificial polyploid clones constructed by genome addition of common carp (*Cyprinus carpio*) and crucian carp (*Carassius auratus*). Sci. China Life Sci. 46: 595–604.

Yamaha, E., M. Murakami, K. Hada, S. Otani, T. Fujimoto, M. Tanaka, S. Sakao, S. Kimura, S. Sato and K. Arai. 2003. Recovery of fertility in male hybrids of a cross between goldfish and common carp by transplantation of PGC (primordial germ cell)-containing graft. Genetica 119: 121–131.

Yarzhombek, A.A. and B.I. Gomelsky. 1993. Solubilization of methyltestosterone by carp. J. Ichthyol. 33: 147–148.

Ye, Y., Z. Wang, J. Zhou and Q. Wu. 2009. Genetic stability of progeny from an artificial allotetraploid carp using sperm from five fish species. Biochem. Genet. 47: 533–539.

Yousefian, M., C. Amirinia, M. Bercsenyi and L. Horvath. 1996. Optimization of shock parameters in inducing mitotic gynogenesis in the carp *Cyprinus carpio* L. Can. J. Zool. 74: 1298–1303.

Zhang, P.J., M. Hayat, C. Joyce, V.L. Gonzalez, C.M. Lin, R.A. Dunham, T.T. Chen and D.A. Powers. 1990. Gene transfer, expression and inheritance of pRSV-rainbow trout-GH cDNA in the common carp, *Cyprinus carpio* (Linnaeus). Mol. Reprod. Devel. 25: 3–13.

Zhong, C., Y. Song, Y. Wang, Y. Li, L. Liao, S. Xie, Z. Zhu and W. Hu. 2012. Growth hormone transgene effects on growth performance are inconsistent among offspring derived from different homozygous transgenic common carp (*Cyprinus carpio* L.). Aquaculture 356-357: 404–411.

Zhu, Z., G. Li, L. He and S. Chen. 1985. Novel gene transfer into fertilized eggs of goldfish (*Carassius auratus*). J. Appl. Ichthyol. 1: 31–33.

6

Traditional Feeding of Common Carp and Strategies for Replacement of Fish Meal

Gert Füllner

Introduction

Aquaculture plays an increasingly important role in securing mankind's rising demand for high-value protein. Consequently, aquaculture production has increased substantially in the last few years. The most important taxonomic group of fish in the world of aquaculture are cyprinids. Worldwide, 5.35 Mio t silver carp (*Hypophthalmichthys molitrix*), 4.57 Mio t of grass carp (*Ctenopharyngodon idella*), and 3.73 Mt of common carp (*Cyprinus carpio*) were produced in 2011 (FAO 2013). These fish species rank among the first three places in global finfish aquaculture statistics. Further cyprinid species such as bighead carp (*Hypophthalmichthys nobilis*) with 2.71 Mt and some of the Indian major carps, including catla (*Catla catla*) and rohu (*Labeo rohita*), rank next. The production of cyprinid species has doubled since 1992 (FAO 2013). Hence, farming of cyprinids in fish ponds is the most important form of aquaculture in the world today.

Common carp is a fast growing fish, which is one of the reasons why carp became one of the most important species for aquaculture. In addition, carp farming in warm water ponds is a simple form of aquaculture that only requires a flat piece of land with adequate water supply. Ponds are simple constructions which are discussed in detail by Hartman et al. in Chapter 3

Saxon State Authority for Environment, Agriculture and Geology, Fisheries Department, P.O. Box 1140, D-02697 Königswartha, Germany.
Email: gert.fuellner@smul.sachsen.de

of this book. Construction and operation of fish farms does not need highly qualified personnel. In many cases, the fishponds can be integrated with surrounding agriculture activities. This means that livestock farming such as production of pigs or poultry can be combined with warm water fish ponds, as it is common in Asian rural aquaculture systems. Another way to boost fish production, as well as production of agricultural products, is rice-fish culture, in which fish and rice production takes place in the same pond or field one after another or even at the same time. In both methods, with input of wastes from animal production or with decomposing plant matter as the remaining wastes of rice-fish culture or from other plant farming wastes, the carp pond receives organic material and hence nutrient input comparable to a cheap fertilizer. This has a strong feed saving effect in such integrated systems.

Another common way to increase yield in carp ponds without increasing the amount of feed is the 'polyculture' approach. In this traditional form of Asian aquaculture, different fish species with different food requirements are produced together in the same pond. An overview on species that are used in polyculture together with carp is given in Chapter 3 (this volume) by Hartmann et al. The 'classic' form of carp polyculture in China is the combination of silver carp (*Hypophthalmichthys molitrix*) and/or bighead carp (*Hypophthalmichtys nobilis*) as phytoplankton feeders, grass carp (*Ctenopharyngodon idella*) as consumers of macrophytes and common carp (*Cyprinus carpio*) as zooplankton and benthos feeders. Each species has different food requirements in polyculture, this property can be leveraged to increase yield (see Chapter 3 this volume for a treatise on polyculture). Moreover, the total yield is not only the sum of all yields of the different species in an ideal polyculture, but is considerably higher than that. The reasons for this are positive effects by repeated cycles of nutrient flows within the system. Different food stuffs ingested by fish are digested differently. Remains of not entirely digested food can be used by other aquatic organisms or function as nutrient input that accelerates primary and secondary production in the ponds.

Compared to other cultured species, common carp is a non-sensitive fish which can thrive even in waters of poor quality. Carp can be reproduced in ponds or in hatcheries without much technological assistance from humans; moreover, the brood stock can be collected in natural waters as well. Reproduction and raising of carp are reviewed in Chapter 4 (this volume) by G. Schmidt.

There is another fundamental reason for the current increase in carp farming. Many species in aquaculture such as salmon or trout are predators, and thus need a high intake of animal protein in their diet, which is normally supplied by fish meal. These raw materials are available in limited supply and thus restrict the development of aquaculture. Carp farming in warm water ponds, however, requires only marginal supply of fish meal and fish oil (Cochrane et al. 2009). Primarily for this reason, the relevance of farming common carp and other cyprinids as a resource-efficient form of aquaculture in ponds is expected to further increase in the future.

Basics of Metabolism

To understand the principles of digestion, it is necessary to describe some basics of the metabolism of carp. The fundamental principles of carp metabolism have been explained by Vinberg (1956). Digestion and metabolization of food components by common carp depend on many environmental factors. Since carp is a poikilothermic organism, its metabolism is strongly influenced by temperature. The time of intestinal passage takes 14–19 hours at 10°C, but only 5–7 hours at 25°C. Intestinal passage is also influenced by frequency of feeding (Schade 1982). Not only the intestinal passage but even the metabolic rate is dependent on temperature, increasing exponentially with increasing temperature (Fig. 6.1).

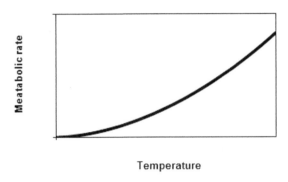

Figure 6.1. General dependency of metabolic rates of fish on temperature.

Furthermore, the metabolic rate is influenced by oxygen supply. In contrast to a gradual temperature dependence the metabolic rate decreases only under suboptimal oxygen conditions. If oxygen supply exceeds optimum level for carp, metabolism increases no further (Fig. 6.2).

Metabolism strongly depends on the live weight of the fish. Vinberg (1956) calculated an equation for the basic metabolism of carp at 20°C:

$$Q = 0.343 \times w^{0.85}$$

In which Q = oxygen consumption in mg O_2 fish^{-1} x h

w = individual weight in g

In practice, the active metabolism rate including activities such as feeding and growing is more relevant. At 23°C the total active metabolism can be described by the following equation:

$$Q = 1.23 \times w^{0.80}$$

This means that small fish have a much higher relative oxygen demand than large fish (Fig. 6.3). In other words, the metabolic rate of smaller carp

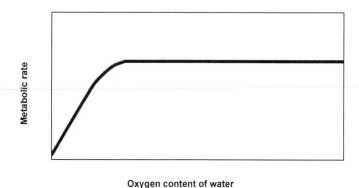

Figure 6.2. General dependency of the metabolic rates of fish on oxygen supply.

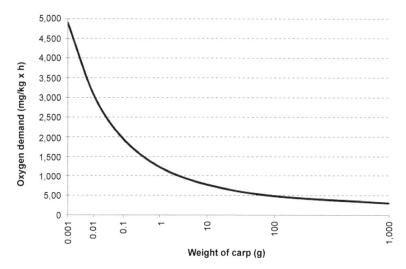

Figure 6.3. Oxygen demand per kg body weight in common carp of different size at 23°C under aquaculture conditions.

is much higher than that of larger individuals. In the same way, the rate of metabolism is strongly tied to growth rate, digestion rate, and finally the amount of feed that can be applied to fish per day.

The effects of external factors on metabolism are considered in feeding tables that relate the daily amount of feed to average body mass, water temperature and sometimes to oxygen supply.

Nutrition of Carp

Carp are opportunistic feeders and live on a broad range of food organisms. They use the concept of 'optimal foraging' which means that carp use the food that is available in the shortest time and with the lowest input of energy, as has been shown by Ivlev's seminal work on juvenile carp (Ivlev 1961).

Adult carp are primarily bottom feeding fish. As secondary food, carp can use zooplankton. If feasible, adult carp can be fed mussels or small fish.

Carp have barbels to detect food in the sediment. In order to find sediment-dwelling food items carp dig intensively in the bottom, screening the bottom by using the nose, separating stones and other non-digestible parts and then put their mouth over prey organisms, and suck them up. The resulting 'feeding holes' in case of production of marketable carp are visible after draining the ponds in fall (Fig. 6.4). For a detailed account on carp feeding ecology please refer to Chapter 9 in this volume.

Figure 6.4. 'Feeding holes' in a carp pond with a depth of more than 20 cm after fishing and draining shows how intensive carp dig the pond bottom during growing season, picture by G. Füllner.

The Process of Digestion in Carp

Similar to other cyprinids, common carp have no stomach. For this reason, their digestion does not work at acidic pH levels and there is no pepsin digestion. Nevertheless, there are other enzymes that are active in the gut of carp (Table 6.1).

The digestive enzymes of carp have optimum activity levels at temperatures of 38–39°C. Consequently, feeding carp at high temperatures and other favorable environmental conditions is possible and even beneficial.

The amount and the composition of digestive enzymes are controlled by the amount of food consumed. If there are no carbohydrates in the feed, for example, only a low amylase activity can be detected. The enzyme activity in carp does not decrease at the same rate as water temperature. In fact, enzyme

Table 6.1. Enzymes in the digestive system of carp (according to Jančarik 1964). The number of asterisks shows the amount of enzymes active in the respective organ or tissue.

	Enzyme	pH-optimum	Substrate	Location and amount	Decomposition product
Proteases	Trypsin	8.0	Protein	Pancreas ✻ ✻ ✻ Intestinal mucosa (IM) ✻ ✻ ✻	Polypeptides
	Dipeptidase	7.8	Dipeptides	IM ✻ ✻ ✻	Amino acids (AA)
	Aminopeptidase	7.8	Polypeptides	IM ✻ ✻ ✻	AA with free amino groups
	Carboxypeptidase	7.8	Polypeptides	IM ✻	AA with free carboxyl groups
Esterases	Lipase	7.0...7.5	Lipids	IM ✻ ✻ ✻ Gut lumen ✻ Gall ✻ Pancreas ✻	Glycerin and fatty acids
Carbo-hydrates	Amylase	6.8...7.0	Starch/ Glycogen	Pancreas ✻ ✻ ✻ IM ✻	Maltose

activity can lag behind a drop in temperature for a couple of days. Hence, it is not necessary to limit the daily food intake, if there are short cold spells in carp ponds in summer according to the feed demand (Füllner 1987). Such short cold spells will not affect the overall enzyme activity.

There is another special reason that influences the digestion of food in the earliest life stages of carp. Postlarvae have a very short and only marginally developed gut. Therefore they are not able to produce enough enzymes to digest nutrients of food. In the first days of their lives, carp rely on the enzymes already present in their food or at least they need a supplement for processing nutrients. Carp larvae need time to adapt the type and amount of their digestive enzymes to the amount and composition of feed intake. Additionally, it takes carp fry some time to produce an appreciable amount of enzymes after changing alimentation from yolk sac to external feeding. Although adult carp can produce all necessary enzymes, small juvenile post-larval carp need a supply from external enzymes for digestion of food. Therefore, it is impossible to raise carp fry with 'simple' artificial feeds in the first days of their life.

A third factor in the development of postlarvae is that the microflora has not yet developed sufficiently in the digestive tract of carp fry. In all stages of life, the microflora of the intestine is very important for digestion because carp do not possess a stomach. It was observed in rohu (*Labeo rohita*) that the supplementation with cultured *Bacillus circulans* improved growth and feed efficiency of larvae fed commercial feed (Mondal et al. 2008). This may be true for larvae of all carp species. Consequently, feeding of postlarvae is possible with special diets, but this requires very expensive microencapsulated feeds that also contain an essential amount of enzymes. However, the easiest and best way is to feed carp fry with natural food in the first days of their life. This high quality natural food can be produced in accurately prepared ponds. High quality natural food comprises the right organisms at the right time which means that particles in a suitable size for small but fast growing carps are available and without containing predacious copepods. Diet choice of carp fry is further described in Chapter 9 (this volume) and the management of ponds and their preparation before stocking are reviewed in Chapter 3 (this volume).

Natural Food as Basis of Carp Farming

Natural food is the key for successful carp production in ponds. Natural food has a high amount of water and it is, therefore, easy to digest by all age groups of carp. It is of particular importance that natural food contains internal digestive enzymes of the consumed organisms.

It is possible to produce carp by using only natural food. Nevertheless, the efficiency of this form of carp farming is low, especially because of a dissipation of food protein into other uses than biomass gain. As can be seen in Table 6.2, the digestible part of natural food organisms almost entirely consists of protein and fat. Because carp can digest native starch, carbohydrate-rich feed can be added as supplement to natural food. In this form of production, the protein of natural food delivers the protein that is necessary for growth and the repair of tissues. The carbohydrates of the supplementary feed deliver the energy for all other vital metabolic processes, such as energy metabolism, locomotion of the fish, or for storage of metabolites, e.g., in lipogenesis. This form of carp farming allows for a more efficient use of the limited protein resources provided by natural food. Simultaneously, the yield of fish in the pond can be doubled or even tripled.

Table 6.2. Nutrient composition of major carp prey items (according Albrecht and Breitsprecher 1969). All values in percent of dry matter.

	Water % fresh matter	Crude protein	Crude fat	NFE*	Crude ash
Chironomids	85.8 (79.0–92.7)	52.1 (52.6–59.6)	7.7 (3.6–11.3)	26.7 (16.6–34.1)	7.7 (5.3–8.0)
Daphnia	90.6 (87.8–93.8)	43.6 (14.3–58.1)	9.6 (2.9–21.3)	23.4 (10.9–44.1)	17.0 (10.4–28.9)
Copepods	90.0 (87.9–92.1)	42.0 (25.1–70.2)	33.0 (11.7– 4.9)	20.0 (10.9–31.7)	6.0 (4.1–7.3)
Tubificides	83.0 (77.8–87.2)	51.8 (46.1–66.6)	17.6 (6.3–19.8)	20.6 (14.6–28.2)	5.9 (4.0–9.0)
Ephemerids	80.5 (74.9–7.2)	60.5 (44.0–71.6)	15.9 (5.1–26.3)	15.4 (8.2–23.2)	8.7 (5.1–10.2)
Caddisfly larvae	76.7 (70.4–87.3)	56.7 (48.5–68.6)	17.2 (3.5–12.1)	19.7 (12.7–31.5)	6.4 (3.5–12.1)

*Nitrogen-free extracts (carbohydrates which consist almost completely of indigestible chitin in case of carp prey)

Value of Natural Food

Natural food contains high amounts of protein, fat and chitin and has a high water content. It features a near-optimal composition of essential amino acids for carp; although, the nutritive value differ depending on the sources or species that make for the natural food items (Table 6.2).

There is another important characteristic of common carp that needs to be mentioned in this context. As shown in Table 6.1, common carp has the ability to digest native starch. This is different to other teleosts as well as other cyprinids. According to recent research the closely related tench (*Tinca tinca*) has an amylase activity that is considerably lower than that of carp (Hidalgo et al. 1999). It is not clear, whether the capability of carp to digest native starch is a natural trait or whether it has been acquired by genetic fixation through anthropogenic selection. Wild carp in nature can even digest plants, e.g., seeds of grasses (Nikolski 1957, Balon 2004). This suggests that the ability to use starch is a plausible result of natural selection leading to a genetically fixed ability to use native starch. This process might well be independent of later selection for high amylase activity strains in aquaculture. However, for more than 150 years grains have been used as supplementary feed and the according selection of fast growing individuals as breeding stock has probably amplified the strength of selection and led to the genetically fixed high amylase activity in aquaculture strains of carp. Regardless of its evolutionary origin, the high amylase activity in carp makes it possible to use grains as feed for carp in ponds if there is also enough natural food as a source of protein.

Benthos

As shown earlier, carp are optimal foragers. Hence sediment-dwelling animals are the primary food for adult carp. The main food consists of chironomids and tubificides with chironomids constituting the major food in the sediment of ponds (Fig. 6.5). They offer a food source of a fairly high biomass per prey item with a nearly optimal nutrient content for the carp (Table 6.3).

The filling of water in a drained pond in spring results in a quick and large development of insect larvae. In carp aquaculture ponds in Germany (Central Europe) that are prepared well for carp production, especially by draining in winter, the amount of chironomids can reach up to 1,000 individuals or 12 grams per square meter soon after filling with water in spring (Merla 1960). This amount of benthos food is much higher than in ponds that are continuously filled with water during winter. In contrast, the proportion of oligochaeta only accounts for about 30–80 individuals or 0.1–0.2 g/m², and is uninfluenced by filling or draining of ponds.

The quantity of ephemerids and trichopterais commonly much lower in pond soils as well as in the diet for carp. Due to the opportunistic type of feeding (see Chapter 9, this volume), carp with body mass higher than

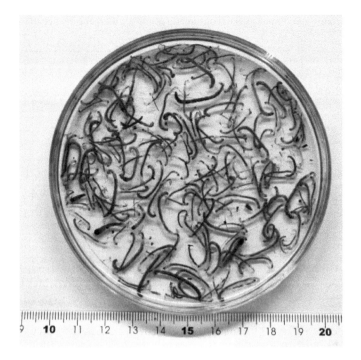

Figure 6.5. Chironomids isolated from pond soil in the experimental pond station Königswartha (Saxony, Germany).

Table 6.3. Nutrient composition of benthos in carp ponds (%). According to Epler et al. (2010).

	Dry matter	Crude protein	Crude fat	NFE*	Crude ash
In fresh matter	13.9–14.9	6.6–7.5	0.8–1.0	2.0–4.2	3.2–3.6
In dry matter	100	44.0–53.8	5.7–6.4	14.5–28.2	21.4–25.9

*Nitrogen-free extracts (nearly 100% indigestible chitin)

200 grams primarily feed on chironomids that are longer than 12 mm. Only when the amount of food organisms is lower than approximately 0.5 g/m² in the sediment of ponds (at which point they become a very hard to exploit food resource), larger carp can be more effective consumers of zooplankton. Therefore, zooplankton is probably regarded as a secondary food for bigger carp with an average body mass higher than 500 grams (Merla 1960). But even if carp use plankton as food, the largest plankton organisms are consumed first.

Zooplankton

Carp fry need the smallest zooplankton as food. The size of the zooplankton organisms depends on the size of the growing carp, and finally on the mouth size of the juvenile fish (Table 6.4).

Table 6.4. Groups and dimensions of consumed natural food organisms depend on the size of carp fry.

Size of carp fry		Natural food organism	Size (mm) of food items
Total length (mm)	Average body mass (mg)		
6.2	1.5	Rotatoria (e.g., *Brachionus* species)	< 0.1
6.2–7.4	1.5–2.95	Rotatoria (e.g., *Chydorus sphaericus*)	0.1–0.3
		Ceriodaphnia reticulata	0.6–0.8
		Bosmina longirostris	0.4–0.6
> 7.2	> 2.08	*Moina retirostris*	1.0–1.6
> 7.5	> 3.02	Rarely rotatoria, primarily bigger zooplankton	
< 12.5	< 34	Not yet ingested: *Daphnia magna, D. pulex, D. longispina*, and bigger cyclops	

Carp of a length of more than 2 cm have a nearly identical demand for any specific type of natural food. Carp larvae smaller than 2 cm prefer to consume protozoa and rotifers. They feed exclusively on microcrustacea (cladocera and copepoda) and rotatoria. These fish are yet too small to effectively dig for pond soil animals. At a mean total length of >2 cm, carp larvae shift their diet to include benthic food resources, but microcrustacea still dominate the larval diet. Small carp (≤15 cm) show a high preference for microcrustacea and tend to avoid benthic macroinvertebrate food resources. As carp size increases, the proportion of macroinvertebrates in gut contents increases as well. Not before the average body mass of carp is higher than 200 g (or 15 cm total length) they can forage effectively on pond bottom food. This is one of the reasons that the production stage of two-year-old carp reaches the highest yield per hectare: At the beginning of the season, carp are still small and can only consume zooplankton. In the second half of the production period they can feed on pond soil animals as well. In contrast, carp fingerlings in the first year of life can only feed on zooplankton as just shown above. Because smaller fish have a higher metabolic rate and a higher growing potential, the yield of two-year-old carp is higher than the production stage for marketable carp in the third year that cannot effectively use zooplankton any more.

Natural Food as Basis for Carp Production

The possible fish yield in carp ponds based on natural food is several times higher than in natural water bodies. One major reason for this is the pond construction. Due to its shallowness a pond can be considered a littoral zone across its entire area. The littoral is the most productive area of a water body for benthic primary production. Besides the adjustment of stock to the harvest capacity, other human management strategies for ponds, e.g., winter-drying play an important role to maximize fish yield. In the early post-winter season, the onset of the reproductive cycles leads to an increased abundance of natural food items such as benthos and zooplankton resulting in high food biomass for fish particularly in freshly filled ponds. In ponds for production of market size carp, the useable bottom feed organisms can already diminish at the beginning of July. Major zooplankton organisms can be over-grazed by carp after shifting to this food source one month later. In ponds for production of smaller carp, over-grazing starts even earlier, or the spring-'peak' of zooplankton (a typical phase of above-average population growth of zooplankton in lakes and ponds of the temperate zone) does not occur at all due to constant grazing by carp (Fig. 6.6). However, over-grazing in ponds directly depends on stocking density of fish.

This over-grazing effect is less of a problem with respect to the growth and production of adult carp, but it is a huge problem for production of one-year-old carp fingerlings. Based on the high intra-specific competition for food resulting from a high density of individuals of the same species, the natural food and with it the only source of high-value protein is depleted by the beginning of August in the Northern Hemisphere (Fig. 6.7). Hence it is occasionally necessary to give one-year old carp not only carbohydrate-rich supplementary feeds in the second part of the season, but rather feeds with high-value protein and a complete dietary profile. In temperate climates this procedure should be a standard practice, at least for the last feeding month (typically in September in the Northern Hemisphere) before wintering. This is necessary to reach a high average body mass and a satisfactory condition of carp for the following overwintering season. Otherwise carp fingerlings

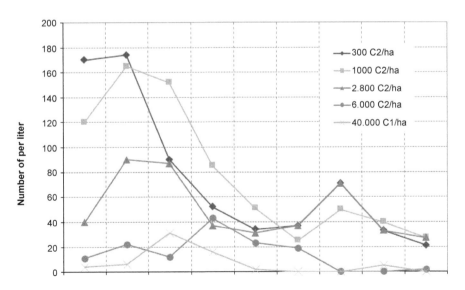

Figure 6.6. Number of daphnia per liter in the course of the growing season in Central Europe in carp ponds stocked with different age groups and stocking densities. C1 = one-year-old carp; C2 = two-year-old carp.

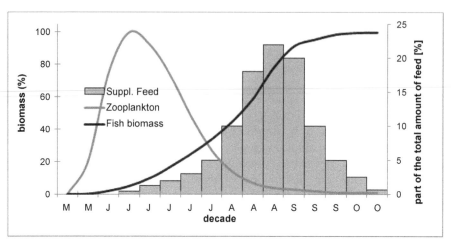

Figure 6.7. Principle of natural food development in a carp pond for production of carp fingerlings in the first year growing season, depending on fish biomass, and application of supplementary feed. X-axis: Ten-day periods during the growing season (May = M to October = O) in Central Europe.

can develop a protein deficit even though they were fed with a high amount of supplementary feed. Most supplementary feeds consist only of grains which deliver only carbohydrates for metabolism. Consequently, fingerlings become fat. This fat, however, is only a sign of an 'apparent good condition' because the carp lacks well-balanced feed. The cause of this 'apparent good condition' is the lack of adequate amounts of nutrients for growth (particularly of high value food protein). Therefore, carp generate excessive fat reserves from the carbohydrates of the food. Intriguingly, carp seem to deposit a lot of fat rather than investing energy into somatic growth. It can be suggested that if more high-value food was available, the fish would become bigger and less fat. Simply expressed, such excessive fat deposition is an expression of malnutrition. This phenomenon is well-described in carp aquaculture and can lead to high winter-kills, because carp are not able to use these 'bad' fat reserves efficiently in case of detrimental environmental conditions (Müller 1966). The fat content of carp can reach 28.5% in the muscle tissue under those conditions (Amlacher 1960). Higher overwintering losses caused by protein deficit can be avoided by feeding supplementary feeds with carbohydrates and a high protein content.

Nutrient and Energy Requirements

The composition of carp feed first has to satisfy the requirement for energy and essential nutrients of the organism. It also has to allow an effective gain of flesh from an economic point of view.

The most important feed element is protein. It is the only 'construction material' of growing organisms and its metabolic role cannot be replaced by other nutrients. The requirement for available protein for carp amounts to 3.0–3.2 g/kg metabolic [$kg^{0.8}$] body mass (Eckhardt et al. 1983, Table 6.5).

Naturally, the quality of proteins and primarily their amino acid composition has to satisfy the nutritional demands of carp. This can be ensured with an adequate demand of high-quality animal components in the feed, especially fish meal. In practice, at high water temperatures and correspondingly high daily feed rations, a crude protein content of 25–30% in the feed is sufficient for older carp. For smaller carp (in the first season) higher crude protein contents of up to 40% are recommended.

Table 6.5. Requirement of available protein by carp with different body masses, calculated with the results of Eckhardt et al. (1983).

Average body mass of carp (g)	Requirement of available protein per day (g/kg fish x d)
1	11.9–12.7
10	7.5–8.0
100	4.8–5.1
1000	3.0–3.2
2000	2.6–2.8

Besides the essential protein which meets the requirements as construction material for tissue growth, there has to be a supply of a more metabolizable source of energy. Energy suppliers in feeds are predominantly carbohydrates but also fats. In fact, for carp it is irrelevant if metabolizable energy sources are fat or carbohydrates, as long as the amount of protein in the food is high enough (Ogino et al. 1976, Takeuchi et al. 1979, Koch 1984). It is also possible for the fish to gain energy from protein, but this is physiologically and hence eventually also economically inefficient. The energy gain from protein is accompanied by an increasing excretion of ammonia by carp. This pollutes the water of the pond, thus degrading environmental conditions. Ammonia can lead to autointoxication of carp and may result in gill necrosis (Gongnet 1984). Thus, high value feed proteins should be used for growth and thus for synthesis of fish protein, wherever possible. Eckhardt et al. (1983) specified an energy supply of 230–260 kJ/$kg^{0.8}$/d as a guiding value, if a high weight gain is sought under optimal temperature conditions for carp growth. The optimal protein to energy relation in food for carp is 12–13 g available protein per MJ metabolizable energy. This results in a recommended protein supply of 2.8–3.4 g/$kg^{0.8}$/d as a guiding value.

Feeding Carp with Supplementary Feeds

In contrast to other fish species, carp is able to digest native starch for example from cereals (Takeuchi et al. 1979). This allows raising fish yield substantially above natural productivity only by adding cereals and without any use of fish meal. However, the supplementary feeds used remain an addition to, rather than a substitution of natural food (Table 6.6).

Table 6.6. Nutrient content of important supplemental feeds for carp farms (in percent of raw matter). Data according to Nehring (1970).

	Water	Crude protein	Crude fat	NFE	Crude ash
Wheat	12.0	12.1–15.2	2.1–2.3	78.3–81.0	1.9–2.2
Rye	12.0	9.9–11.3	1.7–3.5	79.6–83.9	1.9–2.2
Maize	12.0	10.7–11.0	2.5–4.7	80.2–84.4	0.9–1.7
Barley	12.0	11.4–14.5	1.9–2.5	77.6–78.9	2.1–2.8
Oats	12.0	13.0–16.3	5.5–8.2	67.2–71.1	2.1–3.5
Pea	12.0	24.6–40.3	0.9–1.5	47.9–63.4	3.3–3.8
Lupine	12.0	36.7–46.0	4.9–9.5	28.3–39.0	3.9–5.1

Particularly a sufficient protein amount, and in addition, the essential amino acids required for the synthesis of own body proteins are frequently absent from the supplementary feeds. Arginine, histidine, isoleucine, leucine, lysine, methionine, phenylalanine, threonine, tryptophan and valine are essential amino acids for carp (Nose et al. 1974). These amino acids need to be provided through external feeding, but in case of supplementary feeding of carp only natural food can deliver these amino acids in adequate composition. Their total amount in grain is too low and the composition of the amino acids is only suboptimal. Therefore yields still stay limited by natural food if fish are fed only with supplementary feeds.

In moderate temperate climates a weight gain between 800 and 1,200 kg/ha can be reached in carp ponds by feeding with supplementary feeds. This triplicating or quadruplicating of the yield can be reached with protein supply from natural food only. The carbohydrates that are given by supplementary feeds serve as energy supplier for the metabolism. They reduce the allocation of protein from natural food to metabolic uses other than biomass gain to a physiologically acceptable minimum. The production of carp based on a diet of natural food with a high amount of carbohydrate feeding supplement corresponds to the principle of pasture management in livestock farming. Both from an economic perspective as well as for reasons of conservation of natural resources and aquatic pollution this farming method is semi-extensive, which is amongst the most recommendable practices in carp farming.

Supplementary feeding of carp compared to the rearing without additional feed supply increases the land productivity by using low-cost food, because:

- natural food is not depleted as quickly as it is without additional feeding,
- ponds retain their fertility without additional fertilization by the (moderate) input of nutrients via feed, metabolism and excretion,
- the pond's free area (i.e., space without vegetation) can be conserved most economically, because more intensively used ponds have a lesser tendency towards siltation as compared to pure natural food ponds. The

higher biological activity and nutrient availability in intensively used ponds decreases water transparency and therefore reduces macrophyte growth (e.g., reed) that onsets siltation.

Grain is a typical supplementary feed. Even weed-polluted grains are used for carp feeding. Low-priced offers from farmers, like grains that are contaminated with weed seeds or clammy grains, may be used. Even grains containing ergot are not considered to be a problem because carp avoid the intake of ergot, but feeding of ergot-contaminated grain is restricted by law in some countries, e.g., in the European Union (EC 2005). Due to consideration of global market prices wheat or triticale (a hybrid between rye and wheat) are applied as supplementary feeds in temperate climate regions. But the same rearing results can be reached with rye, barley or oat, irrespective of their higher proportion of husks. It is also possible to use other feed stuffs as supplementary feed for carp, such as oilseed cakes or other by-products from the food industry or legumes. However, proteins of legumes are not appropriate as single food for carp and for this reason yield is limited when using pea or lupine, although the protein content of these legumes is much higher than that of grain. This is due to the fact that these protein sources have no well-balanced composition of the essential amino acids and vegetable feeds frequently contain some anti-nutritive substances (which are reviewed in Chapter 13 "Natural Toxins" by C. Pietsch, this volume). Feeding trials during a complete growing season using different supplementary feeds at the experimental pond station Königswartha (Germany) showed good body mass gain of carp as well as gain per hectare and feed conversion efficiency (Fig. 6.8). However feeding of marketable carp only with protein-rich vegetable feeds like lupine, pea or a cheap pelleted high protein Vegetable

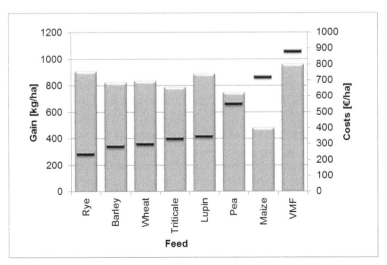

Figure 6.8. Weight gain and feeding costs for rearing carp with different vegetable supplementary feeds for one growing season at the pond station in Königswartha, Germany. VMF = Self-composed pelleted, fully vegetable mixed feed.

Mixed Feed (VMF) instead of grain did not show better yields than feeding with low-protein feeds like wheat, rye or barley. The reasons for this result remain unclear. The different amino acid composition of natural food and protein-rich vegetable feeds or/and the anti-nutritive factors of the vegetable feeds are probably responsible for these results.

For reasons already mentioned, feeding with protein-rich vegetable feeds accomplished no higher weight gain than feeding with grains. Only feeding a pelleted self-composed, fully-vegetable mixed feed (VMF) achieved slightly better results. The reason for this may be explained by the manufacturing process in the pelleting machine. The heating of the pellets could have deactivated some of the anti-nutritive factors in the vegetable raw materials. Other possibilities to remove anti-nutritive components from plant-derived feed ingredients are reviewed in Chapter 13 by C. Pietsch, this volume.

In the above mentioned experiments, the lowest yield was reached by feeding carp with maize which confirms previous studies (Kaushik 1995). Besides, feeding with maize resulted in the worst flesh quality of carp. This is probably due to a high content of oleic acid in maize which resulted in a negative effect on the ratio of polyunsaturated fatty acids (ω3 to ω6 ratio) in carp which received maize (Ćirković et al. 2011). In addition, carp fed with maize at the pond station in Königswartha showed a high content of fat and especially a lot of yellow-colored fat in the visceral part of the fillets and had the worst taste. Consequently, feeding marketable carp with maize is not recommended. In case of need it is possible to use maize as feed for stocking material, but the corn has to be crushed for smaller fish, and crushing maize results in a high amount of un-usable fine particles.

Surprisingly, the highest weight gain and best meat yields were not in accordance with high flesh quality. In carcass examinations differences between feeding groups were identified, but first of all between sexes. Females showed better slaughtering results with respect to meat yield, as they still had lower gonad mass than males after a three summer production time.

Feeding different vegetable supplementary feeds generally resulted indifferent meat yields and flesh qualities of carp (Table 6.7). The flesh quality was better in trials with feeding legumes than with feed prepared from grain. The flesh quality of females was always better than in males. The difference between sexes decreased with larger body mass.

Although several physico-chemical parameters of carp fillets had been analyzed, only luminance showed a significant difference (parameter L in the CIE L*a*b* color space) between the feeding groups barley and wheat, but not between the other groups or sexes. Color values a and b were different between sexes as well as between some of the feeding groups (Table 6.8). The color was measured with Spectrophotometer AUTOVISION SPECTROCAM75 RE (Los Angeles, CA) at 10–20 points on randomized locations across the entire fish fillet. Measurement points were selected so that no values were taken at the bright whitish fat between the myomeres.

Table 6.7. Carcass quality of marketable carp (average of 25 fish per feeding group) after feeding different supplementary feeds for one growing season in ponds. Supplementary feed was given three times a week with application rates according to Table 6.11. Stocking density ca. 800 C2/ha; target yield 1,000 kg/ha. Each trial was conducted with 2–5 replications (i.e., ponds). Slaughtering results are from 25 carp per trial from rearing experiments at the experimental pond station in Königswartha (for more details see Pfeifer and Füllner 2008).

Feed	Average body mass (kg)	Condition factor K**	Mass of fillets without skin (% of total body mass)	Mass of gonads (% of total body mass)	Fat content in the fillet (%)
Natural food	2.07	2.25	38.0	4.9	1.18
Wheat	2.22	2.29	38.4	3.2	6.33
Rye	2.33	2.53	39.8	3.2	9.43
Maize	3.05	2.47	38.9	5.1	8.66
Barley	2.09	2.26	38.8	2.8	6.41
Triticale	3.25	2.49	40.6	3.9	8.31
Pea	3.16	2.32	41.0	4.2	4.83
Lupine	2.33	2.22	38.8	3.5	3.22
VMF*	3.66	2.59	41.9	4.6	7.69

*Self-composed vegetable mixed feed (VMF), made from colza cake, triticale, sunflower extract, toasted soybeans, soybean oil, linseed and mineral nutrients premix (no vitamins). Proximate analysis: 23.2% crude protein, 14.1% crude fat, 9.1% crude fiber, 6.4% ash (calculated as % of fresh matter).

**$K = 100 \frac{W}{L^3}$ with W = entire body wet weight in g, L = length in cm

Table 6.8. Differences in CIE-luminance (L) and CIE-color values a and b of carp fillets derived from fish fed with different supplementary feeds (fish from Table 6.7). If value 'a' is negative, the color is greener, if 'a' is positive, the color is redder. If value 'b' is negative, the color is bluer, if 'b' is positive, the color is more yellow. 'L' can differ between 0 (black) and 100 (white), mean ± standard deviation.

Feed	L	a	b
Natural food	29.04 ± 2.03	−0.27 ± 1.10	1.67 ± 1.44
Wheat	32.09 ± 3.58	−0.58 ± 1.01	1.09 ± 1.25
Rye	30.70 ± 3.36	−0.63 ± 0.72	0.88 ± 1.18
Maize	29.34 ± 4.81	0.29 ± 1.33	1.62 ± 1.59
Barley	28.49 ± 1.66	−0.47 ± 0.81	1.05 ± 1.24
Triticale	29.75 ± 2.97	−0.28 ± 1.39	1.30 ± 1.53
Pea	31.04 ± 4.10	−0.33 ± 1.25	0.90 ± 1.67
Lupine	31.61 ± 2.28	−0.78 ± 1,21	1.19 ± 1.56
VMF	30.85 ± 2.27	−0.44 ± 0.60	0.75 ± 1.11

VMF = Self-composed vegetable mixed feed (see Table 6.7)

In standardized taste testings, carp fed with lupine and peas reached best results, even better than fish that were fed with natural food only (Table 6.9). Maize-fed carp were evaluated as the worst by all testers. Therefore, maize

Table 6.9. Results of tasting tests of marketable carp, fed different supplementary feeds (average of 8 examiners judging samples that have been double encoded so that examiners were blind-testing; repeated testing of the same sample was allowed; samples were not salted or flavored, and refined in a standardized way using a microwave oven for 6 minutes at 400 W; fish from Table 6.7). Possible score range for evaluation between 1 and 5 with 1 being the worst and 5 being the best score.

Feed	Appearance	Odor fresh sample	Odor cooked sample	Texture	Taste	Total Score
Natural food	3.88	4.63	3.81	4.13	4.13	4.12
Wheat	3.88	5.00	3.56	3.88	3.44	3.74
Rye	2.94	4.63	3.63	3.38	3.56	3.58
Maize	4.44	4.69	3.25	3.50	2.94	3.41
Barley	3.69	4.25	3.94	4.38	4.06	4.09
Triticale	3.38	4.50	3.88	4.06	3.94	3.96
Pea	4.75	4.81	3.94	4.56	4.38	4.45
Lupine	4.69	4.94	4.25	4.31	4.44	4.47
VMF	4.25	4.63	3.44	4.19	4.13	4.13

should not be used as supplementary food for marketable carp if good taste is a goal. In addition, carp fed with maize received the worst evaluation not only for taste but also for yield in pond, food conversion and slaughtering parameters in despite the good visual appearance of the fillet (Tables 6.7, 6.9).

The presented tests showed that the fat content is the most important parameter for taste evaluation of carp and a low fat content resulted in a better scoring. Fat content of carp fillet can also vary within a feeding treatment group, especially in groups that were fed with grain. Feeding grain always resulted in high fat content. Carp fed mainly on legumes achieved a low fat content. However, a slightly higher fat content was noted in the carp that were fed exclusively with natural food. There was no relationship between flesh luminance and fat content.

The results indicate that the higher the fat contents of carp, the worse the product quality in terms of taste. Therefore, a moderate fat content should be aimed for. This can be reached by selecting management strategies to yield maximum harvests per hectare by application of high protein vegetable feeds or by choosing lower yields per hectare and a limited supplementary feeding with grain. The latter corresponds to better revenue; additionally, it also meets the goals of a more natural form of aquaculture. Less intense production corresponds to a lower amount of nutrients in pond water and is followed by higher biodiversity of animals and plants in and around ponds.

From an economical point of view, rye was the best supplementary feed for carp. However, the best flesh quality of carp was reached by feeding legumes, which causes a low but acceptable fat content of carp fillets produced in this way. Feeding costs from high protein vegetable feeds like lupine, peas or cheap mixed feed from vegetable components are just as or double as high as the cost of grain and do not deliver a similar increase of yields per hectare. The higher feed price results in higher production costs, but not in a better feed conversion. This is why the use of high protein vegetable feeds appears to be too expensive for the production of marketable carp.

Feeding Techniques for Supplementary Feeds

Grains are water-stable due to their cellulosic cell walls, and so they are very well protected against loss of nutrients. Therefore, feeding of older carp with unbroken grains is only necessary two to three times a week. A feeding of supplementary feed to carp fingerlings three times per week is sufficient too, although grains have to be crushed before feeding. The grist should be as rough as possible, because fine parts of the grist will not be consumed and are lost as fish feed. The dusty part of the grist is then only an expensive fertilizer, which unnecessarily increases the nutrient input into the water. The best technique for preparing crushed carp feeds are older roller-grinders. Modern pebble-mills grind the grains too fine.

Unbroken grain as well as crushed grain have to macerate, otherwise carp cannot absorb the hard feed stuff. This happens very easily in the water of the pond some minutes after application of the feed. This makes a swelling of the grain in a tank or a bucket prior to feeding unnecessary. The requirements for supplementary feed differ considerably with the situation to which the feed is to be applied. It depends on the supply of natural food, on fish biomass and on water temperature. Because of the high amount of natural food supplementary feeding is often not necessary before June (in the Northern Hemisphere). Nevertheless, many recommendations exist for carp farming in moderate climates. They differ depending on the date of publication and geographical region (Table 6.10).

The monthly and daily amounts of feed that should be distributed can be calculated with help of such a table (Table 6.11) or a software tool based on a feeding chart such as Fig. 6.9. In addition to considering the changing feed demand during the course of the year, a software can allocate the amount of feed to the individual feeding days or even the feed tours. The calculations

Table 6.10. Distribution of supplementary feed for production of marketable carp; planned quantity in percent of the total feed amount (according to historical and recent authors).

Author	Geographical region	March	April	May	June	July	Aug.	Sept.
Vogel (1905)	Germany			10	25	30	25	10
Walter (1934)	Bavaria			10	20	30	30	10
Michaels (1988)	U.K.			10	15	25	30	20
Horvath et al. (1992)	Hungary	2	5	10	20	25	28	10
W. Schäperclaus (1998)	Upper and Lower Lusatia			-	15	25–30	40–45	15–20
Steffens (2008)	Upper and Lower Lusatia			5	15	25	40	15

Table 6.11. Recommended monthly feed distribution for carp ponds in Central Europe according to the age group of carp (planned amounts of supplementary feed, in percent of total feed amount).

Month Age group	C_{0-1}	C_{1-2}	C_{2-3}
May	-	maximum 5	5
June	maximum 5	10	15
July	10	20	25
August	50	45	40
September	30	20	15
October	minimum 5	0–5	-
Sum	100	100	100

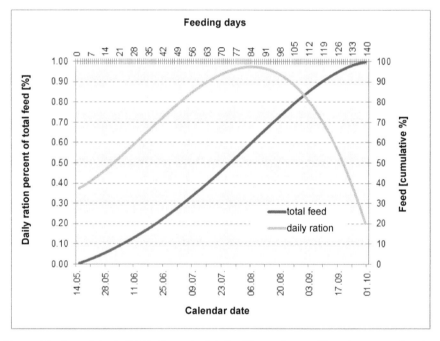

Figure 6.9. Annual feed distribution calculated with a software. The advantage of such a computer program over the calculation with a table is the consistent course of the curve.

are made on the basis of a planned fish yield and an expected feed conversion ratio (FCR). For grains and a yield near the maximum potential limit an FCR of 2.0 for example is acceptable. Feeding a fish meal-containing diet will result in a lower FCR.

In ponds, areas with a hard bottom should be selected as feeding places. The feeding should be in deeper water to prevent birds (e.g., swans or ducks) from consuming the fish feed. However, the feeding should not happen too deep such that the feed does not sink to parts of the pond bottom where the oxygen concentration is low. In ponds with an area less than two hectares it is possible to feed from the levee or near the pond outlet. In bigger ponds it is necessary to feed from boats, following the same 'feeding street' each time (Fig. 6.10). This is necessary for controlling the feed uptake. The feeding street has to be marked with stacks or fixed floaters that simultaneously serve as control points.

Figure 6.10. Feeding street in a traditional carp pond. Picture by G. Füllner.

For distribution of grains special barges can be used, whereas on larger ponds the barges need to be engine-powered to allow for time-efficient administration of feed (Fig. 6.11).

Prior to feeding it must be ensured that the last ration of feed has been entirely consumed by carp. This happens with a feed scraper or with a barge stack. If there is remaining feed from the previous feed application, further feeding has to be stopped. The reason why the fish refused the feed should be explored. If feed remains in the same pond repeatedly, the feeding street or the feeding point has to be changed, because older feed acidifies and is avoided by carp. If there is no more feed from the last portion at the control points, and the environmental conditions are good (e.g., high water temperature, high oxygen content), the amount of feed can be raised. Finally, the feeding with supplementary feeds, such as grains, takes place *ad libitum*, i.e., more than the fish consume within the first feeding day.

It is important to note that the recommended stocking density must not be exceeded during use of supplementary feed. The highest yield can be reached with stocking densities of about 30,000 carp fry, 2,000–3,000 C1 or 800–1,000 C2 per hectare with a FCR around 2.0. Lower stocking density

Figure 6.11. Special feeding boat with swimming chambers and slot bottom for distribution of grains. Picture by G. Füllner.

results in lower FCR and in better fish flesh quality. Thus it is possible to achieve a FCR clearly below 2.0 depending on stocking density (Figs. 6.12, 6.13).

Finally, it should be emphasized that grain is not an alternative for missing natural food. Feeding too high amounts of supplementary feed results in a poor FCR and excessive adsorption of fat in musculature in case of marketable carps. It can also lead to a bad condition of carp fingerlings and carp stocking material.

Feeding Carp with Complete Mixed Feeds

If yields per hectare are higher than the level that is achievable with natural food plus supplementary feed, the use of complete feed will be necessary. This shows that carp might consume fish meal like all carnivorous fishes.

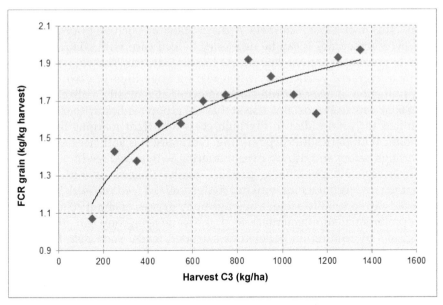

Figure 6.12. Achievable Feed Conversion Ratio (FCR) in carp of the age group C_{1-2} by additional feeding with grain in carp ponds. Data from commercial fish farms in Saxony (Germany).

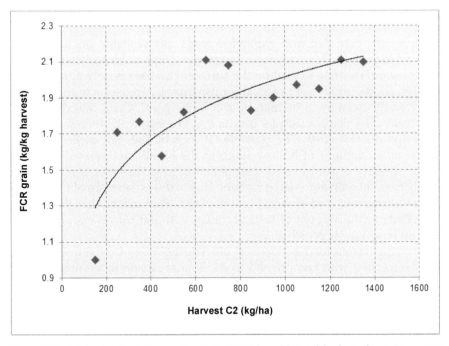

Figure 6.13. Achievable Feed Conversion Ratio (FCR) by additional feeding of carp (age group C_{2-3}) with grains in carp ponds. Data from commercial fish farms in Saxony (Germany).

Pelleted or extruded feeds are five to seven times more expensive than grain. This fact alone hampers a dissipation of such high-priced feeds. However, sometimes it can be necessary to feed carp with adequate mixed feeds to guarantee animal welfare. As has already been shown, natural food can be overgrazed particularly when rearing fingerlings of carp. Therefore, the production of one-year-old stocking material is mostly realized by using full-fledged mixed feeds. But also at this age group feeding with mixed feeds is limited to certain, due to the high costs and tight margins that set the economic boundaries for carp farming. Normally, it is sufficient to feed carp fingerlings before and during over-wintering with mixed feed.

It is certainly possible to produce bigger fingerlings with a rearing technique in which feeding with full-fledged mixed feeds is used. While one-year-old carp normally achieve average body masses of about 30 to 50 grams with preferential supplementary feed, the same age group can attain 100 to 200 grams by feeding on pelleted or extruded feeds. Such carp fingerlings are in a much better condition and almost no losses occur in the following overwintering period.

There are a lot of feeds with different prices on the market depending on the components that are available for feed production and the production process. This raises the question of which complete feed should be used in carp farms. The composition of complete feed has to satisfy the nutritional requirements of carp. But the price differs not only with respect to the components but also with the different manufacturing processes. Today compressed, extruded and expanded feeds are available on the market. Expanded feeds (with a manufacturing process comparable to that of popcorn) are buoyant, i.e., able to float on the surface. This feed is only applicable for garden ponds, and is not practical for fish farms, because swimming feed is more easily eaten by birds than by fish in the pond. Compressed feeds can be obtained by using a simpler production process which therefore leads to lower production costs. The possible fat content does not exceed 12% due to the small amount of hollow space in the highly compressed stuff. The water stability of compressed feed pellets is low. Hence, it is necessary to apply the feed with automated feeders. Due to its tendency for abrasive wear the movement of feed should be minimized. That means that the number of reloads or the amount of transportation with flat conveyors or elevators should be as low as possible.

In carp farms food conversion efficiencies (FCE) between 1.5 and 1.8 can be reached with compressed feeds. In comparison to salmonids this is a relatively low food conversion. The reason for this is the method of food consumption by carp. Carp are sloppy feeders. During feed uptake fast swimming can be observed and crushing of feed is achieved with their pharyngeal teeth. Thereby fine particles are lost as water is pressed through the operculum due to breathing. Pelleted feeds disaggregate clearly faster than extruded feeds, and feed losses in practical pond production are larger because of the carps feeding behavior. Therefore FCE can clearly be improved with extruded feeds. This is one of the reasons why the use of the very expensive extruded mixed feed can be profitable even if the composition of the feed is the same (Table 6.12). One conclusion of these findings is that higher feed quality leads to 50% higher survival rates, and thus one third less carp should be stocked in order not to overstock the ponds and thus slow down fish growth.

Extruded feed has higher water stability and in praxis no abrasive wear. Due to the higher amount of hollow space a fat content of up to 30% in the feed is possible. This high fat content helps to save protein by the same way as by carbohydrates of supplementary feeds. The feed protein can be used for growth (or synthesis of fish protein) and the fat serves as energy source for all other metabolic processes.

Generally, the overwintering period of one-year-old carp is another period where it makes sense to regularly offer mixed feeds. At times when the ice cover is absent, mixed feed has to be applied to one-year-old carp. Due to higher stocking densities in wintering ponds carp do not receive enough foodstuff from natural food. Under these conditions the amount of mixed feed should be 0.5% of the fish biomass per week at maximum. Preferably, water-stable extruded feed with high energy content should be used for feeding during the winter season, since the intake of feed by carp happens very slowly at low water temperatures. Complete grains have high water stability and do not decompose if they rest on the bottom of the pond for a longer time. Nevertheless, feeding of grains in winter is not recommended, because the digestibility of carbohydrates and fat is more temperature-dependent than protein digestibility. Already at temperatures

Table 6.12. Rearing of carp fingerlings (C_1) with pelleted or extruded feeds, similar feeding rates and feed amounts. Due to the different production processes the pelleted and extruded feeds differed in their protein and fat content up to a certain extend but most importantly the digestibility of nutrients and stability in water between the two feed types is different. Therefore, extruded feeds are recommended for C1 carp. Mean of results from experimental pond station in Königswartha, Germany 1999/2000 (for more details see Füllner 2001).

Variant	1999		2000		Index extruded/ Compressed (Mean of 1999 and 2000) in %
	Pelleted	Extruded	Pelleted	Extruded	
Protein contend in feed (%)	30	30	30	36	
Fat content in feed (%)	10	20	15	18	
Replications (n)	4	8	2	2	
Stocking density [n/ha]	25,000	25,000	25,000	25,000	
Harvest [n/ha]	9,510	14,385	6,272	12,534	176
Harvest [kg/ha]	1,850	2,554	1,453	2,029	139
Final weight [g]	204	179	232	162	79
Losses [%]	64	43	75	50	67
FCR	1.60	1.25	1.80	1.35	76
WGF*	8.67	8.58	9.08	8.82	98

*WGF= Weight gain factor = $\frac{\text{Weight gain}}{\text{Initial weight}}$

beneath 20°C the digestibility of carbohydrates and fat decreases sharply, whereas it stays almost constant with decreasing winter-like temperatures for protein (Schade 1982).

Feeding Technique for Mixed Feeds

Mixed feeds should generally be fed only when the nutritional status of the carp makes it a necessity. If there is a sufficient amount of usable natural food, feeding with mixed feed should be avoided. For this reason the monitoring of natural food should precede any application of mixed feeds in carp ponds. Furthermore, it is necessary to control the most important environmental parameters such as water temperature, dissolved oxygen content and pH. Feeding of mixed feed is associated with an input of nutrients into the water, especially phosphorus (P) and nitrogen (N). Therefore, additional fertilization of ponds should be avoided during the period of mixed feed application. The necessary P-supply is already provided through feeding and additional nutrients can increase the growth of phytoplankton. Decomposition of high phytoplankton densities can lead to oxygen deficits which can lead to losses of fish.

Pelleted mixed feeds have to be applied solely by automated feeders because of their short-term water stability. The use of automated feeders can minimize feed losses and reduces labor. Mixed feed should not be sprinkled on a part or the whole pond area with the intention to apply all feed evenly distributed, such that all fish will have the same chance to acquire feed and swimming effort of carp can be minimized. When feed is applied to a wide area of the pond, carp would have to search longer to find the wide spread feed particles. This certainly requires more energy for swimming, and even more importantly requires time. In this time the feed is likely to be used by other predators like water birds, and in addition water-sensitive feeds will decompose within a few minutes. Easy constructed feed dispensers (or self-feeders) are optimal for use in carp ponds (Fig. 6.14). Though not quite as suitable are timed feeders, such as most common belt feeders.

The amount of mixed feed provided depends on the composition of the feed. Pelleted feeds with low crude protein (below 30%) and low fat content can be fed *ad libitum* with self-feeders (pendulum feeder or demand fish feeder) if the temperature and the oxygen content are high and the pH is in normal ranges.

Balanced feeds with higher protein content or higher energy have to be applied in strict adherence to the feeding tables of the producer. These feeding tables consider the most important environmental conditions and the average body mass of the fish. Smaller carp need a higher percentage of feed per kilogram biomass than bigger ones. But a daily ration of more than 5% of the current stocking biomass cannot be fed even under optimal

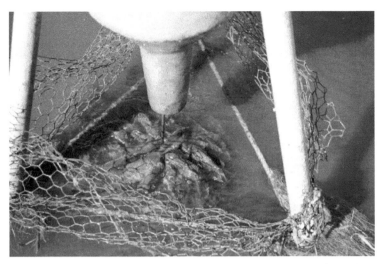

Figure 6.14. Carp fingerlings use a feed dispenser in the experimental pond station in Königswartha, Germany.

environmental conditions. If the daily ration rises to more than 5%, food conversion decreases dramatically. Under such conditions carp waste the feed by sloppy feeding, they do not absorb the total amount of feed and feed digestibility also declines.

Under unfavorable environmental conditions, the amount of feed has to be reduced. For the same reason the filling of the self-feeders should be done late in the morning, after the oxygen content has risen due to photosynthetic activity of algae. On sunny days this should be done past 10 a.m. Generally the feed should be consumed rapidly. Nevertheless, if carp consume the feed entirely it is not a reason to increase the dose of feed.

The modification of the amount of feed should follow a new assessment of the fish biomass. This should be done every 10 to 14 days. To determine the fish biomass an adequate amount of fish has to be caught (more than 50, although it is better to have a 100 specimens). With regard to an estimated percentage of losses it is possible to calculate the biomass precise enough to properly adjust the amount of food.

As has been shown, farming of carp with both supplementary feeds as well as mixed feeds is always possible. Carp production with mixed feeds is expensive. For this reason mixed feeds should be given to carp only for special purposes. The best and recommendable kind of carp farming is the production in ponds with a medium stocking density and feeding fish with good value supplementary feed. This is not only economically recommended, it also conserves the globally limited resources of fish meal.

References

Albrecht, M.L. and B. Breitsprecher. 1969. Untersuchungen über die chemische Zusammensetzung von Fischnährtieren und Fischfuttermitteln. Z. Fischerei N.F. 17: 143–163 (in German).

Amlacher, E. 1960. Das Verhalten der inneren Organe und der Muskulatur dreisömmriger Karpfen aus dem Teich des Dresdner Zwingers bei extremer Kohlenhydratfütterung. 2. Teil. Quanititative Fettbestimmung an Muskulatur, Leber und Niere; Histologie der Haut und Muskulatur; Fettdepots in der Leibeshöhle. Z. Fischerei N.F. 9: 749–762 (in German).

Balon, E.K. 2004. About the oldest domesticates among fishes. Journal of fish Biology 65: 1–27.

Ćirković, M., D. Trbović, D. Ljubojević and V. Đorđević. 2011. Meat quality of fish farmed in polyculture in carp ponds in Republic of Serbia. tehnologijamesa. UDK: 39.2.04:637.56.04(497.11). 106–121.

Cochrane, K., C. De Young, D. Soto and T. Bahri. 2009. Climate change implications for fisheries and aquaculture: overview of current scientific knowledge. FAO Fisheries and Aquaculture Technical Paper. No. 530 Rome.

EC 2005. Regulation (EC) No. 183/2005 of the European Parliament and of the Council of 12 January 2005 laying down requirements for feed hygiene. Official Journal of the European Union 48: 1–22.

Eckhardt, O., K. Becker and K.D. Günther. 1983. Zum Protein- und Energiebedarf wachsender Karpfen (Cyprinus carpio L.). IV. Ableitung des Bedarfs an verfügbarem Protein und an umsetzbarer Energie, wenn hohe Wachstumsraten realisiert werden sollen. Z. Tierphysiol., Tierernähr. u. Futtermittelkd. 49(4/5): 260–265 (in German).

Epler, P., F. Borowiec, Sokołowska-Mikołajczyk and M. Fałowska, B. 2010. Content of basic nutrients, amino acids and fatty acids in the benthos of carp ponds. AACL Bioflux 3(2): 125–131.

Food and Agriculture Organization (FAO) of the United Nations 2013. Fish Stat J. Release: 2.0.0. Dataset Global aquaculture production—Quantity (1950–2011). FAO Rome.

Füllner, G. 1987. Untersuchungen zum Einsatz von Fütterungsautomaten zur Optimierung der Fütterung in der Karpfenteichwirtschaft. Dissertation Humboldt University, Berlin, Germany (in German).

Füllner, G. 2001. Pelletiertes oder extrudiertes Mischfutter für die Fütterung von einsömmrigen Karpfen in der Teichwirtschaft? Fischer & Teichwirt 52(10): 374–376 (in German).

Hidalgo, M.C., E. Urea and A. Sanz. 1999. Comparative study of digestive enzymes in fish with different nutritional habits. Proteoloytic and amylase activities. Aquaculture 170: 267–283.

Horvath, L., G. Tamas and C. Seagrave. 1992. Carp and pond fish culture. Fishing New Books, Oxford U.K., 158 pp.

Ivlev, V.S. 1961. Experimental Ecology of the Feeding of Fishes. New Haven. CT. Yale University Press USA, 302 pp.

Jančarik, A. 1964. Die Verdauung der Hauptnährstoffe beim Karpfen. Z. Fischerei N.F. 12: 601–684 (in German).

Kaushik, S.J. 1995. Nutrient requirements, supply and utilization in the context of carp culture. Aquaculture 129: 225–241.

Khan, T.A. 2003. Dietary studies on exotic carp (Cyprinus carpio L.) from two lakes of western Victoria, Australia: Aquat. Sci. 65: 272–286.

Koch, H.L. 1984. Untersuchungen zur Nährstoff- und Energieverwertung junger wachsender Spiegelkarpfen (Cyprinus carpio L.) bei unterschiedlichen Proteinquellen und –gehalten in gereinigten Versuchsdiäten. Dissertation Georg-August-Universität Göttingen, Germany, 157 pp. (in German).

Merla, G. 1960. Beiträge zur Kenntnis des Wachstums und der Ernährung des Karpfens (Cyprinus carpio L.). Z. Fischerei N.F. 9: 659–734 (in German).

Michaels, V.K. 1988. Carp Farming. Fishing New Books Ltd., Farnham, Surrey, U.K., 207 pp.

Mondal, S., T. Roy, S. Sen and A.K. Ray. 2008. Distribution of enzyme-producing bacteria in the digestive tracts of some freshwater fish. Acta Ichthyologicaet Piscatoria 38: 1–8.

Müller, W. 1966. Zwei häufige Formen schlechter Kondition bei Satzkarpfen. Deutsche Fischerei Zeitung 13: 47–49.

Nehring, K. (ed.). 1970. Futtermitteltabellenwerk. Berlin 1970, 2. Aufl. 1972 (in German).

Nikolski, G.W. 1957. Spezielle Fischkunde. VEB Deutscher Verlag der Wissenschaften Berlin, 632 pp.

Nose, T., D. Lee and Y. Hashimoto. 1974. A note on amino acids essential for growth of young carp. Bull. Jap. Soc. Sci. Fisheries 40: 903–908.

Ogino, C., J.Y. Chiou and T. Takeuchi. 1976. Protein nutrition in fish. VI. Effects of dietary energy sources on the utilization of proteins on rainbow trout and carp. Bull. Jap. Soc. Sci. Fisheries 42: 213–218.

Pfeifer, M. and G. Füllner. 2008. Einfluss der Fütterung unterschiedlicher pflanzlicher Futtermittel auf die Produktqualität von Speisekarpfen. Rundschau für Fleischhygiene und Lebensmittelüberwachung 60(4): 146–152 (in German).

Schade, R. 1982. Untersuchungen zur Nahrungsausnutzung im Darm von Karpfen (*Cyprinus carpio* L.). Arch. Hydrobiol. Suppl. 59: 377–415 (in German).

Steffens, W. 2008. Der Karpfen. Westarp Wissenschaften-Verlagsgesellschaft mbH Hohenwarsleben, Germany, 228 pp. (in German).

Takeuchi, T., E. Watanabe and C. Ogino. 1979. Availability of carbohydrate and lipid as dietary energy sources for carp. Bull. Jap. Soc. Sci. Fisheries 45: 977–982.

Tucker, J.K., F.A. Cronin, D.W. Soergel and C.H. Theiling. 1996. Predation on zebra mussels (*Dreissena polymorpha*) by common carp (*Cyprinus carpio*). Journal of Freshwater Ecology 11(3): 363–372.

Vinberg, G.G. 1956. Rate of metabolism and food requirements of fishes. Fish. Res. Bd. Canada 194.

Vogel, P. 1905. Dritter Band zum Ausführlichen Lehrbuch der Teichwirtschaft. E. Hübner Bautzen, Germany, 902 pp. (in German).

Walter, E. 1934. Grundlagen der allgemeinen fischereilichen Produktionslehre. pp. 483–661. *In*: Demoll-Maier: Handbuch für die Binnenfischerei Mitteleuropas Band IV, Lieferung 4-5. E.Schweizerbart'sche Verlagsbuchhandlung (Erwin Nägele) G.m.b.H. Stuttgart, Germany (in German).

Part III

Diseases and Immune Response to Carp

Carp in the wild as well as in aquaculture facilities are constantly exposed to a variety of environmental stressors. The reader has already learned how different abiotic variables such as temperature and oxygen content affect carp growth and well-being. This part of the book introduces the biological agents that can cause harm to carp. Chapter 7 (disease agents of carp) provides a detailed but accessible account for all the virus, bacteria and eukaryotic single and multi-cell organisms that cause disease in carp. Many of these disease agents have significant economic effects by causing mortality or loss of growth in aquaculture. Others mainly affect carp in the wild and can play an important role for individual and population growth of carp also in natural systems. When available, the author presents topical information on therapies or preventive measures. The most effective measure against any disease agent in carp is the individual's own immune system which is discussed in detail in Chapter 8. The authors present a structured account of carp's lines of defense against pathogens ranging from the mode of entry of pathogens to their recognition, subsequent inflammatory responses and the detailed processes involved in the cellular and a-cellular immune response. The cross-linkage between Chapters 7 and 8 is exemplified by the study of the complexity of parasitic diseases and the immune response against them.

7

Disease Agents and Parasites of Carp

Jasminca Behrmann-Godel

Parasites and pathogens are a ubiquitous part of the environment of every organism. Hereafter called parasites, these invasive species themselves represent a full range of taxa, including viruses, bacteria, protists and several metazoan groups including cestodes and crustaceans. Some parasites develop directly on or in a host, others have extremely complex life cycles, requiring several sequential obligate life cycle hosts in order to complete their development. The infobox for this chapter provides definitions of the important parasitological terms used in this chapter. Ecologically speaking, host organisms can be seen as the environment for the parasite providing a number of distinct habitats for parasites to colonize (Bush et al. 1997). Most parasite species occupy specific tissues or organs on or within their host, and there may often be a whole community of different parasite species colonizing a single host individual. However, the parasite community of one host species may differ between localities and geographical regions because in addition to the immediate surroundings of the host's body, parasites are also influenced by abiotic parameters of the wider environment, such as the temperature, salinity or pH of the wider environment in which the host lives. Furthermore, many parasite species have free-living stages or depend on the occurrence of specific intermediate hosts or vectors (see infobox) that may vary in abundance or occurrence from place to place.

There are numerous ways in which parasites can infect their hosts, and most are particular to species and life cycle stage. Parasites may infect their host directly, for example by penetrating the host's skin, like trematode

Limnological Institute, University of Konstanz, Mainaustrasse 252, 78457 Konstanz.
Email: Jasminca.Behrmann@uni-konstanz.de

Infobox: Parasitological terminology

Parasites are organisms that benefit at the expense of another organism belonging to another species called the **host**, mostly by trophic exploitation. **Endoparasites** live inside the host, **ectoparasites** on the host body surface. The **microparasites** are unicellular organisms such as bacteria and protozoans. Viruses may also be called microparasites. **Macroparasites** are metazoan forms, including flatworms, nematodes, arthropods and several others.

Life strategies of parasites:

Parasites that depend on a host for development are called **obligate parasites**. In contrast, a **facultative parasite** can complete its life cycle without infecting a host. Organisms that are not typically parasitic but can become so under certain specific conditions (such as immune deficiency of the host) are known as **opportunistic parasites**.

Parasite life cycles:

A **direct (monoxen) life cycle** can be found in parasites that exploit only one host for development. An **indirect or complex (heteroxen) life cycle** incorporates several hosts, which are required for development of the parasite. Sexual maturity and reproduction always take place in the **final (definitive, primary) host**. One or several **intermediate (secondary) hosts** are needed for interim phases of development, which may include asexual reproduction or metamorphosis to the next developmental stage. If the intermediate host is used as a carrier to the next host it is typically called a **vector**. In terrestrial systems, most vectors are blood feeding arthropods such as insects like mosquitoes, they transmit parasites (such as *Plasmodium* sp. which causes malaria) to the next host of the parasite's life cycle. Typical vectors in the aquatic environment (especially for fish) are blood-feeding arthropods or annelids like leeches. Some parasites may infect a host but do not undergo development. These hosts are called **paratenic (transport)** hosts, they can be used for dispersal or to reach the next trophic level, thereby raising the parasite's chance of being transmitted to the next host in the life cycle.

---

cercariae. Other infection routes may involve active transmission by a vector or indirect infection through the food chain, either as a result of direct predation on the parasite or when an infected prey organism is consumed by a host.

Parasitic organisms are thought to have developed from primarily non-parasitic, free-living forms, in which natural selection favored specific adaptations for a parasitic way of life. The major benefits of a parasitic lifestyle are transportation to new areas, transmission to new life cycle hosts, shelter, readily accessible food resources, the relatively stable environment provided by the host, and the minimization of the energetic costs of foraging. However there are also disadvantages for parasites, the most troublesome being the immune response of the host to invasion of its body tissues. The exhausting interaction in which both protagonists struggle to adapt to one another is known as the 'parasite-host arms race' and can be seen as the basis for constant co-evolution of both protagonists. Parasites must continually try to overcome the host´s immune attacks by evolving new infection, evasion- or immune modulation strategies, whereas the host tries endlessly to overcome these and either fight off the parasite or develop a tolerance to it (see more details in the next chapter). Under normal circumstances this never-ending battle between the host and parasite can be considerably inconspicuous. As a general rule, parasites do not kill or even seriously harm their hosts, but there may be situations where the balance is disturbed. From a human perspective, the results of a parasite gaining even temporary advantage are usually unwelcome—for example the headache, snuffles and fever associated with an influenza virus triggering our own immune system, or the sight of a pond full of moribund koi, struck down by spring viraemia following infection by *Rhabdovirus carpio*.

Anthropogenic environmental change is increasingly implicated in outbreaks of parasite disease. Factors such as increasing temperature, environmental pollution or the introduction of new species can all unbalance parasite-host interactions. The cumulative effect of multiple environmental stressors maybe particularly damaging, with negative implications for immune function and animal health (Vidal-Martínez et al. 2010, Marcogliese and Pietrock 2011). For example, a rise in temperature might prolong the period in which transmission of particular parasites can take place, allowing them to develop more infective stages or reduce the time required to complete the life cycle, thereby increasing abundance and virulence (determined as the pathogenicity of a parasite or its ability to cause disease) (Aho et al. 1982, Marcogliese 2001, Hakalahti et al. 2006, Poulin 2006, Paull and Johnson 2011). Rising temperatures may also favor alien parasites, removing a temperature barrier and allowing them to enter new systems either directly or as passengers of non-native intermediate hosts. Introduced parasites may cause epizootic outbreaks in naïve host populations that lack adaptations to reduce pathogenicity or to defend against the invaders (Marcogliese 2001, Britton et al. 2011, Behrmann-Godel et al. 2014).

The spread of parasitic disease is a major concern in modern aquaculture. Aquaculture is the world's fastest growing food production sector, with carp (including common carp, grass carp and silver carp) being the most widely consumed type of cultured fish (Naylor et al. 2000). Parasitic disease is a primary constraint on the culture of many aquatic species, and the economic costs of disease control and research are immense (Bondad-Reantaso et al. 2005). The intensification of aquaculture, the expansion of the ornamental fish trade, the increasing globalization of the industry including the transport of aquatic organisms and their products between countries and continents, and the intensive stocking of marine and freshwater habitats with hatchery-reared animals all help facilitate the spread of diseases worldwide.

Parasites of Common Carp

The widespread distribution and adaptability of the common carp, not to mention the attention resulting from its commercial importance, mean that a vast number of associated parasitic species have been identified. The most comprehensive list compiled so far includes 310 parasite species, including representatives of all major taxonomic phyla ranging from myxozoa to arachnidae (Baruš et al. 2002). Like other host organisms, individual carp may be infected by a whole range of parasitic species, known collectively as the parasite infra community. Guegan and Kennedy (1993) have found a clear relationship between the length of time a host species has been present in a region and the species richness of its associated community of helminth parasites. For common carp, therefore, we can expect the parasite community be richer in its native Asian range than in parts of its introduced range such as Europe, where it was brought in the mediaeval period.

Instead of reiterating a list of known parasites for common carp (as can be found, e.g., in Baruš et al. 2002), the focus here will be on selected examples from all major groups, including a description of the most important associated diseases in feral and farmed fish and, where available, suggestions for prevention and cure.

Viral Diseases

The majority of viral fish diseases recognized today have been identified within the last few decades, not because they are new, but because developments in molecular diagnostics have revolutionized the study of fish diseases in the last 50 years (Cunningham 2002). New tools such as the amplification of nucleid acids via Polymerase Chain Reaction (PCR) and subsequent sequencing of informative genes, restriction enzyme digestion and probe hybridization have facilitated research into many previously mysterious viral fish diseases worldwide.

Rhabdoviruses

Of special importance among carp viruses are *Rhabdovirus carpio* (Fijan 1972), the agent of Spring Viraemia of Carp (SVC) and certain fish herpesviruses, which result in significant economic losses in aquaculture and increasingly also in wild fish (Jeney and Jeney 1995, Lepa and Siwicki 2012).

SVC can occur if spring temperature rises above 7°C, causing serious mortality in carp of all ages. Maximum mortality occurs between 10–15°C but the virus may survive at temperatures up to 23°C (Fijan 1999). It can infect all varieties of *Cyprinus carpio* (mirror-, leather-, koi-, grass-, silver-, bighead- and crucian carp and goldfish) as well as other species including pike (*Esox Lucius*), tench (*Tinca tinca*), roach (*Rutilus rutilus*) and wels catfish (*Siluris glanis*). Clinical signs of disease are somewhat unspecific and can be further confused by accompanying bacterial infections; they include lethargy, dark body coloration, decreased ventilation rate, exophthalmia (pop-eye), dropsy and bleeding from gills, fins and internal organs. SVC is transmitted via feces but it has been shown that blood sucking parasites such as leeches (*Piscicola geometra*) and the carp louse (*Argulus foliaceus*) can act as vectors for transmission to uninfected fish (Ahne 1985, Ahne et al. 2002). Suspected cases of SVC should be properly diagnosed by an appropriate laboratory. Unambiguous identification can be achieved either by direct PCR-based methods such as sequencing of certain genes or by indirect methods such as ELISA, which uses antibodies and a color change as a diagnostic tool to detect an infecting agent (Rodák et al. 1993, Koutná et al. 2003). Once the disease is properly diagnosed, control is best achieved by maintaining the water temperature above 20°C (Fijan 1999). Currently no antiviral drugs are available to treat SVC, but recent research has found promising results in the development of a vaccine (Dhar et al. 2014, Kanellos et al. 2006, Ahne et al. 2002).

Herpesviruses

The herpesviruses are a group of highly host-specific large DNA viruses, of which three recognized taxa are known to infect common carp including koi and goldfish. The carp pox herpesvirus CyHV-1 causes a disease known as carp pox (CHV), which typically causes white or gray ulcers in the epithelium of common carp and koi. Cyprinid herpesvirus 2(CyHV-2) causes hematopoietic necrosis in goldfish, and koi herpesvirus CyHV-3 (KHV), affects carp and its varieties including koi and ghost carp (Lepa and Siwicki 2012).

KHV, Fig. 7.1 was first reported in 1998 by Ariav et al. (1998), but is now endemic in Eurasia, Africa and North America (Pokorova et al. 2005), where it is responsible for enormous economic losses in carp industries. Problems with KHV are increasingly reported from wild carp populations where infection can cause mass mortality events such as recently documented in Canada (Garver et al. 2010). Outbreaks of KHV occur seasonally when the water temperature rises above 13°C. At lower temperatures, the virus remains dormant in experimentally infected fish, but causes rapid mortality within 7–12 days following a temperature shift to 23°C (Gilad et al. 2004). Afflicted fish are lethargic and exhibit pale patches on the skin. The gills initially appear pale, with irregular coloration and eventually exhibit epithelium necrosis. Eyeballs are sunken (enopthalmus) (Fig. 7.1a) and mucus production is increased. Fry and juveniles are also affected and infection can result in high rates of mortality (Hedrik et al. 2000, Pokorova et al. 2005). The virus is transmitted horizontally, via viral particles excreted into the water along with feces. There is evidence that the virus can persist as a latent infection in the host and thus be spread even by seemingly healthy fish, then activated by temperature stress (Gilad et al. 2004, Eide et al. 2011). As with SVC, the most efficient methods for diagnosing KHV are based on PCR or ELISA testing (Pokorova et al. 2005).

The severe threat posed by KHV to global carp production has resulted in intensive research and the development of efficient methodologies for prevention and treatment (see Pokorova et al. 2005 and citations herein). Shapira et al. (2002) have identified potentially KHV-resistant local carp

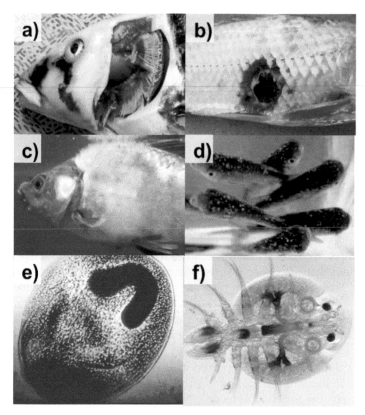

Figure 7.1. a. Koi infected with KHV showing irregular gill colour, necrotic tissue and enopththalmus (photo: kindly provided by Dirk Kleinfeld); b. Ulcer on fish body caused by an *Aeromonas* sp. infection (courtesy of the American Fischeries Society, AFS); c. Carp with *Aeromonas* sp. Infection (foto: http://www.gronau.net/Teich/Koi-Krankheiten.htm); d. grass carp fingerlings with white spot, *Ichthyophtiriusmultifilis* (courtesy of the AFS); e. microscopic image of *Ichthyophtiriusmultifilis*, showing the horseshoe-shaped macronucleus (courtesy of the AFS); f. The carp louse *Argulusfoliaceus* (photo: Behrmann-Godel).

lineages and tried to select for this trait in crossbreeding experiments. Ronen et al. (2003) took an induced resistance approach, exposing carp to the virus for 3–5 days at 23°C, then increasing the water temperature to 30°C. Following this procedure, high levels of specific antibodies were found in the sera of carp, indicating resistance to KHV. Perelberg et al. (2003) found that carp that survive at least two natural outbreaks of KHV became resistant to the disease, and went on to develop an inactivated vaccine (Perelberg et al. 2008). However the protective effects were short-lived and treated fish required regular repeat immunizations. The newest approach is the development of a promising DNA vaccine by Zhou et al. (2013), which may offer long-term protection against KHV.

Bacterial Diseases

Vast numbers of bacteria are associated with fish diseases, but, not all of the species involved are primarily fish pathogens. Many are opportunistic pathogens that only infect the host when it is already weakened or immunologically compromised. Most bacterial diseases typically appear as secondary infections in fish that are already stressed or carrying other infections. As with viral diseases, research into bacterial disease has greatly benefited from recent technical developments in molecular genetics.

Flexibacter

One of the most widespread bacterial diseases in both wild and farmed fish is columnaris disease, caused by *Flavobacterium columnare*. First described in 1922 by Davis in fishes from the Mississippi River, the bacterium can infect a wide range of freshwater fishes and causes severe losses in fish farms worldwide. *F. columnare* is a component of the bacterial microbiota of freshwater fish, fish eggs and water. It can survive and remain infective for several months outside its host, in pond sediment or even in sterile water.

    F. columnare occur as both acute and chronic infections, typically affecting the gills, skin and fins of the host. The severity of infection and the progress of the disease depend on the virulence of the bacterial strain involved, and this may vary considerably. Only fish infected with milder strains show external signs of disease, as the more virulent strains tend to be fatal within a few days, before any external symptoms of columnaris become apparent (Declercq et al. 2013). The first signs of chronic columnaris are small lesions on the fish's body, starting as pale discolorations of the skin. These are typically surrounded by a reddish zone. These usually first appear at the proximal basis of the dorsal fin, and progress to the distal edge, contrasting with 'normal' fin rot in which the damage starts at the outer edge of the fin and progresses towards the base. On the fish's body, the discoloration caused by columnaris progresses from the fin base in a saddle-like manner, hence the alternative name of 'saddle-back' disease (e.g., Morrison et al. 1981). As the disease advances, lesions develop on the head and tail and penetrate deeper into the skin and muscle tissue layers resulting in deep ulcers typically covered with white-yellowish mucus. Often secondary infections with other bacteria and/or fungi aggravate the condition. Discoloration can also be seen on the gills, where yellowish-white-areas of degeneration indicate necrosis, which may cause respiratory distress.

    As with viral diseases, bacterial infections such as columnaris are best diagnosed using PCR methods (see, e.g., Darwish et al. 2004). The advantage of PCR lies in the ability to detect *F. columnare* at very low levels, in tissues such as gills or skin, thereby allowing to prevention and/or control measures to be initiated early. As with many other diseases, the best prevention is to reduce the stocking density in aquaculture facilities or rearing ponds, especially during periods of elevated water temperature. Declercq et al. (2013) suggest that since reducing rearing densities may also provide fewer opportunities for the transmission of ectoparasites and penetrating endoparasites, it may be an efficient general tool in the ecological management of fish disease.

Other prevention methods have been described, including treatment with oral antibiotics such as oxytetracycline, which can be administered along with the feed (Thomas-Jinu and Goodwin 2004), and vaccination with bacteria killed by exposure to heat or formalin (Ransom 1975, Fujihara and Nakatani 1971). Orally delivered oxytetracycline has proven an effective cure in both the early and advanced stages of disease outbreaks (Bullock et al. 1986). A very interesting new approach to columnaris treatment may be the use of *F. columnare*-specific phage lysates. Prasad et al. (2011) reported 100% survival of phage-treated walking catfish *Clarius batrachus* in experimental infection trials.

Aeromonas

Several bacteria from the *Aeromonas* species complex are well known pathogens of fish, and many strains new to science have been identified in the last 10 years (Kozinska et al. 2002). *Aeromonas* species normally occur in fresh, brackish and coastal waters and as opportunistic members of the intestinal microflora of many fish species, where they are usually tolerated by healthy individuals. However, in immunocompromised hosts, secondary infection with *Aeromonas* can be pathogenic. The best studied species are *Aeromonas salmonicida*, *A. hydrophila* and *A. sobria*, and all three have the potential to cause sub-acute to chronic disease in carp and other cyprinids (and salmonids in the case of *A. salmonicida*). *A. salmonicida*, a non-motile aeromonad, is the cause of carp erythrodermatitis (Bootsma et al. 1977). In the course of the disease, roundish ulcers form as a result of epithelial necrosis, and fish may lose their appetite. Mortality is normally low (between 20–25% according to Jeney and Jeney 1995). *A. hydrophila* and *A. sobria* are motile aeromonads. Infections occur mainly in spring and summer when the temperatures rise. Fish showing signs of *A. hydrophila* or *A. sobria* infection are often either co-infected with other pathogenic bacteria, suffering from injuries or living under stressed conditions such as intensive aquacultural facilities (Jeney and Jeney 1995). Symptoms of acute *A. hydrophila* or *A. sobria* infection appear as visible damage on the body surface, such as fin or tail rot, large ulcers and skin damage, and bleeding at the base of fins, the anal vent and the gills (Fig. 7.1b). Systemic (whole body) infection results in severe symptoms including damage to all organs, dropsy and death. Although motile aeromonads can be found almost everywhere, they are much more abundant in polluted water and especially in water with a high organic load, as is often found in densely stocked aquaculture ponds. Under such conditions, fish are continually exposed to high numbers of bacteria. Low oxygen levels, high fish densities, high temperatures and other stress factors increase the likelihood of annual outbreaks of the disease, mostly in spring. Below 4°C, *Aeromonas* spp. are thought to be inactive. However activity and infection rates rise exponentially with temperature increases. At temperatures over 12°C, the fish begin to build immunity to infection, but under stressed conditions this immune response may fail. The best protection against *Aeromonas* infections within cultured carp populations is thus a reduction of stress. As for columnaris, the recommended treatment is antibiotics delivered in fish feed, but Jeney and Jeney (1995) question such treatments. Loss of appetite in sick fish may limit the uptake of the antibiotic medication and any benefit should be weighed against the risk of antibiotic resistance developing in bacteria.

Fungal Diseases

Fungal infections can lead to significant losses in both free-living and cultured fish. Most fungal diseases of fish appear as secondary infections in hosts with pre-existing injuries or bacterial infections. The fungi can enter the body via wounds or ulcers (such as those caused by bacteria like *Aeromona* ssp. described above) and proliferate quickly, often either causing or significantly hastening death. The major pathogenic fungi (known as watermolds) of concern for carp include *Saprolegnia* spp. (see below), *Achlya* spp. and *Branchiomyces* spp., which cause necrosis of the gill filaments (gill rot).

Warm water fish hatcheries are particularly susceptible to watermold disease, and of importance for carp are members of the family Saprolegniaceae, especially the genus *Saprolegnia*. Dead or unfertilized fish eggs are very quickly infected by fungi, and left to proliferate, these quickly threaten healthy eggs and fry. The resulting economic losses in intensive aquaculture can be severe. Typical signs of a *Saprolegnia* infection (saprolegniasis) are cotton-wool-like white colonies of the fungus on the integument of the fish (Fig. 7.1c). Fungal hyphae penetrate the epidermis and the dermis, causing degenerative changes including edema and, eventually sloughing of the skin. Left untreated, infected fish die of osmoregulatory failure. Disinfection measures should be taken as soon the first signs of saprolegniasis are detected. It should be noted that the use of malachite green to treat fungal and other ectoparasitic diseases such as ichthyophthiriosis (see chapter about Protozoa) is no longer recommended, or in fact permitted, in fish farms within the European Union (EU). As an alternative, therapeutic baths with NaCl, formaldehyde or kaliumpermanganate are recommended. Other suggested therapeutics to cure saprolegniasis are very similar to those for columnaris disease (Oláh and Farkas 1987).

Other fungal species commonly implicated in carp disease are *Aspergillus*, *Penicillium* and *Fusarium*, which can be ingested with contaminated food. Several pathogenic fungi are known to produce mycotoxins (e.g., aflatoxin from *Penicillium* sp.) that can be detrimental to a wide variety of animals including fish, and the effects of mycotoxins are reviewed in Chapter 13 by C. Pietsch of this volume. However difficulties in detecting and identifying specific mycotoxins mean it is often impossible to prove a mycotoxicological cause for disease, and the picture is further clouded by the fact that fungi may be present without producing any toxins (reviewed in Kumar et al. 2013).

Diseases Caused by Protozoa

The protozoans infecting fish include both ecto- and endoparasites. Among the endoparasitic species, blood flagellates of the genera *Trypanosoma*, *Cryptobia* and *Trypanoplasma* spp. are of special importance and the most troublesome with respect to carp is the trypanosome *Trypanoplasma borelli*. The ectoparasitic protozoans fall into three broad types. Firstly there are the opportunists with only limited persistence as parasites. These are typically secondary infectors that parasitize mainly stressed or immune-depressed fish. Examples include ciliates of the genera *Ophryoglena*, *Tetrahymena*, *Hemiophrys* and *Glaucoma*. Secondly there are the ubiquitous facultative parasites lacking host or site preference, such as the flagellate *Ichthyobodo necator* (formerly *Costia necatrix*), ciliates of the genus *Chilodonella*, *Ichthyophtirius multifilis*, *Crypotcaryon irritans*, several ubiquitous members of the *Trichodina* and *Tripartiella* species complexes such as *Trichodina reticulata* which infects the skin of carp, and parasitic dinoflagellates of the genus *Oodinium* (*Piscinoodinium*). Thirdly, there are numerous specialized facultative parasites exhibiting limited host specificity but with a pronounced preference for a specific infection site. These include the highly specialized trichodines and several *Tripartiella* spp., all of which parasitize the gills of fish (Paperna 1991).

One of the most common protozoan parasites is the ciliate *Ichthyophtirius multifilis*. Outbreaks can develop into devastating epidemics in intensive fish farms and ornamental fish facilities, often resulting in total stock losses. Commonly known as 'Ich' or 'whitespot' the disease is typified by visible white spots on the body surface, and sometimes also on the gills (Fig. 7.1d). Each white spot represents an individual parasite in the trophont stage, during which the ciliate resides in the host's epidermis where it feeds on skin cells and body fluids and develops to maturity. It then leaves the host and falls to the bottom where it transforms into a round reproductive cyst, the tomont. The tomont undergoes repeated divisions until it has divided into 100 to 2000 ciliated, free-swimming infective stages known as theronts. The theronts search actively for a new host and on contact with a suitable fish (Paperna (1991) suggested that almost any freshwater teleost species will do) the theront penetrates the skin, burrows into the epidermis and transforms into the feeding trophont. The time required to close the life cycle is temperature dependent, with generation time decreasing with rising water temperature. At 20°C, the entire life cycle can be completed in just 8 days. In warm water aquaculture conditions or in closed systems such as aquaria, *I. multifilis* can cause 100% mortality within a very short time. Often the first sign of an *I. multifilis* infection is so called 'flashing' of fish. Infected individuals turn onto their sides and rub themselves on the substratum in an attempt to rid themselves of the parasites. Sunlight reflected off their upturned silvery flanks is seen as flashing in the water. However similar scratching behavior can also be a sign of other skin irritations, so care should be taken with diagnosis. The best approach is to search the skin and gills for the characteristic white spots and examine a skin smear under a microscope. In case of infection this will reveal individual ciliates (up to 1 mm in diameter) often rotating slowly due to the beating of their cilia. The nucleus of the trophont stage has a characteristic horseshoe-shaped macronucleus (Fig. 7.1e). Ichthyophthirios is one of the most troublesome parasitic diseases in cold- and warm water fish cultures of carp, eels and catfish (Paperna 1991), and among the most difficult to treat. Fish may sustain low enzootic infections and encysted tomonts may survive in the habitat for a long time. Stressed host fish kept at high densities during rising spring temperatures provide optimal conditions for the parasite, and increase the risk of epizootic outbreaks that result in vast losses in a very short period of time. However spontaneous recovery has been reported, both in natural fish populations and in fish holding facilities. Clearly some fish species are capable of mounting an effective defense against *I. multifilis,* including humoral and non-humoral immune reactions (reviewed in Abowei et al. 2011). In carp, limited immunity was achieved following a controlled exposure to *I. multifilis* tomonts, and fish subsequently survived high doses of the parasite. However this immunity was not stable and was suppressed following the application of corticosteroids (Abowei et al. 2011).

Diseases Caused by Myxozoa

The Myxozoa are spore-forming parasites that infect freshwater and marine fishes. Previously classified as protozoans, recent molecular analyses and observations of specific functional specializations and multicellular states have required their reclassification a separate phylum of metazoans including more than 2100 described species to date (Yokoyama et al. 2012). Most members of the Myxozoa are not harmful to fish but a few individual species are highly fish-pathogenic. A common and economically important disease of carp in intensive aquaculture is swimbladder inflammation (SBI), caused by *Sphaerospora dykovae* (Dyková and Lom 1988). *S. dykovae* belongs to the class Myxosporea, a group of myozoan parasites that form distinctive infective spores easily identifiable under a microscope by their polar capsules containing coiled filaments. A typical myxosporean life cycle includes two obligate hosts, an annelid (oligochaetes in freshwater and polychaetes in marine water) and a vertebrate, typically a fish (Fig. 7.2). In the latter, the typical myxosporean spores develop by sporogony and are released into the water. They are ingested by the annelid host, in which mature actinospores are formed via sporogony and released into the water. The actinospores infect the fish following contact with the skin, fin or gills and subsequent invasion of the sporoplast completes the life cycle.

The pathology of SBI in carp is typically divided into five stages (Jeney and Jeney 1995). During the first stage, blood vessel dilatation leads to hyperaemia and a petechial rash in the swim bladder wall. In the second

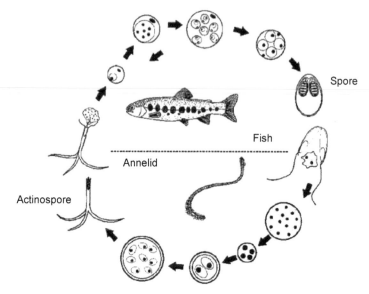

Figure 7.2. Typical myxosporean life cycle with alternating fish and annelid hosts, modified from that of Yokoyama et al. (2012).

stage, hyperaemia decreases and brown or black spots can be seen in the swimbladder wall. In the third stage the swimbladder wall is thickened by inflammation and filled with an exudate. These processes are exacerbated in the fourth stage and layers of the swimbladder wall are necrotized. In the fifth stage cysts are formed in the swim bladder and the organ is filled with serum or pus. At this stage inflammation can spread to other organs and secondary peritonitis may develop. From the third stage onwards, secondary bacterial infections may exacerbate or add to the condition. During the first and second stages of SBI there are no outwards signs of disease and the pathology described above can only be detected by post mortem analysis. It is usually in the third to fifth stages that fish suffering from the disease become noticeably sick. In the chronic form, infected fish may lose their balance and swim on their back or side, or even tail-up. The abdomen is enlarged and the caudal fins may stand out of the water as the fish tries and fails to dive. However these clinical signs typically appear in only 10–20% of infected fish and mortality is relatively rare. Most problems with SBI are in fact caused by accompanying secondary infections for example by bacteria. These can lead to severe losses in hatcheries where the disease can spread rapidly in carp fingerlings (Jeney and Jeney 1995). According to Jeney and Jeney (1995) no treatment against sphaerosporosis is available as yet, and the only effective agent for prevention is fumagillin, which will prevent the development of the disease if added to fish feed (Füllner and Müller 1992, Gomez et al. 2013).

Diseases Caused by Platyhelminthes

The platyhelminthes are a species-rich group with three parasitic classes, the trematodes, monogeneans and cestodes.

Trematoda

Most trematodes have a complex life cycle including several obligate hosts. Depending on the location of infection in the intermediate vertebrate host, the various species are commonly referred to as eye, blood, liver or lung flukes. In most species, the final host is a vertebrate, such as a fish-eating bird like a gull or a cormorant. Within this final host, the trematodes generally inhabit the intestine, where they attain maturity and reproduce sexually. They produce eggs which are voided into the environment along with the feces of the host. Free-living motile miracidia hatch from the eggs and seek a first intermediate host, usually an aquatic snail, which they infect via penetration of the skin. Inside the first intermediate host, the parasite enters its next life stage, known as the redia or sporocyst, which reproduces asexually to produce large numbers of the next infective stage, the cercaria. Cercariae are shed continuously from the snail at a rate of several hundred to several thousand per day for up to several months. The cercariae of most species are motile, and search actively for the next life cycle host, which they usually infect by penetration of the skin. Depending on the species of fluke, this next life cycle host might be a second intermediate host within which the parasites develop into metacercariae that are infective for the next intermediate host, or it might be the final host within which the adult worm develops. The number of obligate intermediate hosts varies between one and six depending on species of fluke. The life cycles of *Diplostomum spathaceum* and *Sanguinicola inermis* are depicted in Fig. 7.3.

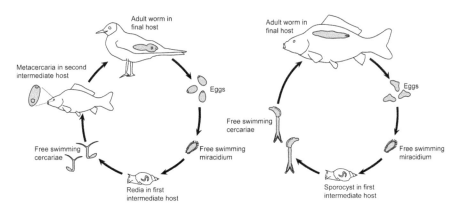

Figure 7.3. Life cycles of two trematode species (a) *Diplostomum spathaceum*, (b) *Sanguinicola inermis*.

Eye flukes belonging to the *Diplostomum* species complex infect a range of fish species including carp and all its varieties (Kennedy and Pojmanska 1996). In the second intermediate fish host, metacercariae are found in the lenses or other parts of the eyes depending on the infecting *Diplostomum* species. In severe infections, the concentration of parasites in the lens can cause cataracts and the fish may go blind, as seen in rainbow trout *Oncorhynchus mykiss* infected with *D. spathaceum* (Shariff et al. 1980). In fish fry, mass infection with cercariae of *Diplostomum* sp. leads to a disease known as diplostomiasis which can result in severe losses (Majoros 1999, Larsen et al. 2005). Another trematode causing serious disease mainly in carp fry is the blood fluke *Sanguinicola inermis.* It infects a broad range of fish hosts, including all varieties of *Cyprinus carpio* and several other cyprinids, and other species such as perch, *Perca fluviatilis,* and pike, *Esox lucius* (Kirk and Lewis 1994). The life cycle is relatively simple, including an aquatic snail intermediate host and a fish final host (Fig. 7.3b). The blood flukes cause gross histopathology in carp including mechanical damage caused by the movement of the flukes through the host's tissues and blood vessels and inflammatory reactions to encapsulated eggs. Inflammation and necrosis of the gill tissues may be a threat to cultured fish (Kirk and Lewis 1998).

Outbreaks of trematode disease in fish rearing ponds are best prevented by controlling the snail host population and by mechanical filtration to remove cercariae from circulation (Larsen et al. 2005). The antihelmintic Praziquantel (also known as Droncit) can be used to cure infected fish (Székely and Molnár 1991).

Monogenea

The monogeneans constitute a very diverse group that infect freshwater and marine fish species almost exclusively. Owing to a long history of host-parasite co-evolution, monogeneans tend to be highly host-specific, and in the wild seldom cause any detriment to the fish they infect (Chubb 1977). However under the dense rearing conditions of aquaculture and the ornamental fish trade, members of the monogenean orders Dactylogyridea and Gyrodactylidea can be seriously problematic. Paperna (1991) recorded annual losses of carp fry to *Dactylogyrus vastator* infections in breeding and nursery carp ponds during the spawning season. Unlike the trematodes, most monogeneans have a simple, direct life cycle and spend their whole life on one single host. Members of the Dactylogyridea are oviparous and infect mainly the gills of their hosts, while the viviparous Gyrodactylidea infect the skin of their host fish (Chubb 1977). A range of effective treatments exist, including Praziquantel and formaldehyde (Schmahl and Taraschewski 1987).

Cestoda

The cestodes or tapeworms constitute a diverse group of approximately 3400 parasitic species with vertebrate hosts (including 800 known species infecting teleosts). Their large size (some species can grow to several meters of length) and conspicuous nature makes cestodes perhaps the best known and most reviled of parasites, and those infecting fish are generally perceived by farmers, anglers and fishermen to be detrimental. In reality however, most tapeworms dwell in the intestine of their hosts, attached by suckers, hooks or other holdfast organs and doing nothing more harmful than absorbing nutriment from the content of the hosts gut via their own skin. Other than a generally marginal reduction in food or vitamin uptake for the host, low numbers of cestodes do little harm and in most cases are probably imperceptible to the host. In commercial terms, cestodes are not usually a serious concern, though those that reside in the host's flesh can reduce demand and profitability. However there are exceptions. Heavy infections with *Bothriocephalus acheilognathi* and *Khawia sinensis* can result in mass mortalities of carp (Hoole et al. 2011). Both species originate from Asia— *B. acheilognathi* from the Amur River (Yukhimenko 1970) and *K. sinensis* from China (Hoole et al. 2011)—and both have been introduced elsewhere along with commercial carp imports. Since arriving in Europe, *B. acheilognathi* has spread along with its carp host and now also infects more than 40 different cyprinids and some further species. In the USA, its presence is a threat to a number of endangered fish species including the woundfin minnow (*Plagopterus argentissimus*), roundtail chub (*Gila robusta*) and speckled dace (*Rhinichthys osculus*) (Hoole et al. 2011).

    B. acheilognathi infects two host species during its life cycle (Fig. 7.4). Eggs are released from adult worms that live in the intestine of the final host—a

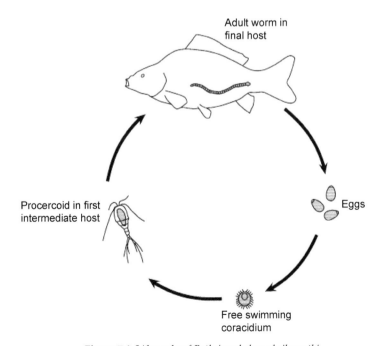

Figure 7.4. Life cycle of *Bothriocephalus acheilognathi*.

fish such as a carp. From these eggs, free-swimming oncomiracidia hatch and are consumed by the intermediate host, a copepod. The procercoid stage develops in the haemocoel of the intermediate host and infects the final host after it has ingested the copepod prey. Clinical symptoms include clumsy movements of fish, lingering at the water surface and emaciation. Parasites usually accumulate in the anterior part of the carp´s intestine where they can cause a blockage, enlargement of the abdomen or in extreme cases rupture of the intestinal wall. The holdfast organ comprises two large sucking grooves, which may engulf parts of the intestinal wall, causing local inflammation and local loss of gut epithelium, sometimes followed by a drastic host immune response. Adult carp infected with tapeworms can be treated in spring and autumn with an antihelmintic (such as Niclosamid). After treatment, ponds should be emptied, dried and treated with unslaked or chlorinated lime.

Diseases Caused by Nematodes

The vast phylum Nematoda includes a number of species that parasitize fish, all of them endoparasites. They occupy a variety of host tissues, but only a few species are known to be pathogenic to carp. These especially include members of the genus *Philometroides* such as *Philometroides cyprini* (formerly known as *P. lusii* or *P. lusiana*) in common carp and *Philometroides sanguineus* in crucian carp. Large infestations of either species may cause a disease known as philometroidosis in both cultured and wild fish (Williams et al. 2012, Moaravec and Cervinka 2005). *P. cyprini* has a complex annual life cycle (Fig. 7.5), in which the carp is the final host and various copepod species can serve as intermediate hosts. Gravid female nematodes occupy the skin under the scales of host fish and release infectious nematode L_1 larvae in spring (May/June in Europe). If ingested by a copepod, the larvae go on to develop

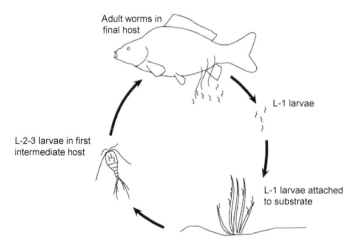

Figure 7.5. Life cycle of *Philometroides cyprini.*

into the fish infective L_3 larvae. When an infected copepod is ingested by a fish, the nematodes penetrate the gut wall and migrate into the body cavity. They aggregate near the swimbladder, gonads and kidney where they grow to 3.5 mm in length and copulate approximately one month post infection. The males then migrate to the swimbladder wall and the females to their position in the skin under the scales, where they continue to grow quickly and may reach up to 110 mm in length by spring. They produce numerous L_1 larvae of approx. 450 μm in length (Schäperclaus 1979).

   P. cyprini parasites are pathogenic to the host. Migrating juveniles cause inflammation of the swimbladder wall and hematophagic feeding habits result in mechanical damage to blood vessels. The movement of female nematodes to the integument leads to the destruction of muscle fibers and increases the risk of secondary infection, for example with *Saprolegnia* spp. For juvenile carp (1+), infection with more than three female nematodes may be fatal (Schäperclaus 1979). Infected hosts exhibit reduced growth, lethargy and abduction of scales, that becomes apparent in spring. Infected fish should not be spawned in hatcheries. The best preventative action against philometroidosis in carp is stringent separation of year classes. Spawners should be removed from the breeding ponds immediately after spawning. Carp fry should be moved from the breeding ponds to nursery ponds no later than 6 days after hatching because the L_1 nematode larvae are not infectious for carp for 7 days after release from the female (Schäperclaus 1979).

Diseases Caused by Crustaceans

The Crustacea (phylum Arthropoda) are an extremely diverse group comprising over 38,000 species. Most crustaceans are primarily aquatic and a large number have a parasitic association with fish. The majority of crustacean fish parasites are ectoparasites that feed on host blood, mucus or skin cells, many also parasitize fish gills. Pathology therefore tends to include mechanical disruption of the epidermis at attachment and feeding sites, often resulting in osmoregulatory or respiratory problems and creating opportunities for secondary infectors such as pathogenic fungi and bacteria. The most well-known crustacean parasite of carp is *Argulus foliaceus* (Fig. 7.1f), which despite its low host specificity is commonly known as the 'carp louse' due to its significant impact on the carp farming industry of Europe and Asia (Paperna 1991). The pathogenicity of the carp louse lies not only in mechanical damage to the fish's skin but also in irritations caused by the secretion of digestive toxins into the epidermis while feeding. Furthermore, it has been shown that *A. foliaceus* can act as a vector for other pathogens such as *Rhabdovirus carpio,* the agent of spring viremia of carp (SVC). *A. foliaceus* has a direct life cycle in which adult females leave the host after copulation and deposit their eggs (up to 500) in a gelatinous string on any suitable surface such as stones. The females may then return to the same host individual or attach to another until ready to deposit the next clutch of eggs. Incubation times for the eggs vary with temperature, but once hatched (within 8 days at 26°C), the parasitic metanauplii search actively for a host fish (Hoole et al. 2011). After attachment, nine larval molts follow in quick succession and the adult stage is reached. In warm summer temperatures, the whole life cycle may be completed in less than 40 days, making it possible for the parasite to realize four generations per year under optimal conditions. Reproductive success peaks at temperatures between 20–25°C, but grinds to a halt at temperatures below 13°C (Paperna 1991). Fish suffering argulosis begin to show behavioral abnormalities such as lethargy and loss of appetite shortly after the onset of infection. Heavy infections may lead to fin damage, increased mucus production and small petechial hemorrhages at parasite attachment sites that may result in skin loss, osmoregulatory distress and secondary infection by fungi and bacteria (Hoole et al. 2011). The symptoms may lead to significant host mortality in fish farms, rearing ponds and natural habitats (Pekmezci et al. 2011). Numerous chemotherapeutics are available to treat argulosis, most of them based on organophosphate insecticides. Ponds with infected stock should be dried and disinfected to eliminate the eggs of the parasite (Hoole et al. 2011).

References

Abowei, J.F.N., O.F. Briyai and S.E. Bassey. 2011. A Review of Some Basic Parasite Diseases in Culture Fisheries Flagellids, Dinoflagellides and Ichthyophthriasis, Ichtyobodiasis, Coccidiosis Trichodiniasis, Heminthiasis, Hirudinea Infestation, Crustacean Parasite and Ciliates. British J. Pharmacol. Toxicol. 2: 213–226.

Ahne, W. 1985. *Argulus foliaceus* L. and *Philometra geometra* L. as mechanical vectors of spring viremia of carp virus (SVCV). J. Fish Dis. 8: 241–242.

Ahne, W., H.V. Bjorklund, S. Essbauer, N. Fijan, G. Kurath and J.R. Winton. 2002. Spring viremia of carp (SVC). Dis. Aquat. Org. 52(3): 261–272.

Aho, J.M., J.W. Camp and G.W. Esch. 1982. Long-term studies on the population biology of *Diplostomum scheuringi* in a thermally altered reservoir. J. Parasitol. 68: 695–708.

Ariav, R., S. Tinman, I. Paperna and I. Bejerano. 1998. First report of newly emerging viral disease of *Cyprinus carpio* species in Israel. Paper read at EAFP 9th international conference, at Rhodes, Greece.

Baruš, V., M. Peňáz and K. Kohlmann. 2002. *Cyprinus carpio* (Linnaeus 1758). *In:* M. Banaresku and H.J. Paepke (eds.). The freshwater fishes of Europa. Cyprinidae 2 Vol. 5/III Part III: Crassius to Cyprinus. Gasterosteidae. AULA-Verlag, Wiebelsheim.

Behrmann-Godel, J., S. Roch and A. Brinker. 2014. Gill worm *Ancyrocephalus percae* (Ergens 1966) outbreak negatively impacts the Eurasian perch *Perca fluviatilis* L. stock of Lake Constance, Germany. J. Fish Dis. Journal of Fish Diseases 37: 925–930.

Bondad-Reantaso, M.G., R.P. Subasinghe, J.R. Arthur, K. Ogawa, S. Chinabut, R. Adlard, Z. Tan and M. Shariff. 2005. Disease and health management in Asian aquaculture. Vet. Parasitol. 132(3-4): 249–72.

Bootsma, R., N. Fijan and J. Blommaert. 1977. Isolation and preliminary identification of the causative agent of carp erythrodermatitis. Vet. Archives 47: 291–302.

Britton, J. Robert, Josephine Pegg and Chris F. Williams. 2011. Pathological and Ecological Host Consequences of Infection by an Introduced Fish Parasite. PLoS ONE 6(10): e26365.

Bullock, G.L., T.C. Hsu and E.B. Shotts. 1986. Columnaris disease of fishes. USFWS Fish Dis. Leaflet 1986: 1–9.

Bush, A.O., K.D. Lafferty, J.M. Lotz and A.W. Shostak. 1997. Parasitology meets ecology on its own terms: Margolis et al. revisited. J. Parasitol. 83(4): 575–83.

Chubb, J.C. 1977. Seasonal Occurrence of Helminths in Freshwater Fishes Part I. Monogenea. Advances in Parasitology 15: 133–199.

Cunningham, C.O. 2002. Molecular diagnosis of fish and shellfish diseases: present status and potential use in disease control. Aquaculture 206: 19–55.

Darwish, A.M., A.A. Ismaiel, J.C. Newton and J. Tang. 2004. Identification of Flavobacterium columnare by a species-specific polymerase chain reaction and renaming of ATCC43622 strain to *Flavobacterium johnsoniae*. Mol. Cell. Probes 18(6): 421–427.

Davis, H.S. 1922. A new bacterial disease in freshwater fishes. United States Bureau Fish. Bull. 38: 37–63.

Declercq, A.M., F. Haesebrouck, W. Van den Broeck, P. Bossier and A. Decostere. 2013. Columnaris disease in fish: a review with emphasis on bacterium-host interactions. Vet. Res. 44: 27.

Dhar, A.K., S.K. Manna and F.C. Thomas Allnutt. 2014. Viral vaccines for farmed finfish. Indian J. Virol. 25(1): 1–17.

Dyková, I. and J. Lom. 1988. Review of pathogenic myxosporeans in intesive culture of carp (*Cyprinus carpio*) in Europe. Folia Parasitologica 35: 289–307.

Eide, K.E., T. Miller-Morgan, J.R. Heidel, M.L. Kent, R.J. Bildfell, S. LaPatra, G. Watson and L. Jin. 2011. Investigation of koi herpesvirus latency in koi. J. Virol. 85(10): 4954–4962.

Fijan, N. 1972. Infectious dropsy in carp—a disease complex. pp. 39–51. *In*: L.E. Mawdesley-Thomas (ed.). Diseases of Fish. Symposium of the Zoological Society of London. Academic Press, London.

Fijan, N. 1999. Spring viraemia of carp and other viral diseases and agents of warmwater fish. *In*: P.K.T. Woo (ed.). Fish Diseases and Disorders. CAB International 3: 177–244.

Fujihara, M.P. and R.E. Nakatani. 1971. Antibody productionand immune responses of rainbow trout and coho salmon to *Chondrococcus columnaris*. J. Fish. Res. Board Can. 28(9): 1253–1258.

Füllner, G. and W. Müller. 1992. The therapy of swimbladder inflammation (*Renicola sphaerosporosis*) of carp. Angewandte Parasitologie 33: 79–90.

Garver, K.A., L. Al-Hussinee, L. Hawley, T. Schroeder, S. Edes, V. LePage, E. Contador, S. Russell, S. Lord, R.M. Stevenson, B. Souter, E. Wright and J.S. Lumsden. 2010. Mass mortality associated with koi herpesvirus in wild common carp in Canada. J. Wildl. Dis. 46(4): 1242–1251.

Gilad, O., S. Yun, F.J. Zagmutt-Vergara, C.M. Leutenegger, H. Bercovier and R.P. Hedrick. 2004. Concentrations of a Koi herpesvirus (KHV) in tissues of experimentally-infected *Cyprinus carpio* koi as assessed by real-time TaqMan PCR. Dis. Aquat. Org. 60(3): 179–187.

Gómez, D., J.O. Sunyer and I. Salinas. 2013. The mucosal immune system of fish: the evolution of tolerating commensals while fighting pathogens. Fish Shellfish Immunol. 35(6): 1729–1739.

Guegan, J.F. and C.R. Kennedy. 1993. Maximum local helminth parasite community richness in British freshwater fish: a test of the colonization time hypothesis. Parasitology 106(Pt1): 91–100.

Hakalahti, T., A. Karvonen and E.T. Valtonen. 2006. Climate warming and disease risks in temperate regions—*Argulus coregoni* and *Diplostomum spathaceum* as case studies. J. Helminthol. 80(2): 93–98.

Hedrick, R.P., O. Gilad, S. Yun, J.V. Spangenberg, G.D. Marty, R.W. Nordhausen, M.J. Kebus, H. Bercovier and A. Eldar. 2000. A Herpesvirus Associated with Mass Mortality of Juvenile and Adult Koi, a Strain of Common Carp. J. Aquat. Anim. Health 12(1): 44–57.

Hoole, D., P. Bucke, P. Burgess and I. Wellby. 2011. Diseases of Carp and other Cyprinid Fishes. Fishing News Books, Blackwell Science, Oxford.

Jeney, Z. and G. Jeney. 1995. Recent achievements in studies on diseases of common carp (*Cyprinus carpio* L.). Aquaculture 129(1–4): 397–420.

Kanellos, T., I.D. Sylvester, F. D'Mello, C.R. Howard, A. Mackie, P.F. Dixon, K.C. Chang, A. Ramstad, P.J. Midtlyng and P.H. Russell. 2006. DNA vaccination can protect *Cyprinus carpio* against spring viraemia of carp virus. Vaccine 24(23): 4927–4933.

Kennedy, C.R. and T. Pojmanska. 1996. Richness and diversity of helminth parasite communities in the common carp and in three more recently introduced carp species. J. Fish Biol. 48(1): 89–100.

Kirk, R.S. and J.W. Lewis. 1994. The distribution and host range of species of the blood fluke *Sanguinicola* in British freshwater fish. J. Helminthol. 68(04): 315–318.

Kirk, R.S. and J.W. Lewis. 1998. Histopathology of *Sanguinicola inermis* infection in carp, *Cyprinus carpio*. J. Helminthol. 72(01): 33–38.

Koutná, M., T. Velelý, J. Piskal and J. Hulová. 2003. Indentification of spring viraemia of carp virus (SVCV) by combined RT-PCR and nested PCR. Dis. Aquat. Org. 55: 229–235.

Kozinska, A., M.J. Figueras, M.R. Chacon and L. Soler. 2002. Phenotypic characteristics and pathogenicity of *Aeromonas genomospecies* isolated from common carp (*Cyprinus carpio* L.). App. Microbio. 93(6): 1034–41.

Kumar, V., S. Roy, D. Barman, A. Kumar, L. Paul and W.A. Meetei. 2013. Importance of mycotoxins in aquaculture feeds. Aquat. Anim. Health 18: 25–29.

Larsen, A.H., J. Bresciani and K. Buchmann. 2005. Pathogenicity of *Diplostomum* cercariae in rainbow trout, and alternative measures to prevent diplostomosis in fish farms. Bull. Eur. Ass. Fish Pathol. 25: 20–27.

Lepa, A. and A.K. Siwicki. 2012. Fish herpesvirus diseases: a short review of current knowledge. Acta Vet. Brno. 81(4): 383–389.

Majoros, G. 1999. Mortality of fish fry as a result of specific and aspecific cercarial invasion under experimental conditions. Acta Vet. Hungarica 47: 433–450.

Marcogliese, D.J. 2001. Implications of climate change for parasitism of animals in the aquatic environment. Can. J. Zool. 79(8): 1331–1351.

Marcogliese, D.J. and M. Pietrock. 2011. Combined effects of parasites and contaminants on animal health: parasites do matter. Trends Parasitol. 27(3): 123–130.

Moravec, F. and S. Cervinka. 2005. Female morphology and systematic status of *Philometroides cyprini* (Nematoda: Philometridae), a parasite of carp. Dis. Aquat. Org. 67(1-2): 105–109.

Morrison, C., J. Cornick, G. Shum and B. Zwicker. 1981. Microbiology and histopathology of "saddle-back" disease of underyearling Atlantic salmon, *Salmo salar*. J. Bacteriol. 78: 225–230.

Naylor, R.L., R.J. Goldburg, J.H. Primavera, N. Kautsky, M.C.M. Beveridge, J. Clay, C. Folke, J. Lubchenco, H. Mooney and M. Troell. 2000. Effect of aquaculture on world fish supplies. Nature 405(6790): 1017–1024.

Oláh, J. and J. Farkas. 1987. Effect of temperature, pH, antibiotics, formalin and malachite green on the growth and survival of *Saprolegnia* and *Achlya* parasitic on fish. Aquaculture 13: 273–288.

Paperna, I. 1991. Diseases caused by parasites in the aquaculture of warm water fish. Ann. Rev. of Fish Dis. 155–194

Paull, S.H. and P.T.J. Johnson. 2011. High temperature enhances host pathology in a snail-trematode system: possible consequences of climate change for the emergence of disease. Freshwater Biol. 56(4): 767–778.

Pekmezci, G.Z., B. Yardimci, C.S. Bolukbas, Y.E. Beyhan and S. Umur. 2011. Mortality due to heavy infestation of *Argulus foliaceus* (Linnaeus 1758) (Branchiura) in pond-reared carp, *Cyprinus carpio* L., 1758 (Pisces). Crustaceana 84: 553–557.

Perelberg, A., H. Smirnov, M. Hutoran, A. Diamant, Y. Bejerano and M. Kotler. 2003. Epidemiological description of a new viral disease afflicting cultured *Cyprinus carpio* in Israel. Israeli J. Aquaculture-Bamidgeh 55(1): 5–12.

Perelberg, A., M. Ilouze, M. Kotler and M. Steinitz. 2008. Antibody response and resistance of *Cyprinus carpio* immunized with cyprinid herpes virus 3 (CyHV-3). Vaccine 26(29-30): 3750–3756.

Pokorova, D., T. Vesely, V. Piackova, S. Reschova and J. Hulova. 2005. Current knowledge on koi herpesvirus (KHV): a review. Vet. Med. 50(4): 139–147.

Poulin, R. 2006. Global warming and temperature-mediated increases in cercarial emergence in trematode parasites. Parasitology 132: 143–151.

Prasad, Y., Arpana, D. Kumar and A.K. Sharma. 2011. Lytic bacteriophages specific to *Flavobacterium columnare* rescue catfish, *Clarias batrachus* (Linn.) from columnaris disease. J. Environ. Biol. 32(2): 161–8.

Ransom, D.P. 1975. Immune reponses of salmonids: (a) oral immunization against *Flexibacter columnaris*. (b) effects of combining antigens in parenterally daministered polyvalent vaccines, M.A. Thesis, Oregon State University, Corvallis, OR.

Rodák, L., Z. Pospisil, J. Tomanek, T. Vesely, T. Obr and L. Valicek. 1993. Enzyme-linked immunosorbent assay (ELISA) for the detection of spring viremia of carp virus (SVCV) in tissue homogenates of the carp. J. Fish Dis. 16: 101–111.

Ronen, A., A. Perelberg, J. Abramowitz, M. Hutoran, S. Tinman, I. Bejerano, M. Steinitz and M. Kotler. 2003. Efficient vaccine against the virus causing a lethal disease in cultured *Cyprinus carpio*. Vaccine 21(32): 4677–4684.

Shapira, Y., A. Perelberg, T. Zak, G. Hulata and B. Levavi-Sivan. 2002. Differences in resistance to koi herpes virus and growth rate between strains of carp (*Cyprinus carpio*) and their hybrids. Israeli J. Aquaculture-Bamidgeh 54: 62–63.

Schäperclaus, W. 1979. Fischkrankheiten, Vol. 4. Akademie-Verlag, Berlin.

Schmahl, G. and H. Taraschewski. 1987. Treatment of fish parasites. Parasitol. Res. 73(4): 341–351.

Shariff, M., R.H. Richards and C. Sommerville. 1980. The histopathology of acute and chronic infections of rainbow trout *Salmo gairdneri* Richardson with eye flukes, *Diplostomum* spp. J. Fish Dis. 6: 455–465.

Székely, C. and K. Molnár. 1991. Praziquantel (Droncit) is effective against diplostomosis of grasscarp *Ctenopharyngodon idella* and silver carp *Hypophthalmichthys molitrix*. Dis. Aquat. Org. 11: 147–150.

Thomas-Jinu, S. and A.E. Goodwin. 2004. Acute columnaris infection in channel catfish, *Ictalurus punctatus* (Rafinesque): efficacy of practical treatments for warmwater aquaculture ponds. J. Fish Dis. 27(1): 23–28.

Vidal-Martínez, V.M., D. Pech, B. Sures, S.T. Purucker and R. Poulin. 2010. Can parasites really reveal environmental impact? Trends Parasitol. 26(1): 44–51.

Williams, C.F., F. Moravec, J.F. Turnbull and H.W. Ferguson. 2012. Seasonal development and pathological changes associated with the parasitic nematode *Philometroides sanguineus* in wild crucian carp *Carassius carassius* (L.) in England. J. Helminthol. 86: 329–338.

Yokoyama, H., D. Grabner and S. Shirakashi. 2012. Transmission biology of the Myxozoa. pp. 3–42. *In*: E.D. Carvalho, G.S. David and R.J. Silva (eds.). Health and Environment in Aquaculture, InTech, Rijeka, Croatia.

Yukhimenko, S.S. 1970. On the occurance of *Bothriocephalus gowkongensis* Yeh, 1955 (Cestoda Pseudophyllidea) in the young of cyprinidae from the Amour river. Parasitologia 4: 480–483.

Zhou, J.X., H. Wang, X.W. Li, X. Zhu, W.L. Lu and D.M. Zhang. 2013. Construction of KHV-CJ ORF25 DNA vaccine and immune challenge test. J. Fish Dis. doi: 10.1111/jfd.12105.

8

The Carp Immune System and Immune Responses to Pathogens

Natalia Ivonne Vera–Jimenez,[1] Geert Frits Wiegertjes[2,*] and Michael Engelbrecht Nielsen[1]

Introduction

The immune system is critical for survival and fitness of any organism, protecting them from infectious agents, toxins (sterile), tissue damage and the impairment that they elicit (Magnadottir 2006, Segner et al. 2011, Uribe et al. 2011). Such protection is based on immune responses against a given threat. The first response launched can be summarized as innate immunity, the earliest barrier to infection and damage. Innate immunity's outcome is based on a general recognition of a group of pathogens or damage signals, rather than on a particular microorganism or molecule and does not comprise long lasting immunity (Medzhitov and Janeway Jr. 2000, Magnadottir 2006). The second response comprises more specific reactions and is collectively called adaptive (or acquired) immunity. The adaptive immune response is developed during the lifetime of an organism as adaptations to infection with any given pathogen; its protection is mediated by B- and T-lymphocytes (Fig. 8.1B) and in many cases leads to immunological memory, conferring long-lasting protection against the specific pathogens that the immune system has been exposed to (Uribe et al. 2011).

---
[1] The Technical University of Denmark, Allergy, Immunotoxicology and Natural Toxins Group, 2860 Søborg, Denmark.
[2] Wageningen University, Cell Biology and Immunology Group, 6700AH Wageningen, Netherlands.
* Corresponding author

The effective protection of the immune system as a whole is achieved through the activation of four more or less sequential main immunological stages. The first is immunological recognition, where the presence of any type of threat is detected (Alvarez-Pellitero 2008, Kawai and Akira 2010). The second stage is the contention and elimination of the pathogen, engaging different immune effect or functions such as phagocytosis, production of reactive oxygen species, activation of complement system proteins, antibody production, etc. (Alvarez-Pellitero 2008, Rieger and Barreda 2011). Since the damage and eradication power launched by the immune system is so strong, it has to be carefully regulated to minimize the self-damage inflicted during an infection; therefore the third stage denotes immune regulation (Tort 2011, Verburg-van Kemenade et al. 2011). And lastly the fourth stage is immunological memory, which allows the induction of a much faster and stronger immune response against previously encountered pathogens (Uribe et al. 2011).

In this chapter, the main components of the carp immune system will be discussed, along with the immune responses elicited by the encounter of carp pathogens and how immune modulators can improve carp health.

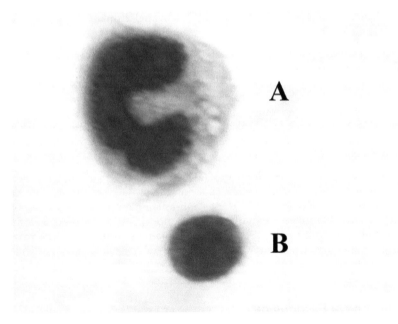

Figure 8.1. A. Carp Leukocytes stained with hematoxylin and Eosin (H&E stain). A) Carp neutrophil. B) Carp lymphocyte. B. Carp Macrophage stained with hematoxylin and Eosin (H&E stain).

Description of the Carp Immune System

Innate Immunity

Carps are constantly threatened by the invasion of microorganisms, and as other vertebrates, they have evolved advanced systems of immune defense to eliminate infective pathogens in their organism. The first line of host defense against pathogens is called innate immunity, it is mediated by phagocytes including granulocytes (Fig. 8.1A), macrophages (Fig. 8.1B) and dendritic cells, and could be considered the most important branch of the immune system in fish. Innate immunity responds quickly to infections and its components are ready to react even before the host gets infected (Medzhitov 2001, Akira et al. 2006, Rauta et al. 2012)

Infobox: macrophage polarization

Mammalian macrophages can be sub-divided in classically activated macrophages, induced in a T helper 1 (T_H1) cytokine environment, and alternatively activated macrophages, induced in a T_H2 cytokine environment. Studying the presence and function of polarized macrophages in teleosts is greatly facilitated by standardized procedures for the isolation and *in vitro* culture of macrophage populations from goldfish and common carp. Differentially activated fish macrophages largely resemble the phenotypes of mammalian macrophages.

Classically activated macrophages (also termed M1 macrophages) can be characterized as macrophages that respond to a microbial stimulus sensed by pattern recognition receptors (PRRs) such as Toll-like receptors (TLRs) in the presence of, in particular, the cytokine interferon-gamma (IFNγ). They typically exhibit high oxygen and nitrogen radical production and may be considered 'inflammatory' macrophages which help combat pathogens but must be kept under tight regulation to prevent them from also causing uncontrolled damage to host tissues.

Alternatively activated macrophages (also termed M2 macrophages) can be characterized as macrophages that respond in the presence of the T_H2 cytokines interleukin (IL)-4 and/or IL-13. These 'anti-inflammatory' macrophages often show increased arginase activity and produce proteins important for the extracellular matrix and increased levels of polyamines (putrescine, spermidine, and spermine) upon activation. M2 macrophages are considered particularly important for wound healing.

Immunological Recognition

Detection of tissue damage in the host is accomplished by the innate immune system, using a series of receptors displayed primarily by immune cells, which identify the nature of the injury and act towards its repair (Zhang and Schluesener 2006, Bianchi 2007, Rebl et al. 2010, Hansen et al. 2011). These receptors are collectively known as Pattern Recognition Receptors (PRRs) (Hansen et al. 2011). Based on molecular signals, PRRs are able to discriminate between infection; open wounds associated with the intrusion of pathogens during injury and pathogen-free trauma such as mechanical tissue damage or sterile inflammation (Schreml et al. 2010, Rock et al. 2010). Infectious agents display a series of specific molecular motifs called Pathogen-Associated Molecular Patterns (PAMPs), which are essential for the survival and pathogenicity of the microorganism (Bianchi 2007). PAMPs are not found in the host and therefore are non-self molecules (Akira et al. 2006). Some examples of PAMPs include lipopolysaccharide (LPS), lipoteichoic acids (LTA), lipoarabinomannan and β-glucans, which are signatures of gram-negative, gram-positive, mycobacterial and fungal pathogens, respectively (Medzhitov and Janeway Jr. 2000, Dalmo and Bøgwald 2008). On the other hand, identification of pathogen-free trauma, such as mechanical damage, is accomplished by the recognition of Damage-Associated Molecular Patterns (DAMPs) (Hansen et al. 2011). DAMPs are endogenous molecules which in healthy tissues are contained inside the cells and hidden from the immune system, but which are released following an injury. DAMPs thus are self-molecules (Lotze et al. 2007, Chen and Nunez 2010). Some examples of DAMPs are Heat Shock Proteins (HSPs), High-Mobility Group Box-1 (HMBG1), ATP, uric acid, hyaluronic acid and mitochondrial DNA (Midwood and Piccinini 2010, Chen and Nunez 2010, Zhang et al. 2010). In fish, the recognition of DAMPs has not been studied in depth, although collagen-derived proteolytic fragments have been reported to indicate the presence of damage (Castillo-Briceño et al. 2009).

The Inflammatory Response

Immediately after the recognition of a threat, an inflammatory response is launched with neutrophils as the first leukocyte cell type that reaches the site of injury. This leads to the local production of highly microbicidal compounds like Reactive Oxygen Species (ROS) and Nitric Oxide (NO) to help eradicate infectious agents (Bianchi 2007). This recruitment takes place in response to a cocktail of cytokines and chemokines produced by local cells (Martin and Leibovich 2005, Gonzalez et al. 2007c). Next, macrophages migrate to the site of injury to clear the wound of dead cells and other debris. Macrophages do not constitutively express high levels of cytokines and growth factors in their resting state. However, they are very sensitive to danger signals and when activated become strong producers of cytokines, which can mediate and sustain the inflammation and activate fibroblasts and endothelial cells (Afonso et al. 1998, Martin and Leibovich 2005). At a later phase, lymphocytes reach the site of injury, marking the progression of the proliferative and tissue remodeling phase (Diegelmann and Evans 2004, Wei 2011). In carp, mobilization of epidermal cells and their phagocytic activity has been reported within the first hours after wounding (Iger and Abraham 1990). This process overlaps with the inflammatory response, which appears to follow the same phases present in mammals, although thrombocytes instead of platelets (fish thrombocytes are the equivalent of platelets found in mammals) are in charge of the coagulation (Jiang and Doolittle 2003).

Immune Effect or Functions of the Innate Immunity

Phagocytosis

Phagocytosis is a fundamental defense mechanism in bony fish. Phagocytic cells such as neutrophils, monocytes and macrophages mediate this process (Nikoskelainen et al. 2005, Rieger and Barreda 2011). Phagocytosis is an active process that is initiated after immunological recognition and binding of the pathogen to the cell surface. Subsequently, the pathogen is surrounded by the phagocyte membrane and then internalized in a membrane-enclosed vesicle called phagosome (Stuart and Ezekowitz 2008). In addition to being phagocytic, macrophages and neutrophils have lysosomes which are membrane-enclosed granules containing enzymes, proteins and peptides detrimental to pathogens. Following internalization, the lysosomes migrate towards the phagosome and fuse with them, generating the so-called phagolysosomes where the lysosomal contents are released to achieve the degradation of the ingested pathogens (Stuart and Ezekowitz 2008, Rieger et al. 2010, Rieger and Barreda 2011).

Respiratory Burst

The production of Reactive Oxygen Species (ROS), also known as respiratory burst, is one of the earliest cellular responses following pathogen recognition. Its initiation is marked by an increase in cellular oxygen uptake, followed by the one electron reduction of molecular oxygen (O_2) to superoxide anions (O_2^-). This reaction is catalyzed by the membrane-associated enzyme Nicotinamide Adenine Dinucleotide Phosphate (NADPH) oxidase, using NADPH as the electron donor (Pick and Mizel 1981, Bellavite 1988, Chung and Secombes 1988, Babior 1999, Madamanchi and Runge 2007). Further reduction of O_2^- produces Hydrogen Peroxide (H_2O_2), which occurs either as a spontaneous dismutation, especially at low pH, or as a catalyzed reaction by a family of enzymes called superoxide dismutase (SOD). Additional reactions of O_2^- and H_2O_2 may lead to the formation of hydroxyl radicals (OH^-), especially in the presence of iron through the Fenton or Haber-Weiss reactions. The interaction of H_2O_2 with myeloperoxidase (MPO) can produce hypochlorous acid and other toxic metabolites if H_2O_2 is not converted to water and molecular oxygen by the enzyme catalase, which can act as a natural scavenger (Fridovich 1978, Jones 1982, Babior 1999, Genestra 2007, Lushchak 2008, Arockiaraj et al. 2012). A schematic representation with the main products of the respiratory burst is shown in Fig. 8.2. Besides the significant function of ROS during infection clearance, their importance during wound healing processes has been established. ROS have been shown to be involved in cell proliferation, but can also cause a detrimental effect on host tissues

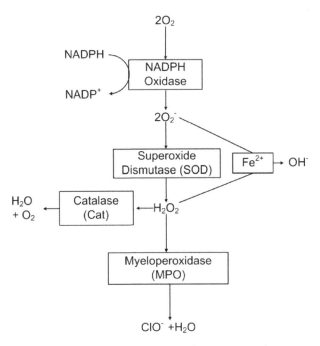

Figure 8.2. Schematic representation of the respiratory burst main products.

due to oxidative stress (Kanta 2011). It has been shown that an extraordinary coordination for the quantities and timing in the production of ROS is needed, and exerts an important impact in the outcome of wound healing in both, mammals and carp (Diegelmann and Evans 2004, Wallach-Dayan et al. 2007, Przybylska et al. 2013, Vera-Jimenez and Nielsen 2013). For example, ROS at low concentrations (1–25 µM) function as second messengers and promote cell proliferation, at higher concentrations (25–50 µM) they have microbicidal effects, and in excessive amounts (> 50 µM) they can impair cell growth and cause apoptosis (Kanta 2011, Vera-Jimenez and Nielsen 2013). Furthermore, although it is conceivable that the majority of the ROS are produced by immune cells during the inflammatory phase, fibroblast and other cell types can contribute to their production at a lesser extent during the wound healing process (Witte and Barbul 2002, Kanta 2011). In brief, ROS play a role in the clearance of pathogens and tissue regeneration processes. Therefore, the respiratory burst and associated ROS constitute important indicators of fish health status (Vera-Jimenez et al. 2013).

Nitric Oxide

Nitric Oxide (NO) is produced by leukocytes during the initial stages of the inflammatory response, and has been proposed as a modulator of the immune response involved in pathogen eradication in mammalian systems and fish in general (Wink et al. 2011, Rieger and Barreda 2011). In addition, some reviews based on wound healing models in mammals have highlighted the importance of NO during wound healing (Witte and Barbul 2002, Soneja et al. 2005), while others focus more on the detrimental effects of the same molecule on cell proliferation (Villalobo 2006). NO is the basis for all reactive nitrogen species, and is formed by the oxidation of L-arginine to L-citrulline by NO synthase (NOS) (Wink et al. 2011). Three forms of NOS have been identified in mammals: neuronal (nNOS), endothelial (eNOS) and inducible (iNOS), from which only iNOS has been shown to be involved in immune defense (Knowles and Moncada 1994). As nitric oxide is detrimental not only for the pathogens but also for the host cells, the activity of NOS is highly regulated. In macrophages this regulation occurs through the regulated expression of the NOS gene, which is induced as a result of signaling cascades generated upon stimulation of pathogen recognition receptors (Saeij et al. 2002, Bilitewski 2008). Large amounts of NO (like the ones produced by iNOS during an inflammatory reaction) are usually complemented by elevated production of ROS. The interaction between nitric oxide and superoxide can lead to the formation of peroxinitrite (ONOO-) a highly bactericidal compound. Besides, it has been suggested that NO can preferentially alter transcription factors sensitive to changes in the cellular redox status (Pfeilschifter et al. 2001, Scharsack et al. 2003, Genestra 2007).

Carp nitric oxide, nitrosative agents and nitrosative enzymes released outside the cells can be detrimental to pathogens and beneficial to the infected fish, as shown during infection with *Ichthyophthirius multifiliis* (Gonzalez et al. 2007b), cyprinid herpesvirus 3-infection (Syakuri et al. 2013), upon microbial activation of macrophages (Forlenza et al. 2011), and in response to fungal cell wall components such as β-glucans (Pietretti et al. 2013). However, they can also contribute to nitrosative stress and tissue injury (Forlenza et al. 2008b). This detrimental effect of NO in carp was demonstrated by Saeij and coworkers who showed that *T. borreli*-infected carp treated with an iNOS inhibitor had higher survival than infected control carp, and that NO production could lead to apoptosis of carp Peripheral Blood Leukocytes (PBL) and inhibited proliferation of blood, spleen and Head Kidney (HK) leukocytes (Saeij et al. 2002, Saeij et al. 2003a, Scharsack et al. 2003).

Subsequently, Forlenza and co-workers studied the contribution of neutrophilic granulocytes and macrophages to nitrosative stress in carp infected with *T. borreli*. They could demonstrate that macrophages restrict the presence of nitrating agents to their phagosomal compartments. Therefore, carp macrophages become nitrated themselves with a minimum to nonexistent contribution to extracellular nitration and tissue injury. On the other hand, neutrophilic granulocytes greatly contribute to tissue nitration likely via peroxynitrite and an MPO-mediated mechanism (Forlenza et al. 2008b).

Lysozyme

Lysozyme is a bactericidal enzyme that hydrolyzes the β-(1,4) glycoside bonds of bacterial cell wall peptidoglycans resulting in bacterial lysis. Lysozyme activity has been primarily associated with defense against Gram-positive bacteria. However, although Gram-negative bacteria are not directly damaged by lysozyme, when complement and other enzymes disrupt the outer cell wall exposing the inner peptidoglycan layer, lysozyme also becomes effective against them. Besides the antibacterial function, lysozyme can promote phagocytosis by direct activation of polymorphonuclear leukocytes and macrophages, or by acting as an opsonin (Magnadottir 2006, Saurabh and Sahoo 2008). Furthermore, it has been reported that lysozyme can damage structures containing muramic acid or glycol chitin and has a restricted detrimental influence on chitin, which is a major component of cell walls of fungi and exoskeletons of certain invertebrates (Saurabh and Sahoo 2008). Two types of lysozyme, the c-type and the g-type lysozyme, have been identified in common and grass carp (Fujiki et al. 2003, Savan et al. 2003, Ye et al. 2010). The distribution of lyzozyme has been reported to be mainly in the head kidney, but also at sites where the risk of bacterial invasion is very high such as the gills, skin, gastrointestinal tract and eggs and in the blood (Ye et al. 2010).

Lysozyme level or activity has been considered an important index of innate immunity of fish. However it has also been shown that several factors can influence its activity, therefore, the use of lysozyme as an indicator of fish health status should be considered carefully. In cultured carp for example, lysozyme activity can vary due to stress conditions such as stocking density, periodic handling, transport, water quality or the use of anesthetics (Saurabh and Sahoo 2008). Variation in lysozyme activity can also be associated with season, water temperature and sexual maturity. Swain and coworkers reported an increase of lysozyme activity in Indian carp (*Labeorohita*) during the rainy and summer seasons, and a decrease during winter (Swain et al. 2007). Furthermore, Studnicka and coworkers described differences on lysozyme activity related to sexual maturity in common carp, obtaining the highest enzyme levels in spawners (Studnicka et al. 1986).

Transferrin and Ferritin

Transferrin (Tf) and Ferritin are the main proteins mediating iron homeostasis. Iron is fundamental to the biology of eukaryotic cells because it plays a key role in functions like oxygen transport, electron transfer and catalysis. Similarly, iron is crucial for pathogens, which must acquire it within their vertebrate hosts in order to replicate and cause disease (Skaar 2010). The importance of Tf for carp blood parasites such as *Trypanoplasma borreli* has been clearly shown (Jurecka et al. 2009a, Jurecka et al. 2009c). Transferrin and ferritin are acute phase proteins that can act as inhibitors of bacterial growth through the regulation of available iron. Therefore they are often up-regulated during infection (Wooldridge and Williams 1993, Jurado 1997). In addition, excessive concentration of iron can cause tissue damage as a result of the formation of highly reactive hydroxyl radicals or oxidants of similar reactivity from hydrogen peroxide via Fenton reaction (H_2O_2 + Fe^{2+} = Fe^{3+} + OH^- + $^*OH$). These radicals are responsible for toxic processes including peroxidation of biological membranes and DNA damage. Hence, Tf and ferritin play an important role in the control of oxidative damage (Meneghini 1997, Magnadottir 2006).

Transferrin is a transport protein of extracellular Fe^{3+}, it binds iron creating therefore a bacteriostatic environment. In addition, it is been claimed that Tf induces neutrophilic end-stage maturation in goldfish macrophages (Stafford and Belosevic 2003, Magnadottir 2006), and that it can activate and modulate the production of NO by goldfish and carp macrophages after its activation through molecular cleavage (Stafford and Belosevic 2003, Jurecka et al. 2009a). Four different alleles of Tf have been cloned and sequenced in carp. In comparison with human and salmonid Tf, the three-dimensional structure of the alleles is well conserved, except for the iron binding sites in one of the Tf lobules. Therefore, it appears, the concentration of Tf in carp serum is twice the concentration found in for example salmonid fish (Jurecka et al. 2009b). Ferritin on the other hand, is an intracellular Fe^{3+} storage protein, which has been shown to release or sequester iron in the event of depletion or overload respectively. Some pathogens can circumvent host-iron withholding through high-affinity iron uptake mechanisms that compete against host-mediated sequestration and the production of chelating agents (Magnadottir 2006, Skaar 2010).

Complement

The complement system plays an important role in the immune defense and involves about 35 soluble and membrane-bound proteins. Normally, the soluble proteins circulate in the body as inactive precursors (pro-proteins). However, after pathogen recognition the pro-proteins are cleaved by proteases in the system. This initiates and amplifies a reaction cascade of further protein cleavages, resulting in the activation of the Membrane Attack Complex (MAC). The MAC eradicates pathogens generating pores in their membrane, this upsets their osmotic balance and ultimately leads to cell death (Holland and Lambris 2002). Three biochemical pathways (illustrated in Fig. 8.3) activate the complement:

i) The classical complement pathway typically requires antigen:antibody complexes to be initiated, therefore it bridges the innate and the adaptive immune systems. These complexes activate the C1 complement factor and initiate the cleavage of C1r and C1s molecules. C1s then cleaves the C4 complement factor into C4b and C4a, enabling the C4b fragment to bind to the activating surface (pathogen membrane). Thereafter, the C2

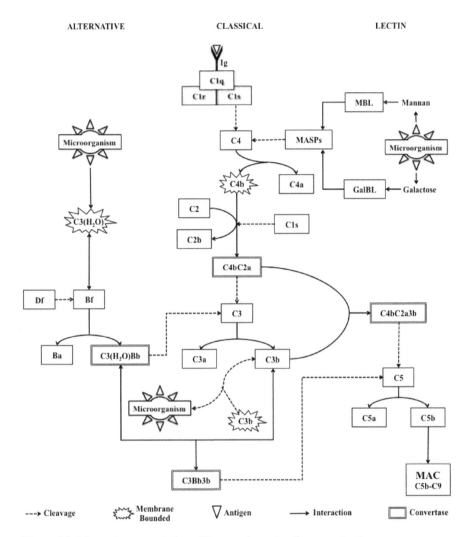

Figure 8.3. Schematic representation of the complement pathways activation.

complement factor binds to C4b, leading to its cleavage into C2a and C2b by the C1s component. The C2a fragment remains bound to C4b, forming an enzymatic unit called the classical C3 convertase (C4b2a). This enzymatic unit can cleave surrounding native C3 complement factor into C3a and C3b fragments. The association of an additional molecule of C3b to the classical C3 convertase, leads to the formation of the classical C5 convertase (C4b2a3b). This convertase is responsible for cleaving C5 into C5a and C5b, allowing the assembly of the MAC (C5b-C9) (Nakao and Yano 1998, Boshra et al. 2006).

ii) The alternative pathway comprises C3 complement factor activation and amplification upon binding of $C3(H_2O)$ to various pathogen surfaces (i.e., viruses, bacteria, fungi, parasites). The bound $C3(H_2O)$ then interacts with Factor B (Bf). Subsequently, factor D (Df) present in plasma, cleaves Bf into Ba and Bb fragments, resulting in the formation of the alternative C3 convertase $C3(H_2O)Bb$. This convertase, although short-lived, is capable of cleaving many surrounding C3 molecules into C3b and C3a. C3b will bind to the pathogen surfaces within proximity, forming additional C3 convertases and forming an amplification loop (amplification of C3 cleavage). This loop rapidly leads to a massive deposition of C3b molecules onto the pathogen surface, which are recognized by complement receptors, promoting phagocytosis. Furthermore, newly generated C3b can also bind to an existing C3 convertase, resulting in the alternative C5 convertase (C3bBb3b). Like the classical C5 convertase (C4b2a3b), the alternative C5 convertase is capable of cleaving C5, which leads to the subsequent assembly of the MAC (Holland and Lambris 2002, Boshra et al. 2006).

iii) The lectin (MBL) pathway is considered the oldest complement pathway. As indicated by its name, it requires the interaction of lectins such as Mannose-Binding Lectin (MBL) and Galactose-Binding Lectin with MBL-like structure (GalBL), with sugar moieties found on the surface of pathogens, like mannan and galactose. Upon binding, the enzymes associated with these lectins, MBL-associated serine proteases (MASPs) are activated. These proteases have been shown to be structurally homologous to C1r and C1s, and to cleave the C4 complement factor (Nakao et al. 2006).

The activation of complement pathways results in various biological functions, including activation of phagocytes, opsonization (enhancement of phagocytosis), respiratory burst, cell lysis and promotion of inflammatory responses. Excessive activation of complement can be harmful for the host, therefore is tightly regulated. Although most of the complement factors have been identified and described in carp, as reviewed by Nakao et al. 2011, only regulatory factor I has been cloned, the rest have yet to be isolated or cloned (Nakao and Yano 1998, Nakao et al. 2011).

Antimicrobial Peptides

Antimicrobial peptides (AMPs) are small, secreted proteins with antimicrobial or bacteriostatic properties against a wide variety of bacterial, fungal, viral and other pathogenic infections. The positively charged amino acids of these peptides bind to negatively charged molecules and substances in the membranes of the microorganisms, creating pores that disrupt the membrane and lyse the pathogen (Noga et al. 2011, Rajanbabu and Chen 2011). In carp, presence of the AMPs apolipoprotein A-I and A-II has been reported in skin (Concha et al. 2003), and β-defensins 1 and 2 in skin and gills (van der Marel et al. 2012).

Neutrophil Degranulation and Extracellular Traps

Neutrophils, as one of the first cells at inflammation sites, are armed with a wide battery of intracellular and extracellular antimicrobial tools. One of these major extracellular defense mechanisms is the Neutrophil Extracellular Trap (NET), which immobilizes and kills pathogens. Although the exact mechanisms involved in the NET formation and its role in infection has not been well determined, *in vitro* experiments in carp showed the release of NET structures by dying neutrophils upon carp neutrophil stimulation with different immunostimulants of bacterial, fungal or viral origin, as a kind of suicide mechanism, called NETosis (Brogden et al. 2012). The NET kinetics and their composition depend on the stimulus used, but it usually leads to the extracellular release of DNA, antimicrobial peptides and stabilizing proteins such as histones as well as the granular enzymes myeloperoxidase (MPO), neutrophil elastase (NE) and matrix metalloproteinase 9 (MMP-9) (Pijanowski et al. 2013). In mammalian systems, some bacteria such as *Streptococcus pyogenes* have been shown to efficiently avoid NET-entrapment and killing by the production and release of NET-degrading nucleases (Buchanan et al. 2006). Brogden and coworkers examined NET formation in common carp challenged with *Aeromonas hydrophila*, and the effect of β-glucan on NETs'stabilization and inhibition of bacterial nucleases. Their results showed that β-glucan was able to induce NET formation and stabilization, protecting them against bacterial nuclease degradation, and thereby reducing the severity of the *A. hydrophila* infection (Brogden et al. 2012). NETs have therefore shown to be a novel important host innate immune defense mechanism of fish neutrophils.

Adaptive Immunity

The adaptive immune system is activated by the innate immune system, and plays a vital role in protection against recurrent infections. Such protection is exerted through the generation of memory cells (cellular immunity) and specific receptors (soluble and membrane-bound) such as T-cell receptors and immunoglobulin (Ig), allowing the fast and efficient elimination of the specific fish pathogens (Rauta et al. 2012).

Cellular Components

Antigen Presenting Cells (APCs)

Phagocytic cells activated during the innate immune response, such as monocytes and macrophages, can display foreign peptides using the major histocompatibility complex (MHC) receptors to activate T cells and elicit an adaptive immune response, two types of MHC receptors are present in teleost fish (Vallejo et al. 1992, Alvarez-Pellitero 2008). In 1990 the first teleostean MHC sequences were described for carp (Hashimoto et al. 1990). The MHC genes of common carp constitute an extended gene family, it includes classical and non-classical class I genes, *B2m* genes, class II *A* and *B* genes (Stet et al. 1993, Rakus et al. 2003). The high variability of the MHC genes results in the presence of numerous alleles and, as a consequence, numerous haplotypes within a population. Each allele has the ability to bind and present different groups of peptides in more or less successful ways. Thus, the response of an organism towards certain pathogens can be influenced by the MHC haplotype (Rakus et al. 2009). Further studies in teleost APCs have demonstrated that the MHC Class I (MHC I) receptor usually displays antigens from intercellular pathogens, while MHC Class II (MHC II) those from extracellular pathogens (Rauta et al. 2012).

T cells

T cells release factors to coordinate responses of other immune cells or cytotoxic factors to directly kill infected or abnormal cells. Numerous genes associated with T cell functioning and signaling have been identified in fish, and a variety of specific immune functions like allograft rejection, mixed leukocyte reaction, and delayed hypersensitivity reaction indicated that T cells are involved in these specific immune activities in a similar manner to that of mammals (Nakanishi et al. 2002, Rauta et al. 2012). However, the functional correlation between postulated T cell associated genes and T cell activity in adaptive immunity in fish is not well understood (Rauta et al. 2012). Two main sets of T cells have been distinguished based on the expression of the membrane bound glycoproteins CD8 or CD4. These molecules act as co-receptors for T cell receptors and help to stabilize the interaction with MHC receptors, enhancing T cell activation. In mammals, CD8 marks cytotoxic T cells, which recognize antigenic peptides associated with MHC class I of the APCs. Their main function is the direct killing of target cells. In contrast, CD4 marks T helper cells, which recognize peptides associated with MHC class II and orchestrate many aspects of the immune responses (Castro et al. 2011). In carp homologous sequences of CD8 have been characterized and have shown to be up-regulated during viral infection (Forlenza et al. 2008a). CD4 cDNA sequences have been found in carp, and although the sequences showed little similarity to that of higher vertebrates, features like gene organization, splicing pattern, binding motifs and key residues are clearly conserved (Sun et al. 2007, Castro et al. 2011, Yamaguchi et al. 2013).

B cells and Immunoglobulins (Ig)

B cells are characterized by the expression of B cell receptors, which are surface immunoglobulin receptors (sig). Fish B cells, like those in mammals have shown to amplify their structural diversity undergoing for instance Ig H-chain rearrangement and allelic exclusion, resulting in an improved specific pathogen recognition (Alvarez-Pellitero 2008). B cells are also responsible for the production of Igs or antibodies (Abs), which are the principal indicators of acquired immunity to pathogens. Igs have evolved to become highly diversified molecules that recognize a remarkably large number of different antigens (Rauta et al. 2012).

Three classes of Ig have been identified in teleost fish, IgM, IgD and IgZ (Fillatreau et al. 2013). IgM can be present in secretions including gut and cutaneous mucus, it can be expressed in the surface of B cells, and its tetrameric form is the most prevalent Ig in fish serum. Furthermore, IgM has been shown to be up-regulated during pathogen exposure in carp (Koumans-van Diepen et al. 1995, Fillatreau et al. 2013). IgD was initially thought to be expressed only in rodents and primates. However, its wide distribution and remarkable plasticity among vertebrates including carp was demonstrated later (Edholm et al. 2011). The IgD direct role in immune defense was suggested by the finding of IgD-only B cells and by the fact that secreted IgD can be bound by granulocytes via an IgD-binding receptor (Xiao et al. 2010). IgZ is a teleost specific Ig isotype that shows variations in gene structure across teleosts (called IgT in rainbow trout). Recently it has been found that IgZ appears to be specialized in mucosal immunity (Zhang et al. 2011). Two subclasses of IgZ have been identified in carp. IgZ1 is expressed in systemic and mucosal organs and is responsive to blood parasites and during wound healing processes in carp larvae, while IgZ2 gene expression is induced during mucosal parasite infection (Ryo et al. 2010, Schmidt and Nielsen 2014).

Immune Responses of Carp to Pathogens

Knowledge about immune mechanisms and immunological traits that can correlate with disease resistance in fish is essential for the development of prophylactic strategies (such as vaccination), and can help during the development of more resistant strains by the use of molecular marker assisted selection (Rakus et al. 2012).

Viral Diseases

Viral diseases are a serious problem in farmed fishes and cause heavy economic losses for the fish farming industry throughout the world (Harikrishnan et al. 2010). As mentioned in Chapter 7, the cyprinid herpesvirus 3 (CyHV-3), also known as Koi Herpes Virus (KHV), is the etiological agent of a highly contagious and extremely virulent disease of common carp (*Cyprinus carpio* L.) and its ornamental koi varieties. The mortality rate in affected ponds is as high as 70–95%, causing very large economic losses in fish farms in many countries worldwide. However, the number of studies on immune responses of carp against CyHV-3 is very limited (Rakus et al. 2012). After exposure of carp to CyHV-3, there is an activation of the type I interferon system. The interferons (IFNs) are a large family of multifunctional secreted proteins involved in immune activation, cell growth regulation and antiviral defense (Goodbourn et al. 2000). The antiviral effect is exerted through the activation of signal transduction pathways, which induces the expression of a large number of IFN-stimulated genes, some of which encode antiviral proteins. Additionally, infection with CyHV-3 prompts cytotoxic T cell (CD8$^+$) activation and the induction of pro-inflammatory molecules like cytokine Interleukin-1β (IL-1β) and Interleukin-12 (IL-12), which can also stimulate the activity of cytotoxic cells. Later during CyHV-3 infection in carp, the cytokine Interleukin-10 (IL-10) has shown to be up-regulated (Rakus et al. 2012). IL-10 has been suggested to be an anti-inflammatory cytokine, therefore reducing the previously elicited inflammatory reaction (Ingerslev et al. 2010). Furthermore, it has been demonstrated that carps develop antibodies against herpesvirus, which in fact can be used as a non-lethal tool for identifying fish populations that may have been previously exposed to the virus and are not showing clinical signs of disease (St-Hilaire et al. 2009).

Bacterial Diseases

The innate defense mechanisms of fish against bacteria include the production of broad-spectrum anti-microbial substances and acute phase proteins, non-classical complement activation, release of cytokines, inflammation and phagocytosis (Ellis 2001). 'Motile *Aeromonas* septicemia' (MAS) is a term used to describe motile aeromonad infections of warm-water fish, including carp. It has been shown that after challenge of carp with *Aeromonashydrophila*, parameters such as cortisol serum concentrations, hemagglutination, hemolytic and bacterial agglutination titers are increased (Swain et al. 2010). Furthermore, increased phagocytic activity and antibody production against *Aeromonas*, as well as an up-regulation of the cytokines IL-1β, TNFα and IL-10 have been reported (Tanekhy et al. 2009, Ardo et al. 2010). However results concerning the production of reactive oxygen species are contradictive; while Swain and coworkers report an increased respiratory burst, Ardo and coworkers report a decrease in the production of ROS after challenge with *A. hydrophila* (Swain et al. 2010, Ardo et al. 2010).

Fungal Diseases

Fungal infections in common carp are serious problems of both natural fisheries and intensive aquaculture. In spite of their importance the knowledge on fungal infections is still very poor (Jeney and Jeney 1995). Some pathological studies of saprolegniasis in fish have been performed, however very few have focused on the fish immune responses. An *in vitro* study based on the immune responses of a monocyte/macrophage cell line from rainbow trout challenged with the water molds *Achlya* and *Saprolegnia*, showed an increase cell adhesion to spores and mycelia from both fungi. Additionally, the pro-inflammatory cytokines IL-1β and TNFα were strongly up-regulated, as well as Interleukin-8 (IL-8), which is involved in chemotaxis of neutrophils and granulocytes (Kales et al. 2007).

Parasitic Diseases

Parasitic infections in teleost fish are often controlled primarily by innate defense mechanisms, which avoid parasite infestation or reduce the severity of infection (Jones 2001). Perhaps one of the best described immune responses to carp parasites is the one elicited by exposure to the blood flagellate *Trypanoplasma borreli*. Studies show that *T. borreli* infection induces TNFα expression in the Head Kidney (HK), followed by an increase in the expression of acute phase proteins, complement factor 3 (C3) and alpha-2-macroglobulin, subsequently IL-1β expression is up-regulated. In the liver however, the expression of acute phase proteins gradually increases during approximately 6 days, followed by a gradual increase in TNFα. The early rapid induction of expression of acute phase proteins in the HK has been related to the rapid increase in TNFα in response to the initial inoculum, while the gradual increase in expression described in the liver is probably triggered by the subsequent infiltration and replication of parasites in tissues (Saeij et al. 2002, Saeij et al. 2003b). Furthermore, *T. borreli* infections in carp are characterized by anemia, splenomegaly and polyclonal B cell activation (Forlenza et al. 2009). *In vitro* studies have shown that *T. borreli* can be effectively eradicated by antibody-dependent complement-mediated parasite lysis. The fact that clearance of *T. borreli* from the carp bloodstream corresponds with the onset of a specific antibody response *in vivo*, and that IgM is required, suggests that the classical complement activation during the later stage of infection is crucial (Forlenza et al. 2009). However, *T. borreli* and other parasites are still able to infect and persist in their host despite the induced immune responses, as they have evolved survival strategies that exploit the host immune system to their own benefit. For example, continuous stimulation of carp phagocytes by *T. borreli* can lead to overproduction of TNFα and Nitric Oxide (NO), which can interfere with the detection and clearance of surface IgM. Furthermore, as described previously in this chapter, NO and related products can cause tissue injury, apoptosis and suppressive effects on lymphocyte proliferation (Saeij et al. 2002, Forlenza et al. 2009).

The immune responses to ectoparasitic infections in carp that are the elicited by exposure to *Ichthyophthirius multifiliis* have also been well studied. Carp infection with *I. multifiliis* up-regulates the expression of the chemokine-a (CXCa) and the chemokine receptor-1 (CXCR1) (Gonzalez et al. 2007b). Chemokines constitute a large multifunctional family of cytokines that induce the migration of cells to the sites of infection or injury. This up-regulation could be related to the presence of pro-inflammatory cytokines, such as IL-1β, and TNFα, which were shown to be up-regulated during *I. multifiliis* infection as well. Furthermore, an up-regulation of the enzymes iNOS and arginase 2 were observed possibly indicating the activation of immune cells at the site of infection (Gonzalez et al. 2007b). The activation of complement by *I. multifiliis* infection was also observed. Gonzalez and coworkers measured a pronounced up-regulation of the carp factor B (Bf/

C2) in skin, blood and liver, and they could also establish a significant up-regulation of complement factor C3 in the liver (Gonzalez et al. 2007a). Studies on rainbow trout have also shown an accumulation of B and T cells at the infected site, and up-regulation of genes encoding IgM, IgT (the trout molecule for IgZ) and IL-8 (Olsen et al. 2011).

Immune Modulation

Immune modulation has risen as an option to improve fish health and welfare. Thus, fish immune modulators are being explored and many agents are currently used by aquaculture industry (Sakai 1999, Moradali et al. 2007). Overcrowding can negatively affect the health of cultured fish, increasing susceptibility to infections and tissue damage. Different chemotherapeutic agents and drugs have been used to treat bacterial infections in cultured fish, but the incidence of drug-resistant bacteria and environmental issues have become major subjects of discussion (Sakai 1999, Douet et al. 2009). Vaccination is also used as a method for prevention of infectious diseases. However, vaccines against several pathogenic agents have not been developed, and the immediate control of all fish diseases and mechanical tissue damage associated with overcrowding is impossible using only vaccines (Ellis 1988, Sakai 1999). Immune-modulators or biological response modifiers are compounds capable of interacting with the immune system and modifying the host response. This interaction can potentially improve fish health by enhancing non-specific defense mechanisms, such as phagocytic activity, cytokine production and complement activation (Sakai 1999, Moradali et al. 2007). The outcome of immune modulation can vary depending of multiple aspects like dose, route and timing of administration, mechanism of action, site of activity and the compound itself (Moradali et al. 2007). Some immune modulators and their effect in carp have been summarized in Table 8.1.

Summary

The carp immune system comprises innate and adaptive immune responses. Innate immunity is the most important branch of the fish immune defense, and is supported by the immunological recognition of pathogens and damage cells, inflammatory responses, as well as by a wide range of immune effect or functions including phagocytosis, production of cell radicals and NETs. In addition, humoral factors such as the complement system, lectins, lysozyme and anti-microbial peptides play an important role on the fish's first line of defense.

The adaptive immune response is the other constituent of the immune system. It is maintained by cellular components such as monocytes, macrophages and B cells (APCs), which have been activated by the innate immune response. APCs present engulfed pathogen moieties to T cells, and in this way, T cells become activated. Signaling immune molecules like cytokines and chemokines, orchestrate the immune responses leading to

Table 8.1. Immunomodulators and their effects in carp.

Species	Substance	Via	Dose	Challenged	Effects	Reference
Common carp	Baker's yeast-derived	Ip injection	500 or 1000 µg/fish	+*Aeromonas hydrophila*	↑ Survival rate ↑ Blood leukocyte ↑ O_2^- ↑ Bactericidal activity	(Selvaraj et al. 2005)
		90 minutes bath	150, 750 or 1000 µg/ml	+*Aeromonas hydrophila*	No effect reported	
		Oral	1, 2, 4%	+*Aeromonas hydrophila*	No effect reported	
Common carp	β-glucan Macro Gard ®	Feed supplement	6 mg/kg/day	+ *Aeromonas salmonicida*	↓ *TNFα1, TNFα2* ↓ *IL-1β, IL-6*	(Falco et al. 2012)
Common Carp	Baker's yeast extract (7% β-glucan, 10% nucleotides)	Oral	50 mg/kg/day	Only à-glucan	↑ *IL-1β, TNFα, IL-12p35, IL12-p40,* CXC chemokines, 1day post treatment ↓ *IL-1β, TNF ζ, IL-12p35, IL12-p40,* at 3, 5, 7 and 10 days post treatment (pt)	(Biswas et al. 2012)
				+*Aeromonas hydrophila*	↑ Phagocytic activity 1 and 3 days pt ↓ Viable pathogen counts	
Common carp	β-glucan Macro Gard®	Bath	0.1 µg/l	Mechanical injure	↑ Wound closure ↑ *IL-1β, IL-6* (1 and 3 d post wound) ↓ *IL-8* (3 and 14 d post wound)	(Przybylska et al. 2013)
Common carp	β-glucan Macro Gard®	Head-Kidney leukocytes stimulation (*in vitro*)	100 µg/ml	Only β-glucan	↑ ROS ↑ NO ↑ *IL-1β, IL-6, IL-11*	(Vera-Jimenez and Nielsen 2013, Vera-Jimenez et al. 2013, Pietretti et al. 2013)

Table 8.1. contd....

Table 8.1. contd.

Species	Substance	Via	Dose	Challenged	Effects	Reference
Common carp	Nucleotides	Oral (intubation)	0.15, 1.5 or 15 mg/fish	+ *Aeromonashydrophila*	↑ complement and lysozyme activity, ↑phagocytosis, and O_2^- ↓ Bacterial cell number	(Sakai et al. 2001)
Common carp	Unmethylated CpG DNA	IP injection	1 and 10 ìg/fish		↑ Phagocytic activity ↑ O_2^- ↑ Lysozyme activity	(Tassakka and Sakai 2002)
Koi carp	Baker's yeast-derived	Oral	0.5%	Only β-glucan	↑ Blood leukocyte counts ↑ Respiratory burst ↑ Phagocytic capacity ↑ Lysozyme activity ↑ Survival rate	(Lin et al. 2011)
Grass carp	Mycelia of *Poriúcocos*-derived	IP injection	10 mg/kg	+ *Aeromonas veronii* + Grass carp haemorrhage virus	↑ Superoxide dismutase and catalase activity ↓ Viral infection	(Kim et al. 2009)
Indian major carp	Oyster mushroom -derived	Head-Kidney leukocytes stimulation (*in vitro*)	1, 10 or 100 µg/ml		↑ Respiratory burst ↑ Phagocytic activity ↑Bactericidal activity	(Kamilya et al. 2006)
Indian major carp	Vitamin C	Oral	1000 mg/kg diet	+ *Aeromonas hydrophila*	↓ Mortality rate Faster infiltration of leukocytes ↓ Necrotic lesions Faster wound healing	(Sobhana et al. 2002)

Indian major carp	β-1,3 glucan	Oral	0.1%	Only β-glucan	↑ Bacterial agglutination titre ↑ Haemagglutinationtitre ↑ Phagocytic activity	(Sahoo and Mukherjee 2001a, Sahoo and Mukherjee 2002)
				+ *Aeromonas hydrophila*	↓ Mortality rate	
				+ *Edwardsiellatarda*	↑ Antibody response ↓ Mortality rate	
Indian major carp	Levamisol	Oral	5 mg/kg	+ *Aeromonas hydrophila*	↓ Mortality rate ↑ Lysozyme activity ↑ ROS	(Sahoo and Mukherjee 2001b)
				+ Aflatoxin $B_1(AFB)_1$ (immunocompromised)	Restored lysozyme activity Restored Phagocytic activity Restored ROS production	
Indian major carp	Garlic	Oral	0.1, 0.5, 1% dry weight	+ *Aeromonas hydrophila*	↑ O_2^- ↑ Lysozyme activity ↓ Mortality rate ↑ Bactericidal activity	(Sahu et al. 2007)

the maturation, activation and recruitment of leukocytes to the infected or injured area, and the further expansion of cytotoxic cells, T helper cells, B memory cells and immunoglobulins. Although the carp adaptive immune system is believed to be less efficient than their counterpart in mammals, it still plays an important role in fish immunity. Furthermore, resistance can be improved with vaccination and the use of immune modulators.

In this chapter immune responses to some of the most common viral, bacterial, fungal and parasitic diseases in carp are described, and immune-modulation is discussed as an option to improve carp health and welfare.

References

Afonso, A., S. Lousada, J. Silva, A.E. Ellis and M.T. Silva. 1998. Neutrophil and macrophage responses to inflammation in the peritoneal cavity of rainbow trout *Oncorhynchus mykiss*. A light and electron microscopic cytochemical study. Dis. Aquat. Org. 34: 27–37.

Akira, S., S. Uematsu and O. Takeuchi. 2006. Pathogen recognition and innate immunity. Cell 124: 783.

Alvarez-Pellitero, P. 2008. Fish immunity and parasite infections: from innate immunity to immunoprophylactic prospects. Vet. Immunol. Immunopathol. 126: 171–198.

Ardo, L., Z. Jeney, A. Adams and G. Jeney. 2010. Immune responses of resistant and sensitive common carp families following experimental challenge with *Aeromonas hydrophila*. Fish Shellfish Immun. 29: 111–116.

Arockiaraj, J., S. Easwvaran, P. Vanaraja, A. Singh, R.Y. Othman and S. Bhassu. 2012. Molecular cloning, characterization and gene expression of an antioxidant enzyme catalase (MrCat) from *Macrobrachium rosenbergii*. Fish Shellfish ImmunolFish Shellfish Immun. 32: 670–682.

Babior, B.M. 1999. The production and use of reactive oxidants by phagocytes. pp. 503–518. *In*: D.L. Gilbert and C.A. Colton (eds.). Reactive Oxygen Species in Biological Systems. Kluwer Academic/Plenum Publishers, New York.

Bellavite, P. 1988. The superoxide-forming enzymatic system of phagocytes. Free Radic. Biol. Med. 4: 225–261.

Bianchi, M.E. 2007. DAMPs, PAMPs and alarmins: all we need to know about danger. J. Leukoc. Biol. 81: 1–5.

Bilitewski, U. 2008. Determination of immunomodulatory effects: focus on functional analysis of phagocytes as representatives of the innate immune system. Anal. Bioanal. Chem. 391: 1545–1554.

Biswas, G., H. Korenaga, H. Takayama, T. Kono, H. Shimokawa and M. Sakai. 2012. Cytokine responses in the common carp, *Cyprinus carpio* L. treated with baker's yeast extract. Aquaculture 356-357: 169.

Boshra, H., J. Li and J.O. Sunyer. 2006. Recent advances on the complement system of teleost fish. Fish Shellfish Immun. 20: 239–262.

Brogden, G., M. von Köckritz-Blickwede, M. Adamek, F. Reuner, V. Jung-Schroers, H.Y. Naim and D. Steinhagen. 2012. β-Glucan protects neutrophil extracellular traps against degradation by *Aeromonas hydrophila* in carp (*Cyprinus carpio*). Fish Shellfish Immun. 33: 1060–1064.

Buchanan, J.T., A.J. Simpson, R.K. Aziz, G.Y. Liu, S.A. Kristian, M. Kotb, J. Feramisco and V. Nizet. 2006. DNase expression allows the pathogen group A *Streptococcus* to escape killing in neutrophil extracellular traps. Curr. Biol. 16: 396–400.

Castillo-Briceño, P., M.P. Sepulcre, E. Chaves-Pozo, J. Meseguer, A. García-Ayala and V. Mulero. 2009. Collagen regulates the activation of professional phagocytes of the teleost fish gilthead seabream. Mol. Immunol. 46: 1409–1415.

Castro, R., D. Bernard, M.P. Lefranc, A. Six, A. Benmansour and P. Boudinot. 2011. T cell diversity and TcR repertoires in teleost fish. Fish Shellfish Immun. 31: 644–654.

Chen, G.Y. and G. Nunez. 2010. Sterile inflammation: sensing and reacting to damage. Nat. Rev. Immunol. 10: 826–837.

Chung, S. and C.J. Secombes. 1988. Analysis of events occurring within teleost macrophages during the respiratory burst. Comp. Biochem. Phys. B 89: 539–544.

Concha, M.I., S. Molina, C. Oyarzún, J. Villanueva and R. Amthauer. 2003. Local expression of apolipoprotein A-I gene and a possible role for HDL in primary defence in the carp skin. Fish Shellfish Immun. 14: 259–273.

Dalmo, R.A. and J. Bøgwald. 2008. β-glucans as conductors of immune symphonies. Fish Shellfish Immun. 25: 384–396.

Diegelmann, R.F. and M.C. Evans. 2004. Wound healing: an overview of acute, fibrotic and delayed healing. Front. Biosci. 9: 283–289.

Douet, D.G., H. Le Bris and E. Giraud. 2009. Environmental aspects of drug and chemical use in aquaculture: An overview. Options Méditerranéennes, A. 86: 105.

Edholm, E., E. Bengten and M. Wilson. 2011. Insights into the function of IgD. Dev. Comp. Immunol. 35: 1309–1316.

Ellis, A.E. 1988. Current aspects of fish vaccination. Dis. Aquat. Org. 4: 159.

Ellis, A.E. 2001. Innate host defense mechanisms of fish against viruses and bacteria. Dev. Comp. Immunol. 25: 827–839.

Falco, A., P. Frost, J. Miest, N. Pionnier, I. Irnazarow and D. Hoole. 2012. Reduced inflammatory response to *Aeromonas salmonicida* infection in common carp (*Cyprinus carpio* L.) fed with β-glucan supplements. Fish Shellfish Immun. 32: 1051.

Fillatreau, S., A. Six, S. Magadan, R. Castro, J.O. Sunyer and P. Boudinot. 2013. The astonishing diversity of Ig classes and B cell repertoires in teleost fish. Front. immun. 4: 28.

Forlenza, M., J.D.A.d.C. Dias, T. Veselý, D. Pokorová, H.F.J. Savelkoul and G.F. Wiegertjes. 2008a. Transcription of signal-3 cytokines, IL-12 and IFNαβ, coincides with the timing of CD8αβ up-regulation during viral infection of common carp (*Cyprinus carpio* L.). Mol. Immunol. 45: 1531–1547.

Forlenza, M., I.R. Fink, G. Raes and G.F. Wiegertjes. 2011. Heterogeneity of macrophage activation in fish. Dev. Comp. Immunol. 35: 1246–1255.

Forlenza, M., M. Nakao, I. Wibowo, M. Joerink, J.A.J. Arts, H.F.J. Savelkoul and G.F. Wiegertjes. 2009. Nitric oxide hinders antibody clearance from the surface of *Trypanoplasma borreli* and increases susceptibility to complement-mediated lysis. Mol. Immunol. 46: 3188–3197.

Forlenza, M., J.P. Scharsack, N.M. Kachamakova, A.J. Taverne-Thiele, J.H. Rombout and G.F. Wiegertjes. 2008b. Differential contribution of neutrophilic granulocytes and macrophages to nitrosative stress in a host-parasite animal model. Mol. Immunol. 45: 3178–3189.

Fridovich, I. 1978. The biology of oxygen radicals. Science 201: 875–880.

Fujiki, K., M. Nakao and B. Dixon. 2003. Molecular cloning and characterisation of a carp (*Cyprinus carpio*) cytokine-like cDNA that shares sequence similarity with IL-6 subfamily cytokines CNTF, OSM and LIF. Dev. Comp. Immunol. 27: 127.

Genestra, M. 2007. Oxyl radicals, redox-sensitive signalling cascades and antioxidants. Cell. Signal. 19: 1807–1819.

Gonzalez, S.F., K. Buchmann and M.E. Nielsen. 2007a. Complement expression in common carp (*Cyprinus carpio* L.) during infection with *Ichthyophthirius multifiliis*. Dev. Comp. Immunol. 31: 576–586.

Gonzalez, S.F., K. Buchmann and M.E. Nielsen. 2007b. Real-time gene expression analysis in carp (*Cyprinus carpio* L.) skin: Inflammatory responses caused by the ectoparasite *Ichthyophthirius multifiliis*. Fish Shellfish Immun. 22: 641–650.

Gonzalez, S.F., M.O. Huising, R. Stakauskas, M. Forlenza, B.M. Lidy Verburg-van Kemenade, K. Buchmann, M.E. Nielsen and G.F. Wiegertjes. 2007c. Real-time gene expression analysis in carp (*Cyprinus carpio* L.) skin: Inflammatory responses to injury mimicking infection with ectoparasites. Dev. Comp. Immunol. 31: 244–254.

Goodbourn, S., L. Didcock and R. Randall. 2000. Interferons: cell signalling, immune modulation, antiviral response and virus countermeasures. J. Gen. Virol. 81: 2341–2364.

Hansen, J.D., L.N. Vojtech and K.J. Laing. 2011. Sensing disease and danger: a survey of vertebrate PRRs and their origins. Dev. Comp. Immunol. 35: 886–897.

Harikrishnan, R., C. Balasundaram and M. Heo. 2010. Molecular studies, disease status and prophylactic measures in grouper aquaculture: economic importance, diseases and immunology. Aquaculture 309: 1–14.

Hashimoto, K., T. Nakanishi and Y. Kurosawa. 1990. Isolation of carp genes encoding major histocompatibility complex antigens. Proc. Natl. Acad. Sci. USA 87: 6863–6867.

Holland, M.C.H. and J.D. Lambris. 2002. The complement system in teleosts. Fish Shellfish Immun. 12: 399–420.

Iger, Y. and M. Abraham. 1990. The process of skin healing in experimentally wounded carp. J. Fish Biol. 36: 421.

Ingerslev, H.C., T. Lunder and M.E. Nielsen. 2010. Inflammatory and regenerative responses in salmonids following mechanical tissue damage and natural infection. Fish Shellfish Immun. 29: 440–450.

Jeney, Z. and G. Jeney. 1995. Recent achievements in studies on diseases of common carp (*Cyprinus carpio* L.). Aquaculture 129: 397.

Jiang, Y. and R.F. Doolittle. 2003. The evolution of vertebrate blood coagulation as viewed from a comparison of puffer fish and sea squirt genomes. Proc. Natl. Acad. Sci. USA 100: 7527.

Jones, D.P. 1982. Intracellular catalase function: Analysis of the catalitic activity by product formation in isolated liver cells. Arch. Biochem. Biophys. 214: 806–814.

Jones, S.R.M. 2001. The occurrence and mechanisms of innate immunity against parasites in fish. Dev. Comp. Immunol. 25: 841–852.

Jurado, R.L. 1997. Iron, infections, and anemia of inflammation. Clin. Infect. Dis. 25: 888–895.

Jurecka, P., I. Irnazarow, J.L. Stafford, A. Ruszczyk, N. Taverne, M. Belosevic, H.F. Savelkoul and G.F. Wiegertjes. 2009a. The induction of nitric oxide response of carp macrophages by transferrin is influenced by the allelic diversity of the molecule. Fish Shellfish Immun. 26: 632–638.

Jurecka, P., I. Irnazarow, A.H. Westphal, M. Forlenza, J.A. Arts, H.F. Savelkoul and G.F. Wiegertjes. 2009b. Allelic discrimination, three-dimensional analysis and gene expression of multiple transferrin alleles of common carp (*Cyprinus carpio* L.). Fish Shellfish Immun. 26: 573–581.

Jurecka, P., G.F. Wiegertjes, K.Ł. Rakus, A. Pilarczyk and I. Irnazarow. 2009c. Genetic resistance of carp (*Cyprinus carpio* L.) to *Trypanoplasma borreli*: influence of transferrin polymorphisms. Vet. Immunol. Immunopathol. 127: 19–25.

Kales, S.C., S.J. DeWitte-Orr, N.C. Bols and B. Dixon. 2007. Response of the rainbow trout monocyte/macrophage cell line, RTS11 to the water molds *Achlya* and *Saprolegnia*. Mol. Immunol. 44: 2303–2314.

Kamilya, D., D. Ghosh, S. Bandyopadhyay, B.C. Mal and T.K. Maiti. 2006. *In vitro* effects of bovine lactoferrin, mushroom glucan and Abrus agglutinin on Indian major carp, catla (*Catla catla*) head kidney leukocytes. Aquaculture 253: 130.

Kanta, J. 2011. The role of hydrogen peroxide and other reactive oxygen species in wound healing. Acta Med. 54: 97–101.

Kawai, T. and S. Akira. 2010. The role of pattern-recognition receptors in innate immunity: Update on toll-like receptors. Nat. Immunol. 11: 373–384.

Kim, Y., F. Ke and Q.Y. Zhang. 2009. Effect of β-glucan on activity of antioxidant enzymes and Mx gene expression in virus infected grass carp. Fish Shellfish Immun. 27: 336.

Knowles, R.G. and S. Moncada. 1994. Nitric oxide synthases in mammals. Biochem. J. 298: 249–258.

Koumans-van Diepen, J.E., E. Egberts, B.R. Peixoto, N. Taverne and J.H. Rombout. 1995. B cell and immunoglobulin heterogeneity in carp (*Cyprinus carpio* L.); an immuno (cyto) chemical study. Dev. Comp. Immunol. 19: 97–108.

Lin, S., Y. Pan and L. Luo. 2011. Effects of dietary beta-1,3-glucan, chitosan or raffinose on the growth, innate immunity and resistance of koi (*Cyprinus carpio koi*). Fish Shellfish Immun. 31: 788–794.

Lotze, M.T., H.J. Zeh, A. Rubartelli, L.J. Sparvero, A.A. Amoscato, N.R. Washburn, M.E. DeVera, X. Liang, M. Tör and T. Billiar. 2007. The grateful dead: damage-associated molecular pattern molecules and reduction/oxidation regulate immunity. Immunol. Rev. 220: 60–81.

Lushchak, V. 2008. Oxidative stress as a component transition metal toxicity in fish. pp. 1–29. *In*: E. Svensson (ed.). Aquatic Toxicology Research Focus. Nova Science Publishers, Inc., New York.

Madamanchi, N. and M. Runge. 2007. Oxidative stress. pp. 549–562. *In*: M.S. Runge and C. Patterson (eds.). Contemporary Cardiology: Principles of Molecular Cardiology. Humana Press Inc., Totowa, NJ.

Magnadottir, B. 2006. Innate immunity of fish (overview). Fish Shellfish Immun 20: 137–151.

Martin, P. and S.J. Leibovich. 2005. Inflammatory cells during wound repair: the good, the bad and the ugly. Trends Cell Biol. 15: 599.

Medzhitov, R. 2001. Toll-like receptors and innate immunity. Nat. Rev. Immunol. 1: 135–145.

Medzhitov, R. and C. Janeway Jr. 2000. Innate immune recognition: mechanisms and pathways. Immunol. Rev. 173: 89.

Meneghini, R. 1997. Iron homeostasis, oxidative stress, and DNA damage. Free Radical Bio. Med. 23: 783–792.

Midwood, K.S. and A.M. Piccinini. 2010. DAMPening inflammation by modulating TLR signalling. Mediators Inflamm. 2010. doi: 10.1155/2010/ 672395. Epub 2010 Jul13.

Moradali, M.F., H. Mostafavi, S. Ghods and G. Hedjaroude. 2007. Immunomodulating and anticancer agents in the realm of macromycetes fungi (macrofungi). Int. Immunopharmacol. 7: 701.

Nakanishi, T., U. Fischer, J.M. Dijkstra, S. Hasegawa, T. Somamoto, N. Okamoto and M. Ototake. 2002. Cytotoxic T cell function in fish. Dev. Comp. Immunol. 26: 131–139.

Nakao, M., T. Kajiya, Y. Sato, T. Somamoto, Y. Kato-Unoki, M. Matsushita, M. Nakata, T. Fujita and T. Yano. 2006. Lectin pathway of bony fish complement: identification of two homologs of the mannose-binding lectin associated with MASP2 in the common carp (*Cyprinus carpio*). J. Immunol. 177: 5471–5479.

Nakao, M., M. Tsujikura, S. Ichiki, T.K. Vo and T. Somamoto. 2011. The complement system in teleost fish: progress of post-homolog-hunting researches. Dev. Comp. Immunol. 35: 1296–1308.

Nakao, M. and T. Yano. 1998. Structural and functional identification of complement components of the bony fish, carp (*Cyprinus carpio*). Immunol. Rev. 166: 27–38.

Nikoskelainen, S., S. Verho, K. Airas and E.M. Lilius. 2005. Adhesion and ingestion activities of fish phagocytes induced by bacterium *Aeromonas salmonicida* can be distinguished and directly measured from highly diluted whole blood of fish. Dev. Comp. Immunol. 29: 525–537.

Noga, E.J., A.J. Ullal, J. Corrales and J.M. Fernandes. 2011. Application of antimicrobial polypeptide host defenses to aquaculture: Exploitation of downregulation and upregulation responses. Comp. Biochem. Physiol. D6: 44–54.

Olsen, M.M., P.W. Kania, R.D. Heinecke, K. Skjoedt, K.J. Rasmussen and K. Buchmann. 2011. Cellular and humoral factors involved in the response of rainbow trout gills to *Ichthyophthirius multifiliis* infections: Molecular and immunohistochemical studies. Fish Shellfish Immun. 30: 859–869.

Pfeilschifter, J., W. Eberhardt and K.F. Beck. 2001. Regulation of gene expression by nitric oxide. Pflugers Arch. 442: 479–486.

Pick, E. and D. Mizel. 1981. Rapid microassays for the measurement of superoxide and hydrogen peroxide production by macrophages in culture using an automatic enzyme immunoassay reader. J. Immunol. Methods 46: 211–226.

Pietretti, D., N.I. Vera-Jimenez, D. Hoole and G.F. Wiegertjes. 2013. Oxidative burst and nitric oxide responses in carp macrophages induced by zymosan, MacroGard[(R)] and selective dectin-1 agonists suggest recognition by multiple pattern recognition receptors. Fish Shellfish Immun. 35: 847–857.

Pijanowski, L., L. Golbach, E. Kolaczkowska, M. Scheer, B.M.L. Verburg-van Kemenade and M. Chadzinska. 2013. Carp neutrophilic granulocytes form extracellular traps via ROS-dependent and independent pathways. Fish Shellfish Immun. 34: 1244–1252.

Przybylska, D.A., J.G. Schmidt, N.I. Vera-Jimenez, D. Steinhagen and M.E. Nielsen. 2013. β-glucan enriched bath directly stimulates the wound healing process in common carp (*Cyprinus carpio* L.). Fish Shellfish Immun. 35: 998–1006.

Rajanbabu, V. and J.Y. Chen. 2011. Applications of antimicrobial peptides from fish and perspectives for the future. Peptides 32: 415–420.

Rakus, K.Ł., I. Irnazarow, M. Adamek, L. Palmeira, Y. Kawana, I. Hirono, H. Kondo, M. Matras, D. Steinhagen, B. Flasz, G. Brogden, A. Vanderplasschen and T. Aoki. 2012. Gene expression analysis of common carp (*Cyprinus carpio* L.) lines during Cyprinid herpesvirus 3 infection yields insights into differential immune responses. Dev. Comp. Immunol. 37: 65–76.

Rakus, K.Ł., G.F. Wiegertjes, P. Jurecka, P.D. Walker, A. Pilarczyk and I. Irnazarow. 2009. Major histocompatibility (MH) class II *B* gene polymorphism influences disease resistance of common carp (*Cyprinus carpio* L.). Aquaculture 288: 44–50.

Rakus, K.Ł., G.F. Wiegertjes, R.J. Stet, H.F. Savelkoul, A. Pilarczyk and I. Irnazarow. 2003. Polymorphism of major histocompatibility complex class II *B* genes in different lines of the common carp (*Cyprinus carpio*). Aquat. Living Resour. 16: 432–437.

Rauta, P.R., B. Nayak and S. Das. 2012. Immune system and immune responses in fish and their role in comparative immunity study: A model for higher organisms. Immunol. Lett. 148: 23–33.

Rebl, A., T. Goldammer and H. Seyfert. 2010. Toll-like receptor signaling in bony fish. Vet. Immunol. Immunopathol. 134: 139–150.

Rieger, A.M., B.E. Hall and D.R. Barreda. 2010. Macrophage activation differentially modulates particle binding, phagocytosis and downstream antimicrobial mechanisms. Dev. Comp. Immunol. 34: 1144–1159.

Rieger, A.M. and D.R. Barreda. 2011. Antimicrobial mechanisms of fish leukocytes. Dev. Comp. Immunol. 35: 1238–1245.

Rock, K.L., E. Latz, F. Ontiveros and H. Kono. 2010. The sterile inflammatory response. Annu. Rev. Immunol. 28: 321–342.

Ryo, S., R.H.M. Wijdeven, A. Tyagi, T. Hermsen, T. Kono, I. Karunasagar, J.H.W.M. Rombout, M. Sakai, B.M.L.V. Kemenade and R. Savan. 2010. Common carp have two subclasses of bonyfish specific antibody IgZ showing differential expression in response to infection. Dev. Comp. Immunol. 34: 1183–1190.

Saeij, J.P.J., W.B. van Muiswinkel, M. van de Meent, C. Amaral and G.F. Wiegertjes. 2003a. Different capacities of carp leukocytes to encounter nitric oxide-mediated stress: a role for the intracellular reduced glutathione pool. Dev. Comp. Immunol. 27: 555–568.

Saeij, J.P.J., W.B. Van Muiswinkel, A. Groeneveld and G.F. Wiegertjes. 2002. Immune modulation by fish kinetoplastid parasites: arole for nitric oxide. Parasitology 124: 77–86.

Saeij, J.P.J., B.J.d. Vries and G.F. Wiegertjes. 2003b. The immune response of carp to *Trypanoplasma borreli*: kinetics of immune gene expression and polyclonal lymphocyte activation. Dev. Comp. Immunol. 27: 859–874.

Sahoo, P.K. and S.C. Mukherjee. 2002. The effect of dietary immunomodulation upon *Edwardsiella tarda* vaccination in healthy and immunocompromised Indian major carp (*Labeo rohita*). Fish Shellfish Immun. 12: 1.

Sahoo, P.K. and S.C. Mukherjee. 2001a. Effect of dietary β-1,3 glucan on immune responses and disease resistance of healthy and aflatoxin B 1-induced immunocompromised rohu (*Labeo rohita* Hamilton). Fish Shellfish Immun. 11: 683.

Sahoo, P. and S. Mukherjee. 2001b. Dietary intake of levamisole improves non-specific immunity and disease resistance of healthy and aflatoxin-induced immunocompromised rohu, *Labeo rohita*. J. Appl. Aquacult. 11: 15–25.

Sahu, S., B.K. Das, B.K. Mishra, J. Pradhan and N. Sarangi. 2007. Effect of *Allium sativum* on the immunity and survival of *Labeo rohita* infected with *Aeromonas hydrophila*. J. Appl. Ichthyol. 23: 80–86.

Sakai, M. 1999. Current research status of fish immunostimulants. Aquaculture 172: 63.

Sakai, M., K. Taniguchi, K. Mamoto, H. Ogawa and M. Tabata. 2001. Immunostimulant effects of nucleotide isolated from yeast RNA on carp, *Cyprinus carpio* L. J. Fish Dis. 24: 433–438.

Saurabh, S. and P. Sahoo. 2008. Lysozyme: an important defence molecule of fish innate immune system. Aquacult. Res. 39: 223–239.

Savan, R., A. Aman and M. Sakai. 2003. Molecular cloning of G type lysozyme cDNA in common carp (*Cyprinus carpio* L.). Fish Shellfish Immun. 15: 263–268.

Scharsack, J., D. Steinhagen, C. Kleczka, J. Schmidt, W. Körting, R. Michael, W. Leibold and H. Schuberth. 2003. The haemoflagellate *Trypanoplasma borreli* induces the production of nitric oxide, which is associated with modulation of carp (*Cyprinus carpio* L.) leucocyte functions. Fish Shellfish Immun. 14: 207–222.

Schmidt, J.G. and M.E. Nielsen. 2014. Expression of immune system-related genes during ontogeny in experimentally wounded common carp (*Cyprinus carpio*) larvae and juveniles. Dev. Comp. Immunol. 42: 186–196.

Schreml, S., R.M. Szeimies, L. Prantl, M. Landthaler and P. Babilas. 2010. Wound healing in the 21st century. J. Am. Acad. Dermatol. 63: 866–881.

Segner, H., A.M. Möller, M. Wenger and A. Casanova-Nakayama. 2011. Fish immunotoxicology: research at the crossroads of immunology, ecology and toxicology. Int. Stud. Environ. Chem. 6: 1–15.

Selvaraj, V., K. Sampath and V. Sekar. 2005. Administration of yeast glucan enhances survival and some non-specific and specific immune parameters in carp (*Cyprinus carpio*) infected with *Aeromonas hydrophila*. Fish Shellfish Immun. 19: 293.

Skaar, E.P. 2010. The battle for iron between bacterial pathogens and their vertebrate hosts. PLoS Pathog. 6: e1000949.

Sobhana, K.S., C.V. Mohan and K.M. Shankar. 2002. Effect of dietary vitamin C on the disease susceptibility and inflammatory response of mrigal, *Cirrhinus mrigala* (Hamilton) to experimental infection of *Aeromonas hydrophila*. Aquaculture 207: 225–238.

Soneja, A., M. Drews and T. Malinski. 2005. Role of nitric oxide, nitroxidative and oxidative stress in wound healing. Pharmacol. Rep. 57: 108–119.

Stafford, J.L. and M. Belosevic. 2003. Transferrin and the innate immune response of fish: identification of a novel mechanism of macrophage activation. Dev. Comp. Immunol. 27: 539–554.

Stet, R.M., S.H. van Erp, T. Hermsen, H.A. Sültmann and E. Egberts. 1993. Polymorphism and estimation of the number of *MhcCyca* class I and class I genes in laboratory strains of the common carp (*Cyprinus carpio* L.). Dev. Comp. Immunol. 17: 141–156.

St-Hilaire, S., N. Beevers, C. Joiner, R.P. Hedrick and K. Way. 2009. Antibody response of two populations of common carp, *Cyprinus carpio* L., exposed to koi herpesvirus. J. Fish Dis. 32: 311–320.

Stuart, L.M. and R.A. Ezekowitz. 2008. Phagocytosis and comparative innate immunity: learning on the fly. Nat. Rev. Immunol. 8: 131–141.

Studnicka, M., A. Siwicki and B. Ryka. 1986. Lysozyme level in car (*Cyprinus carpio* L.). Bamidgeh. 38: 22–25.

Sun, X., N. Shang, W. Hu, Y. Wang and Q. Guo. 2007. Molecular cloning and characterization of carp (*Cyprinus carpio* L.) CD8β and CD4-like genes. Fish Shellfish Immun. 23: 1242–1255.

Swain, P., T. Behera, D. Mohapatra, P.K. Nanda, S.K. Nayak, P.K. Meher and B.K. Das. 2010. Derivation of rough attenuated variants from smooth virulent *Aeromonas hydrophila* and their immunogenicity in fish. Vaccine 28: 4626–4631.

Swain, P., S. Dash, P.K. Sahoo, P. Routray, S.K. Sahoo, S.D. Gupta, P.K. Meher and N. Sarangi. 2007. Non-specific immune parameters of brood Indian major carp *Labeo rohita* and their seasonal variations. Fish Shellfish Immun. 22: 38–43.

Syakuri, H., M. Adamek, G. Brogden, K.Ł. Rakus, M. Matras, I. Irnazarow and D. Steinhagen. 2013. Intestinal barrier of carp (*Cyprinus carpio* L.) during a cyprinid herpesvirus 3-infection: Molecular identification and regulation of the mRNA expression of claudin encoding genes. Fish Shellfish Immun. 34: 305–314.

Tanekhy, M., T. Kono and M. Sakai. 2009. Expression profile of cytokine genes in the common carp species *Cyprinus carpio* L. following infection with *Aeromonas hydrophila*. B. Eur. Assoc. Fish Pat. 29: 198–204.

Tassakka, A.C.M.A.R. and M. Sakai. 2002. CpG oligodeoxynucleotides enhance the non-specific immune responses on carp, *Cyprinus carpio*. Aquaculture 209: 1–10.

Tort, L. 2011. Stress and immune modulation in fish. Dev. Comp. Immunol. 35: 1366–1375.

Uribe, C., H. Folch, R. Enriquez and G. Moran. 2011. Innate and adaptive immunity in teleost fish: A review. Vet. Med. 56: 486–503.

Vallejo, A.N., N.W. Miller and L. William Clem. 1992. Antigen processing and presentation in teleost immune responses. Annu. Rev. Fish Dis. 2: 73–89.

van der Marel, M., M. Adamek, S.F. Gonzalez, P. Frost, J.H.W.M. Rombout, G.F. Wiegertjes, H.F.J. Savelkoul and D. Steinhagen. 2012. Molecular cloning and expression of two β-defensin and two mucin genes in common carp (*Cyprinus carpio* L.) and their up-regulation after β-glucan feeding. Fish Shellfish Immun. 32: 494–501.

Vera-Jimenez, N.I. and M.E. Nielsen. 2013. Carp head kidney leukocytes display different patterns of oxygen radical production after stimulation with PAMPs and DAMPs. Mol. Immunol. 55: 231–236.

Vera-Jimenez, N.I., D. Pietretti, G.F. Wiegertjes and M.E. Nielsen. 2013. Comparative study of β-glucan induced respiratory burst measured by Nitroblue tetrazolium assay and Real-time luminol-enhanced chemiluminescence assay in common carp (*Cyprinus carpio* L.). Fish Shellfish Immun. 34: 1216–1222.

Verburg-van Kemenade, B.M., C.M. Ribeiro and M. Chadzinska. 2011. Neuroendocrine-immune interaction in fish: differential regulation of phagocyte activity by neuroendocrine factors. Gen. Comp. Endocrinol. 172: 31–38.

Villalobo, A. 2006. Nitric oxide and cell proliferation. FEBS J. 273: 2329–2344.

Wallach-Dayan, S.B., R. Golan-Gerstl and R. Breuer. 2007. Evasion of myofibroblasts from immune surveillance: a mechanism for tissue fibrosis. Proc. Natl. Acad. Sci. USA 104: 20460–20465.

Wei, L. 2011. Immunological aspect of cardiac remodeling: T lymphocyte subsets in inflammation-mediated cardiac fibrosis. Exp. Mol. Pathol. 90: 74–78.

Wink, D.A., H.B. Hines, R.Y.S. Cheng, C.H. Switzer, W. Flores-Santana, M.P. Vitek, L.A. Ridnour and C.A. Colton. 2011. Nitric oxide and redox mechanisms in the immune response. J. Leukocyte Biol. 89: 873–891.

Witte, M.B. and A. Barbul. 2002. Role of nitric oxide in wound repair. Am. J. Surg. 183: 406–412.

Wooldridge, K.G. and P.H. Williams. 1993. Iron uptake mechanisms of pathogenic bacteria. FEMS Microbiol. Rev. 12: 325–348.

Xiao, F., Y. Wang, W. Yan, M. Chang, M. Yao, Q. Xu, X. Wang, Q. Gao and P. Nie. 2010. Ig heavy chain genes and their locus in grass carp (*Ctenopharyngodon idella*). Fish Shellfish Immun. 29: 594–599.

Yamaguchi, T., F. Katakura, K. Someya, J.M. Dijkstra, T. Moritomo and T. Nakanishi. 2013. Clonal growth of carp (*Cyprinus carpio*) T cells *in vitro*: Long-term proliferation of Th2-like cells. Fish Shellfish Immun. 34: 433–442.

Ye, X., L. Zhang, Y. Tian, A. Tan, J. Bai and S. Li. 2010. Identification and expression analysis of the g-type and c-type lysozymes in grass carp *Ctenopharyngodon idellus*. Dev. Comp. Immunol. 34: 501–509.

Zhang, Q., M. Raoof, Y. Chen, Y. Sumi, T. Sursal, W. Junger, K. Brohi, K. Itagaki and C.J. Hauser. 2010. Circulating mitochondrial DAMPs cause inflammatory responses to injury. Nature 464: 104–107.

Zhang, Y., I. Salinas and J. Oriol Sunyer. 2011. Recent findings on the structure and function of teleost IgT. Fish Shellfish Immun. 31: 627–634.

Zhang, Z. and H.J. Schluesener. 2006. Mammalian toll-like receptors: from endogenous ligands to tissue regeneration. Cell. Mol. Life Sci. 63: 2901.

Part IV

Ecology

Ecology is described as the manifold interactions between the phenotype and all adjacent and indirectly affected non-adjacent components of the food web. This also includes humans as part of this equation, because it is us who affect food webs the most as, e.g., top predators when fishing.

The consideration of humans as top-predators already implies that food web interactions are trophic in the sense of who eats who. This part of the book therefore begins with a chapter describing the feeding ecology of carp. The reader learns that carp undergo different ontogenetic niche shifts. This means carp change their feeding behavior as they grow. This kind of feeding behavior ranges from consumption of zooplankton by carp larvae to crushing hard-shelled mollusks and bivalves by adult carp. The chapter on feeding ecology of carp builds and expands on the different nutritional requirements that are described for keeping carp in aquaculture (Chapters 4 and 6). Carp do not only affect the food web by consuming prey, also their feeding activity alters the food web structure indirectly. While digging through the sediment adult carp dislocate plants, resuspend nutrient buried in the sediment and increase turbidity in a process called bioturbation. These feeding effects are also elaborated upon in the second chapter of the ecology section (Chapter 10) that describes how carp can become invasive species and alter invaded ecosystems in non-desirable ways. In Oceania carp proliferate as an unwanted addition to the pristine freshwater fauna and their feeding ecology leads to adverse effects. Carp as invasive species are a problem that pose economic and ecological threats and hence there is a great deal of research on how to manage invasive carp population. Chapter 10 reviews this research thus synthesizing topical information on the ecology of carp. Ecosystem managers seek to mitigate the negative effects of carp in Oceania. This undertaking requires a detailed knowledge of the feeding and reproductive ecology of the invasive species. Hence there is a lot to be learnt on reproduction, feeding and growth of wild carp from the study of its effects as an invasive species. Information on growth of carp in the wild is also of great interest to sport fishers which is the topic of the last part of the ecology section (Chapter 11). Fishers want to catch big fish. A management of wild carp populations for a sustainable and economically viable sport fisheries therefore requires a great deal of understanding of factors that influence the stocking success, the population growth, individual growth and the negative effects that can come with the proliferation of carp populations. The chapter on recreational fishing summarizes this knowledge and critically assesses threshold traits above which carp can become a nuisance depending on the ecosystem they proliferate in. Fishing for carp with rod and line is a centuries-old tradition and steadily increases in popularity. It hence epitomizes the strong connection between this species and humans. Carp and humans interact not only as aquaculturists and their market species but also as ecosystem managers dealing with the effect of invasive carp and sport fishers seeking to continuously catch big fish by sustainably managing a fishing stock.

9

Feeding Ecology of Carp

Brian Huser[1,][*] and Pia Bartels[2]

The common carp is generally considered an omnivore, feeding on a wide variety of organic material and prey. With increasing size however, carp usually prefer macroinvertebrates. Adult carp possess several distinct morphological features that facilitate feeding on benthos, especially on organisms that live on and in the sediment. The characteristic feeding style of carp, i.e., probing the sediment at the bottom, can result in multiple important effects on lake and pond ecosystems. In this chapter, we first briefly describe the feeding apparatus of carp. Subsequently, we describe the diet of carp in detail, and present stable isotope analysis as a powerful tool to determine resource utilization. In the last part, we present multiple direct and indirect effects of carp feeding on aquatic communities and ecosystems. These impacts are particularly important for the management of aquatic ecosystems.

Resource Utilization of Common Carp

Feeding Apparatus

Carp display a set of distinct morphological features that facilitate omnivory. One set of features is well adapted to enable the retention and ingestion of small food items, whereas the other set is particularly adapted to feed on larger and hard prey. Zooplankton and other small particles, for example, can be stirred up from the sediment and ingested by carp via gulping. The expansion of the buccal cavity thereby produces a slow suction flow that

[1] Department of Aquatic Sciences and Assessment, Swedish University of Agricultural Sciences, Box 7050, SE-750 07 Uppsala, Sweden.
  Email: brian.huser@slu.se
[2] Department of Ecology and Environmental Science, Umeå University, SE-901 87 Umeå, Sweden.
  Email: pia.bartels.pb@gmail.com
[*] Corresponding author

sucks in particles from all sides around the mouth. Particles are then retained by the branchial sieve; a filter-like structure that is attached to the branchial arches. The mesh width of the branchial sieve determines the particle size that will be most efficiently retained. Sibbing et al. (1986) demonstrated that carp can efficiently retain phytoplankton, large zooplankton and detritus (> 250 μm) with their branchial sieve. Carp fry can consume smaller (> 100 μm) prey (Gisbert et al. 1996). With increasing size of carp however, the ability to retain small food items declines (Sibbing 1988). That is due to the increasing size of the branchial sieve during growth, resulting in less efficient planktivory of carp larger than 30 cm (Sibbing 1988).

Benthivory and the ingestion of large food items is facilitated by a set of special characteristics of the feeding apparatus such as a protusible mouth, toothless jaws and toothless palatine, and palatal and postlingual organs (Sibbing 1991). The protusible mouth is especially well adapted to ingest large prey. Thereby, the carp's upper jaw is extended downwards, forming a rounded suction tube, and the oropharyngeal and opercular cavities are rapidly expanded, resulting in a fast and strong suction flow that efficiently sucks in a target food item. Even attached prey or prey hidden in crevices can be ingested by the strong suction. The size of the prey that can be ingested is limited by the gape width and can be approximately 9% of the standard length of small carp, and approximately 7% of the standard length of carp larger than 25 cm (Sibbing 1988). The ingested food item is then transported by waves of muscular contractions between the palatal and postlingual organs to the chewing cavity. In the chewing cavity, food is then crushed between the pharyngeal jaws and the chewing pad. In particular, hard material, such as plant seeds, mollusks and debris, is efficiently processed by this powerful chewing mechanism (Sibbing 1988, Tucker et al. 1996). These adaptations enable carp to feed on a wide variety of food items, ranging from small to large particles, which facilitates an omnivorous life style.

Ontogenetic Diet Shift and Carp Diet

As many other fish species, carp change their resource utilization during their ontogeny (Fig. 9.1). Carp larvae feed endogenously, from the remains of the egg nutrition. Approximately 20–30 days after hatching, changes in carp at 20–25 mm standard length initiate the juvenile period (Vilizzi and Walker 1999). This switch is reflected in their feeding morphology. The position of the mouth changes from being sub-ventral in the free-living embryo to being terminal in the feeding larvae (Vilizzi and Walker 1999). The terminal mouth position mirrors the onset of the planktivore life stage. Exactly how long this stage persists is unclear. Khan (2003) found that carp fry < 2 cm fed exclusively on zooplankton and started including benthos at a length > 2 cm (see also Kloskowski 2011). In contrast, Vass and Vass Van Oven (1959) showed that carp fry ingested benthos (chironomids) as early as two days after hatching, i.e., at a length of 7.5 mm. Hoda and Tsukahara (1971) report changes in feeding morphology that suggest the transition to benthivory such as a

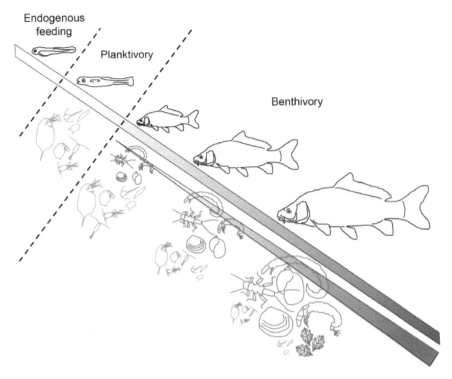

Figure 9.1. Ontogentic diet shift of carp. Carp embryos are supported by endogenous nutrition. Carp fry is planktivorous but preferably feeds on cladocerans. With increasing size, carp increasingly include macrobenthos in their diets (indicated by red arrow). The switch from planktivory to benthivory might happen very early during ontogeny (see text for more details). Large carp can supplement their diet with plant material. Large zooplankton and detritus (not shown) occurs in the diet of carp in all size classes. Different sizes of food items indicate different importance for carp diet, where large items are more important than small items. Different colors represent different resources: green = phytoplankton; blue = zooplankton; red = macroinvertebrates; brown = plant material (seeds and leaves).

downwardly projected mouth as early as in 20 mm standard length juveniles and an increase in mouth protrusibility between 20–25 mm standard length. With these changes, carp resemble the adult form in feeding morphology, feeding habitats and locomotion. Zooplankton is the most important resource for carp fry. Umesh et al. (1999) demonstrated that carp fry grew substantially better (on average 50%) in tanks where zooplankton production was high compared to control tanks where phytoplankton production was higher than zooplankton production. However, tanks with high zooplankton densities also contained biofilms with high quantities of attached algae and other food organisms, suggesting that microbenthos might be an important food source for carp at early life stages. With increasing body size of carp, planktivory becomes less important, and macrobenthos constitutes the major proportion of carp diet (Chapman and Fernando 1994, Garcia-Berthou 2001, Khan 2003,

Britton et al. 2007), although zooplankton is commonly found in the diet of carp of all size classes (Garcia-Berthou 2001, Khan 2003, Britton et al. 2007).

Adult carp diet has been demonstrated to be exceptionally diverse, including phytoplankton, zooplankton, pelagic invertebrates, macrobenthos, plant material such as seeds and leaves, detritus, fish eggs, and in some cases fish. Although animal items are the preferred food when both plant and animal matter are abundantly available (Eder and Carlson 1977), plant material is frequently found in the gut. However, reports on the importance of plant matter and detritus vary substantially in the literature. Some studies demonstrate high proportions of detritus and plant matter in carp diets (e.g., Chapman and Fernando 1994, Michel and Oberdorff 1995, Garcia-Berthou 2001), whereas others report only a minor role of plant material and debris in carp diets (e.g., Khan 2003, Britton et al. 2007). The utilization of plant material can be an important supplement to the carp diet (Hume et al. 1983), in particular when animal resources become scarce. Carp consume a wide variety of macroinvertebrates including insects such as adult and larval chironomids and other diptera, ephemeroptera, zygoptera, hemiptera, trichoptera, mollusca such as gastropoda and bivalvia, small crustacea such as amphipoda and ostracoda, and annelida such as oligochaeta (e.g., Sibbing 1988, Michel and Oberdorff 1995, Marsden 1997, Garcia-Berthou 2001, Khan 2003, Britton et al. 2007, Singh et al. 2010, Zambrano et al. 2010, Kloskowski 2011). Several studies suggest high preference for chironomids (Hruska 1961, Eder and Carlson 1977, Spataru et al. 1980, Hume et al. 1983, Lammes and Hoogenboezem 1991, Garcia-Berthou 2001, Khan 2003). Fish eggs, fish, and crayfish have occasionally been reported in the diets of large carp (Marsden 1997, Khan 2003, Zambrano et al. 2010, Jackson et al. 2012). Britton et al. (2007) found remains of small fish in carp as small as 12 mm. However, they also report that fish and crayfish presented only a minor fraction of carp diets. Adult crayfish are vulnerable to carp predation during ecdysis (Hasan and MacIntosh 1992), but juvenile crayfish may be more accessible to carp. In summary, carp diet mirrors a highly omnivorous feeding behavior including a broad range of available pelagic and benthic resources, suggesting that carp might compete with many native and other introduced species.

Prey Preference and Diet Overlap

The omnivorous feeding style from an early age can result in high overlap in resource use between different age classes of carp (Khan 2003, Kloskowski 2011). Generally, carp fry and very small carp show a more different diet from adult carp due to the high preference for zooplankton. Khan (2003) showed that diet overlap between small and medium carp was intermediate (Schoener's similarity index: 0.55–0.65) and between small and large carp was low (0.36–0.44). An overlap of more than 60% (> 0.60) is considered as substantial (Wallace 1981). Carp fry generally prefer to feed on copepods, rotifers, and small-bodied cladocerans (Kloskowski 2011), whereas larger carp usually prefer large cladocerans (Gisbert et al. 1996, Garcia-Berthou

2001, Khan 2003, Kloskowsi 2011). In adult carp, large cladocerans also dominate the zooplankton diet (Kloskowski 2011), presumably because the size-selective filtering system promotes the retention of large-sized zooplankton (Sibbing 1988).

With increasing size and the onset of benthivory, diet overlap becomes increasingly higher (0.72–0.74, Khan 2003; 0.62–0.89, Kloskowski 2011). Adult carp generally prefer macroinvertebrates over plankton. In an experimental study, Rahman et al. (2010) demonstrated that carp maintained in ponds with no access to benthic resources included 70.9% zooplankton and 29.1% phytoplankton in their diet. In contrast, carp in ponds with access to a bottom substrate usually preferred macroinvertebrates (76.1% of diet content), whereas zooplankton (17.9%) and phytoplankton (6.0%) accounted for only a minor fraction of the diet. This indicates that carp can switch to less preferred resources when preferred resources become scarce or are not available, although growth rates might decline (Rahman et al. 2010). However, specific growth rates were more than 30% higher when bottom substrate was available, suggesting that macroinvertebrates are an essential resource for adult carp. In addition to the general preference for macroinvertebrates, Garcia-Berthou (2001) reported size-dependent variation in diet at the benthivorous stage, with smaller individuals preying on meiobenthos (benthic cladocerans, octracods) and larger carp preying on macrobenthos (*Chaoborus*, large chironomids). This and the fact that small carp change their resource preference from small- to large-bodied zooplankton suggests at least partial resource partitioning between different size/age classes.

Stable Isotope Analysis as a Tool to Estimate Resource Utilization

Estimating a consumer's diet accurately and realistically is challenging. Traditional diet estimates with stomach/gut analysis can provide high resolution of resource utilization, i.e., depending on the digestion level, resources can be determined to species level. However, different prey have differential digestion rates. Rapidly digestible prey such as soft-bodied items might therefore be underestimated or neglected. Khan (2003) found high proportions of digested content or empty stomachs in carp, suggesting that some of the prey might be under-estimated. Furthermore, some prey, for instance algae, plant material and detritus, might be amorphous, and therefore difficult to determine. One of the major difficulties with stomach/gut analyses is however the temporal resolution. Stomach/gut contents merely reflect a 'snapshot' of the consumer's resource utilization, thereby disregarding potential differential assimilation of alternative prey. In the past decades, ecologists have increasingly used stable isotope analyses to estimate resource utilization. Stable isotope analysis reliably mirrors resource use, provides a long-term estimate, and is therefore a powerful complement to traditional stomach/gut analyses.

The most commonly used stable isotopes for resource utilization estimates are ^{13}C and ^{15}N. Trace amounts of the natural carbon and nitrogen pool occur as these 'heavy' isotopes. The slightly heavier atomic mass of ^{13}C and ^{15}N compared to the more abundant ^{12}C and ^{14}N can lead to different biochemical reaction rates. Primary producers preferentially incorporate lighter ^{12}C. This preference results in a change of the isotopic signature of the producer compared to its resource, also known as biofractionation. The ratio of the heavier isotope to the lighter more common isotope (relative to a standard) is expressed as δ. Standard references for C and N are PeeDee limestone and atmospheric nitrogen, respectively (Peterson and Fry 1987). In lake ecosystems, phytoplankton and periphyton often show very different isotopic signatures. Phytoplankton is depleted in ^{13}C and displays $\delta^{13}C$ values ranging between –19‰ and –40‰ (Peterson and Fry 1987, Yoshioka et al. 1994, Schindler et al. 1997). In contrast, periphyton is typically enriched in ^{13}C. Boundary layer effects that occur near solid surfaces can reduce the availability of carbon (Hecky and Hesslein 1995), thus periphyton incorporates higher amounts of less-preferred ^{13}C. This discrepancy in isotopic signature between benthic and pelagic primary producers then offers a convenient method to estimate the contribution of benthic/pelagic resources to higher consumers (Hecky and Hesslein 1995).

Whereas carbon isotopes are typically used to determine the origin of the utilized resources, nitrogen isotopes are mainly used to determine a consumer's trophic position. Consumers preferentially excrete the light ^{14}N, resulting in an accumulation of heavy ^{15}N in the consumer's tissue. This accumulation is further transferred through the food web. Thus, many animals typically show a trophic level shift in $\delta^{15}N$ of about 3–3.5 (DeNiro and Epstein 1978, Checkley and Entzeroth 1985, Peterson and Fry 1987), i.e., the consumer's $\delta^{15}N$ value is 3–3.5‰ higher than its resource. Thus, nitrogen isotopes provide an estimate of a consumer's trophic position in the food web.

Generally, diet studies estimated with stable isotope analyses reflect the omnivorous feeding strategies of carp. For instance, Reid et al. (2012) showed that depending on the hydrological connectivity of different billabongs in the eastern highlands of Australia, different resources were important for carp. In channels, biofilm contributed 44–79% (1 and 99 percentile) to carp diet. In highly connected billabongs, C_3 macrophytes (7–37%), coarse (7–35%) and fine (5–41%) particulate organic matter, and aquatic invertebrates (63–68%) were most important for carp diet, whereas in billabongs with low connectivity, C_4 macrophytes (16–31%) and seston (15–41%) were most important (Reid et al. 2012). In their study, carp therefore relied largely on benthic resources including detritus. In contrast, Matsuzaki et al. (2010) showed that depending on the hybridization levels between native and domesticated individuals, carp relied 48% (41–56%) and 59% (53–66%) on pelagic primary producers for individuals with high and low hybrid index, respectively. Similarly, Soto et al. (2011) indicated high pelagic reliance (≥ 80%) of carp in a reservoir in the northeastern Iberian Peninsula.

Since stable isotope analyses integrate resource utilization over the long-term, they are exceptionally useful for the evaluation of resource partitioning between different species. Because of their omnivorous feeding style, carp can potentially compete with various native and introduced species (Tatrai et al. 1999, Mazumder et al. 2012). Zambrano et al. (2010) indicated that native axolotl salamanders (*Ambystoma mexicanum*) showed high overlap in isotopic niche space between invasive tilapia (*Oreochromis niloticus*) and carp in a large water system in Mexico, partly explaining the observed decrease in axolotl populations in this area. Similarly, Jackson et al. (2012) observed substantial changes in niche space of Louisiana swamp crayfish (*Procambarus clarkii*) coinciding with the introduction of carp into Lake Naivasha, Kenya. With the arrival of carp, crayfish niche space decreased substantially. In years with low carp abundance, crayfish niche space increased notably, suggesting strong competition between these species.

Zambrano et al. (2010) showed that carp relied up to 40.1% on benthic resources determined by stable isotope analyses. In the same study, the authors estimated 69.1% benthic reliance by gut analyses, indicating an over-estimation of benthic resources when determined by gut analyses. Trophic position of carp was estimated to 3.2 by stable isotope analyses, whereas it was estimated to 2.1 with gut analyses. This emphasizes the fact that the gut content can over-or underestimate particular resource use. In addition to being a 'snapshot' of resource utilization, different resources also differ in quality. High quality resources might contribute minor proportions to resource utilization, but due to being efficiently used, contribute higher proportions to the build-up of biomass. For instance, although plant material is often found in carp guts, it is not efficiently used by carp (Sibbing 1988), and thus likely only contributes in minor proportions to carp biomass.

Gut analyses can give high-resolution estimates of resource utilization, whereas stable isotope analyses provide a long-term estimate of resource use. Carp is a highly omnivorous species, and resource utilization might largely be determined by the availability of various resources. Combining both methodologies to determine the diet is a powerful way to study the feeding ecology of fish.

Effects of Carp Feeding on Ecosystems

The ability of a species to regulate lake ecosystem processes is primarily due to abundance (Power et al. 1996), population size structure (Davis et al. 2005), trophic level (Carpenter et al. 1992), and food source or mode of feeding (Matthews 1998) but biotic and abiotic factors, for example planktonic nutrient generation (Hudson et al. 1999) or watershed nutrient loading (Brabrand et al. 1990), are also important. Bottom-feeding fish are considered to be important regulators (or engineers) of aquatic community structure through their direct and indirect influence on water transparency, nutrient cycling, plankton and aquatic biota (Northcote 1988, Parkos et al. 2003, Weber and Brown 2009, Bajer et al. 2009). Cahn (1929) was the first to describe detrimental effects on water quality (i.e., increased turbidity and macrophyte loss) due to common carp invasion. The most commonly noted changes to aquatic ecosystems, sediment resuspension, nutrient concentrations and algal biomass, are associated with increased biomass densities and size of carp and other benthivorous fish species (Meijer et al. 1990a, Richardson et al. 1990, Breukelaar et al. 1994, Roberts et al. 1995, Lougheed et al. 1998, Driver et al. 2005, Badiou and Goldsborough 2010, Weber and Brown 2011, Akhurst et al. 2012). Carp can act as ecosystem engineers via the three main ways they alter both biotic and abiotic properties of aquatic ecosystems: selective consumption of non-benthic and benthic organisms, nutrient translocation and release to the water column, and physical disruption of sediment and macrophytes (Fig. 9.2).

Carp have the ability to alter aquatic systems so strongly that they may induce changes on an ecosystem scale. For example, the alternative equilibria theory, as posited by Scheffer (1990) and Scheffer et al. (1993), states that shallow lake ecosystems primarily exist in one of two alternative equilibria: either a clear water or turbid state. The clear water state is characterized by abundant macrophyte growth and diverse aquatic biota, whereas the turbid state is the opposing condition (Scheffer et al. 1993, Ibelings et al. 2007). Though there are many ecological processes and environmental characteristics that can contribute to the persistence of, or transformation between states (e.g., lake morphology, nutrient loading sources and magnitude, food web structure), the link between turbidity and aquatic macrophyte growth is primary (Scheffer et al. 1993) because macrophtyes contribute to sediment stability, reduce wind-driven sediment resuspension, provide habitat for aquatic biota, and compete directly with phytoplankton for nutrients.

Bioturbation, i.e., the disturbance of sediments through the activity of animals, can increase turbidity through direct sediment loading, increased phytoplankton growth through nutrient translocation, or macrophyte damage leading to increased wind-driven sediment loading. Effects from bioturbation by fish such as carp are generally accepted to be greater than

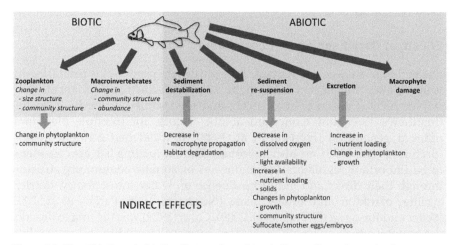

Figure 9.2. Direct biotic and abiotic effects and resulting indirect effects of carp feeding.

those caused by macroinvertebrates (Jeppesen et al. 1998, Svensson et al. 1999), but of course this depends on carp life stage and biomass density (Driver et al. 2005). System-wide effects are not universal, though, and in some cases minimal effects from carp introduction are detected. For example, native fish species persisted at similar biomass after over-stocking of carp at 482 fish ha^{-1} in a small (10.2 ha), dimictic (mean depth 6.5 m) eutrophic lake (Barthelmes and Brämick 2003). Zambrano et al. (2001) also predicted that deeper lakes would show less sever impacts from carp invasion. In general, however, it is clear that the introduction of carp has, and continues to cause dramatic ecological changes at both the community and ecosystem level, especially in shallow systems, and estimating the importance of carp regulation of aquatic ecosystems can help managers improve water quality and biotic community health. The effects carp have, both directly and indirectly, on aquatic ecosystems are described here. However, it is important to look at the basic changes carp cause so we can better understand how these changes relate to and affect biota in aquatic systems.

Sediment Burrowing and Resuspension

The most obvious change to shallow aquatic systems is the increase in suspended solids (Meijer et al. 1990b, Barton et al. 2000), which can be directly related to carp biomass (Meijer et al. 1990a, Driver et al. 2005) and a frequent reduction in submersed plants (Bajer et al. 2009) in aquatic systems affected by common carp. In a summary of available studies on carp and water quality in the literature, the vast majority of cases had increases in turbidity (and/or suspended solids) after carp introduction (Table 9.1). Because adult carp primarily forage on benthic macroinvertebrates, they induce severe mixing or bioturbation of the sediment. The removal of submersed plants alone, however, is enough to increase wind-induced increases in suspended sediment (Dieter 1990). Breukelaar et al. (1994) determined that suspended sediment concentration increased linearly with carp biomass (46 g m^{-2} d^{-1}) inducing a reciprocal 0.38 m^{-1} Secchi disc depth (a measure of water clarity). Dramatic (4-fold) increases in suspended sediment were also detected by Fischer et al. (2013) in small (3.2 m^2), shallow (1m water depth) enclosures. Increased turbidity can also be caused by increases in planktonic biomass (Breukelaar et al. 1994), with the expulsion of sediment particles and pore-water nutrients and excretion contributing directly to the increased growth.

Carp have also been shown to increase turbidity more than native benthic species in many cases (Williams et al. 2002, Parkos et al. 2003). Breukelaar et al. (1994), however, showed that bream could increase suspended sediment more than carp in experimental ponds at biomass densities greater than 300 kg ha^{-1}. Although they can increase turbidity at low abundance levels (Richardson et al. 1990, Drenner et al. 1998, Zambrano et al. 1998), a switch from clear water to turbid state is only like to occur once carp biomass surpasses a critical threshold (Zambrano and Hinojosa 1999). Reported thresholds vary, but range from roughly 200 to 320 kg ha^{-1} (Zambrano and

Table 9.1. Studies showing positive (+), negative (−), or no significant (NS) effects of carp on multiple biotic and abiotic variables. Exp type = experiment type; Phyto = phytoplankton; Zoop = zooplankton; BI = benthic invertebrates; Turb(SS) = turbidity(suspended solids); TN = total nitrogen; TP = total phosphorus; DO = dissolved oxygen.

Study	Location	Substrate	System	Time (d*)	Exp type	Size (ha*)	Depth (m)	Denisty (kg/ha)	Fish size (cm*)	Phyto Biomass(density)	Zoop	BI	Mac	Pisc	Plan	Turb(SS)	TN	TP	DO	pH
Akhurst et al. 2012	Australia		Impoundment	36	Enclosure	0.00031	1	1875	26-27	+	NS		-			+	+	+		
Angeler et al. 2002	Spain		Wetland	42	Enclosure	0.00008	2	5500-6500	52-61	+	-		NS			(+)	+	+		-
Badiou&Goldsborough 2010[a]	Canada		Wetland	2years	Cells	5-7		0-1247					NS	NS		(+)			-	-
Bajer et al. 2009	US		Lake	4years	Whole Lake	508	3.5	10-255	40-50	+				NS			+	+		
Breukalaar et al. 1994	Netherlands	Sandy clay/clay	Pond	180	Pond	0.1	1	0-500	3.6	+						+	+	+		
Chakrabarty &Das 2007			Fish Jar/Tank	12	Fish Jar/Tank	5-150 L				+						+	+	+		
Chumchal&Drenner 2004	US	Sand	Mesocosm	33	Mesocosm	0.0003	2.2	670	Adult	+						+				
Chumchal et al. 2005	US	Clay Loam	Pond	28	Whole Pond	0.36	1.2	0-465	30-32	+	NS					NS	+	+		
Cline et al. 1994	US	Sand, Clay	Tubs	9	Enclosure	0.0002	0.6	4533	Fingerling							+	+	+		
Crivelli 1983	France	Mud	Marsh	71	Enclosure	0.0064	0.43	0-726								NS				-
Drenner et al. 1997	US		Hatchery	1991-95	Whole Pond	0.38		566						NS		+				
Driver et al. 2005	Australia	Fine sediment/day	Pond	69	Enclosure	0.0065	0.35-0.75	0-650	Variable	+			NS				+	+		
Egertson&Downing 2004	Canada		Whole Lake	Variable	Whole Lake	Variable	Variable	Variable	NA	+		-	-	-	+/-	(NS)	+	NS	NS	-
Evelsizer&Turner 2006	Canada		Marsh	89	Exclosure	0.0009	0.56-0.76	NA	35.8	+		-	-	NS	-		NS	+		
Fisher et al. 2013	US		Mesocosm	35	Mesocosm	0.00025	1.2	4098		+		-	-	NS		+	+	+		
Forester&Lawrence 1978	US	Gravel	Pond	570	Whole Pond	0.1		230	NA				-							
Haines 1973	US		Pond	450	Whole Pond	0.0025	1.2	70	5.3				-							
Hinojosa &Zambrano 2004	Mexico		Pond		Whole Pond	0.05	1	5-6												
Jackson et al. 2010	US		Lake	2001-06	Whole Lake	4-2174	1-12	Variable	Variable	+		-				(+)	+	+		
Khan et al. 2003	Australia		Pond	28	Whole Pond	0.0005	0.8	250	14.1-14.5	+						+	+	+		
King &Hunt 1967	US	Aquatic soil	Marsh	1964-65	Exlosure	0.0005	0.02-0.66	448-560	NA	+			-							
King et al. 1997[d]	Australia	Silt, clay, loam	Pond	120	Whole Pond	0.28-0.8	4	100-1181	31-70	+	NS		-			+				
Kloskowski 2011	Poland		Pond	92	Whole Pond	1-69	0.7-1.3	170-240		+	NS		-			+				
Lamarra 1975	US		Pond	60	Enclosure	0.0009	1.6	0-1250	13-59	+			-	-			+			
Lougheed et al. 1998	Canada	Sand, silt	Lake	42	Enclosure	0.005	0.4-0.7	23-2100	16.1	+	NS		-			+		NS		
Matsuzaki et al. 2007	Japan		Pond	36	Enclosure	0.0004	0.7-0.8	369	14.5	+	+		-	NS		+	+	NS		
Matsuzaki et al. 2009	Japan		Pond	180	Enclosure	0.0004	0.7-0.8	160-480	3	+	(+)		-	NS		(+)	+	-		
Meijer et al. 1990a	Netherlands		Pond	150	Pond	1	1.5	320	2	+			-			(+)		NS		
Meijer et al. 1990b	Netherlands		Pond	72	Pond	0.0006	1.3	466	15-19.3		(+)		-			(+)				
Miller&Crowl 2006	US	Mud, clay	Lake	70	Enclosure	0.06	0.5	160	NA	+	NS	-	-			+	NS	+		
Parkos et al. 2003	US	Clay	Pond	91	Enclosure	0.07	1.5	174	43-69	NS	NS	-	-			+		+		
Parkos et al. 2006	US	Clay	Pond	28	Enclosure	0.06	1.6	476		NS	NS(+)	-	-		+	+	+	+		
Rahman et al. 2008	Bangledesh		Fish Tank	86	Enclosure	0.0004	1.6	300	6.8	+	+				+		NS	+	NS	NS
Richardson et al. 1990	US		Tank	72	Tank	0.0004	0.7	634	NA	+						NS	+	NS	NS	NS
Robel 1961[b]	US	Sandy clay	Marsh	20	Enclosure	0.0012	2	18	NA	NS									-	
Roberts et al. 1995[c]	Australia	Pot sediment	Pond	8	Enclosure	0.0012	0.3	0-660	15-25	NS			-			NS	NS	NS	-	-
			Pond		Endosure	0.0012	0.9	226	15-25											
Robertson et al. 1997[d]	Australia		Pond	120	Pond	0.28-0.8	0.9	101-348	NA			-				+(+)	NS	NS		
							4	669-1181	NA											

Study	Country	Substrate	Water body		Type															
Roozen et al. 2007	Netherlands	Silt and/or clay	Lake	11	Enclosure	0.00008	0.9	---	---			+			(+)	NS	NS	-	-
Sidorkewicj et al. 1998	Argentina	Sandy clay loam	Channel	127	Enclosure	0.0175	0.3-0.8	138-275 18-40	23 25 11					-	+				
															NS					
Sidorkewicj et al. 1999	Argentina	Sandy clay loam	Aquaria	30	Enclosure	0.000024	0.5	575	7.9-14.2				-	-	+					
Tapia&Zambrano 2003	Mexico		Ponds	210	Whole Pond	0.5-10	<3	17-25	Fingerling					-	+					
Threinen&Helm 1954	US	Muck and/or sand	Lake	68	Enclosure	0.0037	0.5-0.9	NA	NA					-						
Tryon 1954	US	Peaty/mud/sand	Lake	1950-52	Enclosure	0.004	0.3-0.9	NA	NA				-	-	+					
Wahl et al. 2011	US	Clay	Pond	60	Enclosure	0.06	Variable	277-430	57.3	-	NS			-				+	-
Weber &Brown 2009	US		Lake	Variable	Whole Lake	Variable	Variable	Variable	13-44	+	+			-					
Wilcox&Hornbach 1991	US	Sand, silt, clay	Lake	38	Exclosure	0.0015	0.6	0-700	15		NS		NS						
Williams&Moss 2003	UK		Lake	42	Enclosure	0.0004	1	0-700		NS	NS		(+)	-	(+)				
Williams et al. 2002	UK		Lake	42-63	Enclosure	0.0004	1-1.3	358	15									-	
Wolfe et al. 2009			Pond	60	Enclosure	0.05	1	140-560	5.3			+/-		+/-	+				
Zambrano&Hinojosa 1999	Mexico		Pond	330	Exclosure	0.05	1	0-4	3-5		-		-	NS	+				-
Zambrano et al. 1998[e]	Mexico	Mud	Pond	330	Whole Pond	0.8-8	<2				(-)		NS	-	(NS)				

*Units, unless noted otherwise

aEffects shown in parentheses are only for carp biomass densities >600 kg/ha

bEffects only at >178 kg/ha carp biomass

cExperiments run consecutively created large variation in second experiment (lower biomass density)

dComparison presented only between low and high biomass density of carp

eLow statistical power

Hinojosa 1999). It is possible however, that smaller population densities could also cause a shift in state over longer periods, especially as benthic invertebrate populations decline and carp increase foraging time, resulting in increased sediment disturbance and resuspension (Werner and Anholt 1993). In a study by Ritvo et al. (2004), carp stirred sediment to such a degree that sediment bulk density increased significantly. Although it has been suggested that this was due to compaction, increased density may also be caused by translocation, mineralization, or ingestion of high organic content sediment, which is generally lower in density than sediment composed of mostly inorganic particles (Håkanson and Jansson 1983).

Bottom feeding fish can also alter the active sediment layer by increasing the depth of sediment mixing; the active layer of sediment being regarded as the portion of sediment that interacts directly with the overlying water column. A general rule of thumb is that the active layer of sediment is approximately 10 cm in sediment depth. This will obviously vary and active layer thickness can be affected by the type and abundance of both sediment dwelling organisms and organisms that forage in the sediment, as well as the type of sediment and velocity of water at the sediment-water interface. The effect carp have on sediment mixing depth via foraging has not been well documented. A number of older studies showed that sediment mixing depth in waters inhabited by adult bream (*Abramis brama*) and carp could reach 12 cm (Alikunhi 1966, Panek 1987). Comparisons between disturbed and undisturbed areas are generally lacking, and thus specific changes in active sediment layer and the subsequent effects on, for example nutrient availability, are not possible. In addition, carp may not act as they would in a natural habitat when placed in an enclosure, due to artificial constraints (e.g., food limitation, Miller and Crowl 2006), which makes determining true changes in mixing depth difficult. In a recent study, however, Huser et al. (2013) showed that carp could increase sediment mixing depth by a factor of 3. Mixing depth in exclosures without carp was approximately 4 cm, whereas it averaged over 14 cm in other areas of a soft bottom, shallow lake with an approximate carp biomass density of 200–250 kg ha^{-1} and average fish size of 3.4 kg.

Sediment composition is likely a key factor controlling the magnitude of effects detected after carp invasion or during experimental studies. Unfortunately, of the numerous studies on the common carp, many barely describe (or do not describe at all) sediment characteristics that may explain some of the conflicting results characterizing ecosystem changes caused by carp. Comparisons between such studies thus can be difficult, especially when considering thresholds of carp biomass that may be used as targets for lake management. For example, no correlations were found between carp biomass and turbidity in some enclosure studies (Crivelli 1983, Miller and Crowl 2006), whereas increased turbidity is often found in ponds and lakes (e.g., Sidorkewicj et al. 1998, Parkos et al. 2003, Schrage and Downing 2004). Other studies have not observed significant nutrient effects from carp (e.g., nitrogen, Matsuzaki et al. 2007) and it has been suggested that burrowing effects on sediment may depend on sediment characteristics (Robel 1961, Crivelli 1983, Roberts et al. 1995).

Biomass thresholds for negative ecosystem effects range from approximately 100 to 450 kg ha^{-1} (Fletcher et al. 1985, Williams et al. 2002, Haas et al. 2007, Bajer et al. 2009, Weber and Brown 2009, Hicks et al. 2012). On the other hand, increased turbidity and nutrient mobilization, and thus secondary effects such as increased phytoplankton biomass, may be limited in ecosystems with hard substrates that limit the foraging capabilities of carp (Håkanson and Jansson 1983, Crivelli 1983, Havens 1993, Roberts et al. 1995). Given such a broad range of threshold values, and limited information on other factors, besides carp, that contribute to ecological degradation following carp invasion, it is difficult to provide a precise biomass density representing a threshold for a specific impact. Both Chapter 11 on recreational fishing and growth in natural carp populations by Ragnarsson-Stabo and Chapter 10 on carp as invasive species by Hicks and Ling provide more guidance in this area but clearly more studies specific to carp size and density and critical thresholds, above which ecological quality is degraded, are needed.

Water Chemistry

Carp can also alter the chemistry of aquatic systems, both in the water column and at the sediment water interface. Mixing of sediment by adult carp has two main effects on the hydrochemistry at the sediment surface. First, diffusion/release rates can increase due to the physical movement of sediment bound and pore-water constituents caused by carp and second, aerobic decomposition increases as greater amounts of sediment are exposed to oxygen (Graneli 1979, Kadir et al. 2006, Phan-Van et al. 2008). Aerobic mineralization generally occurs in the uppermost few millimeters of sediment (Håkanson and Jansson 1983) and in the cavities created by animal burrowing (Kristensen and Blackburn 1987), whereas carp likely expand this 'aerated' zone by increasing the sediment mixing depth (Huser et al. 2013) leading to increased mineralization of organic matter previously buried in anoxic sediment layers (Ritvo et al. 2004). The increase in turbidity (i.e., decreased light) and increased organic matter mineralization caused by carp burrowing may be possible drivers behind decreased oxygen concentrations seen in some studies (Table 9.1). This is somewhat counter-intuitive because as we will see later on in this chapter, over 80% of carp studies reported herein describe increases in phytoplankton, which should generally lead to increased production of oxygen during daylight hours. Only three out of the seven studies showing decreased oxygen concentrations also had increased phytoplankton levels; six, however, had elevated turbidity and/or suspended solids showing that, at least in these particular studies, changes in turbidity may be more important with respect to oxygen status than oxygen produced during photosynthetic primary production.

Water column pH also seems to be affected in the presence of carp. Although the number of studies with data is low (N = 5), in all cases pH decreased in the presence of carp. In addition, Czech carp farmers regulate elevated pH levels by increasing stocks of larger (> 2+) carp that increase turbidity causing a decrease in photosynthetic activity (Adamek and Marsalek 2012). As we will see later, in many cases high carp biomass density leads to increases in phytoplankton biomass (and presumably photosynthetic primary productivity), which should lead to higher pH levels due to the consumption of carbonic acid (H_2CO_3), also known as dissolved CO_2. This apparent contradiction may be explained by increased turbidity and light limitation of macrophytes and phytoplankton, increased bacterial degradation/mineralization of sediment organic matter, or a combination of both.

Nutrients

One of the most frequently studied aspects of carp in aquatic ecosystems is the effect carp populations have on nutrients in the water column. And, one of the most difficult aspects in these studies has been shown to be the untangling of exactly how carp cause changes in nutrient fluxes and cycling in aquatic systems. Benthivorous species, unlike those that feed in the water column, have the ability to enhance the amount of nutrients in the water, both via excretion of sediment-derived food sources and through physical 'pumping' of nutrients in the sediment through bioturbation. Planktivorous fish (including early life stage benthivores), on the other hand, transform phosphorus already within the pelagic system and do not alter the total amount of phosphorus available in the water column, except at shorter timescales.

Studies on carp and nutrients have focused on P and N because these nutrients are known to regulate primary productivity in most freshwater systems and elevated concentrations of these two nutrients are consistently found in the presence of carp (see Table 9.1). Some differences in opinion remain on whether excretion (Lamarra 1975), sediment disturbance (Driver et al. 2005), or both are the major drivers of increased nutrients when carp invade aquatic systems. Unfortunately, comparisons of many early studies on carp size and water quality are difficult due to both variability in carp size within a study and inconsistent reporting of actual carp sizes, as detailed by Driver (2002). The idea that there is a continuum of nutrient mobilization, from excretion to sediment disturbance, as carp size increases is now, however, the dominant theory (e.g., Driver et al. 2005, Weber and Brown 2009, Kloskowski 2011). The transition from planktivorous to benthivorous feeding behavior when carp shift from the larval to juvenile state is associated with a negative correlation of P in excretion and wet weight of small carp (Lamarra 1975). Even so, excretion rates for some benthic species can exceed the sum of external loading sources for both N and P (Gido 2002). In addition, Brabrand et al. (1990) concluded that phosphorus supply from benthic feeding fishes could be nearly double that coming from external sources. Small carp (10 to 15 mm), however, may not have a net effect on nutrients (Meijer et al.

1990b) because their diet consists of high amounts of zooplankton, a nutrient source already present in the water column. On the other hand, Driver et al. (2005) showed that larger carp mobilized more P per unit weight than smaller carp, even though larger fish retain more P, have lower mass specific ingestion rates, and excrete less P per unit size. The additional P mobilized was attributed to input from the sediment. Bioavailability of nutrients can also be altered by carp as was shown by Morgan and Hicks (2013) with a model positively relating wet mass of carp to the proportion of ammonia to total N due to increasing excretion rates of ammonia as fish mass increased.

Even though carp and other benthivorous fish species can have strong impacts on nutrient cycling in aquatic systems, changes in nutrient export from sediment to overlying waters are not well documented. Indirect evidence includes increased turbidity (Cline et al. 1994, Roberts et al. 1995) and nutrients (Lamarra 1975, Lougheed et al. 1998) in the overlying water, with the greatest effects detected when high biomass and large, adult carp are present (Driver et al. 2005). Experimental design has also confounded results and interpretations with respect to changes in sediment P availability and carp. Lamarra (1975) suggested excretion was the main driver of increased P in waters inhabited by carp and that there was no physical transfer of sediment P to the water. Although we now know that nutrient release via excretion vs. sediment release is generally dependent on the size/age of carp present, the tests conducted by Lamarra to determine the effect of increased physical sediment mixing on P release (water mixing by paddle once a week) were not a realistic representation of carp burrowing (Qin and Threlkeld 1990). In addition, enclosures giving carp access to the sediment, such as those used by Lamarra, make determination of P source difficult.

Aquaculture studies tend to look at maximizing nutrient availability and productivity, and thus may provide some insight into how carp affect nutrient release from sediment in natural systems. Chakrabarty and Das (2007) showed that bioturbation induced release of soluble reactive phosphate from apatite (a generally insoluble phosphorus source) increased by 65–90% when carp were introduced but phosphorus flux increased by only 6.3–7.2 in the absence of apatite. Similarly, Jana et al. (1992) showed an influx increase between 72 and 100% from sediment including phosphate rock when carp were present whereas P release increased by 7–8% in the absence of phosphate rock.

Indirect effects resulting from carp burrowing can also lead to increased nutrients in aquatic systems. Increased internal nutrient loading has been linked to loss of macrophytes and sediment bed stability (Moss et al. 2002, Parkos et al. 2003, Schrage and Downing 2004, Bajer et al. 2009), potentially leading to increased mixing between sediment and water due to increased water velocity at the sediment surface (Håkanson and Jansson 1983). Destruction and decomposition of macrophytes can lead to increases in water column nutrients (Carpenter and Lodge 1986, Parkos et al. 2003), although this nutrient source is likely to decrease over time as a carp population reduces macrophyte abundance. Carp may also increase nutrients indirectly by physically mixing sediment with oxygenated water, increasing aerobic degradation of organic matter. Ritvo et al. (2004) showed significant decreases in easily oxidized organic matter in the sediment as carp number increased in treatment enclosures, implying substantial export of nutrients that were previously buried in the sediment. Bioturbation by carp can lead to reduced nitrogen and phosphorus accumulation in sediment as well (Jana and Sahu 1993, Rahman et al. 2008), presumably due to both direct consumption and increased aerobic degradation of organic matter that result in export of these previously bound nutrients to the water column. If, as noted previously in this chapter, carp increase sediment mixing depth by up to a factor of three, increased rates of organic matter oxidation may contribute a substantial, previously unavailable, pool of nutrients to the water column. Huser et al. (2013) suggest that the potentially available P pool, that is the pool of P available for release from the sediment, may increase by up to a factor of three in the presence of carp.

Some studies have not observed an effect of common carp on nutrients (e.g., Matsuzaki et al. 2007) and it has been suggested that burrowing effects on sediment nutrient release may depend on sediment characteristics (Robel 1961, Crivelli 1983, Roberts et al. 1995). Although detailed sediment information is lacking in many of the carp studies covered in this chapter, approximately 70% of cases showed increases in TN and/or TP and only one of these studies (Matsuzaki et al. 2007) showed a decline in TP (Table 9.1). Thus, there is a strong body of evidence showing increased nutrient levels in systems affect by carp but clearly more work is needed to describe the specific circumstances that lead to increases in nutrient generation from both fish excretion, sediment disruption and other direct and indirect effects.

Phytoplankton and Bacterioplankton

The increase in water column nutrients associated with common carp can lead to increased phytoplankton and bacterioplankton production (Andersson et al. 1988, Qin and Threkkeld 1990, Breukelaar et al. 1994, Vanni 2002, Chumchal and Drenner 2004, Matsuzaki et al. 2007, Akhurst et al. 2012, Fisher et al. 2013). Studies involving more detailed analysis of phytoplankton (not just chlorophyll a or total biomass) have shown that community response to carp can vary. For example, Williams et al. (2002) showed minimal effect of carp on edible (< 30 µm) and non-edible phytoplankton (> 30 µm or filamentous), possibly due to nitrogen limitation in the study system. Matsuzaki et al. (2007) on the other hand, found that cyanobacteria dominated species composition in carp enclosures whereas cryptophytes dominated in enclosures without carp. Some studies showing changes in phytoplankton community composition and biomass indicate nutrients (generally phosphorus) as a contributing factor. At least 50% of phosphorus excreted by carp is easily available for uptake and production by phytoplankton (Lamarra 1975); this will vary, however, based on carp size, diet and habitat variability. In a review of studies available in the literature,

81% showed a positive relationship between carp and phytoplankton biomass, whereas only 4% showed a decline and 15% showed no trend either way (Table 9.1).

It is well documented that increased nutrient availability can lead to increased biomass of phytoplankton, but physical stirring of the sediment can also lead to increased biomass via recruitment of resting stages of planktonfrom the sediment surface. Brunberg and Blomqvist (2003) showed that up to 50% of *Microcystis* colonies were recruited over the summer from the sediment in a shallow bay and Verspagen et al. (2004) suggested that rather than an active process (buoyancy regulation) initiated by *Microcystis*, sediment recruitment is more likely to be caused by wind driven resuspension or bioturbation. This phenomenon has also been shown to occur with other cyanobacteria (Karlsson-Elfgren and Brunberg 2004) and other studies have shown a shift to cyanobacteria in aquatic systems inhabited by carp (Williams and Moss 2003), in which, increased sediment recruitment likely played a role. Due to the increase of cyanobacteria in some systems domination by carp, reverse consequences of increases in taxa and their toxins on carp are also possible and are reviewed in Chapter 13 on natural toxins's effects on carp by Pietsch (this volume). Resuspension effects from bioturbation are not limited to cyanobacteria. In a study by Havens (1991), motile cryptophytes dominated community composition in fishless enclosures or when nets limited access to sediment, whereas green algae thrived in enclosures where fish had access to the sediment. The effects from increased sediment recruitment are not restricted to shallow lakes or littoral regions, because shallow areas can act as inoculation sites for pelagic zones (Brunberg and Blomqvist 2003). Thus, carp feeding habits may play an important role in both total biomass and community composition of phytoplankton in lakes by favoring those taxa that benefit from sediment resuspension, especially during summer months when carp are actively feeding. As we will discuss later, carp can also exert top down control of phytoplankton by reducing zooplankton abundance and altering community structure, either directly or indirectly, thereby leading to decreased grazing pressure on phytoplankton (Carpenter et al. 1985).

Zooplankton

The effect carp have on zooplankton communities is likely to be variable due to differing carp population sizes and ecosystem structures. Our summary of available studies in the literature confirms this, with no strong trend either way with respect to zooplankton biomass or density in systems affected by carp (Table 9.1). Populations dominated by adult carp are likely to have more of an indirect influence through nutrient remobilization, habitat alteration (i.e., reduced macrophytes, and foraging limitation caused by increased turbidity), whereas juveniles directly affect zooplankton via consumption. Some studies have shown an increase in zooplankton biomass in relation to carp presence (Table 9.1). Increases in food supply (Drenner et al. 1998, Kahn et al. 2003) and decreases in benthic invertebrate predators (Johnson and Crowley 1980) are potential factors behind increasing zooplankton biomass. Parkos et al. (2003) detected an increase in zooplankton abundance in the presence of carp, which contradicts the results from a number of other studies where reductions (Meijer et al. 1990b, Richardson et al. 1990, Lougheed et al. 1998) or no effects (Qin and Threlkeld 1990, Cline et al. 1994, Akhurts et al. 2012) were found. As indicated by Parkos et al. (2003), cascading effects from the consumption of invertebrate predators that use zooplankton as a food source were the likely reason for increased zooplankton abundance. Although both bottom-up (increased nutrients) or top-down (decreased densities of macroinvertebrate predators) were suggested as principal drivers, carp had a substantially stronger effect on benthic macroinvertebrate density in this study, indicating the top-down effect was primarily responsible for the increase in zooplankton abundance. This reasoning was supported by peaks in zooplankton abundance that coincided with decreases in odonates, an invertebrate predator that feeds on zooplankton, in experimental carp enclosures (Parkos et al. 2003).

Carp can also alter zooplankton community structure, causing shifts from large- to small-bodied species, which likely explains some of the contradictory evidence regarding changes in total zooplankton biomass. Because turbidity can limit phytoplankton ingestion by large-bodied zooplankton, a shift in community structure to smaller-bodied zooplankton taxa can occur when inorganic turbidity increases (Kirk 1991). Other studies have revealed that carp effects on zooplankton can be taxa-specific. Matsusuki et al. (2007) detected an increase in rotifer abundance in treatments with carp, which was likely due to grazing on large-bodied zooplankton that then lead to an increase in biomass of smaller-bodied taxa (Richardson et al. 1990). In a study by Kloskowski (2011), no difference in total zooplankton biomass was detected between ponds stocked with consecutively larger size carp, but community structure shifted. During summer, large-size cladocerans were lower in ponds with medium to large size carp when compared to 1-summer fingerlings, which the authors suggested was a result of depredation by ≥ 1+ carp during the spring.

Benthic Invertebrates

The dietary preference of adult carp for macroinvertebrates can have substantial effects on macroinvertebrate community composition and can lead to reduced invertebrate abundance, diversity, evenness and richness (Wilcox and Hornbach 1991, Parkos et al. 2003, Miller and Crowl 2006, Stewart and Downing 2008). Tàtrai et al. (1994) detected a decrease in macroinvertebrate abundance, especially for chironomids, but effects on macroinvertebrate communities have been shown to vary. Parkos et al. (2003) showed that most benthic invertebrate taxa were reduced by both high (476 kg ha^{-1})and low (174 kg ha^{-1}) biomass densities of carp, but under low grazing pressure, chironomid biomass was unaffected by carp. Community composition may also switch to larger sized taxa due to carps general preference for smaller bodied prey (Covich and Knezevic 1978). Carp have been shown to reduce macroinvertebrate abundance, even when prevented access to sediment (Matsuzaki et al. 2007). The authors of this study hypothesized that the reduction of submerged macrophytes, via phytoplankton shading due to increased carp excretion, negatively affected the macroinvertebrate community. Oligochaeta, for example, have been shown to be abundant in the vicinity of macrophyte root systems (Sagova-Mareckova and Kvet 2002), likely due to differing sediment chemistry and root exudates that stimulate bacteria, an important food source for oligochaeta (McMurry et al. 1983). Thus, even younger carp that prefer to feed on zooplankton can negatively affect the benthic macroinvertebrate community through indirect effects primarily due to excretion.

Macrophytes

As with other aquatic species covered earlier, carp can drastically alter the abundance and community composition of macrophytes in surface waters, both directly and indirectly. And, because macrophytes are important for their ability to stabilize sediment, compete with phytoplankton for light and nutrients, and provide food, habitat and prey refugia, changes in abundance and diversity will affect other species in the aquatic ecosystem. Foraging activity of adult carp is commonly associated with uprooted aquatic vegetation (Crivelli 1983, Lougheed et al. 1998, Zambrano et al. 2001) and the effects of carp burrowing can lead to a loss of both abundance and diversity of macrophytes (Threinen and Helm 1954, Crivelli 1983, Parkos et al. 2003, Miller and Crowl 2006, Akhurts et al. 2012). Many studies have documented sparse aquatic vegetation in lakes with abundant carp (Cahn 1929, Cahoon 1953, Schrage and Downing 2004, Bajer et al. 2009) and negative effects are generally seen once a certain density threshold is reached, typically 100 to 200 kg/ha (Bajer et al. 2009). The changes in macrophyte community structure and abundance can be both direct via uprooting or breakage (Robel 1962, Lougheed et al. 1998, Hinojosa-Garro and Zambrano 2004) and decreased re-establishment due to continuous physical disturbance (Hootsmans 1999), or indirect through increased turbidity caused by sediment resuspension (Robel 1962, King and Hunt 1967, Lamarra 1976, Miller and Crowel 2006) and increased phytoplankton biomass (Hellström 1991).

Not all aquatic systems are equally affected by the presence of carp, as some are able to maintain a clear-water state with abundant macrophytes (Lougheed et al. 1998). Shallow systems that allow adequate light penetration to substrata can allow macrophytes to persist in some cases (Zambrano and Hinojosa 1999, Evelsizer and Turner 2006). In addition, not all macrophyte taxa are affected evenly by carp burrowing. Floating leaved species (e.g., white water lily) and some emergents (e.g., American lotus) can resist the effects of carp foraging (Crivelli 1983, Drenner et al. 1998, Bajer et al. 2009). Such species have large rhizomes and substantial overwintering carbohydrate reserves (Wetzel 1983), which can help shoots grow to the surface even in high turbid waters caused by carp feeding activity. The large rhizomes can also help withstand uprooting by carp (Crivelli 1983). Submerged species, on the other hand, have weaker root systems and are more dependent on light penetration and thus are more susceptible to the resulting effects of adult carp foraging in sediments (Roberts et al. 1995). Thus, carp can have a structuring influence on macrophyte communities, via both light limitation and physical sediment disturbance and uprooting of more sensitive species, whereas tolerant species may survive or even thrive in affected systems.

Fish

Ecosystem changes caused by carp have the potential to influence native fish communities but direct evidence is mostly lacking. Changes in macrophyte, invertebrate and planktonic communities, nutrient levels and availability, and water chemistry can all indirectly affect native fish species, which can confound or potentially mask direct effects. It does appear that native fish species, often less tolerant than carp, generally decline as lakes become more eutrophic, whereas carp are adapted to eutrophic systems and their populations tend to increase with increasing aquatic productivity (see Chapter 10, this volume). Common carp are more efficient than some native species at consuming benthic invertebrates (Zambrano et al. 1998, Parkos et al. 2003), which can negatively affect the growth of native species that rely on invertebrate prey as a food source (Wahl et al. 2011). Wolfe et al. (2009) also showed reduced survival of centrarchid species can occur in the presence of common carp, potentially due to food source competition. Large- (*Micropterous salmoides*) and smallmouth bass (*Micropterous dolomieui*), however, do not seem to be as affected by carp (Forester and Lawrence 1978) and some species, such as the white crappie (*Pomoxis annularis*), have increased in abundance in the presence of carp (Egertson and Downing 2004), likely due to the ability to thrive in turbid conditions created by carp.

As noted above, researchers have shown inverse relationships between carp and planktivores (i.e., bluegill (*Lepomis macrochirus*) and crappie (*Pomoxis nigromaculatus*) abundance (Jackson et al. 2010, Weber and Brown 2011). Although it has been suggested that this relationship may be driven by poor water quality conditions caused by carp, other researchers have shown that some predator species may limit carp recruitment via egg or larvae predation when these species are large enough in number. Bajer and Sorensen (2010) noted carp recruitment only occurred when shallow spawning areas were free from potential carp predators (centrarchids, percids and esocids) due to periodic anoxia. Thus, the negative relationship seen between carp and some native fish species may be driven by either the negative effects of degraded water quality on native fish, direct resource competition, or poor recruitment of carp caused by native fish predation on eggs and juvenile carp.

Summary

The somewhat contradictory results found from both experimental and removal studies (see Chapter 11) show that the effect carp have on aquatic ecosystems is complex due to the interrelated direct and indirect changes carp cause to these systems. The combination of top-down and bottom-up effects, and the varying effects on aquatic systems at different life stages, make comparisons between studies and development of adequate study designs difficult. Recent research has shown, however, that there is a continuum of effects caused by carp as they age and switch from juvenile to adult stages. Both fish size and feeding ecology influence the type and magnitude of the disturbance effect but other factors, such as sediment type, lake morphology and magnitudes of other loading/disturbance sources may also play important roles in determining the relative changes between systems invaded by carp. This is why applying specific thresholds, e.g., biomass density or fish per unit area, should be done with care, especially when considering management options to improve or restore water quality in aquatic systems. Given the vast areas that carp inhabit (or have invaded), additional research is warranted to better understand the wide ranging effects carp feeding ecology have on aquatic ecosystems and to develop or improve methods to manage populations and improve the ecological health of affected systems.

References

Adámek, Z. and B. Maršálek. 2012. Bioturbation of sediments by benthic macroinvertebrates and fish and its implication for pond ecosystems: A review. Aquacult. Int. 21(1): 1–17.

Akhurst, D.J., G.B. Jones, M. Clark and A. Reichelt-Brushett. 2012. Effects of carp, gambusia, and australian bass on water quality in a subtropical freshwater reservoir. Lake Reserv. Manage. 28: 212–223.

Alikunhi, K.H. 1966. Synopsis of biological data on common carp (*Cyprinus carpio* (Linnaeus)), 1758: Asia and the far east.

Andersson, G., W. Granéli and J. Stenson. 1988. The influence of animals on phosphorus cycling in lake ecosystems. Hydrobiologia 170: 267–284.

Angeler, D.G., M. Álvarez-Cobelas, S. Sánchez-Carrillo and M.A. Rodrigo. 2002. Assessment of exotic fish impacts on water quality and zooplankton in a degraded semi-arid floodplain wetland. Aquat. Sci. 64: 76–86.

Badiou, P.H.J. and L.G. Goldsborough. 2010. Ecological impacts of an exotic benthivorous fish in large experimental wetlands, Delta Marsh, Canada. Wetlands 30: 657–667.

Bajer, P.G., G. Sullivan and P.W. Sorensen. 2009. Effects of a rapidly increasing population of common carp on vegetative cover and waterfowl in a recently restored midwestern shallow lake. Hydrobiologia 632: 235–245.

Bajer, P.G. and P.W. Sorensen. 2010. Recruitment and abundance of an invasive fish, the common carp, is driven by its propensity to invade and reproduce in basins that experience winter-time hypoxia in interconnected lakes. Biol. Invasions 12: 1101–1112.

Barthelmes, D. and U. Brämick. 2003. Variability of a cyprinid lake ecosystem with special emphasis on the native fish fauna under intensive fisheries management including common carp (*Cyprinus carpio*) and silver carp (*Hypophthalmichthys molitrix*). Limnologica 33: 10–28.

Barton, D.R., N. Kelton and R.I. Eedy. 2000. The effects of carp (*Cyprinus carpio* L.) on sediment export from a small urban impoundment. J. Aquat. Ecosyst. Stress Recovery 8: 155–159.

Brabrand, A., B.A. Faafeng and J.P.M. Nilssen. 1990. Relative importance of phosphorus supply to phytoplankton production—fish excretion versus external loading. Can. J. Fish. Aquat. Sci. 47: 364-372.

Breukelaar, A.W., E.H.R.R. Lammens, J.G.P.K. Breteler and I. TÁTrai. 1994. Effects of benthivorous bream (*Abramis brama*) and carp (*Cyprinus carpio*) on sediment resuspension and concentrations of nutrients and chlorophyll a. Freshwater Biol. 32: 113–121.

Britton, J.R., R.R. Boar, J. Grey, J. Foster, J. Lugonzo and D.M. Harper. 2007. From introduction to fishery dominance: The initial impacts of the invasive carp *Cyprinus carpio* in Lake Naivasha, Kenya, 1999 to 2006. J. Fish Biol. 71 (Supplement D): 239–257.

Brunberg, A.K. and P. Blomqvist. 2003. Recruitment of microcystis (cyanophyceae) from lake sediments: The importance of littoral inocula. J. Phycol. 39: 58–63.

Cahn, A.R. 1929. The effect of carp on a small lake: The carp as a dominant. Ecology 10: 271–274.

Cahoon, W.G. 1953. Commercial carp removal at lake mattamuskeet, north Carolina. J. Wildlife Manage. 17: 312–317.

Carpenter, S.R., J.F. Kitchell and J.R. Hodgson. 1985. Cascading trophic interactions and lake productivity. Bioscience 35: 634–639.

Carpenter, S.R. and D.M. Lodge. 1986. Effects of submersed macrophytes on ecosystem processes. Aquat. Bot. 26: 341–370.

Carpenter, S.R., C.E. Kraft, R. Wright, H. Xi, P.A. Soranno and J.R. Hodgson. 1992. Resilience and resistance of a lake phosphorus cycle before and after food web manipulation. Am. Nat. 140: 781–798.

Chakrabarty, D. and S.K. Das. 2007. Bioturbation-induced phosphorous release from an insoluble phosphate source. Biosystems 90: 309–313.

Chapman, G. and C.H. Fernando. 1994. The diets and related aspects of feeding of Nile tilapia (*Oreochromisniloticus* L.) and common carp (*Cyprinus carpio* L.) in lowland rice fields in northeast Thailand. Aquacult. 123: 281–307.

Checkley, D.M. and L.C. Entzeroth. 1985. Elemental and isotopic fractionation of carbon and nitrogen by marine, planktonic copepods and implications to the marine nitrogen cycle. J. Plankt. Res. 7(4): 553–568.

Chumchal, M.M. and R.W. Drenner. 2004. Interrelationships between phosphorus loading and common carp in the regulation of phytoplankton biomass. Arch. Hydrobiol. 161: 147–158.

Chumchal, M.M., W.H. Nowlin and R.W. Drenner. 2005. Biomass-dependent effects of common carp on water quality in shallow ponds. Hydrobiologia 545: 271–277.

Cline, J., T. East and S. Threlkeld. 1994. Fish interactions with the sediment-water interface. Hydrobiologia 275-276: 301–311.

Covich, A.P. and B. Knezevic. 1978. Size-selective predation by fish on thin-shelled gastropods (Lymnaea): the significance of floating vegetation (Trapa) as a physical refuge. Verh. Internat. Verein. Limnol. 20: 2172–2177.

Crivelli, A.J. 1983. The destruction of aquatic vegetation by carp—a comparison between southern France and the United-States. Hydrobiologia 106: 37–41.

Davis, M.B., R.G. Shaw and J.R. Etterson. 2005. Evolutionary responses to changing climate. Ecology 86: 1704–1714.

DeNiro, M.J. and S. Epstein. 1978. Influence of diet on the distribution of carbon isotopes in animals. Geochimica et Cosmochimica Acta 42(5): 495–506.

Dieter, C.D. 1990. The importance of emergent vegetation in reducing sediment resuspension in wetlands. J. Freshwater Ecol. 5: 467–473.

Drenner, R.W., K.L. Gallo, C.M. Edwards, K.E. Rieger and E.D. Dibble. 1997. Common carp affect turbidity and angler catch rates of largemouth bass in ponds. N. Am. J. Fish. Manage. 17: 1010–1013.

Drenner, R.W., K.L. Gallo, R.M. Baca and J.D. Smith. 1998. Synergistic effects of nutrient loading and omnivorous fish on phytoplankton biomass. Can. J. Fish. Aquat. Sci. 55: 2087–2096.

Driver, P.D. 2002. The role of carp (*Cyprinus carpio* L.) size in the degradation of freshwater ecosystems. University of Canberra, Australia.

Driver, P.D., G.P. Closs and T. Koen. 2005. The effects of size and density of carp (*Cyprinus carpio* L.) on water quality in an experimental pond. Arch. Hydrobiol. 163: 117–131.

Eder, S. and C.A. Carlson. 1977. Food habits of carp and white suckers in the south Platte and St. Vrain Rivers and Goosequill Pond, Weld Country, Colorado. Trans. Am. Fish. Soc. 106: 339–346.

Egertson, C.J. and J.A. Downing. 2004. Relationship of fish catch and composition to water quality in a suite of agriculturally eutrophic lakes. Can. J. Fish. Aquat. Sci. 61: 1784–1796.

Evelsizer, V.D. and A.M. Turner. 2006. Species-specific responses of aquatic macrophytes to fish exclusion in a prairie marsh: a manipulative experiment. Wetlands 26: 430–437.

Fischer, J.R., R.M. Krogman and M.C. Quist. 2013. Influences of native and non-native benthivorous fishes on aquatic ecosystem degradation. Hydrobiologia 711: 187–199.

Fletcher, A.R., A.K. Morison and D.J. Hume. 1985. Effects of carp, *Cyprinus carpio* L., on communities of aquatic vegetation and turbidity of waterbodies in the lower Goulburn River basin. Aust. J. Mar. Fresh. Res. 36: 311–327.

Forester, T.S. and J.M. Lawrence. 1978. Effects of grass carp and carp on populations of bluegill and largemouth bass in ponds. T. Am. Fish. Soc. 107: 172–175.

Garcia-Berthou, E. 2001. Size- and depth-dependent variation in habitat and diet of the common carp (*Cyprinus carpio*). Aquat. Sci. 63: 466–476.

Gido, K.B. 2002. Interspecific comparisons and the potential importance of nutrient excretion by benthic fishes in a large reservoir. T. Am. Fish. Soc. 131: 260–270.

Gisbert, E., L. Cardona and F. Castelló. 1996. Resource partitioning among planktivorous fish larvae and fry in a Mediterranean coastal lagoon. Estuar. Coast. Shelf S. 43: 723–735.

Graneli, W. 1979. Influence of chironomus-plumosus larvae on the oxygen-uptake of sediment. Arch. Hydrobiol. 87: 385–403.

Haas, K., U. Köhler, S. Diehl, P. Köhler, S. Dietrich, S. Holler, A. Jaensch, M. Niedermaier and J. Vilsmeier. 2007. Influence of fish on habitat choice of water birds: a whole system experiment. Ecology 88: 2915–2925.

Haines, T.A. 1973. Effects of nutrient enrichment and a rough fish population (carp) on a game fish population (smallmouth bass). T. Am. Fish. Soc. 102: 346–354.

Hasan, M.R. and D.J. MacIntosh. 1992. Optimum food particle size in relation to body size of common carp, *Cyprinus carpio* L., fry. Aquacult. Fish. Manage. 23: 315–325.

Havens, K.E. 1991. Fish-induced sediment resuspension: effects on phytoplankton biomass and community structure in a shallow hypereutrophic lake. J. Plankton Res. 13(6): 1163–1176.

Havens, K.E. 1993. Responses to experimental fish manipulations in a shallow, hypereutrophic lake: The relative importance of benthic nutrient recycling and trophic cascade. Hydrobiologia 254: 73–80.

Hecky, R.E. and R.H. Hesslein. 1995. Contributions of benthic algae to lake food webs as revealed by stable isotope analysis. J. North. Am. Benthol. Soc. 14(4): 631–653.

Hellström, T. 1991. The effect of resuspension on algal production in a shallow lake. Hydrobiologia 213: 183–190.

Hicks, B.J., N. Ling and A.J. Daniel. 2012. Common carp (*Cyprinus carpio*). pp. 247–260. *In*: A handbook of global freshwater invasive species, Earthscan, London.

Hinojosa-Garro, D. and L. Zambrano. 2004. Interactions of common carp (*Cyprinus carpio*) with benthic crayfish decapods in shallow ponds. Hydrobiologia 515: 115–122.

Hoda, S.M.S. and H. Tsukahara. 1971. Studies on the development and relative growth in the carp, *Cyprinus carpio* (L.). Journal of the Faculty of Agriculture, Kyushu University 16(4): 387–509.

Hootsmans, M.J.M. 1999. Modelling potamogeton pectinatus: For better or for worse. Hydrobiologia 415: 7–11.

Hruska, V. 1961. An attempt at a direct investigation of influence of the carp stock on the bottom fauna of two ponds. Verh. Int. Ver. Theor. Angew. Limnol. 14: 732–736.

Hudson, J.J., W.D. Taylor and D.W. Schindler. 1999. Planktonic nutrient regeneration and cycling efficiency in temperate lakes. Nature 400: 659–661.

Hume, D.J., A.R. Fletcher and A.K. Morison. 1983. Carp program final report. Carp program publication no. 10, Fisheries and Wildlife Division, Victoria, Australia.

Huser, B.J., P.G. Bajer, C.J. Chizinski and P.W. Sorenson. 2015. Effects of common carp (*Cyprinus carpio*) on sediment mixing depth and mobile phosphorus mass in the active sediment layer of a shallow lake. Hydrobiologia, accepted.

Håkanson, L. and M. Jansson. 1983. Principals of Lake Sedimentology. Springer-Verlag, Berlin.

Ibelings, B.W., R. Portielje, E.H.R.R. Lammens, R. Noordhuis, M.S. van den Berg, W. Joosse and M.L. Meijer. 2007. Resilience of alternative stable states during the recovery of shallow lakes from eutrophication: Lake veluwe as a case study. Ecosystems 10: 4–16.

Jackson, M.C., I. Donohue, A.L. Jackson, J.R. Britton, D.M. Harper and J. Grey. 2012. Population-level metrics of trophic structure based on stable isotopes and their application to invasion ecology. PLoS One 7(2): e31757.

Jackson, Z.J., M.C. Quist, J.A. Downing and J.G. Larscheid. 2010. Common carp (*Cyprinus carpio*), sport fishes, and water quality: Ecological thresholds in agriculturally eutrophic lakes. Lake Reserv. Manage. 26: 14–22.

Jana, B.B. and S.N. Sahu. 1993. Relative performance of 3 bottom grazing fishes (*cyprinus-carpio*, cirrhinus-mrigala, heteropneustes-fossilis) in increasing the fertilizer value of phosphate rock. Aquaculture 115: 19–29.

Jeppesen, E., J.P. Jensen, M. Sondergaard, T. Lauridsen, F.P. Moller and K. Sandby. 1998. Changes in nitrogen retention in shallow eutrophic lakes following a decline in density of cyprinids. Arch. Hydrobiol 142: 129–151.

Johnson, D.M. and P.H. Crowley. 1980. "Odonate 'hide and seek': habitat-specific rules?" pp. 569–579. *In*: Evolution and ecology of zooplankton communities, New England University Press, Hannover.

Kadir, A., R.S. Kundu, A. Milstein and M.A. Wahab. 2006. Effects of silver carp and small indigenous species on pond ecology and carp polycultures in Bangladesh. Aquaculture 261: 1065–1076.

Karlsson-Elfgren, I. and A.K. Brunberg. 2004. The importance of shallow sediments in the recruitment of anabaena and aphanizomenon (cyanophyceae). J. Phycol. 40: 831–836.

Khan, T.A. 2003. Dietary studies on exotic carp (*Cyprinus carpio* L.) from two lakes of western Victoria, Australia. Aquat. Sci. 65: 272–286.

Khan, T., M. Wilson and M. Khan. 2003. Evidence for invasive carp mediated trophic cascade in shallow lakes of Western Victoria, Australia. Hydrobiologia 506-509: 465–472.

King, D.R. and G.S. Hunt. 1967. Effect of carp on vegetation in a lake erie marsh. J. Wildlife Manage. 31(1): 181–188.

King, A.J., A.I. Robertson and M.R. Healey. 1997. Experimental manipulations of the biomass of introduced carp (*Cyprinus carpio*) in billabongs. 1. Impacts on water-column properties. Mar. Freshwater Res. 48: 435–443.

Kirk, K.L. 1991. Inorganic particles alter competition in grazing plankton—the role of selective feeding. Ecology 72: 915–923.

Kloskowski, J. 2011. Differential effects of age-structured common carp (*Cyprinus carpio*) stocks on pond invertebrate communities: Implications for recreational and wildlife use of farm ponds. Aquacult. Int. 19: 1151–1164.

Kristensen, E. and T.H. Blackburn. 1987. The fate of organic-carbon and nitrogen in experimental marine sediment systems—influence of bioturbation and anoxia. J. Mar. Res. 45: 231–257.

Lamarra, V.A. 1975. Digestive activities of carp as a major contributor to the nutrient loading of lakes. Verh. Internat. Verein. Limnol. 19: 2461–2468.

Lamarra, V.A. 1976. Experimental studies of the effects of carp (*cyprinus carpio* L.) on the chemistry and biology of lakes.University of MN, Mineapolis, MN.

Lammes, E.H.R.R. and W. Hoogenboezem. 1991. Diet and feeding behavior. pp. 353–372. *In*: I.J. Winfield and J.S. Nelson (eds.). Cyprinid Fishes: Systematics, Biology, and Exploitation. Chapman and Hall, London.

Lougheed, V.L., B. Crosbie and P. Chow-Fraser. 1998. Predictions on the effect of common carp (*Cyprinus carpio*) exclusion on water quality, zooplankton, and submergent macrophytes in a great lakes wetland. Can. J. Fish. Aquat. Sci. 55: 1189–1197.

Marsden, J.E. 1997. Common carp diet includes zebra mussels and lake trout eggs. J. Fresh. Ecol. 12(3): 491–492.

Matsuzaki, S.-i.S., N. Usio, N. Takamura and I. Washitani. 2007. Effects of common carp on nutrient dynamics and littoral community composition: roles of excretion and bioturbation. Fund. Appl. Limnol./Arch. Hydrobiol. 168: 27–38.

Matsuzaki, S.S., K. Mabuchi, N. Takamura, B.J. Hicks, M. Nishida and I. Washitani. 2010. Stable isotope and molecular analyses indicate that hybridization with non-native domesticated common carp influence habitat use of native carp. Oikos 119: 964–971.

Matsuzaki, S.S., N. Usio, N. Takamura and I. Washitani. 2009. Contrasting impacts of invasive engineers on freshwater ecosystems: An experiment and meta-analysis. Oecologia 158: 673–686.

Mazumder, D., M. Johansen, N. Saintilan, J. Iles, T. Kobayashi, L. Knowles and L. Wen. 2012. Trophic shifts involving native and exotic fish during hydrologic recession in floodplain wetlands. Wetlands 32: 267–275.

Matthews, W.J. 1998. Patterns in Freshwater Fish Ecology. Chapman and Hall, New York.

Meijer, M.L., M.W. Haan, A.W. Breukelaar and H. Buiteveld. 1990a. Is reduction of the benthivorous fish an important cause of high transparency following biomanipulation in shallow lakes? Hydrobiologia 200-201: 303–315.

Meijer, M.L., E.H.R.R. Lammens, A.J.P. Raat, M.P. Grimm and S.H. Hosper. 1990b. Impact of cyprinids on zooplankton and algae in ten drainable ponds. Hydrobiologia 191: 275–284.

Michel, P. and T. Oberdorff. 1995. Feeding habitat of fourteen European freshwater fish species. Cybium 19: 5–46.

Miller, S.A. and T.A. Crowl. 2006. Effects of common carp (*Cyprinus carpio*) on macrophytes and invertebrate communities in a shallow lake. Freshwater Biol. 51: 85–94.

Morgan, D.K.J. and B.J. Hicks. 2013. A metabolic theory of ecology applied to temperature and mass-dependence of N and P excretion by common carp. Hydrobiologia 705: 135–145.

Moss, B., L. Carvalho and J. Plewes. 2002. The lake at llandrindod wells—a restoration comedy? Aquat. Conserv. 12: 229–245.

Northcote, T.G. 1988. Fish in the structure and function of freshwater ecosystems: A "top-down" view. Can. J. Fish. Aquat. Sci. 45: 361–379.

Panek, F.M. 1987. Biology and ecology of carp. Bethesda, MD: American Fisheries Society.

Parkos, J.J., V.J. Santucci and D.H. Wahl. 2003. Effects of adult common carp (*Cyprinus carpio*) on multiple trophic levels in shallow mesocosms. Can. J. Fish. Aquat. Sci. 60: 182–192.

Parkos, J.J., V.J. Santucci and D.H. Wahl. 2006. Effectiveness of a plastic mesh substrate cover for reducing the effects of common carp on aquatic ecosystems. N. Am. J. Fish. Manage. 26: 861–866.

Peterson, B.J. and B. Fry. 1987. Stable isotopes in ecosystem studies. Ann. Rev. Ecol. Syst. 18: 293–320.

Phan-Van, M., D. Rousseau and N. De Pauw. 2008. Effects of fish bioturbation on the vertical distribution of water temperature and dissolved oxygen in a fish culture-integrated waste stabilization pond system in vietnam. Aquaculture 281: 28–33.

Power, M.E., D. Tilman, J.A. Estes, B.A. Menge, W.J. Bond, L.S. Mills, G. Daily, J.C. Castilla, J. Lubchenco and R.T. Paine. 1996. Challenges in the quest for keystones. Bioscience 46: 609–620.

Qin, J.G. and S.T. Threlkeld. 1990. Experimental comparison of the effects of benthivorous fish and planktivorous fish on plankton community structure. Arch. Hydrobiol. 119: 121–141.

Rahman, M.M., Q. Jo, Y.G. Gong, S.A. Miller and M.Y. Hossain. 2008. A comparative study of common carp (*Cyprinus carpio* L.) and calbasu (*Labeo calbasu hamilton*) on bottom soil resuspension, water quality, nutrient accumulations, food intake and growth of fish in simulated rohu (*Labeo rohita hamilton*) ponds. Aquaculture 285: 78–83.

Rahman, M.M., S. Kadowaki, S.R. Balcombe and M.A. Wahab. 2010. Common carp (*Cyprinus carpio* L.) alters its feeding niche in response to changing food resources: direct observations in simulated ponds. Ecol. Res. 25: 303–309.

Reid, M.A., M.D. Delong and M.C. Thoms. 2012. The influence of hydrological connectivity on food web structure in floodplain lakes. River Res. Applic. 28: 827–844.

Richardson, W.B., S.A. Wickham and S.T. Threlkeld. 1990. Foodweb response to the experimental manipulation of a benthivore (*Cyprinus-carpio*), zooplanktivore (*Menidia-beryllina*) and benthic insects. Arch. Hydrobiol. 119: 143–165.

Ritvo, G., M. Kochba and Y. Avnimelech. 2004. The effects of common carp bioturbation on fishpond bottom soil. Aquaculture 242: 345–356.

Robel, R.J. 1961. The effects of carp populations on the production of waterfowl food plants on a western waterfowl marsh. Pages 147–159. Transactions of the North American Wildlife and Natural Resources Conference.

Robel, R.J. 1961. The effects of carp populations on the production of waterfowl food plants on a western waterfowl marsh. Trans. N. Am. Wildl. Nat. Resour. Conf. 26: 147–159.

Roberts, J., A. Chick, L. Oswald and P. Thompson. 1995. Effect of carp, *Cyprinus carpio* L., an exotic benthivorous fish, on aquatic plants and water quality in experimental ponds. Mar. Freshwater Res. 46: 1171–1180.

Robertson, A.I., M.R. Healey and A.J. King. 1997. Experimental manipulations of the biomass of introduced carp (*Cyprinus carpio*) in billabongs. I. Impacts on benthic properties and processes. Mar. Freshwater Res. 48: 445–454.

Roozen, F.C.J.M., M. LÜRling, H. Vlek, E.A.J. Van Der Pouw Kraan, B.W. Ibelings and M. Scheffer. 2007. Resuspension of algal cells by benthivorous fish boosts phytoplankton biomass and alters community structure in shallow lakes. Freshwater Biol. 52: 977–987.

Sagova-Mareckova, M. and J. Kvet. 2002. Impact of oxygen released by the roots of aquatic macrophytes on composition and distribution of benthic macroinvertebrates in a mesocosm experiment. Arch. Hydrobiol. 155: 567–584.

Scheffer, M. 1990. Multiplicity of stable states in fresh-water systems. Hydrobiologia 200: 475–486.

Scheffer, M., S.H. Hosper, M.L. Meijer, B. Moss and E. Jeppesen. 1993. Alternative equilibria in shallow lakes. Trends Ecol. Evol. 8: 275–279.

Schindler, D.E., S.R. Carpenter, J.J. Cole, J.F. Kitchell and M.L. Pace. 1997. Influence of food web structure on carbon exchange between lakes and the atmosphere. Science 277(5323): 248–251.

Schrage, L.J. and J.A. Downing. 2004. Pathways of increased water clarity after fish removal from ventura marsh; a shallow, eutrophic wetland. Hydrobiologia 511: 215–231.

Sibbing, F.A., J.W.M. Osse and A. Terlouw. 1986. Food handling in the carp (*Cyprinus carpio*), its movement patterns, mechanisms, and limitations. J. Zool. London (A) 210: 161–203.

Sibbing, F.A. 1988. Specializations and limitations in the utilization of food resources by the carp, *Cyprinus carpio*: a study of oral food processing. Envir. Biol. Fish. 22(3): 161–178.

Sibbing, F.A. 1991. Food capture and oral processing. pp. 377–412. *In*: I.J. Winfield and J.S. Nelson (eds.). Cyprinid Fishes: Systematics, Biology, and Exploitation. Chapman and Hall, London.

Sidorkewicj, N.S., A.C. López Cazorla, O.A. Fernández, G.C. Möckel and M.A. Burgos. 1999. Effects of *Cyprinus carpio* on potamogeton pectinatus in experimental culture: The incidence of the periphyton. Hydrobiologia 415: 13–19.

Sidorkewicj, N.S., A.C.L. Cazorla, K.J. Murphy, M.R. Sabbatini, O.A. Fernandez and J.C.J. Domaniewski. 1998. Interaction of common carp with aquatic weeds in argentine drainage channels. J. Aquat. Plant Manage. 36: 5–10.

Singh, A.K., A.K. Pathak and W.S. Lakra. 2010. Invasion of an exotic fish—common carp, *Cyprinus carpio* L. (Actinopterygii: Cypriniformes: Cyprinidae) in the Ganga River, India and its impacts. Acta Ichthyologicaet Piscatoria 40(1): 11–19.

Soto, D.X., L.I. Wassenaar, K.A. Hobson and J. Catalan. 2011. Effects of size and diet on stable hydrogen isotope values (dD) in fish: implications for tracing origins of individuals and their food source. Can. J. Fish.Aquat. Sci. 68: 2011–2019.

Spataru, P., B. Hepher and A. Halevy. 1980. The effect of the method of supplementary feed application on the feeding habits of carp (*Cyprinus carpio* L.) with regard to natural food in ponds. Hydrobiologia 72: 171–178.

Stewart, T.W. and J.A. Downing. 2008. Macroinvertebrate communities and environmental conditions in recently constructed wetlands. Wetlands 28: 141–150.

Svensson, J.M., E. Bergman and G. Andersson. 1999. Impact of cyprinid reduction on the benthic macroinvertebrate community and implications for increased nitrogen retention. Hydrobiologia 404: 99–112.

Tapia, M. and L. Zambrano. 2003. From aquaculture goals to real social and ecological impacts: Carp introduction in rural central mexico. AMBIO: A Journal of the Human Environment 32: 252–257.

Tatrai, I., K. Matyas, J. Korponai, G. Paulovits, P. Pomogyi and M. Presing. 1999. Stable isotope analysis of food webs in wetland areas of Lake Balaton, Hungary. Archiv für Hydrobiologie 146(1): 117–128.

Tàtrai, I., E.H. Lammens, A.W. Breukelaar and J.G.P. Klein Breteler. 1994. The impact of mature cyprinid fish on the composition and biomass of benthic macroinvertebrates. Archiv für Hydrobiologie 131: 309–320.

Threinen, C.W. and W.T. Helm. 1954. Experiments and observations designed to show carp destruction of aquatic vegetation. J. Wildlife Manage. 18: 247–251.

Tryon, C.A., Jr. 1954. The effect of carp exclosures on growth of submerged aquatic vegetation in pymatuning lake, pennsylvania. J. Wildlife Manage. 18: 251–254.

Tucker, J.K., F.A. Cronin and D.W. Soergel. 1996. Predation on zebra mussel (*Dreissena polymorpha*) by common carp (*Cyprinus carpio*). J. Freshwater Ecol. 11: 363–372.

Umesh, N.R., K.M. Shankar and C.V. Mohan. 1999. Enhancing growth of common carp, rohu and Mozambique tilapia through plant substrate: the role of bacterial biofilm. Aquacult. Int. 7: 251–260.

Vanni, M.J. 2002. Nutrient cycling by animals in freshwater ecosystems. Annu. Rev. Ecol. Syst. 33: 341–370.

Vass, K.F. and A. Vass Van Oven. 1959. Studies on the production and utilization of natural food in Indonesian carp ponds. Hydrobiologia 12: 308–392.

Verspagen, J.M.H., E.O.F.M. Snelder, P.M. Visser, J. Huisman, L.R. Mur and B.W. Ibelings. 2004. Recruitment of benthic microcystis (cyanophyceae) to the water column: Internal buoyancy changes or resuspension? J. Phycol. 40: 260–270.

Vilizzi, L. and K.F. Walker. 1999. The onset of the juvenile period in carp, *Cyprinus carpio*: a literature survey. Environ. Biol. Fish. 56: 93–102.

Wahl, D., M. Wolfe, V. Santucci, Jr. and J. Freedman. 2011. Invasive carp and prey community composition disrupt trophic cascades in eutrophic ponds. Hydrobiologia 678: 49–63.

Wallace, R.K. 1981. An assessment of diet-overlap indexes. T. Am. Fish. Soc. 110: 72–76.

Weber, M.J. and M.L. Brown. 2009. Effects of common carp on aquatic ecosystems 80 years after "carp as a dominant": Ecological insights for fisheries management. Rev. Fish. Sci. 17: 524–537.

Weber, M.J. and M.L. Brown. 2011. Relationships among invasive common carp, native fishes and physicochemical characteristics in upper midwest (USA) lakes. Ecol. Freshw. Fish 20: 270–278.

Werner, E.E. and B.R. Anholt. 1993. Ecological consequences of the trade-off between growth and mortality-rates mediated by foraging activity. Am. Nat. 142: 242–272.

Wetzel, R.G. 1983. Limnology. Saunders College Publishing, New York.

Wilcox, T.P. and D.J. Hornbach. 1991. Macrobenthic community response to carp (*Cyprinus carpio* L.) foraging. J. Freshwater Ecol. 6: 171–183.

Williams, A. and B. Moss. 2003. Effects of different fish species and biomass on plankton interactions in a shallow lake. Hydrobiologia 491: 331–346.

Williams, A.E., B. Moss and J. Eaton. 2002. Fish induced macrophyte loss in shallow lakes: Top-down and bottom-up processes in mesocosm experiments. Freshwater Biol. 47: 2216–2232.

Wolfe, M.D., V.J. Santucci, L.M. Einfalt and D.H. Wahl. 2009. Effects of common carp on reproduction, growth, and survival of largemouth bass and bluegills. T. Am. Fish. Soc. 138: 975–983.

Yoshioka, T., E. Wada and H. Hayashi. 1994. A stable isotope study on seasonal food web dynamics in a eutrophic lake. Ecology 75(3): 835–846.

Zambrano, L. and D. Hinojosa. 1999. Direct and indirect effects of carp (*Cyprinus carpio* L.) on macrophyte and benthic communities in experimental shallow ponds in central Mexico. Hydrobiologia 408-409: 131–138.

Zambrano, L., M. Scheffer and M. Martínez-Ramos. 2001. Catastrophic response of lakes to benthivorous fish introduction. Oikos 94: 344–350.

Zambrano, L., M. Perrow, C. Macías-García and V. Aguirre-Hidalgo. 1998. Impact of introduced carp (*Cyprinus carpio*) in subtropical shallow ponds in central mexico. J. Aquat. Ecosyst. Stress Recovery 6: 281–288.

Zambrano, L., E. Valiente and M.J. Vander Zanden. 2010. Food web overlap among native axolotl (*Ambystoma mexicanum*) and two exotic fishes: carp (*Cyprinus carpio*) and tilapia (*Oreochromis niloticus*) in Xochimilco, Mexico City. Biol. Invasions 12: 3061–3069.

10

Carp as an Invasive Species

Brendan J. Hicks[a,*] and Nicholas Ling[b]

Introduction

The common carp is listed among the eight most invasive fish species on the list of "One hundred of the world's worst invasive alien species" (http://www.issg.org/database/species/; Cambray 2003). The major impacts of common carp are typically regarded as reduced water quality and destruction of aquatic vegetation by uprooting it, but this explains only a small fraction of the problems posed by common carp as an invasive species. In many countries, carp are considered a noxious pest due to their disruptive benthic feeding behaviour, which can resuspend sediments, undermine river banks and dislodge macrophytes (Roberts et al. 1995, Williams et al. 2002, Bajer et al. 2009, Hicks et al. 2012). Resuspension of benthic sediments also increases turbidity, which reduces the amount of light available to macrophytes for photosynthesis, which leads to macrophyte collapse (Gehrke and Harris 1994, Rowe 2007). The potential impacts that carp may have on water bodies is exacerbated because they are long-lived, highly fecund and can exist at extremely high densities (e.g., 3,000 kg/ha) (Harris and Gehrke 1997). Accordingly, concerns have also been raised regarding the potential impact high densities of carp may have on native fish populations through mechanisms such as interspecific competition and habitat destruction (Gilligan and Rayner 2007). Paradoxically, common carp are themselves at risk in their native range from other invaders, e.g., *Carassius gibelio* in the Czech Republic (Lusk et al. 2010).

School of Science, Faculty of Science and Engineering, The University of Waikato, Private Bag 3105, Hamilton 3240, New Zealand.
[a] Email: b.hicks@waikato.ac.nz
[b] Email: n.ling@waikato.ac.nz
* Corresponding author

Infobox: Terminology for carp as an invasive species

Common carp as a species is a **super invader** because it has characteristics that enable it to have strong effects in all habitats under certain conditions. Carp are a super invader in some temperate countries with mild climates such as South Africa, Australia, and New Zealand, partly because they are **limnophilic** (thrive in lakes and ponds), **euryhaline** (tolerate a wide range of salinities) **eurythermal** (tolerate of a wide range of temperatures), and are resistant to **hypoxia** (i.e., low concentrations of dissolved oxygen). Carp are **benthivores**, feeding by filtering bottom sediments, and juvenile common carp live among **emergent macrophytes**, i.e., aquatic plants that protrude above the water surface. **Propagule pressure** is a composite measure of the number of individuals of a species released into a region to which they are not native and is a combination of the number released plus the **fecundity**, or reproductive potential, of those individuals. **Winterkill**, or mortality caused by ice-cover induced deoxygenation, reduces competitor and predator numbers, allowing carp to exploit nursery habitat that is relatively free of competitors and predators. Sufficient **predator pressure** from egg predators in shallow marginal nursery habitat can reduce the spawning success of carp. The **life history traits** of carp, i.e., age and size at maturity, number of eggs per female, sex ratio, migratory tendencies, and longevity all contribute to their success as a super invader.

Invasion of common carp in southeastern Australia followed **logistic growth**, where populations increase slowly at first over time, followed by a phase of rapid expansion and then slower population growth as it approaches the carrying capacity of the environment. Environmental fluctuations and the 2–3 year delay in the onset of maturity cause **delayed logistic population growth**, where population size oscillates as it approaches carrying capacity. **Semi-arid** regions such as southeastern Australia receive precipitation that is a little below potential evapotranspiration, and are intermediate between deserts and humid climates. **Ricker stock–recruitment models** relate the number of recruits to the spawners that produced them and show that common carp are highly successful at expanding their populations at low spawner densities because of their high reproductive potential.

Female common carp have large numbers of **oocytes** (immature egg cells) in their **gonads** (ovaries); during reproduction in **spawning aggregations,** several males compete to spawn with a female, usually among **submerged aquatic macrophytes**. These water plants are necessary to support the sticky fertilized eggs until they mature. **Pheromones**, or chemicals secreted by carp that influence the behaviour of other individuals, can function as an attractant of the opposite sex.

The age of fish is usually determined from **sagittal otoliths**, which are one of the three pairs of ear bones composed of calcium carbonate. In carp, however, the sagittal otolith is less readable than the smaller **astericus otoliths**, which are more commonly used in age determination in this species. Structures used in aging also include **opercles** (a bone in the gill cover) and **vertebrae** (skeletal elements of the backbone). Age as the dependent variable is related to fish size as the independent variable in the **von Bertalanffy growth model. Otolith microchemistry** is used to investigate the chemical composition of the ear bones and can be related to the chemistry of the water body where the fish grew to determine spawning origin.

Infobox: contd....

Infobox: contd....

Rotenone is a chemical compound derived from plant roots that is used widely as a toxicant to kill fish and insects. It is applied to lakes and ponds as a restoration technique to reverse the environmental degradation caused by carp and other benthivores. Another control measure is the use of radiotagged **Judas fish**, usually males that are known as **Judas males**, to track aggregations of carp for more effective control by netting or other capture techniques.

Carp **resuspend** soft bottom sediments so that fine particles are mixed into the water column, greatly reducing water clarity. Carp nutrient excretion fuels the growth of **cyanobacteria** (i.e., small floating bacteria containing the green photosynthetic pigment **chlorophyll *a***) that further reduce water clarity. Cyanobacteria can produce toxins that limit their consumption by **zooplankton**, small free-swimming, filter-feeding animals, and these toxins can be poisonous to animals and humans that drink or bathe in the water. **Secchi depth** is a measurement of water clarity that is used to monitor changes in light penetration in lakes. Decrease in water clarity is an environmental effect that particularly affects fish species native to the southern hemisphere such as **galaxiids** in Australia and New Zealand; **galaxiids** are small, migratory fish many species of which are in decline.

Common carp have complex genetic relations with goldfish, producing **triploid** progeny (with one extra set of chromosomes than normal carp) when carp-goldfish hybrids cross with carp. Normal carp are **diploid**, with only two sets of chromosomes. The **F$_1$-generation** is the first hybrid generation resulting from a cross mating of carp and goldfish, and the **F$_2$-generation** is the second generation from hybrids breeding with each other or either parental type. **Introgression** is the flow of genetic material from one species into the gene pool of another, and goldfish control is as important as common carp control to limit yet another environmentally degrading invader, the carp-goldfish hybrid.

To understand the nature of the risk posed by common carp to aquatic ecosystems, the context of the species' biology and origins is important. The species probably evolved in Central Asia in the area of the Caspian Sea at the end of the Pliocene about 2.6 million years ago. After its emergence as a species, a first wave of expansion led to a continuous distribution of common carp in Eurasia from the Don and Danube Rivers in the west to the Amur River drainage basin and China in the Far East, which was broken up into eastern and western populations during multiple Pleistocene glaciations (Berg 1964, Kohlmann and Kersten 2013). Carp spread into the Black and Aral Seas, and subsequently into Europe as far west as the Danube River 8,000 to 10,000 years ago (Balon 1995). The spread to China, Japan and south-east occurred later, but the species' biogeography has been obscured by extensive translocation and mixing of different genetic stocks to improve aquacultural and ornamental strains. As a species, the common carp can now be divided into three subspecies: *C. c. carpio* in Europe, *C. c. viridiviolaceus*[1] in south-east Asia, and *C. c. haematopterus* in the Far East (Vilizzi 2012).

---

[1] *C. c. viridiviolaceus* is synonymous to *C. c. rubrofuscus*, which is the name that is more often used in Chapter 1 on carp genetics.

One of the major problems with common carp as an invasive species is its desirability as a sports fish, a food fish and an ornamental fish. Common carp in its native Europe and Asia is a highly valued food source and recreational game species (Gilligan and Rayner 2007), and probably because of this is now among the most translocated vertebrate species in the world. Carp have been spread to parts of Europe, Asia, Africa, Australia, New Zealand and North and South America (Parameswaran et al. 1972, Howes 1991, Vilizzi 2012). The intentional introduction into Britain the 1490s is the earliest documented carp introduction to Britain (Britton et al. 2010a). As a sports fish, carp have been introduced into ponds, lakes and rivers by coarse anglers, who are so-called because of the large scales of their primary target fish species. Secondly, use of common carp as a food fish has resulted in their intensive culture in ponds, with the almost inevitable consequence that at some point floods or poor management would result in accidental escapes of carp from containment ponds into the wild. Such was the scenario in the South-east of Australia, where carp from a fish farm near Boolara, Victoria, escaped into the River Murray system in the early 1960s (Shearer and Mulley 1978).

Carp have been introduced to more than 100 countries outside their native range (FishBase 2010) (Fig. 10.1). Common carp were introduced into Chile between 1895 and 1920 (Prochelle and Campos 1985), and into India for aquaculture in 1959, and escapees have recently been found in the Ganga River (Ganges) in northern India, posing a risk to the native fish biodiversity (Singh et al. 2010). The range expansion of common carp is clearly far from over, as witnessed by their recent discovery in the Wilderness Lakes System, a wetland of international importance in the Western Cape Province, South

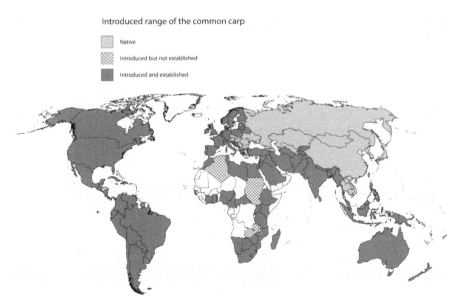

Figure 10.1. The global distribution of common carp (*Cyprinus carpio* L.) Source: www.fishbase. org.

Africa (Olds et al. 2011). Thus we need to understand the risks posed by common carp in areas where they are not native.

This chapter will focus on the characteristics that predispose common carp to be a super invader of aquatic habitats, and summarize the most recent understanding of the effects on ecosystems and the environment.

Risk Assessment

Risk assessments usually revolve around life history traits that predispose a species to become invasive and an understanding of its habitat requirements. Risk assessment models for establishment of exotic vertebrates rank common carp as the most invasive freshwater fish in Australia and New Zealand, considering factors such as climate matching, introduction success, and taxon risk score (Bomford 2008). Common carp are naturally limnophilic, inhabiting slow flowing downstream reaches of large rivers, river deltas, backwaters, wetlands and floodplain lakes throughout their native range (Nikolski 1933). They are euryhaline, tolerating 10‰ salinity for at least three months and surviving in water up to 14‰ salinity (Al-Hamed 1971, Crivelli 1981), and eurythermal, surviving from near-freezing to an upper lethal temperature of 43°C to 46°C (Opuszński et al. 1989). They also tolerate poor water quality and low dissolved oxygen (P_{crit} ~15% saturation at 15°C; Ultsch et al. 1980) but are far less tolerant to hypoxia than other closely related cyprinids such as goldfish and crucian carp.

In an analysis of risk assessment and ecological predictions for alien fishes in North America, Kolar and Lodge (2002) suggested that the main attribute that determines the spread of fish species is its ability to tolerate a wide range of temperatures. Once spread, several factors determine establishment: 1) growth to greater than 68% of its average adult size by two years of age, 2) tolerance of a wide range of temperatures, 3) tolerance of a wide range of salinities, and 4) a history of invasiveness. Factors that contribute to the impact of a fish species are 1) small egg diameter (indicating high fecundity), 2) the ability to survive low water temperatures, and 3) tolerance of a wide range of salinities (Table 10.1). These factors have been used to evaluate invasion risk for a number of freshwater species in the Great Lakes region using a classification and regression tree (CART) approach (Fig. 10.2). Considering these attributes, as we will do below, common carp are almost the perfect invader. Marchetti et al. (2004) used a range of attributes such as adult trophic status, size of native range, parental care, maximum fecundity, maximum adult size, maximum lifespan, physiological tolerance, and propagule pressure to determine the invasiveness of non-native fish in California, USA. Common carp scored high for all these attributes. Using a Fish Invasiveness Scoring Kit, Copp et al. (2009) ranked common carp as the most invasive fish in Britain. A recent summary of risk assessment procedures and current knowledge on non-native freshwater fish introductions suggests combining the economic values of species on a global scale with their associated level of ecological risk (Gozlan et al. 2010). Fish with high

Table 10.1. Variables contributing to a discriminant function describing the establishment, spread and impact analyses by providing according function coefficients. Positive coefficients are associated with becoming established, with spreading quickly and with being perceived as a nuisance; negative coefficients are associated with failing to become established, with spreading slowly and with being perceived as a non-nuisance. After: Table S6, supplementary material, Kolar and Lodge (2002).

Variables contributing to discriminant analysis model	Function coefficients
ESTABLISHMENT	
Closer to mature length by 2 years	0.0704
Wide temperature range tolerance	0.1430
Wide salinity range tolerance	1.5844
Species has no history of invasiveness	−1.6402
SPREAD	
Survive high temperatures	−1.5254
Wide temperature range tolerance	0.9551
Closer to mature length by 2 years	< −0.0001
IMPACT	
Egg diameter	−2.4176
Survive low temperatures	0.1790
Wide salinity range tolerance	0.7360

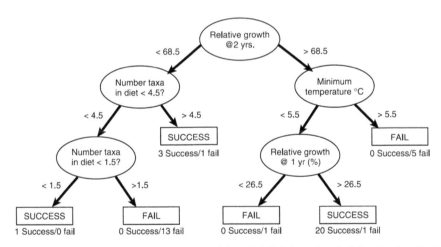

Figure 10.2. CART decision tree of successful and failed introduced fishes in the Great Lakes. Ovals represent decision points; rectangles are terminal points in the tree resulting in classification. The numbers of known successful and failed alien species categorized into each terminus are given, illustrating that 2 of 45 species were misclassified. Source: Fig. 2, Kolar and Lodge (2002). Reproduced with permission.

ecological risk and low economic value are problematic, and common carp fit into this category. Despite their importance as a recreational and aquaculture species in highly developed regions of the world, the economic value per fish is low, which makes a commercial harvest from the wild uneconomic or only marginally profitable. Some authors have taken an economic evaluation approach to the justification for invasive species control when considering common carp removal (Homans and Smith 2013). There are also issues of palability of common carp in some highly developed regions where they are not widely regarded as an acceptable table fish (e.g., USA, Australia, and New Zealand).

Some Recent Examples of Spread

One of the best documented instances of common carp expanding their population size in a previously carp-free environment comes from the Murray-Darling River system in Australia. The release of the Boolara strain of common carp in 1961/1962 into Lake Hawthorn, which drains into the Murray River, was followed by expansion of their geographic distribution in southeastern Australia. Logistic and delayed logistic population growth models best explained the population growth between 1962 and 2002. Annual catches increased rapidly from almost zero between 1962 and 1970 to a maximum of about 570 tonnes/year in 1978 (Forsyth et al. 2013).

The introduction of koi carp (*C. c. haematopterus*, Smith and McVeagh 1987) into New Zealand is undocumented but it is most likely that they were brought in as juveniles, inadvertently or intentionally, in one or more consignments of goldfish, *Carassius auratus*, during the 1960s. Further importation of goldfish into New Zealand was prohibited in 1972 (Kilford 1984, Pullan 1984a, Smith and Pullan 1987, McDowall 1990). Koi were widespread in ornamental fish ponds in the 1970s and 1980s and between 1978 and 1983 koi were caught in several locations in the wild (Pullan 1984a,b, Hanchet 1990). Eight carp were caught in November 1983 and February 1984 in the Waikato River system (Kilford 1984, Pullan 1984a). The relative ease with which these carp were caught (four fish in three seine net hauls), together with anecdotal reports of eel fishermen catching 50 or more fish suggests that koi had already established a large population size in the Whangamarino River and its extensive wetland system by 1983. Koi carp abundance increased dramatically in the wider Waikato River basin between 1990 and 2004 according to bowfisher catch rates (see later) and the species was estimated to contribute up to 69% of total fish biomass in the river by 2005 (Hicks et al. 2005).

In Kenya, Eastern Africa, following the accidental introduction of common carp into Lake Naivasha during 1999, a self-sustaining population became rapidly established and in early 2004 became the principal species exploited in the commercial fishery. Over 9,000 kg of carp were harvested from the lake between October 2005 and 2006, when fish between fork lengths (FL) of 200 and 800 mm (> 8 kg) were captured. Carp of 100 mm FL fed primarily on benthic macroinvertebrates, which were previously unexploited by the indigenous fish community (Britton et al. 2007). Low wage rates probably enhance the profitability of carp harvest in less developed countries such as Kenya compared to highly developed economies.

The Life History and Ecological Niche of Common Carp

Habitat Requirements

Carp have a complex life history that can be divided into spawning, foraging and over-wintering phases, each having different habitat associations. Habitat suitability curves have been developed for modelling key life history phases (Edwards and Twomey 1982). In particular, carp move into interconnected shallow lakes, floodplains and wetlands to spawn (Stuart and Jones 2006, Bajer and Sorensen 2010) and juveniles develop in shallow floodplains and marshes (King 2004). Adult carp tend to forage in warmer shallower waters in summer (Penne and Pierce 2008), but over-winter in large aggregations in deeper water (Johnsen and Hasler 1977).

However, the relationship of common carp relative abundance to lake depth is complicated by winter temperature regimes. Shallow lakes in areas where winter ice-over is common, such as Eastern South Dakota, are prone to periodic winter-time hypoxia (i.e., winterkill) that may limit common carp recruitment and decrease extreme carp abundance (Weber et al. 2010). In these areas, carp abundance and size increases with increasing mean and maximum lake depth. These results contrast with previous work in more temperate locations where common carp abundance decreased with increasing lake depth (Egertson and Downing 2004).

Life History Characteristics

Reproduction

Adult wild carp typically become sexually mature at two to three years old for females and one to two years for males (Crivelli 1981). Relative fecundity ranges from 19,300 to 216,000 oocytes/kg total body mass, with a mean of 97,200 oocytes/kg. Adults average around 400 to 500 mm FL and 2 to 3 kg but may grow much larger, so adult females typically shed around 300,000 eggs per spawning. Tempero et al. (2006) found that maximum gonad weight as a percentage of total body weight was about 20%, and occurred in wild New Zealand koi in early spring, with females averaging 100,000 eggs/kg total body mass. Common carp in Victoria, Australia, had higher fecundity (120,000 to 1,540,000 eggs per fish or 163,000 eggs/kg; Sivakumaran et al. 2003) and in culture, fecundities may be up to 300,000 eggs/kg (Billard 1999), but can vary considerably in different geographical areas (Weber and Brown 2012a).

Environmentally determined sex ratios of carp populations are typically skewed slightly towards females, with male:female ratios generally from 1:1.33 to 1:1.54 (Mauck and Summerfelt 1971, Crivelli 1981). However, spawning aggregations in temporary marshes can have more males than females with male:female ratios of 1:0.56 to 1:0.60 (Crivelli 1981). These observations do not necessarily reflect sex ratios at recruitment. Carp in temperate latitudes begin to spawn in spring as water temperatures rise above about 17°C and depending on the duration of favourable conditions further spawning events may occur throughout summer (Tempero et al. 2006). We observed carp spawning in November (austral spring) in Lake Sorell, Tasmania, when the main body of the lake was 12°C but solar heating of the lake shallows raised littoral temperatures to 26°C. Egg and larval development appears to be limited by temperatures > 26°C as development declined at 28 and 30°C (Sapkale et al. 2011). In the tropical climate of West Bengal, Guha and Mukherjee (1991) observed two distinct reproductive cycles per year in winter and summer.

Despite the high fecundity of carp and the ability to spawn more than once a year under favourable conditions, recruitment appears to be highly variable and strongly dependent on climatic conditions such as ice-over in winter and spring water temperatures. Carp spawn in warm shallow marginal zones of rivers and lakes. The eggs are adhesive and are laid among submerged aquatic macrophytes or on inundated terrestrial plants. Ideal spawning habitat has water temperatures > 17°C, high water levels to inundate marginal vegetated spawning habitat, and sheltered conditions so that wind-driven waves cannot dislodge eggs laid on submerged vegetation or smother eggs with sediment (Phelps et al. 2008). Carp in the Murray-Darling Basin typically sought warm, slow-flowing waters for spawning, such as the floodplain waters and river backwaters abundant in lowland alluvial river catchments (Driver et al. 2005), similar to previous findings from other countries (McCrimmon 1968, Lubinski et al. 1986, Balon 1995).

Flow regulation has an effect on common carp spawning success (Driver et al. 2005). Unregulated rivers have more flood events, especially medium-sized flows (Walker et al. 1995), and provide more opportunities for carp spawning and access to nursery habitat in shallow vegetated floodplain habitats (McCrimmon 1968). Spawning and recruitment of common carp can occur in the absence of flooding (e.g., King 2004, Cheshire 2010), but carp recruitment is much increased when increased river flows inundate floodplains (King et al. 2003, Crook and Gillanders 2006, Stuart and Jones 2006, Humphries et al. 2008, Bice and Zampatti 2011, Beesley et al. 2012).

An analysis of 18 lakes in South Dakota, USA, showed that carp recruitment fluctuated synchronously throughout the region, apparently in response to climate (known as the Moran Effect; Phelps et al. 2008). A further demonstration of the influence of climate on carp recruitment was provided by Bajer and Sorensen (2010), who found that successful carp recruitment in shallow lakes followed severe winter hypoxia, possibly in response to reduced egg-predator pressure in shallow marginal nursery habitat. Spawning success is linked to water level, and strong year classes are associated with spring flooding (Inland Fisheries Service 2010, Taylor et al. 2012). It is likely that spawning aggregations are stimulated, at least in part, by the behavioural influence of pheromones. Sisler and Sorensen (2008) demonstrated that carp can discriminate between the odour of conspecifics and goldfish.

Age and Growth

Although captive carp are reputed to be among the longest living vertebrates, with claims of some individuals exceeding 200 years, such cases appear unverified and are probably exaggerated. Although captive fish may live considerably longer than wild fish, wild carp typically average four to five years old, with few individuals surviving past age 15. The oldest individuals examined by Bajer and Sorensen (2010) were 17 and 34 years of age in Lakes Echo and Susan, respectively, in Minnesota, USA. Carp exhibit a range of growth rates, with some evidence that males are generally smaller than females (Table 10.2), which is consistent with the earlier maturity and shorter life span of males. Growth of juveniles is inversely related to carp density, and Ricker stock-recruitment models indicated that peak production of age-0 common carp occurred when adult abundance was low, suggesting that density-dependent intraspecific competition and environmental conditions regulate recruitment and first-year growth of carp in shallow lakes (Weber and Brown 2013).

Koi carp in New Zealand exceed the growth rates of carp from Europe and Chile after age four and achieve greater maximum size, although their growth rate is always less than that of fish in Australia (Tempero et al. 2006). Males rarely lived in excess of 8 years, whereas females lived to 12 years (Tempero et al. 2006). However, aging carp presents some problems because the larger sagittal otoliths are less readable than the small astericus otoliths and most studies use scales instead (Vilizzi and Walker 1995). Phelps et al. (2007) compared age estimations of carp using scales, fin rays, opercles, vertebrae and astericus otoliths and found that all structures other than otoliths consistently underestimated the ages of carp older than 10 to 13 years.

Temperature has a significant and positive effect on the foraging and growth of juvenile common carp (90–105 mm FL) between 16 and 28°C, with an optimum for foraging and growth between 24 and 28°C (Oyugi et al. 2012). Temperature influences carp size; increasing annual temperature range increases the maximum size of carp (L_∞) and decreases the curvature term (K) of the von Bertalanffy growth model (Fig. 10.3).

Diet

Common carp are highly efficient suctorial substrate feeders, which is the cause of most of their environmental effects. They are omnivores, feeding on a wide range of aquatic invertebrates, plants, seeds, algae and detritus, but focussing on chironomids where they are available (Bartels and Huser, this volume). Such broad food preferences suggests that common carp will compete with many native and other introduced species, as suggested by stable isotope analyses (Kelleway et al. 2010, Hicks et al. 2010).

Table 10.2. von Bertalanffy growth curve parameters of common carp (*Cyprinus carpio*) from Australia, France, Spain, Slovakia, Kenya, and Chile, including koi carp from New Zealand. Length method refers to the use of fork, standard, or total length. – = not given.

Country	L_∞ (cm)	K	t_0	Length method	Source
Australia, Murray River, males	48.9	0.249	-0.519	Fork	Brown et al. 2005
Spain	50.0	0.300	0.200	Fork	Fernández-Delgado 1990
New Zealand, Hikutaia Cut	50.0	0.215	0.000	Fork	Tempero 2004
Chile	51.5	0.320	0.150	Total	Prochelle and Campos 1985
Australia, Murray River	51.5	0.236	-0.542	Fork	Brown et al. 2005
France	51.6	0.270	-0.400	Fork	Crivelli 1981
Australia, Murray River, females	59.4	0.177	-0.609	Fork	Brown et al. 2005
South Africa, Lake Gariep, males	60.7	0.350	0.160	Fork	Winker et al. 2011
Europe, Lesser Danube above Kolárovo, females	61.0	0.270	-0.300	Standard	Balon 1995
Kenya, Lake Naivasha, males	61.8	0.680	–	Fork	Oyugi et al. 2011
Lesser Danube above Kolárovo, males	65.0	0.200	-0.500	Standard	Balon 1995
Australia	65.5	0.260	-0.400	Total	Vilizzi and Walker 1999
South Africa, Lake Gariep, females	66.2	0.390	0.160	Fork	Winker et al. 2011
New Zealand, Waikato River	67.5	0.210	0.150	Fork	Tempero et al. 2006
Croatia, Lake Vransko	75.2	0.122	-0.550	Total	Treer et al. 2003
Kenya, Lake Naivasha, females	75.5	0.750	–	Fork	Oyugi et al. 2011

Note: The equation for the von Bertalanffy growth is $L_t = L_\infty(1-e^{-K(t-t_0)})$, where L_t = length in cm at time t (age in years), L_∞ = asymptotic maximum length, K = characteristic curvature. Both sexes are combined unless otherwise stated.

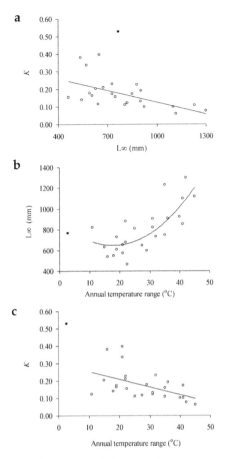

Figure 10.3. a. Relationship of maximum fork length (L_∞) and the curvature (K) of the von Bertalanffy growth model for populations of *Cyprinus carpio*; b. relationship of L_∞ to the annual temperature range experienced by populations; c. relationship of K to the annual temperature range experienced by populations. ○ = their invasive range and • = Lake Naivasha. Source: Oyugi et al. (2011). Reproduced with permission.

Movements and Migration

Movements and migration in invaded habitats are key aspects of the invasive nature of common carp, and mirror those in their native range, where severe winters and spring flooding are common. Carp live in rivers and lakes but move to floodplains and flooded ephemeral marshes in spring to spawn (Koblitskaya 1977, Bajer and Sorensen 2010). Although carp appear to show little habitat selectivity in Australian and New Zealand rivers they also show a remarkable degree of site fidelity. Jones and Stuart (2007) observed that sub adult carp (284–328 mm) occurred equally in mainstream and off-stream areas among woody debris or aquatic vegetation, although they avoided high flows. The majority of carp moved little (< 5 km), but some fish occasionally undertook extensive migrations to access suitable spawning, feeding or over-wintering habitats. The home range of carp in Australia was similarly generally limited, with a mean of 525 m, but one out of 15 fish had a maximum home range of 2.1 km (Crook 2004). In the Waikato River, New Zealand, 85% of 76 recaptured dart-tagged wild koi were caught within 5 km of their release site (mean time at liberty of 519 days; Osborne et al. 2009). However, monitoring with a combination of acoustic and radio telemetry in the Waikato River system showed that recaptures of dart-tagged carp revealed an incomplete view of movement, and that carp actually undertook pre-spawning movements of hundreds of kilometres, associated with changes in water levels or temperature, although these fish often returned to near their site of origin (Daniel et al. 2011), consistent with the appearance of limited movement from dart tagging studies.

Carp are very sensitive to water temperature; changes of as little as ~0.1°C per hour have been sufficient to alter activity patterns of common carp in heated effluent water (Cooke and Schreer 2003), and an increase of 2–3°C over one to two days triggered movements in the Waikato River (Daniel 2009, Daniel et al. 2011).

Almost identical results were obtained from movement studies of carp in the Murray-Darling River system in south-eastern Australia. Eighty per cent of the 293 recaptured externally-tagged carp were caught < 5 km from their release site (mean time at liberty 442 days; Stuart and Jones 2006). A prolonged drought in the River Murray changed normal carp behaviour, restricting the expected seasonal off-channel carp movements into floodplain spawning habitats (Conallin et al. 2011). Conversely, deliberate flow releases intended to maintain floodplain wetland habitat function and native fish communities had the unintended consequence of facilitating off-channel movement and spawning of common carp in the River Murray (Conallin et al. 2012). A separate Australian study showed that 65% of radio-tagged carp displayed long-term site fidelity to within 100 m, whereas two fish travelled more than 650 km downstream (Jones and Stuart 2009). However, most fish moved into adjacent floodplain habitat upon flooding. In the highlands of Tasmania, common carp in Lake Sorell displayed increased mobility during spring–summer periods, moving into shallow habitat rich in macrophytes, particularly during years of high lake levels. During years of low lake levels, this pattern was altered, with frequent use of a rocky 'secondary' habitat. During winter, carp congregated in deeper habitat (Taylor et al. 2012).

In the Fox River, Illinois, radio-tagged common carp used areas with flowing water most frequently in spring, but their use of flowing and mid-reach habitats declined through summer and fall and was lowest in winter (Fig. 10.4) (Butler and Wahl 2010).

A study of radio-tagged fish in a 40-ha lake in Minnesota revealed that carp typically occupied a highly restricted home range (~100 m × 70 m) but quickly learned the location of a supplemental food supply and henceforth undertook substantial night-time movements (> 300 m) to the food source, returning to within their home range during daytime hours (Bajer et al. 2010).

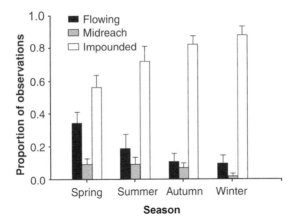

Figure 10.4. Proportion of observations of individual common carp that occurred in flowing, midreach, and impounded habitat zones of the Fox River, Illinois, during each season. Bars represent mean (± 1 SE) proportions of observations on individual fish in each habitat zone. Source: Butler and Wahl (2010). Reproduced with permission.

In conclusion, common carp appears to be a 'partial migrator', in which only a portion of the population moves; movement patterns do not fit classical definitions of 'true' migration in which all of a cohort migrate.

Lateral Movements

Movements from rivers into lateral floodplain wetlands for spawning seem to be a common characteristic of carp life history, with subsequent export of larvae and juveniles from wetlands. Crook et al. (2013) used otolith microchemistry to identify recruitment hotspots for carp in the Lachlan River system, Australia. They showed that particular wetland and floodplain nursery areas dominated carp recruitment, which declined with increasing distance from wetland nurseries. Recruitment success was strongly influenced by river flows and water management, offering the possibility of reducing carp recruitment by targeted water management at nursery sites. The possibility of excluding adult carp from entering wetland habitats also has the potential to reduce the direct impacts of carp in such systems. Differences in carp recruitment and residency in regulated and unregulated rivers in the Murray-Darling catchment suggest that while floods enhance opportunities for dispersal, regulated flows reduce the likelihood of high-flow mortality (Driver et al. 2005). The seasonal activities of carp in the floodplain of the Rhine River included temperature-dependent lateral spawning migrations from the river into oxbows and subsequent spawning (Scharbert and Borcherding 2013).

Carp also undergo seasonal movements in cold, temperate lakes, aggregating in deeper, warmer water during winter. Carp tagged with ultrasonic transmitters in Lake Mendota (Wisconsin, USA) aggregated in two areas of deeper water in the lake from late autumn and were able to be successfully targeted by commercial fishers (Johnsen and Hasler 1977). Adult carp over-winter in deep lakes that are not subject to winterkill, but aggressively move into winterkill-prone shallow regions in the spring to spawn. This accounts for recruitment peaks in years following severe winter hypoxia, which allows carp to exploit nursery habitat that is relatively free of predators (Bajer and Sorensen 2010). Similar winter aggregations have been observed in Lake Sorell (Tasmania, Australia) by following the movements of radio tagged fish (Taylor et al. 2012).

Population Density

Few studies have examined carp population density within the species' native range. One study of the fishery of Lake Balaton in Hungary showed that carp did not dominate the fish biomass in this multispecies ecosystem despite supplemental stocking for recreational angling (Biro 1997). However, many studies outside their native range have demonstrated that common carp quickly dominate fish biomass and cause significant density-dependent ecological impacts. In a survey of 20 major river basins throughout the US, carp were the most commonly collected introduced fish species (Meador et al. 2003). In Australia (King et al. 2012) and New Zealand (Hicks et al. 2006) common carp came to dominate the fish community within 20 years of first establishment. In the Waikato River system, koi carp comprise on average 69% of the total fish biomass. At some lower river sites koi carp are 80–90% of the fish biomass (Hicks et al. 2005).

In the channelized upper Missouri River, a density of 1,224 carp/km was estimated, dropping to 607 carp/km in a reach with 9,369 channel catfish/km (Hesse and Newcombe 1982). Carp outside the natural range typically dominate fish populations and usually exceed 500 kg/ha in biomass. Carp populations can expand rapidly to attain a biomass of 3,144 kg/ha with densities exceeding 1,000 individuals/ha (Koehn 2004). Driver et al. (2005) estimated carp biomass of up to 692 kg/ha in some regions of the Murray-Darling Basin, with localized densities exceeding 2,000 kg/ha. In New Zealand, spawning biomass estimated from boat electrofishing may reach 4,030 kg/ha (Hicks et al. 2006). In the Camargue, France, biomasses of 8–335 kg/ha were estimated by sampling with rotenone and carp was the dominant fish by biomass (mean 62%; Crivelli 1981). Introductions in this area of the south of France (Etang de Maugiuo) associated with dyke construction extend historically back to 1695. In small midwestern lakes of the USA, carp biomass estimated by boat electrofishing was 65–495 kg/ha, and electrofishing catch per unit effort (carp/hour) related well to mark-recapture estimates of abundance (r^2 = 0.83; Bajer and Sorensen 2012).

Predators

Carp quickly outgrow the exploitable size range for most piscivorous fish (Crivelli 1981, Liao et al. 2001), complicating their management by biological control. The ability of carp to degrade water quality in shallow lakes probably reduces the efficiency of visually feeding piscivorous fish, and an increase in carp abundance in Lake Naivasha, Kenya, coincided with a decline in largemouth bass (*Micropterus salmoides*), an introduced piscivore (Britton et al. 2010b).

Ecological Effects

Concerns about the environmental consequence of carp introduction date back to the 1850s in the US (McCrimmon 1968, Weber and Brown 2009). The detrimental effects of common carp on the water quality of soft-bottomed, shallow lakes was first reported in Southern Wisconsin, US, by Cahn (1929) with a change from clear to turbid water, loss of aquatic vegetation and an implied loss of fish diversity, raising serious concerns about the implications of carp proliferation for ecosystem management. A large number of studies in aquaria, exclosures that exclude fish and lakes have since confirmed the ability of carp to resuspend sediments and nutrients due to their vigorous benthic feeding activity (e.g., Cahoon 1953, Zambrano et al. 1998, Zambrano and Hinojosa 1999, Zambrano et al. 2001, Driver et al. 2005, Matsuzaki et al. 2007, Weber and Brown 2009, Kloskowski 2011a,b, Wahl et al. 2011, Weber and Brown 2011, Akhurst et al. 2012). Carp can also reduce the abundance of aquatic macrophytes (Cahoon 1953, Zambrano et al. 1998, Zambrano and Hinojosa 1999, Williams et al. 2002, Hinojosa-Garro and Zambrano 2004, Wahl et al. 2011) and benthic invertebrates (e.g., Zambrano et al. 1998, Zambrano and Hinojosa 1999, Hinojosa-Garro and Zambrano 2004, Wahl et al. 2011, Fischer et al. 2013), and increase abundance of cyanobacteria (Williams and Moss 2003) and chlorophyll *a* concentrations (Fischer et al. 2013). The ecological consequences of common carp invasions include reduced water clarity and increases in algal blooms, stimulated by nutrient release from carp. Further information on the feeding activity of carp and ecological effects may be found in Chapter 9 on the feeding ecology of carp by Bartels and Huser (this volume). Enclosure experiments, modelling and direct measurements of excretion rates have shown that fish can be important in the cycling of nutrients in freshwater systems (e.g., Schaus et al. 1997, Mehner et al. 2005, Post and Walters 2009, Morgan and Hicks 2013). When the effects of sediment resuspension were included, phosphorus concentrations and chlorophyll-*a* concentrations increased with increasing carp biomass between 300 and 800 kg/ha, and with increasing fish size (Driver et al. 2005). Elevated nitrogen and phosphorus concentrations in the presence of carp are a consistent finding (King et al. 1997, Zambrano et al. 1999, Badiou and Goldsborough 2010, Akhurst et al. 2012).

Biomass Thresholds

Biomass thresholds for ecological effects have been the focus of considerable research, and a critical carp biomass of 450 kg/ha was originally identified in Australia (Fletcher et al. 1985). However, loss of macrophytes in soft-bottomed water bodies can occur at biomasses of > 200 kg/ha (Williams et al. 2002). Weber and Brown (2009) summarize a consensus threshold of 450 kg/ha from an extensive literature review, but Bajer et al. (2009) found that the ecological integrity of a shallow lake in North America was jeopardized at densities of ~100 kg/ha (Fig. 10.5). Careful interpretation of the effects of carp on vegetation and waterfowl suggests that safe biomasses of carp are < 50 kg/ha, and that significant impacts occur at biomasses ≥ 100 kg/ha (Bajer et al. 2009, Hicks et al. 2012). A carp biomass of 120–130 kg/ha was sufficient to depress macroinvertebrate and plant biomass, and to elevate chlorophyll

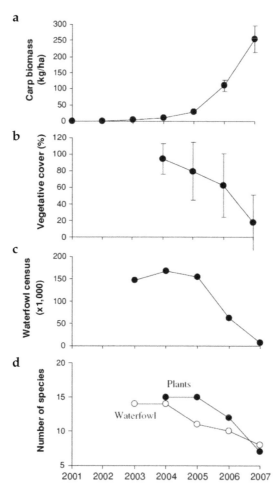

Figure 10.5. Reduction of aquatic macrophyte cover, number of wildfowl and species richness with increasing biomass of common carp in two shallow lakes, Hennepin and Hopper, Illinois: a. Biomass of carp 2001–2007 (mean ± 1 SD); b. per cent vegetative cover (mean ± 1 SD); c. cumulative waterfowl (dabbling and diving ducks) count; and d. number of aquatic plant and duck species. Source: Bajer et al. (2009). Reproduced with permission.

a concentration (Haas et al. 2007). The response of suspended sediment to increasing carp biomasses was approximately linear with increasing carp biomass, but was much stronger in the presence of large carp (mean weight 4.35 kg) than smaller carp (mean weight 2.79 kg) at the same biomass (Fig. 10.6) (Badiou and Goldsborough 2010).

Common carp contribute to eutrophication and algal blooms directly through nutrients released by excretion and defaecation. Recent short-term measurements show that rates of nutrient release are related to biomass, fish body size and water temperature (Morgan and Hicks 2013). At a fish biomass of 500 kg/ha, the total N released by 500-g carp at 20°C (30 mg N/m^2/day) would approximately equal the net internal N loading rate estimated for a New Zealand lake (34 mg N/m^2/day at 7 m; Burger et al. 2007). The same carp loading would release 4.8 mg P/m^2/day.

Because of the difficulty of estimating the absolute biomass of carp, some authors have suggested using a threshold catch per unit effort (CPUE). In Iowa, for instance, 2 kg/fyke net/night was suggested as a threshold CPUE, above which carp have damaging effects on the environment (Jackson et al. 2010). This threshold for ecological effect of common carp was confirmed in South Dakota (Weber and Brown 2011).

Competition with Wildfowl and Fish

Carp are highly efficient harvesters of benthic invertebrate communities, and because of this trait can compete with other benthivores, reducing breeding success and abundance of wildfowl (Haas et al. 2007, Bajer et al. 2009, Broyer and Calenge 2010). Conditions where the biomass of common carp dominates the local fish fauna have common characteristics (e.g., the floodplain habitats of Barmah-Millewa Forest on the Murray River system, Stuart and Jones 2002). In contrast, backwater lake habitats in the Illinois River had fewer larval common carp but more native fish species than main channel habitats, suggesting that off-channel floodplain habitats were refugia for native fish (Nannini 2012). In South Dakota, juvenile common carp (mean total length 49–75 mm) showed a strong habitat preference for emergent macrophytes in summer (Weber and Brown 2012b).

A strong dietary overlap has been demonstrated with tench (*Tinca tinca*) (Adàmek et al. 2003), which may explain the disappearance of tench from the Whangamarino Wetland, North Island, New Zealand, following carp invasion in the early 1980s. In October 1980, tench were caught in the Whangamarino River (Strickland 1980, 1981). In August 2007 and March 2008, extensive boat electrofishing and netting found no tench, with koi carp dominating the fish biomass at all sites. Mean captured biomass of koi carp was 356 kg/ha (Hicks et al. 2008); accounting for capture efficiency (Hicks et al. 2006), actual biomass might be about 700 kg/ha. Koi carp biomass was 93% of the total fish biomass (Hicks et al. 2008).

In 129 Iowa lakes, high densities of common carp were related to undesirable water quality conditions and low abundance of important sport

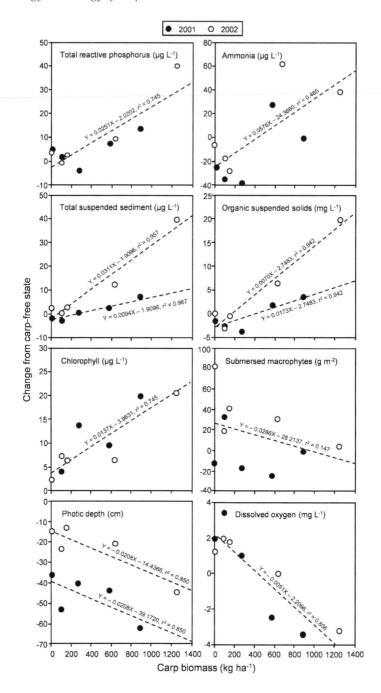

Figure 10.6. Relationship between the change in water quality parameters and common carp biomass (kg/ha) between the baseline study year (2000) and two experimental stocking years (2001 and 2002) in large experimental wetlands, Delta Marsh, Manitoba, Canada. Source: Badiou and Goldsborough (2010). Reproduced with permission.

fishes (Jackson et al. 2010). Shallow lakes had higher densities of common carp compared to deeper lakes, suggesting that shallow lakes are most sensitive to the effects of common carp and that removal of common carp will have the greatest ecological benefits in shallow lakes. In floodplain wetlands of the Murray-Darling Basin, Australia, low flow conditions intensify competition between carp and native fish (Mazumder et al. 2012).

Genetic Effects

An ecological effect of invading common carp is their hybridization with other closely related cyprinids such as the native subspecies (most likely *C. c. haematopterus*; Vilizzi 2012) in Lake Biwa, Japan's largest lake. More than 80% of specimens from the deep off-shore waters (30–70 m) were the native *C. c. haematopterus* (Mabuchi et al. 2006), but *C. c. carpio* introduced for aquaculture have escaped and are now interbreeding with the native subspecies (Matsuzaki et al. 2010). Common carp also interbreed with goldfish (*Carassius auratus* L.). There has been speculation about whether the offspring of such pairings are fertile, and Hänfling et al. (2005) found some evidence of back-crossed hybrids. Hybrids of female koi carp (an ornamental variant of *C. c. haematopterus*) and male goldfish *Carassius auratus auratus* produced by artificial spawning produced F_1 hybrid males that were sterile at age 3, but some F_1 hybrid females were fertile, producing eggs after hormonal injection (Gomelsky et al. 2012). A back-cross obtained by using intact koi carp sperm to inseminate eggs from F_1 hybrid females produced triploid progeny without application of any treatment to the eggs, demonstrating that the koi carp × goldfish hybrid females produced diploid eggs. In Australia, most hybrids were inferred to be F_1-generation, but some F_2-generation and back-crossed individuals were detected, indicating that gene flow occurs between carp and goldfish in Australia. Gene flow was biased in favour of male carp mating with female goldfish, as 19 of the 20 F_1-generation hybrids had goldfish maternal ancestry. Control programmes for common carp should consider controlling goldfish to prevent the risks posed by introgression with this related species (Haynes et al. 2012).

In New Zealand, koi carp-goldfish hybrids (e.g., Fig. 10.7) have been recognized (Pullan and Smith 1987) and comprise about 1% of the carp catch. In Australia, three separate carp strains have been identified, and one hybrid with goldfish (Shearer and Mulley 1978) (Fig. 10.8). Supporting evidence for significant genetic differences between carp strains is further provided by experimental studies of hybridization; chromosomal number of experimental carp x rosy barb (*Puntius conchonius* Ham., synonym *Barbus conchonius*) hybrids varied significantly depending on whether the carp strain was European or koi carp (Váradi et al. 1995).

Figure 10.7. Wild koi carp x goldfish hybrid (355-mm fork length) from New Zealand. Photo: Brendan Hicks.

Figure 10.8. Boolara strain common carp x goldfish hybrid (330-mm fork length) from Australia. Photo: Michael Lake.

Common Carp as a Parasite Reservoir

Common carp in the USA harbour one of the most well-known and dangerous fish parasites, the Asian tapeworm (*Bothriocephalus opsarichthydis*), a pseudophyllaeid cestode that was first discovered in the intestines of grass carp (*Ctenopharyngodonidella*). The invasiveness and spread of common carp, and its transplantation along with bait fishes, has allowed the Asian tapeworm to spread to the endangered desert fishes of the United States (Cucherousset and Olden 2011). For a more detailed account on the parasites of carp please refer to Chapter 7 on parasites of carp by Behrmann-Godel (this volume).

Management Efforts Employed and Their Effectiveness

Population Control Methods

Alternative control strategies have been well summarized by Britton et al. (2011), but most removal programmes are ineffective in the long term, and to date only small lakes have a high probability of successful removal (Koehn et al. 2000). Removal efforts generally fall into one of three categories: one-time removals, on-going or annual removals and eradication attempts. One-time removals of common carp can yield important scientific data but are of little long-term value as water quality will return to the pre-removal state as fish biomass increases post removal (Meijer et al. 1999). Annual removals can be effective for reducing biomass of common carp and can potentially improve water quality, but are time consuming and costly. Population modelling using CARPSIM suggests that size-selective removal might be useful to reduce biomass to < 60% of the biomass before exploitation (B_0), but that there is little prospect of reducing biomass to < 10% of B_0 unless fishing mortality $F > 1.4$, where F is a function of catchability, selectivity and fishing effort (Brown and Walker 2004).

There are a host of potential removal methods for reducing or eradicating common carp including netting (Cahoon 1953), trapping (Stuart et al. 2006), virus introduction (Matsui et al. 2008), exclusion (Lougheed et al. 1998, Lougheed and Chow-Fraser 2001), water level manipulation (Yamamoto et al. 2006), boat electrofishing (Hicks et al. 2006) and poisoning (Frederieke et al. 2005). Complete removal of common carp is often the goal of carp control programmes but is generally only considered feasible for relatively small water bodies that can be emptied or poisoned. However, complete eradication is the goal of on-going radio telemetry assisted removal in the large, shallow Lake Sorell in the Tasmanian highlands, Australia (Diggle et al. 2004, Taylor et al. 2012).

Emerging technologies such as pheromone attraction (Sisler and Sorensen 2008) and gene modification (i.e., daughterless carp; Grewe et al. 2005) show promise but may be decades away from being available as management tools because of the research effort required.

Biological Control and Winterkill

Some predatory fish in South Dakota, such as largemouth bass *Micropterus salmoides* Lacépède, smallmouth bass *Micropterus dolomieui* Lacepède, walleye *Sander vitreus* (Mitchill), and northern pike *Esoxlucius* L., show a preference for eating juvenile common carp, which offers some hope of biological control by effectively managed predators (Weber and Brown 2012c) although Meronek et al. (1996) concluded that this was the least effective control method for nuisance fish because of its low mortality rate.

Winter mortalities can occur naturally in ice-covered lakes due to oxygen deficiencies, e.g., in Canada. Following the breakup of ice in Lake Scucog,

Ontario, in 1959–1960, 80,000 dead carp were found (McCrimmon 1968). One recent finding suggests that the native bluegill sunfish *Lepomis macrochirus* might be an effective predator of common carp eggs in interconnected lakes in the Upper Mississippi River Basin that do not experience winter hypoxia (winterkill) (Bajer et al. 2012).

Commercial Harvest

Although there is no single method for controlling common carp that is effective in all situations, commercial harvest has been used frequently for removing carp. Carp are caught commercially with a wide range of fishing gear, including large-meshed gill nets, fyke nets, or a combination of seine, trammel or fyke nets. Boat electrofishing is used for commercial carp fishing in Australia (Bell, K., K and C Fisheries, pers. comm.). In Ontario, Canada, 29,262 tonnes were caught between 1908 and 1966 (a mean of 496 tonnes/year; McCrimmon 1968). A total of 32 tonnes/year were caught commercially in the Camargues, but 84% of the catch was released because of low commercial value (Crivelli 1981). Boat electrofishing at night was the method of choice for removing rainbow trout and common carp from the Colorado River in Grand Canyon, Arizona (Coggins et al. 2011). Removal of 13.6 tonnes annually from the artificial Lake Scucog (area 68 km^2, mean depth 1.4 m) in Ontario, Canada, had no apparent effect on the standing stock of carp (McCrimmon 1968). Because of mercury contamination, adverse public reaction to fishing methods, an abundance of small bones and variable taste, carp has fallen from favour as a table fish in North America (Fritz 1987), making commercial fishing unprofitable in some markets, and therefore unlikely to control carp abundance. In South Africa, a combination of recreational angling and subsistence fish harvest removed 79 tonnes/year of fish from the 360 km^2 Lake Gariep, South Africa's largest freshwater reservoir (Ellender et al. 2010). Common carp were 78.5% of the catch by weight, which suggests that 62 tonnes of carp were removed annually, with no evidence of overfishing from declining fish size or CPUE.

Harvest of a majority of annual production is required to achieve successful biomass control. Readily available biomass data and tools can be used in an adaptive management framework to successfully control common carp and other nuisance fishes by pulsed commercial fishing (Colvin et al. 2012). Continued carp removal is unlikely to be justifiable unless fishing is economically viable to fund the on-going effort. Given that eradication is rarely an option, especially in large water bodies, biomass reduction to a predefined target is helpful. Colvin et al. (2012) use the Ecotrophic Coefficient (EC) as proposed by Ricker (1946) as a simple, biomass-based metric that may be useful for setting harvest targets based on a single year of data. The EC is calculated as the ratio of biomass harvested to biomass produced over an annual period, and varies from 0 in the case of no harvest to greater than 1 under a scenario of very heavy harvest (Colvin et al. 2012). A suitable objective, based on observed ecosystem effects, is a biomass of no more than

100 kg/ha (Bajer et al. 2009). A commercial fishery is used to limit the impacts of common carp on the ecosystem in Clear Lake, Iowa, and more than 1000 tonnes of common carp have been harvested from the system over the past 70 years (Wahl 2001, Colvin et al. 2010).

Novel Carp Removal Methods

Floating Bait

In New Zealand, a floating bait has been developed that can be deployed in feeding stations with or without a poison such as rotenone (Morgan et al. 2013). The objective is to minimize poison use, to reduce by-kill and target carp specifically, and to allow removal of uneaten bait. Such baits targeting carp have been produced before (Fajt 1996), and although baits were effective in experimental conditions, they have proved problematic in field situations, causing considerable collateral mortality of non-target species (Gerke 2003).

Separation Traps

In Australia, considerable work has been done into designing a trap capable of selectively removing adult carp based on their propensity to either jump over or push past objects in their desire to access floodplain spawning habitat (Stuart et al. 2006). The latest trap design uses weighted one-way bars that allow carp to push into a trap but not escape (Thwaites et al. 2010). Design of the bar weight and spacing optimizes capture of carp and minimizes capture of local non-target species. The trap is then lifted regularly for emptying and disposal of the catch (Thwaites 2011). A further sophistication of a recently installed version of this trap on a fish-way providing access to a large Waikato River floodplain lake in New Zealand (Lake Waikare) is to add a macerator and biodigester to the lift trap (Fig. 10.9). Automation of the system allows the trap to be emptied regularly and the catch is composted and pelletised as fertiliser. In its first 6 weeks of operation this system removed around 13 tonnes of carp migrating upstream into the lake with the production of around 6 tonnes of fertilizer (Bruno David, Waikato Regional Council, pers. comm.).

Bowfishing

A novel method that has been used in the USA is bowfishing (Quinn 2010). In Arkansas, bowfishing tournaments are popular, and five species accounted for 84% of fish harvested: spotted gar *Lepisosteus oculatus* Winchell, common carp, shortnose gar *L. platostomus* Rafinesque, spotted sucker *Minytrema melanops* Rafinesque, and smallmouth buffalo *Ictiobus bubalus* Rafinesque. Mean harvest rate for tournament participants was 3.8 ± 1.1 fish/hour (mean ± 1 standard deviation). Among the tournament winners, the harvest rate was 7.7 ± 2.8 fish/hour, which is a highly effective catch rate compared with other recreational capture methods (Quinn 2010).

Figure 10.9. Automated carp separation trap installed on a fish-way entering Lake Waikare, New Zealand. The separation trap (right) can be lifted to discharge the catch into a macerator and auger feed (left) which supplies a biodigestor installed in the shipping container at left. Photo: Bruno David.

In New Zealand, a bowfishing tournament is held in early November (spring) and is organized by the New Zealand Bowhunters Association (http://nzbowhunters.co.nz/public_html/NEW2012/82-2/koi-carp-the-classic/). Carp are caught in the rivers, lakes and wetlands of the lower Waikato River Basin. The tournament has run since 1990, and a combination of teams, generally of two people, and individual hunters have caught a total of 61 tonnes of carp (almost 25,000 fish) between 1990 and 2012, averaging 2.66 kg in individual weight (Fig. 10.10). Numbers of contestants at each tournament have varied between 32 and 88 hunters. Average weight of carp fell from around 4 kg before 1996 to about 2 kg from 2006 to 2012, reflecting density-dependent growth reduction as the population expanded rapidly between 1990 and about 2004. Assuming that hunters bowfished for about 10 hours during each tournament, the average catch rate per hunter since 2004, when the carp population appears to have reached its maximum, has been 2.73 ± 0.76 fish/hour (mean ± 1 standard deviation). Despite the best efforts of the bowfishers, this method is not likely to result in effective population control in New Zealand because of the seasonal nature of the hunting, which occurs mainly in spring when the carp are spawning in shallow water near lake and river margins of the Waikato River and floodplain lakes and wetlands, small numbers of bowhunters, and the widespread nature of the carp. There are also ethical concerns about bowfishing because of the

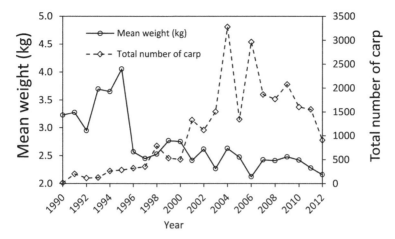

Figure 10.10. Total number and mean weight of koi carp caught in the New Zealand Bowhunters Association's annual 2-day tournament, the World Koi Carp Classic. Source: unpublished data from the New Zealand Bowhunters Association.

likelihood of suffering caused to the animals by non-lethal penetration by the arrows. In many countries with strict animal welfare laws, bowfishing is not legal and hence cannot be a management option.

Carp Exclusion

In water bodies that dry periodically, or have artificial water level controls, exclusion devices and traps have been used to prevent reinvasion of carp when reflooding occurs. Water level manipulation, where it is possible, offers significant advantages for carp control. Summer drawdown reduces recruitment of carp and goldfish (Yamamoto et al. 2006), and outflow manipulation can be used to attract carp to traps. Carp exclusion alone can lead to significant improvements in habitat quality; exclusion via construction of the Cootes Paradise Fishway, protecting Cootes Paradise Marsh, a large urban wetland of western Lake Ontario, Canada, from carp invasion, became operational in 1997. Since that time, turbidity and chlorophyll *a* concentrations in the marsh have declined dramatically (Thomasen and Chow-Fraser 2012). Exclusion methods are being trialled in New Zealand (Fig. 10.11) based on methods developed in Australia (Thwaites et al. 2010, Thwaites 2011).

Effectiveness of Control Measures

Because carp can attain very high biomasses, reductions of carp abundance to below threshold biomasses can improve water quality even where eradication is not possible. A notable success has been carp removal from the Botany Wetlands, New South Wales, Australia, which comprises 11

Figure 10.11. a. One-way gate installed over culvert entrance to a floodplain lake, showing carp aggregating outside the gate. b. Carp can leave the lake downstream via the hinged, weighted push bars, but the same bars prevent upstream movement of adult carp back into the lake. Photo: Adam Daniel.

interconnected ponds and adjacent land covering an area of 58 ha. Ten tonnes of cyprinid biomass (4,073 common carp and 261 goldfish) were removed over nine years, and during this time water clarity determined as Secchi depth improved from 0.4 m to 1.2 m over the removal period, and cyanobacterial density decreased (Pinto et al. 2005). However, environmental gains can be easily reversed by reinvasion and natural reproduction. Following initial near-eradication of carp by rotenone treatment, water clarity and macrophyte abundance decreased as carp proliferated in Hennepin and Hopper Lakes, Illinois, USA (Bajer et al. 2009).

The use of poisons such as rotenone to control carp has many unwanted side effects such as by-kill of non-target aquatic species (Ling 2009) and public health concerns (Ling 2003). Public concern over the use of poisons is a barrier to their use in many countries, despite low toxicity to humans and avian or mammalian wildlife (Ling 2003).

Although there has been limited success eradicating common carp on a large scale, there has been progress that gives hope to future advancements in removal strategies. Radio-tagged Judas fish have been used in South-Central Minnesota, USA, to track coldwater aggregations of carp in winter for removal (Bajer et al. 2011). Once located, these aggregations could be removed using seine nets with an efficiency of up to 94%.

A particularly determined and long-term removal effort is the programme of the Inland Fisheries Service to remove common carp from two large lakes in the high country of the island of Tasmania, Australia. Common carp were discovered in Lake Sorell (4,770 ha) and Lake Crescent (2,365 ha) in 1995 (Diggle et al. 2004, Taylor 2012). These lakes now hold the last common carp in Tasmania after eradication of populations in farm dams on the northwest coast in the 1970s. The lakes also have an endemic galaxiid species, the endangered golden galaxias (*Galaxias auratus* Johnston) and state-renowned brown trout (*Salmo trutta* L.) fisheries. Since the discovery of the unwanted carp population, the state government has used a host of removal techniques including applying hydrated lime to kill carp eggs and fish removal guided by radio telemetry at a cost of AUS$ 300,000 to $400,000 per year (Diggle et al. 2004, Inland Fisheries Service 2010). A total of 17,307 carp have been removed from Lake Sorell (including 14,517 juveniles from the 2009 spawning) and 7,797 from Lake Crescent since 1995 (Inland Fisheries Service 2010, Donkers et al. 2012). In addition, fine-mesh screens were installed at the lakes' outlet to prevent fish and eggs from leaving the system.

Lake Crescent is now believed to be carp-free. Removal of carp from Lake Crescent stands as a landmark of persistence and the strategic use of radio-tagged fish to target the remaining carp at low densities. The key to the success achieved by the Inland Fisheries Service has been the ability to almost eliminate recruitment by blocking littoral spawning habitat with barriers. Unfortunately, even physical exclusion is not completely effective because common carp reportedly spawned on the barrier netting used to prevent them from accessing more suitable spawning habitat.

In semi-arid areas, one-way pumping has been used to retain native fish populations while excluding invasive exotics such as carp. One-way pumping involves moving water but not carp into floodplain habitats to sustain native fish. Inundation managed through pumping in Hattah Lakes, a Ramsar wetland site on the regulated Murray River floodplain, Victoria, Australia, was seen as a viable option for the creation of fish habitat, for promoting native fish recruitment, and to rehabilitate native fish communities (Vilizzi 2013b).

Response to Climate Change

As global temperatures warm, temperate regions with fish communities that are currently dominated by cold-water species with physiological optima < 20°C are vulnerable to the effects of warming temperatures caused by climate change. This includes displacement of native species by common carp, which have a physiological optimum of 20–28°C, and warm-water fishes with physiological optima > 28°C that are able to establish and invade as the thermal constraints on the expression of their life history traits diminish (Britton et al. 2010a).

Conclusions and Summary

Common carp are very well adapted to take advantage of suitable habitats and establish easily. They have behavioural traits allowing them to spread widely and expand their populations rapidly. Characteristics of carp that make them liable to being a super invader are: early maturity (considering their ultimate size), very high fecundity, large size at maturity, wide temperature and salinity tolerances, omnivorous diet exploiting most available foods, and virtual immunity from predation once they pass the earliest juvenile stages. A key characteristic to the success of carp in large rivers with extensive floodplains containing lakes and wetlands is their flexible migration strategy. While a proportion of the population stays in off-channel habitats such as shallow lakes year-round, another proportion of the population migrates back to the main river, and some of these individuals undertake migrations of up to hundreds of kilometres in search of suitable spawning habitat. This presents something of a contradiction, given their generally small home ranges and high site fidelity, and suggests that carp have a sophisticated internal map of a river system that allows them to navigate long distances but also to return accurately to small home ranges. Further, our studies have shown that fish that are tagged together are often recaptured together, suggesting that there is strong social cohesion among small groups of fish (Osborne et al. 2009).

The environmental destruction that common carp can inflict on habitats where they are not native is well documented, and includes significant increases in suspended sediment from their feeding activities, increased nutrients from their excretion, and the natural tendency for populations to grow to high biomasses exceeding 1,000 kg/ha where they completely dominate the fish community. Related impacts include virtual elimination of aquatic macrophytes, competition with other fish species by reducing macroinvertebrate abundance, reduced waterfowl abundance caused by lack of macrophytes and invertebrates, and genetic impacts on related cyprinid species where they are present.

Simulated population responses of common carp to commercial exploitation have shown that moderate exploitation rates of about 40% are optimal for population control. At this rate, maximum lifetime egg production was reduced to 77–89% of starting levels, indicating the potential for recruitment overfishing. Thus commercial fishing plus an integrated approach that removes all sizes of common carp has a greater chance of controlling population abundance (Weber et al. 2011).

Integrated management, such as that developed at Lake Sorell, Tasmania (Taylor et al. 2012), combines repeated net captures, blocking carp from spawning habitat and using radio-tracked Judas males to remove residual fish. Other holistic approaches to management, where flow regulation is possible, include using inundation strategies to benefit native wetland fish and to disadvantage invasive populations of common carp (e.g., Vilizzi 2013a). Innovative combinations of techniques, incorporating decision support tools where possible (Vilizzi 2013b) are the most promising forms of control of common carp in locations where they are not wanted.

Acknowledgments

This study was funded by contract UOWX0505 from the New Zealand Ministry for Business, Innovation and Employment. We gratefully acknowledge the provision of data by the New Zealand Bowhunters Association. Without their dedicated efforts we would have no clear view of the increasing numbers of carp in the Waikato River basin.

References

Adàmek, Z., D. Kortan, P. Lepic and J. Andreji. 2003. Impacts of otter (*Lutra lutra* L.) predation on fishponds: A study of fish remains at ponds in the Czech Republic. Aquacul. Int. 11: 389–396.

Akhurst, D.J., G.B. Jones, M. Clark and A. Reichelt-Brushett. 2012. Effects of carp, gambusia, and Australian bass on water quality in a subtropical freshwater reservoir. Lake Reservoir Manage. 28: 212–223.

Al-Hamed, M.I. 1971. Salinity tolerance of common carp (*Cyprinus carpio* L.). Bulletin of the Iraq Natl. Hist. Mus. 7: 1–16.

Badiou, P.H.J. and L.G. Goldsborough. 2010. Ecological impacts of an exotic benthivorous fish in large experimental wetlands, Delta Marsh, Canada. Wetlands 30: 657–667.

Bajer, P.G., C.J. Chizinski and P.W. Sorensen. 2011. Using the Judas technique to locate and remove wintertime aggregations of invasive common carp. Fish. Manag. Ecol. 18: 497–505.

Bajer, P.G., C.J. Chizinski, J.J. Silbernagel and P.W. Sorensen. 2012. Variation in native micro-predator abundance explains recruitment of a mobile invasive fish, the common carp, in a naturally unstable environment. Biol. Invasions 14: 1919–1929.

Bajer, P.G., H. Lim, M.J. Travaline, B.D. Miller and P.W. Sorensen. 2010. Cognitive aspects of food searching behavior in free-ranging wild common carp. Environ. Biol. Fishes 88: 295–30.

Bajer, P.G. and P.W. Sorensen. 2010. Recruitment and abundance of an invasive fish, the common carp, is driven by its propensity to invade and reproduce in basins that experience winter-time hypoxia in interconnected lakes. Biol. Invasions 12: 1101–1112.

Bajer, P.G. and P.W. Sorensen. 2012. Using boat electrofishing to estimate the abundance of invasive common carp in small midwestern lakes. N. Am. J. Fish. Manage. 32: 817–822.

Bajer, P.G., G.S. Sullivan and P.W. Sorensen. 2009. Effects of a rapidly increasing population of common carp on vegetative cover and waterfowl in a recently restored Midwestern shallow lake. Hydrobiologia 632: 235–245.

Balon, E.K. 1995. Origin and domestication of the wild carp, *Cyprinus carpio*: From Roman gourmets to the swimming flowers. Aquaculture 129: 3–48.

Beesley, L., A.J. King, F. Amtstaetter, J.D. Koehn, B. Gawne, A. Price, D.L. Nielsen, L. Vilizzi and S.N. Meredith. 2012. Does flooding affect spatiotemporal variation of fish assemblages in temperate floodplain wetlands? Freshwater Biol. 57: 2230–2246.

Berg, L.S. 1964. Freshwater fishes of the U.S.S.R. and the adjacent countries. Israel Program for Scientific Translations Ltd., Jerusalem, vol. 2, 4th edition. (Russian version published 1949).

Bice, C.M. and B.P. Zampatti. 2011. Engineered water level management facilitates recruitment of non-native common carp, *Cyprinus carpio*, in a regulated lowland river. Ecol. Eng. 37: 1901–1904.

Billard, R. 1999. Carp Biology and Culture. Springer-Verlag, Berlin.

Biro, P. 1997. Temporal variation in Lake Balaton and its fish populations. Ecol. Freshw. Fish 6: 196–216.

Bomford, M. 2008. Risk assessment models for establishment of exotic vertebrates in Australia and New Zealand. Invasive Animals Cooperative Research Centre, Canberra.

Britton, J., D. Harper, D. Oyugi and J. Grey. 2010b. The introduced *Micropterus salmoides* in an equatorial lake: a paradoxical loser in an invasion meltdown scenario? Biol. Invasions 12: 3439–3448.

Britton, J.R., R.R. Boar, J. Grey, J. Foster, J. Lugonzo and D.M. Harper. 2007. From introduction to fishery dominance: the initial impacts of the invasive carp *Cyprinus carpio* in Lake Naivasha, Kenya, 1999 to 2006. J. Fish Biol. 71, Sup. sd: 239–257.

Britton, J.R., J. Cucherousset, G.D. Davies, M.J. Godard and G.H. Copp. 2010a. Non-native fishes and climate change: predicting species responses to warming temperatures in a temperate region. Freshwater Biol. 55: 1130–1141.

Britton, J.R., R.E. Gozlan and G.H. Copp. 2011. Managing non-native fish in the environment. Fish. Fish. 12: 256–274.

Brown, P., K.P. Sivakumaran, D. Stoessel and A. Giles. 2005. Population biology of carp (*Cyprinus carpio* L.) in the mid-Murray River and Barmah Forest Wetlands, Australia. Mar. Freshwater Res. 56: 1151–1164.

Brown, P. and T.I. Walker. 2004. CARPSIM: Stochastic simulation modelling of wild carp (*Cyprinus carpio* L.) population dynamics, with applications to pest control. Ecol. Modell. 176: 83–97.

Broyer, J. and C. Calenge. 2010. Influence of fish-farming management on duck breeding in French fish pond systems. Hydrobiologia 637: 173–185.

Burger, D.F., D.P. Hamilton, C.A. Pilditch and M.M. Gibbs. 2007. Benthic nutrient fluxes in a eutrophic, polymictic lake. Hydrobiologia 584: 13–25.

Butler, S.E. and David H. Wahl. 2010. Common carp distribution, movements, and habitat use in a river impounded by multiple low-head dams. Trans. Am. Fish. Soc. 139: 1121–1135.

Cahn, A.R. 1929. The effect of carp on a small lake: The carp as a dominant. Ecology 10: 271–275.

Cahoon, W.G. 1953. Commercial carp removal at Lake Mattamuskeet, North Carolina. J. Wildl. Manage. 17: 312–316.

Cambray, J.A. 2003. Impact on indigenous species biodiversity caused by the globalisation of alien recreational freshwater fisheries. Hydrobiologia 500: 217–230.

Carey, M.P. and D.H. Wahl. 2010. Native fish diversity alters the effects of an invasive species on food webs. Ecology 91: 2965–2974.

Cheshire, K.J.M. 2010. Larval fish assemblages in the Lower River Murray, Australia: examining the influence of hydrology, habitat and food. D. Phil. Thesis, University of Adelaide, Adelaide.

Coggins, Jr., L.G., M.D. Yard and W.E. Pine III. 2011. Nonnative fish control in the Colorado River in Grand Canyon, Arizona: an effective program or serendipitous timing? Trans. Am. Fish. Soc. 140: 456–470.

Colvin, M.E., E. Katzenmyer, T.W. Stewart and C.L. Pierce. 2010. The Clear Lake ecosystem simulation model (CLESM) project. Annual report to Iowa Department of Natural Resources, Ames. Unpublished report.

Colvin, M.E., C.L. Pierce, T.W. Stewart and S.E. Grummer. 2012. Strategies to control a common carp population by pulsed commercial harvest. N. Am. J. Fish. Manage. 32: 1251–1264.

Conallin, A.J., K.A. Hillyard, K.F. Walker, B.M. Gillanders and B.B. Smith. 2011. Offstream movements of fish during drought in a regulated lowland river. River Res. Appl. 27: 1237–1252.

Conallin, A.J., B.B. Smith, L.A. Thwaites, K.F. Walker and B.M. Gillanders. 2012. Environmental water allocations in regulated lowland rivers may encourage offstream movements and spawning by common carp, *Cyprinus carpio*: implications for wetland rehabilitation. Mar. Freshwater Res. 63: 865–877.

Cooke, S.J. and J.F. Schreer. 2003. Environmental monitoring using physiological telemetry—a case study examining common carp responses to thermal pollution in a coal-fired generating station effluent. Water, Air, Soil Pollut. 142: 113–136.

Copp, G.H., L. Vilizzi, J. Mumford, M.J. Godard, G. Fenwick and R.E. Gozlan. 2009. Calibration of FISK, an invasiveness screening tool for non-native freshwater fishes. Risk Anal. 29: 457–467.

Crivelli, A.J. 1981. The biology of the common carp, *Cyprinus carpio* L. in the Camargue, southern France. J. Fish Biol. 18: 271–290.

Crook, D.A. 2004. Is the home range concept compatible with the movements of two species of lowland river fish? J. Anim. Ecol. 73: 353–366.

Crook, D.A. and B.M. Gillanders. 2006. Use of otolith chemical signatures to estimate carp recruitment sources in the mid-Murray River, Australia. River Res. Appl. 22: 871–879.

Crook, D.A., J.I. Macdonald, D.J. McNeil, D.M. Gilligan, M. Asmus, R. Maas and J. Woodhead. 2013. Recruitment sources and dispersal of an invasive fish in a large river system as revealed by otolith chemistry analysis. Can. J. Fish. Aquat. Sci. 70: 953–963.

Cucherousset, J. and J.D. Olden. 2011. Ecological impacts of non-native freshwater fishes. Fisheries 36: 215–230.

Daniel, A.J. 2009. Detecting exploitable stages in the life history of koi carp (*Cyprinus carpio*) in New Zealand. Ph.D. Thesis, University of Waikato, Hamilton, New Zealand.

Daniel, A.J., B.J. Hicks, N. Ling and B.O. David. 2011. Movements of radio- and acoustic-tagged adult koi carp in the Waikato River, New Zealand. N. Am. J. Fish. Manage. 31: 352–362.

Diggle, J., J. Day and N. Bax. 2004. Eradicating European carp from Tasmania and implications for national European carp eradication. Inland Fisheries Service, Moonah, Report no. 2000/182, Moonah, Australia. Unpublished report.

Donkers, P., J.G. Patil, C. Wisniewski and J.E. Diggle. 2012. Validation of mark-recapture population estimates for invasive common carp, *Cyprinus carpio*, in Lake Crescent, Tasmania. J. Appl. Ichthyol. 28: 7–14.

Driver, P.D., J.H. Harris, G.P. Closs and T.B. Koen. 2005. Effects of flow regulation in the Murray-Darling Basin, Australia. River Res. Appl. 21: 327–335.

Edwards, E.A. and K.A. Twomey. 1982. Habitat suitability index models: Common carp. US Dept. Int. Fish Wildl. Serv. FWS/OBS-82/10.12, Fish and Wildlife Service, US Department of the Interior, Washington DC. Unpublished report.

Egertson, C.J. and J.A. Downing. 2004. Relationship of fish catch and composition to water quality in a suite of agriculturally eutrophic lakes. Can. J. Fish. Aquat. Sci. 61: 1784–1796.

Ellender, B.R., O.L.F. Weyl, H. Winker and A.J. Booth. 2010. Quantifying the annual fish harvest from South Africa's largest freshwater reservoir. Water SA 36: 45–52.

Fajt, J.R. 1996. Toxicity of rotenone to common carp and grass carp: respiratory effects, oral toxicity, and evaluation of a poison bait. Ph.D. Thesis, Auburn University, Alabama.

Fischer, J.R., R.M. Krogman and M.C. Quist. 2013. Influences of native and non-native benthivorous fishes on aquatic ecosystem degradation. Hydrobiologia 711: 187–199.

FishBase. 2010. R. Froese and D. Pauly (eds.). Animalia–Chordata–Actinopterygii, Cephalaspidomorphi, Elasmobranchii, Holocephali, Myxini, Sarcopterygii. http://www.catalogueoflife.org/annual-checklist/2010/details/database/id/10.

Fernández-Delgado, C. 1990. Life history patterns of the common carp, *Cyprinus carpio*, in the estuary of the Guadalquivir river in south-west Spain. Hydrobiologia 206: 19–28.

Fletcher, A.R., A.K. Morison and D.J. Hume. 1985. Effects of carp, *Cyprinus carpio* L., on communities of aquatic vegetation and turbidity of waterbodies in the lower Goulburn River basin. Aust. J. Mar. Freshwater Res. 36: 311–327.

Forsyth, D.M., J.D. Koehn, D.I. MacKenzie and I.G. Stuart. 2013. Population dynamics of invading freshwater fish: common carp (*Cyprinus carpio*) in the Murray-Darling Basin, Australia. Biol. Invasions 15: 341–354.

Frederieke, J., P. Kroon, C. Gehrke and T. Kurwie. 2005. Palatability of rotenone and antimycin baits for carp control. Ecol. Manage. Restor. 6: 228–229.

Fritz, A. 1987. Commercial fishing for carp. pp. 17–30. *In*: E.L. Cooper (ed.). Carp in North America. American Fisheries Society, Bethesda, Maryland.

Gehrke, P.C. 2003. Preliminary assessment of oral rotenone baits for carp control in New South Wales. pp. 149–154. *In*: Managing Invasive Freshwater Fish in New Zealand. Proceedings of a workshop hosted by Department of Conservation, 10–12 May 2001, Hamilton. Department of Conservation, Wellington.

Gehrke, P.C. and J.H. Harris. 1994. The role of fish in cyanobacterial blooms in Australia. Mar. Freshwater Res. 45: 905–915.

Gilligan, D. and T. Rayner. 2007. The distribution, spread, ecological impacts and potential control of carp in the upper Murray River. NSW Department of Primary Industries, Fisheries Research Report Series: No. 14. NSW Department of Primary Industries, Cronulla, New South Wales, Australia. Unpublished report.

Gomelsky, B., K.J. Schneider and D.A. Plouffe. 2012. Koi × goldfish hybrid females produce triploid progeny when backcrossed to koi males. N. Am. J. Aquac. 74: 449–452.

Gozlan, R.E., J.R. Britton, I. Cowx and G.H. Copp. 2010. Current knowledge on non-native freshwater fish introductions. J. Fish Biol. 76: 751–786.

Grewe, P., B. Natasha, J. Beyer, J. Patil and R. Thresher. 2005. Sex-specific apoptosis for achieving daughterless fish. p. 59. *In*: J. Parkes, M. Statham and G. Edwards (eds.). Proceedings of the 13th Australasian Vertebrate Pest Conference, 2–6 May, 2005, Wellington, New Zealand. Landcare Research, Lincoln, New Zealand.

Guha, D. and D. Mukherjee. 1991. Seasonal cyclical changes in the gonadal activity of common carp, *Cyprinus carpio* Linn. Indian J. Fish. 38: 218–223.

Haas, K., U. Köhler, S. Diehl, P. Köhler, S. Dietrich, S. Holler, A. Jaensch, M. Niedermaier and J. Vilsmeier. 2007. Influence of fish on habitat choice of water birds: a whole system experiment. Ecology 88: 2915–2925.

Hanchet, S. 1990. The effect of koi carp on New Zealand's aquatic ecosystems. New Zealand Freshwater Fisheries Report No. 117. Freshwater Fisheries Centre, MAF Fisheries, Ministry of Agriculture and Fisheries, Rotorua, New Zealand. Unpublished report.

Hänfling, B., P. Bolton, M. Harley and G.R. Carvalho. 2005. A molecular approach to detect hybridisation between crucian carp (*Carassius carassius*) and non-indigenous carp species (*Carassius* spp. and *Cyprinus carpio*). Freshwater Biol. 50: 403–417.

Harris, J.H. and P.E. Gehrke. 1997. Fish and rivers in stress: the NSW river survey. New South Wales Fisheries Office of Conservation and the Cooperative Research Centre for Freshwater Ecology, Sydney.

Haynes, G.D., J. Gongora, D.M. Gilligan, P. Grewe, C. Moran, F.W. Nicholas. 2012. Cryptic hybridization and introgression between invasive cyprinid species *Cyprinus carpio* and *Carassius carassius* in Australia: implications for invasive species management. Anim. Conserv. 15: 83–94.

Hesse, L.W. and B.A. Newcomb. 1982. On estimating the abundance of fish in the upper channelized Missouri River. N. Am. J. Fish. Manage. 2: 80–83.

Hicks, B.J., J. Brijs and D.G. Bell. 2008. Electrofishing survey of the fish community in the Whangamarino Wetland. CBER Contract Report No. 67. Centre for Biodiversity and Ecology Research, Department of Biological Sciences, School of Science and Engineering, The University of Waikato, Hamilton.

Hicks, B.J., N. Ling and M.W. Osborne. 2006. Quantitative estimates of fish abundance from boat electrofishing. pp. 104–111. *In*: M.J. Phelan and H. Bajhau (eds.). A guide to Monitoring Fish Stocks and Aquatic Ecosystems, 11–15 July 2005, Darwin, Australia, Australian Society for Fish Biology, Darwin, Australia.

Hicks, B.J., N. Ling, M.W. Osborne, D.G. Bell and C.A. Ring. 2005. Boat electrofishing survey of the lower Waikato River and its tributaries. CBER Contract Report No. 39. Client report prepared for Environment Waikato. Centre for Biodiversity and Ecology Research, Department of Biological Sciences, The University of Waikato, Hamilton.

Hicks, B.J., N. Ling and B. Wilson. 2010. Introduced fish. pp. 209–228. *In*: K.C. Collier and D.P. Hamilton (eds.). The waters of the Waikato: ecology of New Zealand's longest river. Environment Waikato, Hamilton. Environment Waikato and the Centre for Biodiversity and Ecology Research (University of Waikato), Hamilton.

Hicks, B.J., N. Ling and A.J. Daniel. 2012. Common carp (*Cyprinus carpio*). pp. 247–260. *In*: R.A. Francis (ed.). A Handbook of Global Freshwater Invasive Species, Earthscan, London.

Hinojosa-Garro, D. and L. Zambrano. 2004. Interactions of common carp (*Cyprinus carpio*) with benthic crayfish decapods in shallow ponds. Hydrobiologia 515: 115–122.

Homans, F.R. and D.J. Smith. 2013. Evaluating management options for aquatic invasive species: concepts and methods. Biol. Invasions 15: 7–16.

Howes, G.J. 1991. Systematics and biogeography: an overview. pp. 1–33. *In*: I.J. Winfield and J.S. Nelson (eds.). Cyprinid Fishes—Systematics, Biology and Exploitation. Chapman and Hall, London, England.

Humphries, P., P. Brown, J. Douglas, A. Pickworth, R. Strongman, K. Hall and L. Serafini. 2008. Flow-related patterns in abundance and composition of the fish fauna of a degraded Australian lowland river. Freshwater Biol. 53: 789–813.

Inland Fisheries Service. 2010. Carp management program: Annual report 2009–2010. Inland Fisheries Service, New Norfolk, Tasmania, Australia. Unpublished report.

Jackson, Z.J., M.C. Quist, J.A. Downing and J.G. Larscheid. 2010. Common carp (*Cyprinus carpio*), sport fishes, and water quality: Ecological thresholds in agriculturally eutrophic lakes. Lake Reservoir Manage. 26: 14–22.

Johnsen, P.B. and A.D. Hasler. 1977. Winter aggregations of carp (*Cyprinus carpio*) as revealed by ultrasonic tracking. Trans. Am. Fish. Soc. 106: 556–559.

Jones, M.J. and I.G. Stuart. 2007. Movements and habitat use of common carp (*Cyprinus carpio*) and Murray cod (*Maccullochella peelii peelii*) juveniles in a large lowland Australian river. Ecol. Freshw. Fish. 16: 210–220.

Jones, M.J. and I.G. Stuart. 2009. Lateral movement of common carp (*Cyprinus carpio* L.) in a large lowland river and floodplain. Ecol. Freshw. Fish. 18: 72–82.

Kelleway, J., D. Mazumder, G.G. Wilson, N. Saintilan, L. Knowles, J. Iles and T. Kobayashi. 2010. Trophic structure of benthic resources and consumers varies across a regulated floodplain wetland. Mar. Freshwater Res. 61: 430–440.

Kilford, B. 1984. Koi carp spread alarms scientists. Catch 84(11): 7.

King, A.J. 2004. Ontogenetic patterns of habitat use by fishes within the main channel of an Australian floodplain river. J. Fish Biol. 65: 1582–1603.

King, A.J., P. Humphries and P.S. Lake. 2003. Fish recruitment on floodplains: the roles of flooding and life history characteristics. Can. J. Fish. Aquat. Sci. 60: 773–786.

King, A.J., A.I. Robertson and M.R. Healey. 1997. Experimental manipulations of the biomass of introduced carp (*Cyprinus carpio*) in billabongs. I. Impacts on water-column properties. Mar. Freshwater Res. 48: 435–443.

King, A.J., Z. Tonkin and J. Lieshcke. 2012. Short-term effects of a prolonged blackwater event on aquatic fauna in the Murray River, Australia: considerations for future events. Mar. Freshwater Res. 63: 576–586.

Koblitskaya, A.F. 1977. The succession of spawning communities in the Volga Delta. J. Ichthyol. 17: 534–547.

Kloskowski, J. 2011a. Differential effects of age-structured common carp (*Cyprinus carpio*) stocks on pond invertebrate communities: implications for recreational and wildlife use of farm ponds. Aquacult. Int. 19: 1151–1164.

Kloskowski, J. 2011b. Impact of common carp *Cyprinus carpio* on aquatic communities: direct trophic effects versus habitat deterioration. Fundam. Appl. Limnol. 178: 245–255.

Koehn, J.D. 2004. Carp (*Cyprinus carpio*) as a powerful invader in Australian waterways. Freshwater Biol. 49: 882–894.

Koehn, J.D., A. Brumley and P. Gehrke. 2000. Managing the impacts of carp. Bureau of Rural Sciences, Canberra. Unpublished report.

Kohlmann, K. and P. Kersten. 2013. Deeper insight into the origin and spread of European common carp (*Cyprinus carpio carpio*) based on mitochondrial D-loop sequence polymorphisms. Aquaculture 376–379: 97–104.

Kolar, C.S. and D.M. Lodge. 2002. Ecological predictions and risk assessment for alien fishes in North America. Science 298: 1233–1236.

Liao, H., C.L. Pierce and J.G. Larscheid. 2001. Empirical assessment of indices of prey importance in the diets of fish. Trans. Am. Fish. Soc. 130: 583–591.

Ling, N. 2003. Rotenone—a review of its toxicity and use for fisheries management. Science for Conservation 211, Department of Conservation, Wellington, New Zealand.

Ling, N. 2009. Management of invasive fish. pp. 185–204. *In*: M.N. Clout and P.A. Williams (eds.). Invasive Species Management: A Handbook of Principles and Techniques. Oxford University Press, Oxford.

Lougheed, V.L., B. Crosbie and P. Chow-Fraser. 1998. Predictions on the effect of common carp (*Cyprinus carpio*) exclusion on water quality, zooplankton, and submergent macrophytes in a Great Lakes wetland. Can. J. Fish. Aquat. Sci. 55: 1189–1197.

Lougheed, V.L. and P. Chow-Fraser. 2001. Spatial variability in the response of lower trophic levels after carp exclusion from a freshwater marsh. J. Aquat. Ecosyst. Stress Recovery 9: 21–34.

Lubinski, N.S., A. van Vooren, G. Farabee, J. Janacec and J.D. Jackson. 1986. Common carp in the Upper Mississippi River. Hydrobiologia 136: 141–154.

Lusk, S., V. Lusková and L. Hanel. 2010. Alien fish species in the Czech Republic and their impact on the native fish fauna. Folia Zool. 59: 57–72.

Mabuchi, K., M. Miya, H. Senou, T. Suzuki and M. Nishida. 2006. Complete mitochondrial DNA sequence of the Lake Biwa wild strain of common carp (*Cyprinus carpio* L.): further evidence for an ancient origin. Aquaculture 257: 68–77.

Marchetti, M.P., P.B. Moyle and R. Levine. 2004. Invasive species profiling? Exploring the characteristics of non-native fishes across invasion stages in California. Freshwater Biol. 49: 646–661.

Matsui, K., M. Honjo, Y. Kohmatsu, K. Uchii, R. Yonekura and Z. Kawabata. 2008. Detection and significance of koi herpesvirus (KHV) in freshwater environments. Freshwater Biol. 53: 1262–1272.

Matsuzaki, S.S., N. Usio, N. Takamura and I. Washitani. 2007. Effects of common carp on nutrient dynamics and littoral community composition: roles of excretion and bioturbation. Fundam. Appl. Limnol. 168/1: 27–38.

Matsuzaki, S.S., K. Mabuchi, N. Takamura, B.J. Hicks, M. Nishida and I. Washitani. 2010. Stable isotope and molecular analyses indicate that hybridization with non-native domesticated common carp influence habitat use of native carp. Oikos 119: 964–971.

Mauck, P.E. and R.C. Summerfelt. 1971. Sex ratio, age of spawning fish, and duration of spawning in the carp, *Cyprinus carpio* (Linnaeus) in Lake Carl Blackwell, Oklahoma. Trans. Kans. Acad. Sci. 74: 221–227.

Mazumder, D., M. Johansen, Saintilan, N.J. Iles, T. Kobayashi, L. Knowles and L. Wen. 2012. Trophic shifts involving native and exotic fish during hydrologic recession in floodplain wetlands. Wetlands 32: 267–275.

McCrimmon, H.R. 1968. Carp in Canada. Bull. Fish. Res. Board Can. 165: 1–89.

McDowall, R.M. 1990. New Zealand freshwater fish: a guide and natural history. Heinemann Reed, Auckland.

Meador, M.R., L.R. Brown and T. Short. 2003. Relations between introduced fish and environmental conditions at large geographic scales. Ecol. Indic. 3: 81–92.

Mehner, T., J. Ihlau, H. Dārner and F. Hālker. 2005. Can feeding of fish on terrestrial insects subsidize the nutrient pool of lakes? Limnol. Oceanogr. 50: 2022–2031.

Meijer, M.-L., I. De Boois, M. Scheffer, R. Portielje and H. Hosper. 1999. Biomanipulation in shallow lakes in The Netherlands: An evaluation of 18 case studies. Hydrobiologia 408–409: 13–30.

Meronek, T.G., P.M. Bouchard, E.R. Buckner, T.M. Burri, K.K. Demmerly, D.C. Hateli, R.A. Klumb, S.H. Schmidt and D.W. Coble. 1996. A review of fish control projects. N. Am. J. Fish. Manage.16: 63–74.

Morgan, D.K.J. and B.J. Hicks. 2013. A metabolic theory of ecology applied to temperature and mass-dependence of N and P excretion by common carp. Hydrobiologia 705: 135–145.

Morgan, D.K.J., C.J.R. Verbeek, K.A. Rosentrater and B.J. Hicks. 2013. The palatability of flavoured novel floating pellets made with brewer's spent grain to captive carp. N. Z. J. Zool. 40: 170–174.

Nannini, M.A., J. Goodrich, J.M. Dettmers, D.A. Soluk and D.H. Wahl. 2012. Larval and early juvenile fish dynamics in main channel and backwater lake habitats of the Illinois River ecosystem. Ecol. Freshw. Fish. 21: 499–509.

Nikolski, G.V. 1933. On the influence of the rate of flow on the fish fauna of the rivers of Central Asia. J. Anim. Ecol. 2: 266–281.

Olds, A.A., M.K.S. Smith, O.L.F. Weyl and I.A. Russell. 2011. Invasive alien freshwater fishes in the Wilderness Lakes System, a wetland of international importance in the Western Cape Province, South Africa. Afr. Zool. 46: 179–184.

Opuszyński, K., A. Lirski, L. Myszkowski and J. Wolnicki. 1989. Upper lethal and rearing temperatures for juvenile common carp, *Cyprinus carpio* L., and silver carp, *Hypophthalmichthys molitrix* (Valenciennes). Aquacult. Res. 20: 287–294.

Osborne, M.W., N. Ling, B.J. Hicks and G.W. Tempero. 2009. Movement, social cohesion, and site fidelity in adult koi carp, *Cyprinus carpio* L. Fish. Manag. Ecol. 16: 169–176.

Oyugi, D.O., J. Cucherousset, M.J. Ntibab, S.M. Kisia, D.M. Harpere and J.R. Britton. 2011. Life history traits of an equatorial common carp *Cyprinus carpio* population in relation to thermal influences on invasive populations. Fish. Res. 110: 92–97.

Oyugi, D.O., J. Cucherousset, D.J. Baker and J.R. Britton. 2012. Effects of temperature on the foraging and growth rate of juvenile common carp, *Cyprinus carpio*. J. Therm. Biol. 37: 89–94.

Parameswaran, S., K.H. Alikunhi and K.K. Sukumaran. 1972. Observations on the maturation, fecundity and breeding of the common carp, *Cyprinus carpio* Linnaeus. Indian J. Fish. 19: 110–124.

Penne, C.R. and C.L. Pierce. 2008. Seasonal distribution, aggregation, and habitat selection of common carp in Clear Lake, Iowa. Trans. Am. Fish. Soc. 137: 1050–1062.

Phelps, Q.E., K.R. Edwards and D.W. Willis. 2007. Precision of five structures for estimating age of common carp. N. Am. J. Fish. Manage. 27: 103–105.

Phelps, Q.E., B.D.S. Graeb and D.W. Willis. 2008. Influence of the Moran Effect on spatiotemporal synchrony in common carp recruitment. Trans. Am. Fish. Soc. 137: 1701–1708.

Pinto, L., N. Chandrasena, J. Pera, P. Hawkins, D. Eccles and R. Sim. 2005. Managing invasive carp (*Cyprinus carpio* L.) for habitat enhancement at Botany Wetlands, Australia. Aquat. Conserv.: Mar. Freshwat. Ecosyst. 15: 447–462.

Post, D.M. and A.W. Walters. 2009. Nutrient excretion rates of anadromous alewives during their spawning migration. Trans. Am. Fish. Soc. 138: 264–268.

Prochelle, O. and H. Campos. 1985. The biology of the introduced carp, *Cyprinus carpio* L., in the River Cayumpu, Valdivia, Chile. Stud. Neotrop. Fauna Environ. 20: 65–82.

Pullan, S.G. 1984a. Japanese koi (*Cyprinus carpio*) in the Waikato River system. Report Number 1 April 1984. Unpublished internal report, Ministry of Agriculture and Fisheries, Auckland.

Pullan, S.G. 1984b. Japanese koi (*Cyprinus carpio*) in the Waikato River system. Report Number May 1984. Unpublished internal report, Ministry of Agriculture and Fisheries, Auckland.

Pullan, S. and P.J. Smith. 1987. Identification of hybrids between koi (*Cyprinus carpio*) and goldfish (*Carassiusauratus*). N. Z. J. Mar. Freshwater Res. 21: 41–46.

Quinn, J.W. 2010. A survey of bowfishing tournaments in Arkansas. N. Am. J. Fish. Manage. 30: 1376–1384.

Ricker, W.E. 1946. Production and utilization of fish populations. Ecol. Monogr. 16: 373–391.

Roberts, J., A. Chick, L. Oswald and P. Thompson. 1995. Effect of carp, *Cyprinus carpio* L., an exotic benthivorous fish, on aquatic plants and water quality in experimental ponds. Mar. Freshwater Res. 46: 1171–1180.

Rowe, D.K. 2007. Exotic fish introductions and the decline of water clarity in small North Island, New Zealand lakes: a multi-species problem. Hydrobiologia 583: 345–358.

Sapkale, P.H., R.K. Singh and A.S. Desai. 2011. Optimal water temperature and pH for development of eggs and growth of spawn of common carp (*Cyprinus carpio*). J. Appl. Anim. Res. 39: 339–345.

Scharbert, A. and J. Borcherding. 2013. Relationships of hydrology and life-history strategies on the spatio-temporal habitat utilisation of fish in European temperate river floodplains. Ecol. Indic. 29: 348–360.

Schaus, M.H., M.J. Vanni, T.E. Wissing, M.T. Bremigan, J.E. Garvey and R.A. Stein. 1997. Nitrogen and phosphorus excretion by detritivorous gizzard shad in a reservoir ecosystem. Limnol. Oceanogr. 42: 1386–1397.

Shearer, K.D. and J.C. Mulley. 1978. The introduction and distribution of the carp, *Cyprinus carpio* Linnaeus, in Australia. Aust. J. Mar. Freshwater Res. 29: 551–563.

Singh, A.K., A.K. Pathak and W.S. Lakra. 2010. Invasion of an exotic fish—common carp, *Cyprinus carpio* L. (Actinopterygii: Cypriniformes: Cyprinidae) in the Ganga River, India and its impacts. Acta Ichthyol. Piscatoria 40: 11–19.

Sisler, S.P. and P. Sorensen. 2008. Common carp and goldfish discern conspecific identity using chemical cue. Behaviour 145: 1409–1425.

Sivakumaran, K.P., P. Brown, D. Stoessel and A. Giles. 2003. Maturation and reproductive biology of female wild carp, *Cyprinus carpio*, in Victoria, Australia. Environ. Biol. Fishes 68: 321–332.

Smith, P.J. and S.M. McVeagh. 1987. Genetic analyses of carp, goldfish, and carp–goldfish hybrids in New Zealand. DOC Research and Development Series 219. Science and Technical Publishing, Department of Conservation, Wellington, New Zealand.

Smith, P.J. and S. Pullan. 1987. Identification of hybrids between koi (*Cyprinus carpio*) and goldfish (*Carassiusauratus*). N. Z. J. Mar. Freshwater Res. 21: 41–46.

Strickland, R. 1981. Whangamarino swamp. Freshwater Catch 14: 11.

Strickland, R.R. 1980. Fisheries aspects of the Whangamarino Swamp. Fisheries Environmental Report No. 7. NZ Ministry of Agriculture and Fisheries. Turangi.

Stuart, I. and M. Jones. 2002. Ecology and management of common carp in the Barmah-Millewa Forest: final report of the point source management of carp project to Agriculture Fisheries and Forestry Australia, Arthur Rylah Institute, Heidelberg, Victoria, Australia.

Stuart, I.G. and M. Jones. 2006. Movement of common carp, *Cyprinus carpio*, in a regulated lowland Australian river: Implications for management. Fisheries Manag. Ecol. 13: 213–219.

Stuart, I.G., A. Williams, J. McKenzie and T. Holt. 2006. Managing a migratory pest species: a selective trap for common carp. N. Am. J. Fish. Manage. 26: 888–893.

Taylor, A.H., S.R. Tracey, K. Hartmann and J.G. Patil. 2012. Exploiting seasonal habitat use of the common carp, *Cyprinus carpio*, in a lacustrine system for management and eradication. Mar. Freshwater Res. 63: 587–597.

Tempero, G.W. 2004. Population biology of koi carp in the Waikato region. M.Sc. Thesis, University of Waikato, Hamilton.

Tempero, G.W., N. Ling, B.J. Hicks and M.W. Osborne. 2006. Age composition, growth, and reproduction of koi carp (*Cyprinus carpio* L.) in the lower Waikato, New Zealand. N. Z. J. Mar. Freshwater Res. 40: 571–583.

Thomasen, S. and P. Chow-Fraser. 2012. Detecting changes in ecosystem quality following long-term restoration efforts in Cootes Paradise Marsh. Ecol. Indic. 13: 82–92.

Treer, T., B. Varga, R. Safner, I. Aničić, M. Piria and T. Odak. 2003. Growth of the common carp (*Cyprinus carpio*) introduced into the Mediterranean Vransko Lake. J. Appl. Ichthyol. 19: 383–386.

Thwaites, L.A., B.B. Smith, M. Decelis, D. Fleer and A. Conallin. 2010. A novel push trap element to manage carp (*Cyprinus carpio* L.): a laboratory trial. Mar. Freshwater Res. 61: 42–48.

Thwaites, L.A. 2011. Proof of concept of a novel wetland carp separation cage at Lake Bonney, South Australia. A summary report for the Invasive Animals Cooperative Research Centre and the South Australian Murray-Darling Basin Natural Resources Management Board (PDF 1.2 MB). SARDI Publication No. F2011/000086-1. SARDI Research Report Series No. 530. South Australian Research and Development Institute (Aquatic Sciences), Adelaide.

Ultsch, G.R., M.E. Ott and N. Heisler. 1980. Standard metabolic rate, critical oxygen tension, and aerobic scope for spontaneous activity of trout (*Salmo gairdneri*) and carp (*Cyprinus carpio*) in acidified water. Comp. Biochem. Physiol. A 67: 329–335.

Váradi, L., A. Hidas, E. Várkonyi and L. Horváth. 1995. Interesting phenomena in hybridization of carp (*Cyprinus carpio*) and rosy barb (*Barbus conchonius*). Aquaculture 129: 211–214.

Vilizzi, L. and K.F. Walker. 1995. Otoliths as a potential indicator of age in common carp, *Cyprinus carpio* L. (Cyprinidae, Teleostei). Trans. R. Soc. South Aust. 119: 97–98.

Vilizzi, L. and K.F. Walker. 1999. Age and growth of the common carp, *Cyprinus carpio*, in the River Murray, Australia: validation, consistency of age interpretation, and growth models. Environ. Biol. Fishes 54: 77–106.

Vilizzi, L. 2012. The common carp, *Cyprinus carpio*, in the Mediterranean region: origin, distribution, economic benefits, impacts and management. Fish. Manag. Ecol. 19: 93–110.

Vilizzi, L., A. Price, L. Beesley, B. Gawnea, A.J. King, J.D. Koehn, S.N. Meredith and D.L. Nielsen. 2013a. Model development of a Bayesian Belief Network for managing inundation events for wetland fish. Environ. Modell. Softw. 41: 1–14.

Vilizzi, L., B.J. McCarthy, O. Scholz, C.P. Sharpe and D. B. Wood. 2013b. Managed and natural inundation: benefits for conservation of native fish in a semi-arid wetland system. Aquatic Conserv.: Mar. Freshw. Ecosyst. 23: 37–50.

Wahl, J. 2001. An analysis of the fishery of Clear Lake, Iowa. Iowa Department of Natural Resources, Des Moines.

Wahl, D.H., M.D. Wolfe, V.J. Santucci Jr. and J.A. Freedman. 2011. Invasive carp and prey community composition disrupt trophic cascades in eutrophic ponds. Invasive carp and prey community composition disrupt trophic cascades in eutrophic ponds. Hydrobiologia 678: 49–63.

Walker, K.F., F. Sheldon and J.T. Puckridge. 1995. A perspective on dryland river ecosystems. Regul. River. 11: 85–104.

Weber, M.J. and M.L. Brown. 2009. Effects of common carp on aquatic ecosystems 80 years after "Carp as a dominant": ecological insights for fisheries management. Rev. Fish. Sci. 17: 524–537.

Weber, M.J. and M.L. Brown. 2011. Relationships among invasive common carp, native fishes and physicochemical characteristics in upper Midwest (USA) lakes. Ecol. Freshw. Fish. 20: 270–278.

Weber, M.J. and M.L. Brown. 2012a. Maternal effects of common carp on egg quantity and quality. J. Freshwater Ecol. 27: 409–417.

Weber, M.J. and M.L. Brown. 2012b. Effects of predator species, vegetation and prey assemblage on prey preferences of predators with emphasis on vulnerability of age-0 common carp. Fish. Manag. Ecol. 19: 293–300.

Weber, M.J. and M.L. Brown. 2012c. Diel and temporal habitat use of four juvenile fishes in a complex glacial lake. Lake Reservoir Manage. 28: 120–129.

Weber, M.J. and M.L. Brown. 2013. Density-dependence and environmental conditions regulate recruitment and first-year growth of common carp in shallow lakes. Trans. Am. Fish. Soc. 142: 471–482.

Weber, M.J., M.L. Brown and D.W. Willis. 2010. Spatial variability of common carp populations in relation to lake morphology and physicochemical parameters in the upper Midwest United States. Ecol. Freshw. Fish. 19: 555–565.

Weber, M.J., M.J. Hennen and M.L. Brown. 2011. Simulated population responses of common carp to commercial exploitation. N. Am. J. Fish. Manage. 31: 269–279.

Williams, A.E. and B. Moss. 2003. Effects of different fish species and biomass on plankton interactions in a shallow lake. Hydrobiologia 491: 331–346.

Williams, A.E., B. Moss and J. Eaton. 2002. Fish induced macrophyte loss in shallow lakes: top–down and bottom–up processes in mesocosm experiments. Freshwater Biol. 47: 2216–2232.

Winker, H., O.L.F. Weyl, A.J. Booth and B.R. Ellender. 2011. Life history and population dynamics of invasive common carp, *Cyprinus carpio*, within a large turbid African impoundment. Mar. Freshwater Res. 62: 1270–1280.

Yamamoto, T., Y. Kohmatsu and M. Yuma. 2006. Effects of summer drawdown on cyprinid fish larvae in Lake Biwa. Jpn. J. Limnol. 7: 75–82.

Zambrano, L. and D. Hinojosa. 1999. Direct and indirect effects of carp (*Cyprinus carpio* L.) on macrophyte and benthic communities in experimental shallow ponds in central Mexico. Hydrobiologia 408–409: 131–138.

Zambrano, L., M.R. Perrow, C. Macías-García and V. Aguirre-Hidalgo. 1998. Impact of introduced carp (*Cyprinus carpio*) in subtropical shallow ponds in Central Mexico. J. Aquat. Ecosyst. Stress Recovery 6: 281–288.

Zambrano, L., M. Scheffer and M. Martínez-Ramos. 2001. Catastrophic response of lakes to benthivorous fish introduction. Oikos 94: 344–350.

11

Recreational Fishing for Carp—Implications for Management and Growth of Carp Populations

Henrik Ragnarsson-Stabo

Few fish species give rise to as conflicting opinions as carp. On the one hand, carp are often considered an invasive species and a pest. On the other hand, many consider carp the most desirable species on the planet, and devote their lives to catching them. Carp fishing is increasing in popularity and also the demand for carp fisheries, for example more than 1.9 million carp were stocked in England and Wales in 2002–2003 (Environment agency 2004). In Europe, carp fishing has a high socioeconomic value (e.g., Arlinghaus and Mehner 2003). In England and Wales alone, 26 million days were spent coarse angling. The angler gross expenditure was in excess of 1 billion euro and 30,000 jobs were generated (Radford et al. 2007). Considering the increasing demand for carp fishing, it is important that there are guidelines for managing carp lakes sustainably without jeopardizing the ecosystems of these lakes. Here the attraction of carp to anglers, especially emphasizing growth rates and biomasses of carp in nature and how to combine carp fishing and healthy lakes are explored.

Science has yet to explain why so many find fishing for carp addictive. It is a challenging fish to catch, and one that has fascinated anglers for centuries (Walton 1653, Mascall 1590, Bartlett 1903). Carp fishers have been described as "big 'still' men, slow of movement, soft-footed and low-voiced, many

Swedish University of Agricultural Sciences.
Email: henrik.ragnarsson-stabo@slu.se

have nagging lean wives and it is by the calm secluded waters that they have found peace and quietness for their troubled lives" (BB 1950). However, carp fishing has a more general appeal than that and in later years more women have taken up carp fishing. In 2013 an all-female team won the World Carp Challenge (WCC). Carp fishing is one of the most popular branches of recreational fishing world-wide, and increasing in popularity. There is a rapidly developing industry surrounding the sport, with specialized equipment, bait firms and commercially managed fisheries. There are hundreds of books dedicated to carp fishing, in many languages, though the majority are in English. The plethora of dedicated carp angling magazines in Europe reflect its popularity—the list here is long but far from complete: Carpworld, Big Carp, Carpology, Crafty Carper, Total Carp, Advanced Carp Fishing, Carp Talk, Carp Addict, Svenskt Karpfiske, Spiegel, Karper, Monkeyclimber, Carpe Passion, Carpe Record, Carpe Nature, CarpaXTutti, Pescare, Karpfen, Carpmirror, Carp in Focus, Carp and Fun and Carp Hunters Magazine. In some countries carp are caught to be eaten, but in many countries the fishing is strictly catch-and-release, and this type of fishing is becoming more and more widespread. Catch-and-release means that all fish are returned alive to the lake. The anglers advocating catch-and-release often value the carp so highly that there can be conflicts with those who fish and kill (Arlinghaus 2007). Many measures are taken to minimize the stress and risk of serious injury for the fish, landing nets with a mesh of knotless knots are used, soft unhooking mats, solutions to treat wounds, and so on. It is a good example of where the fishermen work hard at "reducing injury, stress and mortality without ceasing recreational angling" (Arlinghaus et al. 2007). Still, there is a need for more studies to improve these practices, similar to the study on the effect of hook size on damage to carp mouths by Rapp et al. (2008). In Switzerland and Germany there are laws that limit catch and release angling, based on animal welfare issues.

Most of the sport fishing is targeted at rare large specimen-sized carp that are found in relatively low stocks. As Izaak Walton put it 360 years ago carp grow to these exceptional sizes where they are few and far between. "The carp, if he have water-room and good feed, will grow to a very great bigness" (Walton 1653). In exceptional cases they can grow up to over 30 kilos in 18–20 years (Danau 2013). Many of the popular sport fishing lakes have clear water and an abundance of macrophytes, rather than the turbid waters that the high density populations can cause, as described elsewhere in this book.

Sport Fishing for Carp, Now and Then

Carp fishing is varied, as they can be caught using a range of methods, from float fishing, free-lining, bottom fishing with several rods (Paisley 2002), even fly-fishing (e.g., Reynolds et al. 1997) and ice-fishing. Good fishing, challenging fishing for big fish, can be found in many different waterbodies: artificial lakes, such as gravel pits, clay pits, chalk pits and reservoirs (e.g.,

Gibbinson 1989, 1999), tiny waters (Yates 1986, BB 1950), large lakes (Briggs 2009) and rivers (Wayte 2000). Several books describe the exploits of carp globetrotters (e.g., Bursell 1999, Davies-Patrick 2004).

Historically, carp was considered to be almost impossible to catch with a rod and line. Sheringham (1912) suggested that luck was the most vital equipment for the carp angler, and his best tip for the would-be angler was to touch wood and be on friendly terms with a black cat. However, advances in tackle and tactics have made them more catchable. Some important advances are bite alarms and specialized rods (Walker 1953), the invention of the hair-rig (Maddocks 1981) where the bait is attached to the hook using a very thin line, and bait that was acceptable by carp but no other species, so called boilies. This development was further progressed by fishing deliberately for big fish, termed specimen hunting, i.e., fishing targeting larger than average specimens (Walker 1953, Gibbinson 1983), and the formation of specialist clubs for sharing ideas and findings, such as the British carp study group. Targeting individual known specimens that sometimes are given names is common (e.g., Bailey and Page 1986, Hearn 2000). Individual fish can, in some cases, be caught several times per year for many years without serious injuries or retarded growth (Fig. 11.1). A single fishing session may continue for several weeks, in exceptional cases months (Langley-Hobbs 2013). Therefore a lot of specialized equipment has been developed—for example bedchairs and tents (bivvies), to make the wait more comfortable. In some of the very low stocked waters an angler might not expect more than a couple of bites per year—the low catch per unit effort is compensated with patience (Fig. 11.2). These hard-to-catch fish are often ascribed human intellect, sometimes even superhuman. But then, how else could the angler live with being outwitted by a fish, time and time again? To increase the

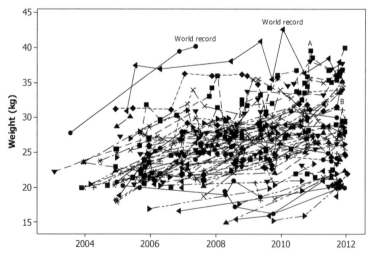

Figure 11.1. 119 individual carp over 20 kg from Lac de Curton, France. The data was compiled by Alijn Danau and Arjen Uitbeijerse.

Figure 11.2. Catch per unit effort is low but compensated for with patience.

chances several rods are often used (2–4 is common, where allowed), together with electronical bite alarms that wake the angler, should he be asleep when the fish takes the bait. Sometimes the most important step is forgotten, to find the fish and fish where they are. Finding them sometimes calls for a bit of detective work (Bailey 1994), for example looking for small gas bubbles rising to the surface as a carp gently disturbs the silty bottom, or a quivering lily-pad as something disturbs it in passing (Sharman 1980). Bait amount, fishing effort, and experience have been shown to lead to more carp caught (Arlinghaus and Mehner 2003).

In some circumstances carp quickly learn to avoid getting hooked (Beukema 1969), and therefore the angler has to keep varying his tactics (Maddocks and Cundiff 1996). In other circumstances, carp seem to learn not at all (Linfield 1980) probably due to hunger caused by over-stocking and subsequent high competition. Pre-baiting with similar baits that are used for fishing is a common method to overcome the carps' caution, and an essential practice for successful fishing (Arlinghaus and Mehner 2003). Pre-baiting is by no means a new phenomenon, it is mentioned in many early works including "the Compleat Angler" (Walton 1653). Carp fishing legend, Dick Walker wrote: "First, brainwash your carp" (Walker 1981), meaning that you should pre-bait the intended fishing spot with the same bait as you plan to fish with. This encourages the carp to treat your bait with less caution than they would normally do, increasing the chance of hooking one. This practice is essential in catch and release fishing, as the fish otherwise learn to avoid bait very quickly.

Carp can be caught throughout the year, even when most of the lake is frozen with ice, though their activity and feeding is decreased in cold water. There are many reports of carp caught in water down to, and below, 4 degrees Celsius (e.g., Briggs 2009, Maddocks 1981, Gibbinson 1989). During the warmest parts of the year, sometimes the carp seem to become more prone to eating food off the surface of the lake (e.g., Crow 2006). Thus, the fishing style and strategy need to be adapted to the situation (e.g., Hughes and Crow 1997).

Baits Used for Carp

Carp are omnivores and can be caught on a wide range of different baits (Hutchinson 1989, Pullen 1988, Townley 1998, Venables 1949). Boilies, maggots, bread, chironomid larvae, nuts, beans, seeds, pellets, luncheon meat, cheese, dried fruit, wheat, hemp, maize, sweetcorn, worms, prawns, mussels, tares, fish, potatoes, worms, sweet-corn, maize, bread, dog food to name a few. Still, the majority of carp are caught on hard boiled paste baits, i.e., boilies. Izaak Walton (1653) advocated the use of "flesh of a Rabbet or Cat cut small" in carp baits. The modern versions are generally pet-free, and tend to be composed of semolina, soy meal, bird foods, fishmeal, milk powder, spices, flavors and feeding stimulants, that are mixed with eggs then hardened by boiling to withstand the attention of less wanted species, so called nuisance species. The nuisance species are generally smaller cyprinids, and cannot take the big and hard baits back to their pharyngeal teeth for crushing, making the baits very selective. Boilies tend to be divided into different types depending on what the dominant ingredients are, e.g., fishmeal boilies and birdfood boilies (Townley 1998). A range of additives are available to boost the effectiveness of the bait—and the angler's confidence. Carp respond positively to a range of smells and tastes (Kasumyan and Døving 2003), and anglers are constantly trying to find new additives to take advantage of this. Commercially available additives include esters, essential oils, fish oils, nut oils, butyric acid, crustacean/fish extracts, molasses; the list is all but endless.

Carp at the Northern Border of their Distribution

Carp grow big when they have a lot to eat, and they have a lot to eat when there are few competitors. Low stocks, no spawning, and little competition from other species seem to be more important for growth than the lake type. Izaak Walton (1653) pinpointed the problem with successful management of carp: "in some ponds Carps will not breed, especially in cold ponds; but where they will breed, they breed innumerably". And there's the catch—to be interesting for sport fishing, the carp must not breed. It is not known what determines where carp can reproduce successfully—and cause problems—and where they cannot—and cause no problems.

Carp was introduced to Sweden in 1560 by Peder Oxe (see Filipsson 1994). They have been stocked into lakes and ponds since, even stocking of a bay in the largest lake of Sweden, Lake Vänern (5,650 km²) in a moderately successful attempt to catch them a couple of years later (Olsson 1939). Since the 1980s more than 2,000 water bodies have been stocked with carp, according to the only carp farm in Sweden, Aneboda Fiskodling, Lammhult. Here, successful reproduction in the wild is very rare indeed. The only documented cases are from lakes that lack piscivorous fish such as pike (*Esox luscious*), perch (*Perca fluviatilis*) and zander (*Sander lucioperca*). Examples include former put-and-take lakes that have been treated with rotenone prior to stocking with salmonid fish such as rainbow trout (*Oncorhynchus mykiss*), and formerly fishless lakes, such as gravel pits. The carp in the natural lakes are observed to spawn, but there is no recruitment. This phenomenon also occurs in crucian carp (*Carassius carassius*), that was long believed to be two different species—the small stunted pond-crucian and the large lake-crucian (see Rolfe 2010). In crucian carp this is due to them being sensitive to predation. In the ponds where there is no piscivorous fish, they breed successfully and grow stunted due to intraspecific competition for food. This means that because there is no mortality due to predation there is an extremely high density of individuals that compete for limited resources. Competition for resources leads to a reduced growth of the individuals resulting in a water body densely populated with small-sized carp, i.e., a stunted population. In lakes with piscivores the handful that survive are released from competition, resources are not limited and the few survivors quickly grow out of the predation window, i.e., the size range where they are sensitive to predation (Holopainen et al. 1997). Similarly, even in the cold Swedish climate, when released from predation common carp will 'breed innumerably' which both increases the risk of detrimental effects on the environment and leads to a population of small fish that is undesirable for recreational angling. The key, then, to managing carp lakes with both anglers and the environment in mind, is to ensure that there is a low density of carp, by effective management of predatory fish and a low stocking density of carp and to keep the number of competitors for limited food resources low enough to allow for a fast growth of few individuals.

There is a widely held belief that mortality during the first winter is what keeps the carp from successful recruitment in Sweden. However, if this was the case then we would expect to find young-of-the-year (YOY) in the standardized fish sampling with multi-mesh gillnets that is conducted in many lakes in late summer and early autumn. An analysis of the National Register of Survey test-fishing (Swedish University of Agricultural Sciences 2014) reveals that in the 7,837 fishing events in 3,494 lakes young-of-the-year carp has only been caught in one lake. Larger carp have been caught on nine occasions in six different lakes (Table 11.1). This strongly suggests that other factors such as, e.g., predation control the recruitment of carp in the tested lakes.

Table 11.1. Carp was caught in six different lakes. Young of the year (YOY) was only caught in the rotenone-treated Lake Träsksjön. Data obtained with permission/courtesy of from the Institute of Freshwater Research, Swedish University of Agricultural Sciences.

Lake	Altitude	Area (ha)	Max depth	NPUE	Notes
Ellestadsjön	38	278	6	0.17; 0.17; 0.17	
Rammsjön	81	8	7	0.12	
Vesljungasjön	111	57	3	0.5	
Surtesjön	98	92	12	0.06	
Råckstaträsk	11	3	2	0.25	
Träsksjön	10	10	4	1.25; 1.5	Treated with rotenone

The catch per unit effort, i.e., carp per net and night, was low in all lakes (min 0.16, max 1.5, median 0.5). Lake Ellestadsjön was stocked with carp in 1917 (Filipsson 1994), and carp were caught now and then in the decades afterwards.

The lake with the YOY-carp (Lake Träsksjön) was treated with rotenone and used as a put-and-take fishery for trout before the carp were introduced. So it is safe to assume that there were no piscivorous fish present. In later years the piscivorous perch (*Perca fluviatilis*) has been introduced into the lake. Perch thrive in the lake and have grown extremely fast compared to other lakes in the region (Fig. 11.3)

No perch older than 7 years was found in L. Träsksjön in 2012. After the establishment of perch no YOY-carp has been caught (Fig. 11.4). The carp that remain have started growing rapidly and are now much larger (mean total length = 379 mm) than before the addition of a piscivorous fish (mean total length = 99 mm). The number of fish caught per unit effort (NPUE) is the same before and after (NPUE = 76 in 2001 and 76 in 2012) but the weight per

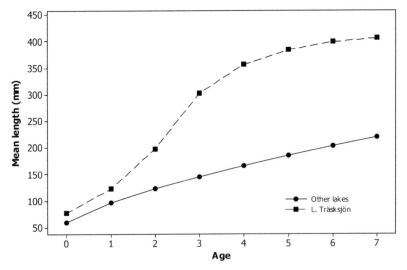

Figure 11.3. Growth of the perch in Lake Träsksjön compared to other lakes in the same region of Sweden. Data courtesy of the Institute of Freshwater Research, Swedish University of Agricultural Sciences.

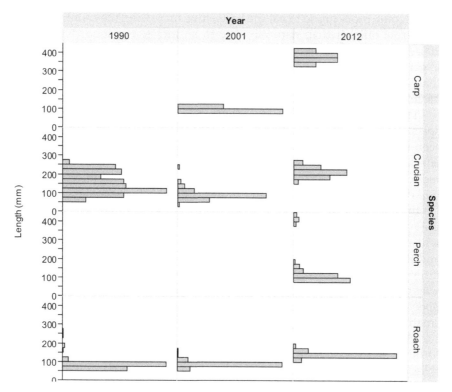

Figure 11.4. The length distribution of carp (*Cyprinus carpio*), crucian (*Carassius carassius*), roach (*Rutilus rutilus*) and perch (*Perca fluviatilis*) caught in gill nets during test fishing in Lake Träsksjön, Sweden before and after piscivorous perch established a population there. All cyprinids smaller than 10 cm have disappeared from the catch test-fishing with standardized gill nets in Lake Träsksjön after perch was introduced. Data from national register of survey test-fishing, Institute of Freshwater Research, Swedish University of Agricultural Sciences (2014).

unit effort (WPUE) has almost increased by a factor of 10 (WPUE = 6299 g in 2012 and 677 g in 2001). The increased WPUE is due to a larger mean weight of the cyprinids (Fig. 11.4) and a lot of large perch in the catch (several over 1 kg each per net). Also the other cyprinids in the lake have been affected by the perch. Before the introduction the mean length was less than 10 cm, now no cyprinid < 10 cm is caught (Fig. 11.4). The mean length of crucian carp and roach has increased dramatically (Fig. 11.4). The cyprinids have also increased in mean weight with a factor of between 4 (roach) and 50 (carp).

Consequences of Carp Fishing

In contrast to fishing for most other species, carp fishing can actually benefit from a high fishing pressure because catch and release fishing minimizes mortality due to fishing (e.g., Rapp et al. 2012) and the habit of pre-baiting with feed that is of high nutritional value (Arlinghaus and Niesar 2003). Carp can become more prone to eat the baits, thus increasing the number of bites, and grow quicker than naturally due to an increase in food (Niesar et al. 2004). The high quality of the feed used during fishing, means that the feed is generally eaten, digested and incorporated into the fish biomass, rather than leading to high feed loss rate and low oxygen levels due to increased bacterial breakdown (Arlinghaus and Niesar 2005).

High carp densities are undesirable both from an ecological and a specialist carp angler point of view. When dominant they have, as can be seen elsewhere in this book, highly density-dependent negative ecological effects, such as increased resuspension, increased eutrophication effects, reductions in macrophytes. In the high density populations the effect of carp will depend on the size structure, as they undergo ontogenetic niche shifts. On the one hand, an abundance of very small carp will affect the structure of the zooplankton community. On the other hand, an abundance of large carp will affect turbidity and macrophyte cover. In addition, for optimal growth—and thus worth to the angling community—the carp are dependent on low intra- and interspecies competition (e.g., Lorenzon 1996). In over-crowded water bodies they often grow stunted and are not as interesting for fishing. Specialized carp anglers generally prefer catching a few large fish than many small ones (Arlinghaus and Mehner 2003).

While phosphorus input from anglers' baits is generally small compared to other sources, it could be significant in lakes that attract many anglers (Niesar et al. 2004) and should be taken into account especially in sensitive lakes, such as oligotrophic lakes with a small water volume or a long retention time and a high angling pressure (Arlinghaus and Mehner 2003). The simplest solutions in these lakes are to either limit the number of fishing licenses and/or limit the amount of bait used daily, especially the cheaper baits such as particles and pellets. The use of boilies tend to be self-regulating, as they are expensive—often costing around 15 € per kilo. However, a total ban on ground baiting and pre-baiting is not recommended, as that would—as mentioned earlier—almost make carp fishing impossible, and therefore decrease the positive effects on the socioeconomy, the regional economy and the general well-being of fishermen.

Thus it should not be hard to educate anglers and fisheries managers that carp stocks should be kept low. This can be achieved, in practice, by stocking the lakes at low levels when a new fishery is started, by ensuring that there is little or no recruitment by managing the piscivorous predators correctly and by avoiding stocking carp in predator-free lakes. Carp are notoriously hard to catch with nets, so in the cases where the densities are too high, cooperation between lake managers and fisheries managers is advised and carp fishermen recruited to reduce the carp population to below the thresholds for impact on the environment and allow maximum growth (e.g., Arlinghaus and Mehner 2003).

Biomass thresholds for ecological effects have been the focus of considerable research, reviewed in the Chapter 10 Carp as an invasive species. To summarize the chapter's findings: often detrimental effects have been observed where carp biomass exceeds 450 kg/ha (Fletcher et al. 1985, Weber and Brown 2009), but some loss of macrophytes can in some cases occur already at biomasses of > 200 kg/ha (Williams et al. 2002). In very sensitive habitats effects on macrophytes can occur at ~100 kg/ha (Haas et al. 2007, Bajer et al. 2009). The authors who are experts on invasive carp suggest that safe biomasses are < 50 kg/ha. To be on the safe side, low stocking densities (< 5–10 carp per hectare depending on lake type) are advised so that the final biomass will not exceed the threshold densities. It is much easier to do an additional stocking at a later date than to remove fish.

Currently, very few bait firms consider the environment when formulating baits. Rather, the baits are designed to maximize nutritional value and attraction while keeping the price acceptable. In some cases, even preservatives, which have been shown to be toxic to carp are used to extend the shelf-life of commercial baits (Rapp et al. 2008). Carp and some other cyprinids find feed with a large amount of fish meals attractive (Kasumyan and Døving 2003), and these meals are high in phosphorus (P) concentration—the nutrient that generally limit phytoplankton growth in fresh water (Wetzel 2001). To minimize the P leakage, baits should be coarse grained, unsolvable in water, have a high P digestability and a low P-level (Arlinghaus and Niesar 2005). However, this is hard to combine with a bait that is acceptable to fish and fishermen. Clearly more effort should be put into formulating ecological baits.

Local anglers should be educated on the risk of negative impact from excessive baiting. Besides the issue with eutrophication, over-feeding might even lead to obesity in carp (Hoole et al. 2001). One possibility to reduce the impact of angling, where there are documented impacts, is to limit the number of licenses, to decrease the angling pressure. Arlinghaus and Mehner (2003) published a simple model that can be used as a guideline for the maximum number that should be allowed depending on the lake size and sensitivity. The recommended number of anglers per hectare ($10^4\,m^2$) varied between approximately 0.1 (oligotrophic lake with 50 years water retention time) and 0.5 (eutrophic lake with 0.1 years retention time). A successful communication about the potential of faster growth for the carp with less competition could lead to fishermen taking part in biomanipulation-type fishing after for example bream, a fish that causes more resuspension than carp (Breukelaar 1994).

Low stocks of large-sized carp can be an asset rather than a problem in areas where they are not invasive. The interest in carp fishing is continually growing globally. In order to combine carp fishing and a healthy ecosystem it is of importance that these water bodies be well managed.

Sustainable Management of a Carp Lake

How then, should a carp lake be managed sustainably to avoid the potentially detrimental effects of the carp, and the carp angler?

1. Carp density should be kept low. A maximum final biomass of 50–100 kg/ha in sensitive lakes. The biomass should never be allowed to exceed 450 kg/ha.
2. The amount of bait used should be regulated in sensitive lakes. Either through maximum amounts allowed or a maximum number of anglers.
3. Only bait of good quality with known ingredients should be allowed. Any seeds and other particles should be boiled to be easily assimilated by the fish which reduces nutrient pollution via baits.
4. It is strongly recommended that the macrophyte cover and water chemistry be monitored to alert managers if excessive baiting or too high carp density affects the water body in unwanted ways.
5. Piscivorous fish stocks should be kept high to keep the density of carp and their competitors low.

Examples of Well-managed Lakes

"Any fisherman who dreams of the perfect pool is always hoping that his imagined paradise really exists and that one day he might actually find it" thus begins a classic carp fishing book by Chris Yates (1992). This search for the perfect lake has taken some carp anglers around the globe.

Rainbow Lake—Lac de Curton

One of these perfect lakes—the most popular carp lake in the world now, 2014, is Lac de Curton, close to Hostens, Bordeaux, France (Fig. 11.5). To fish there you need to book many years in advance as the lake is fully booked, allowing ~25 anglers every day of the year, even during the winters. The reason is, of course, that the fish grow exceptionally large and that the lake is challenging. The lake is a 46 ha old water filled open pit mine. Before filling with water the area was covered with pine trees, which means the lake is now full of dead wood. Due to the acidic nature of the soil, the lake is limed each winter. Despite a long theoretical retention time, and constant heavy baiting, the total phosphorus level in the lake is not high (Table 11.2). Likely because relatively few anglers are allowed to fish, approximately 0.5 fisherman/ha, which is in line with the recommendations based on the model in Arlinghaus and Mehner (2003). There are a lot of predatory fish in the lake, which keeps the reproduction down. Only a few 'new' fish are recruited each year. There are many large pike (*Esox lucius*), some even over 15 kg, perch (*Perca fluviatilis*) and small mouth bass (*Micropterus dolomieu*). There is also very little competition, in the form of bream (*Abramis brama*), tench (*Tinca tinca*) and roach (*Rutilus rutilus*), likely due to the high predation pressure as there is no active management of those species.

Alijn Danau, Belgium, and Arjen Uitbeijerse, the Netherlands, have gathered reports of catches of carp over 20 kilos from the lake (Fig. 11.6). Many of which they have caught themselves (Danau 2007, 2013, Uitbeijerse 2013).

Figure 11.5. Dawn at the perfect pool. Photo: Henrik Ragnarsson Stabo.

Table 11.2. Water chemistry data from Lac de Curton, surface water during autumn mixing conditions. The sample was taken by the author and analysis commissioned to the Erken laboratory, Uppsala University.

Date	pH (25°C)	Cond. (mS/m)	Alk. (meqv/l)	Phosphate (µg/l)	TP (µg/l)	TN (µg/l)
2012-12-01	6.8	14	0.19	< 1	22	650

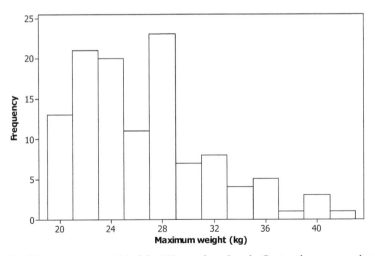

Figure 11.6. The maximum weight of the 119 carp from Lac du Curton that were analyzed. The data was compiled by Alijn Danau and Arjen Uitbeijerse.

Each carp has been identified and assigned an id-number. The mirror carp are easy to identify as their scale patterns are unique, similar to fingerprints in humans. The common carp are harder to identify from pictures, but it is almost always possible to identify some unique scale pattern, old scar or similar marks (Huntingford et al. 2013). This allows us to follow the growth of the individual carp. In total 380 carp over 20 kilos have been identified. A subset of 119 carp has been analyzed here. One of the carp was caught by the author of this chapter in 2011 (Fig. 11.7). It lost a scale in the landing net, which was taken to the age-reading laboratory at the Institute of Freshwater Research, Drottningholm, and aged by counting the number of annuli (rings) on a scale (http://www.slu.se/en/departments/aquatic-resources/contact/ifr/laboratory-for-fish-age-analysis/). The carp was found to be 18–20 years (Fig. 11.8). That carp was likely hatched in 1991–1993. Some research revealed that Ken Townley, an author of carp fishing literature from the UK, had witnessed a stocking of 3-year old carp in April 1995 (Fig. 11.9).

Some extraordinary growth has been recorded in individual fish, such as the first of the two world records from the lake (Fig. 11.10). The mean growth is more moderate. The common carp in Lac de Curton grow slightly faster than the mirrors t (54) = 2.07, p < 0.05. The mean weight increase per year is 1.9 kg for the commons and 1.6 for the mirrors. The biggest common that has been caught in the lakes weighed 41.3 kilos (2012) and was a world record for a short time. Interestingly, some fish vary up to 5 kg within years between post-spawning and winter-condition (Fig. 11.10), one of these is a mirror carp that was caught by the author in 2010 (Fig. 11.11).

Figure 11.7. Common carp 31.2 kg caught 2011-11-08 in Lac de Curton, France. Picture taken by Stephan Nielsen.

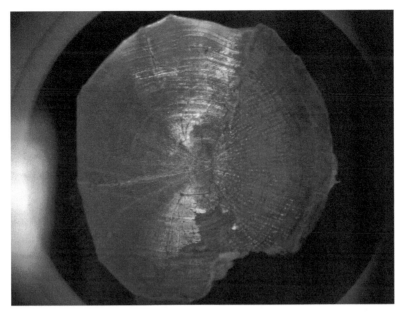

Figure 11.8. Scale from a 31.2 kilo common carp caught in Lac de Curton 2011. Bordeaux. Age reading estimated the age of 18–20 years. Photo: Magnus Kokkin, Institute of Freshwater Research.

Figure 11.9. Picture from stocking of 3+ carp in Lac de Curton 1995. Photo: Ken Townley, UK.

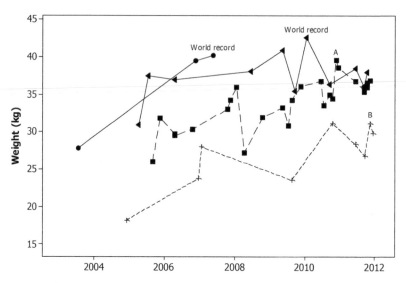

Figure 11.10. The growth of two former world record carp from Lac de Curton, France. A and B denote two carp caught by the author. The data was compiled by Alijn Danau and Arjen Uitbeijerse.

Figure 11.11. 39.6 kg mirror carp caught 2010-11-21 in Lac de Curton, France. Picture taken by Stephan Nielsen.

Lake Tegelbruksdammen

Even in Sweden, with a very short growing season, carp can grow well if the lake is well managed. The growth in Fig. 11.12. is from three different Swedish lakes, Lake Tegelbruksdammen, Lake Kroppkärrsjön and Lake

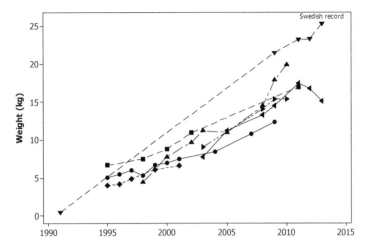

Figure 11.12. Growth in three Swedish lakes. The data was compiled by Jon Sköld, Anders Idarsson and Mikael Sandström.

Snuggan. All of them have low stocking densities (5–100 kg/ha), low fishing pressure (0.5–0.005 anglers per hectare) and a high predation pressure. All carp died in Lake Snuggan in the winter of 2010–2011, probably due to anoxic conditions during that time.

Lake Tegelbruksdammen holds the Swedish record, a fish named Rebecca (25.36 kg). The lake had suffered from anoxic conditions historically and lacked predators. Only high densities of stunted crucian carp were present in the lake. The local fishing club removed macrophytes and stocked the lake with 300 pike of 0.2–3 kg in the winter of 1997. By 1999 almost all crucian carp were gone. Once released from competition the growth of carp, and the remaining crucian carp, increased drastically. The carp there are now the biggest in Sweden, with several individual fish taking turns of being the national record.

References

Arlinghaus, R. 2007. Voluntary catch-and-release can generate conflict within the recreational angling community: a qualitative case study of specialised carp, *Cyprinus carpio*, angling in Germany. Fish. Manag. Ecol. 14: 161–171.

Arlinghaus, R., S.J. Cooke, A. Schwab and I.G. Cowx. 2007. Fish welfare: a challenge to the feelings-based approach, with implications for recreational fishing. Fish Fish. 8: 57–71.

Arlinghaus, R. and M. Niesar. 2005. Nutrient digestibility of angling baits for carp, *Cyprinus carpio*, with implications for groundbait formulation and eutrophication control. Fish. Manag. Ecol. 12: 91–97.

Arlinghaus, R. and T. Mehner. 2003. Socio-economic characterisation of specialised common carp (*Cyprinus carpio* L.) anglers in Germany, and implications for inland fisheries management and eutrophication control. Fish. Res. 61: 19–33.

Bailey, J. 1994. The Fishing Detective. Collins Willow, London.

Bailey, J. and M. Page. 1986. Quest for the Queen. The Crowood Press, Wiltshire.

Bartlett, S.P. 1903. Angling for Carp, and Some Hints as to Best Mode of Cooking. Trans. Am. Fish. Soc. 32: 47–57.

BB. 1950. Confessions of a Carp Fisher. Eyre & Spottiswoode, London.

Beukema, J.J. 1969. Angling experiments with carp (*Cyprinus carpio* L.). Neth. J. Zool. 20: 81–92.

Briggs, S. 2009. Cassien and Beyond. Freebird Publishing, UK.

Breukelaar, A.W., E.H.R.R. Lammens, J.P.G. Klein Breteler and I. Tatrai. 1994. Effect of benthivorous bream (*Abramis brama*) and carp (*Cyprinus carpio*) on resuspension. Verh. Internat. Verein. Limno. 25: 2144–2147.

Bursell, J. 1999. Specimen Mede. Forlaget Mosegaard, Hammershoj, DK.

CEN (European Committee for Standardisation). 2005. EN 14757. Water Quality—Sampling of Fish with Multi-mesh Gillnets, Brussels. CEN, Brussels.

Clifford, K. 2011. A History of Carp Fishing Revisited. Sandholme Publishing Ltd., Newport.

Crow, S. 2006. Carp Fishing. Advanced Tactics. The Crowood Press Ltd., Marlborough.

Danau, A. 2007. Karper and Andere Muzen. Westerlaan Publishers, Lichtenvoorde, Holland.

Danau, A. 2013. Het Heilige Beest. Westerlaan Publishers, Lichtenvoorde, Holland.

Davies-Patrick, T. 2004. Globetrotters Quest: A Worldwide Search for Carp and other Giant Fish. Westerlaan Publishers, The Netherlands.

Environment agency. 2004. Our nations' fisheries. Environment agency, Bristol. http://www.environment-agency.gov.uk/static/documents/Research/fisheries_eng_765655.pdf.

Filipsson, O. 1994. Nya fiskbestånd genom inplantering eller spridning av fisk. Inf. Sötvattenslab. Drottningholm 2: 1–65.

Gibbinson, J. 1983. Modern Specimen Hunting. Beekay Publishers, Henlow.

Gibbinson, J. 1989. Big Water Carp. Beekay Publishers, Henlow.

Gibbinson, J. 1999. Gravel Pit Carp. Laneman Publishing, Dorking.

Hearn, T. 2000. In Pursuit of the Largest. Bountyhunter Publications, Hampshire.

Holopainen, I.J., W.M. Tonn and C.A. Paszkowski. 1997. Tales of two fish: the dichotomous biology of crucian carp (*Carassius carassius* (L.)) in northern Europe. Ann. Zool. Fennici 34: 1–22.

Hoole, D., W. Buckle, P. Burgess and I. Wellby. 2001. Diseases of carp and other cyprinid fishes. Blackwell Science, Oxford.

Hughes, R. and S. Crow. 1997. Strategic Carp Fishing. The Crowood Press Ltd., Marlborough.

Huntingford, F.A., F.L. Borçato and F.O. Mesquita. 2013. Identifying individual common carp *Cyprinus carpio* using scale pattern. J. Fish Biol. 83: 1453–1458.

Hutchinson, R. 1989. Guide to Carp Baits. Wonderdog Publishers, Louth.

Kasumyan, A.O. and K.B. Døving. 2003. Taste preferences in fishes. Fish and Fisheries 4: 289–347.

Langley-Hobbs, J. 2013. http://www.thebigcarphunter.com.

Linfield, R.S.J. 1980. Catchability and stock density of common carp, *Cyprinus carpio* L. in a lake fishery. Aquac. Res. 11: 11–22.

Lorenzen, K. 1996. A simple von Bertalanffy model for density-dependent growth in extensive aquaculture, with an application to common carp (*Cyprinus carpio*). Aquaculture 142: 191–205.

Maddocks, K. 1981. Carp Fever. Beekay Publishers, Henlow.

Maddocks, K. and J. Cundiff. 1996. Carp Rigs. Beekay International, Henlow.

Mascall, L. 1590. The Carpe—in A booke of Fishing with Hooke & Line, and of All Other Instruments Thereunto Belonging. John Wolfe, London.

Niesar, M., R. Arlinghaus, B. Rennert and T. Mehner. 2004. Coupling insights from a carp, *Cyprinus carpio*, angler survey with feeding experiments to evaluate composition, quality and phosphorus input of groundbait in coarse fishing. Fish. Manag. Ecol. 11: 225–235.

Olsson, F. 1939. Om inplantering av karp i sjöar. Svensk fiskeritidskrift 11: 347–348.

Paisley, T. 2002. Carp! Angling Books Ltd., Sheffield.

Pullen, G. 1988. Freshwater Fishing Baits. The Oxford Illustrated Press, Somerset, England.

Radford, A.F., G. Riddington and H. Gibson. 2007. Economic Evaluation of Inland Fisheries: The Economic Impact of Freshwater Angling in England & Wales. Environment Agency, Bristol, 165 pp.

Radford, A.F., G. Riddington and D. Tingley. 2001. Economic Evaluation of inland fisheries. Environment Agency R&D Project W2-039/PR/1 (Module A), pp. 166.

Rapp, T., S.J. Cooke and R. Arlinghaus. 2008. Exploitation of specialised fisheries resources: The importance of hook size in recreational angling for large common carp (*Cyprinus carpio* L.). Fish. Res. 94: 79–83.

Rapp, T., J. Hallermann, S.J. Cooke, S.K. Hetz, S. Wuertz and R. Arlinghaus. 2012. Physiological and behavioural consequences of capture and retention in carp sacks on common carp (*Cyprinus carpio* L.), with implications for catch-and-release recreational fishing. Fish. Res. 125–126: 57–68.

Rapp, T., T. Meinelt, A. Krüger and R. Arlinghaus. 2008. Acute toxicity of preservative chemicals in organic baits used in carp, *Cyprinus carpio*, recreational fishing. Fish. Manag. Ecol. 15: 163–166.

Reynolds, B., B. Befus and J. Berryman. 1997. Carp on the Fly. Johnson Books, Boulder, USA.

Rolfe, P. 2010. Crock of Gold: Seeking the Crucian Carp. M Press, Romford.

Sharman, G. 1980. Carp and the Carp Angler. Century Hutchinson, London.

Sheringham, H.T. 1912. Coarse Fishing. A. & C. Black, London.

Swedish University of Agricultural Sciences. 2014. National Register of Survey test-fishing (NORS). http://www.slu.se/en/departments/aquatic-resources/databases/national-register-of-survey-test-fishing-nors/.

Spurgeon, J., G. Colurullo, A.F. Radford and D. Tingley. 2001. Evaluation of inland fisheries. Environment Agency. R&D Project W2-039/PR/2 (Module B), pp. 162.

Townley, K. 1998. Carp Baits. Beekay International, Withy pool, UK.

Uitbeijerse, A. At the end of the Rainbow. International carper 118: 183–187.

Yates, C. 1986. Casting at the Sun. Pelham Books, London, pp. 232.

Yates, C. 1992. The Secret Carp. Merlin Unwin Books, Ludlow.

Venables, B. 1949. Mr. Crabtree Goes Fishing. The Daily Mirror, London. UK.

Walker, R. 1953. Stillwater Angling. Macgibbon & Kee, London.

Walker, R. 1981. Catching Fish, Knowing their Feeding Habits. David & Charles, London.

Walton, I. 1653. The compleat angler. http://www.gutenberg.org/files/9198/9198-h/9198-h.htm.

Wayte, N. 2000. River Carping. The Crowood Press Ltd., Ramsbury.

Wetzel, R.G. 2001. Limnology Lake and River Ecosystems. Academic Press, San Diego.

Part V

Toxicology

Over the past century and up to the present time humans have introduced a large number of chemical substances into the environment. These come from industrial and agricultural activities or households and although they often successfully fulfill a certain purpose, many of them have undesirable toxic and harmful effects to the environment. Fish are sentinels of their aquatic environment as they are in constant contact and hence have a biochemical exchange to the medium they live in. Thus fish are usually used in environmental risk assessments in order to elucidate the biochemical fate and potential hazard of environmental contaminants. One of the most prominent and most widely used fishes are common carp. Carp have been used as model species in aquatic toxicology for many years and the knowledge on effects of widely distributed pesticides is summarized in Chapter 12. Since for most pesticides the environmental concentrations are known, this chapter compares the outcomes of laboratory studies to environmental concentrations in order to assess the relevance of the toxicological effects for carp in the wild.

Carp are not only exposed to anthropogenic substances but also natural substances originating from fungi, plants, algae or mollusks and can result in an exposure of carp to toxic compounds. Chapter 13 therefore focuses on the effects of natural toxins on carp and provides an in-depth description of the state of art in this field of research. The presence of mycotoxins and other anti-nutritional factors in fish feed is highly relevant in nutrition of cyprinids in aquaculture and still needs further research in the future.

12

Effects of Pesticides on Carp

Radka Dobsikova[1,*] and Josef Velisek[2]

Introduction

The pesticide contamination of aquatic ecosystem has increased, mainly in the last few decades, due to their extensive use in agricultural, chemical and industrial processes. Because of their widespread distribution and toxic nature, pesticides have recently become an important class of water pollutants.

Pesticides are used to control, pests' which means to destroy harmful organisms that infect agricultural fields. Pesticides can contaminate both surface and ground water which leads to pollution in aquatic environments (Kuivila and Foe 1995). Pesticides are introduced into water systems from various sources, e.g., industrial effluents, agricultural runoff and chemical spills (Eichelberger and Lichtenberg 1971). They are among the most frequently occurring organic pollutants in agricultural soils (Benitez et al. 2006) and ground and surface waters (Rebich et al. 2004), causing harmful ecological effects on natural ecosystems, especially aquatic systems (Glusczak et al. 2011, Rossi et al. 2011). They cause damage to target organisms as well as non-target ones, including fish. Fish kills caused by various pesticides have been reported on a global scale (Eichelberger and Lichtenberg 1971).

Fish species play a major ecological role in aquatic food webs because of their function as a carrier of energy from lower to higher trophic levels (van

[1] Department of Veterinary Public Health and Animal Welfare, Faculty of Veterinary Hygiene and Ecology, University of Veterinary and Pharmaceutical Sciences Brno, Palackeho tr. 1/3, 612 42 Brno, Czech Republic.
Email: dobsikovar@vfu.cz
[2] Research Institute of Fish Culture and Hydrobiology Vodnany, Faculty of Fisheries and Protection of Waters, South Bohemian Research Center of Aquaculture and Biodiversity of Hydrocenoses, University of South Bohemia Ceske Budejovice, Zatisi 725/II, 389 25 Vodnany, Czech Republic.
Email: velisek@frov.jcu.cz
* Corresponding author

der Oost et al. 2003). For this reason, fish species are widely used to evaluate the health of aquatic ecosystems (Whyte et al. 2000).

The species common carp (*Cyprinus carpio* L.) is one of the most common freshwater fish species in waters of Central Europe. It is also a species reared for commercial purposes. Carp is an omnivorous, benthopelagic, freshwater fish, widely used in biomonitoring programs (Huang et al. 2007, Falfushynska and Stolyar 2009). It is often used as a sentinel organism due to its ease of sampling, good adaptability to environmental conditions, and high ecological and economic relevance achieved in studies that use carp as a model organism. OECD (Organisation for Economic Co-operation and Development) guidelines usually use carp as a standard test organism for chemical testing, e.g., in OECD 203 (Fish, Acute Toxicity Test, 1992) or OECD 212 (Fish, Short-term Toxicity Test on Embryo and Sac-fry Stages, 1998).

The effects of pesticides on carp, including their ethiology, mechanism of toxic action, toxicity, clinical symptoms, pathological and morphological picture in the fish exposed to various chemical groups of pesticides are reviewed.

Organochlorine Pesticides

Ethiology, Mechanism of Toxic Action and Toxicity

Organochlorines are a group of chemicals composed of carbon, chlorine and hydrogen. As pesticides, they are also referred to as chlorinated organics, chlorinated insecticides or chlorinated synthetics. Although the first chlorinated hydrocarbon was synthesized in 1874, its insecticidal properties were not discovered until 1939. It was introduced as DDT in 1942 during World War II and it subsequently helped to control vectored diseases such as typhus and malaria. The advantages of these synthetic chemicals over previously used botanical (naturally occurring chemicals extracted from plants) or other natural insecticides were, e.g., improved efficacy, lower use rates, lower costs and greater persistence (Smith 2000, Brundage and Barnett 2010).

Organochlorine pesticides are neurotoxins affecting sodium and potassium channels in neurons. A decrease in potassium permeability and inhibition of calmodulin, Na^+/K^+- and Ca^{2+}-ATPase activity occur in organisms poisoned by organochlorine insecticides (Narahashi 2010). Furthermore, these insecticides interfere with estrogenic hormones (Kelce et al. 1995, Tilson 1998) and calcium metabolism (Peakall 1969).

Organochlorine pesticides are frequently detected in organisms on higher trophic levels. They are highly persistent and accumulate in fatty tissues of organisms and lipoproteins of cell membranes (Chowdhury et al. 1990). DDT and its breakdown products are transported from warmer regions of the world to the Arctic area by a phenomenon of global distillation and the substances accumulate in the food web (Smith 1991). Organochlorine

Infobox: Pesticide and Pathological Terminology

Multifold detrimental effects of toxic substances may occur in fish. **Necrosis** is commonly observed after exposure to cell-toxic chemicals and is defined as a form of cell injury that results in premature death of cells in living tissue by autolysis. Necrosis is caused by the actions of external factors (such as infection, toxins, or trauma) on cells or tissues that result in unregulated digestion of cell components. Consequently **necrotic foci** occur as numerous, relatively small or tiny, fairly well-circumscribed, usually spheroidal portions of tissue that manifest as coagulative, caseous, or gummatous necrosis and are characteristically associated with agents that are hematogenously disseminated. Accordingly, **lesions** are abnormalities in the tissue of an organism (in general terms: 'damage'), usually caused by disease or trauma. Lesions can occur in parenchyma that is the functional part of an organ in the body. **Parenchymal organs** are all glands, liver, lymphatic organs, kidney, spleen, ovary, and testis. On the outer body surface **desquamation** may occur which is also called skin peeling—the shedding of the outermost membrane or layer of the skin. If **cytotoxic effects** in blood cells occur, **anemia** (i.e., a decrease in number of red blood cells, RBCs, or less than the normal quantity of hemoglobin in the blood) may be observed. Anemia may also be diagnosed where there is a decreased oxygen-binding ability of each hemoglobin molecule due to deformity or lack in numerical development as in some other types of hemoglobin deficiency. **Extravasation** is the leakage of fluid out of its container. In the case of inflammation, it refers to the movement of white blood cells from the capillaries to the tissues surrounding them (diapedesis). **Hyperaemia** is the increase of blood flow to different tissues in the body. Clinically, hyperaemia in tissues is manifested as **erythema**. This, for example, may lead to **eye hemorrhage** which is a condition where blood vessels inside the eye rupture and bleed, leaving red splotches on the white of the eye, in the retina, or between the retina and the lens. In addition, **congestion of visceral vessels** may be observed that is an excessive or abnormal accumulation of blood in visceral vessels. In contrast the term **ischemia** is used for a restriction in blood supply to tissues, causing a shortage of oxygen and glucose needed for cellular metabolism. Ischemia is generally caused by problems with blood vessels, with resultant damage to or dysfunction of tissue. If small dilated blood vessels near the surface of the skin or mucous membranes occur this is called **telangiectasias** or **angioectasias**. In addition, **edema** formerly known as dropsy or hydropsy, occur due to an abnormal accumulation of fluid in the interstitium, which is a location beneath the skin or in one or more cavities of the body. It is clinically shown as swelling which can be observed after exposure to many chemicals. Histological examinations then often also show **vacuolization** which is the formation of vacuoles within or adjacent to cells. As a response to cellular damages **hyperplasia** may occur which manifests as an increase in number of cells/proliferation of cells. Often also changes of the activities of metabolic enzymes can be observed including enzymes important for **transamination** which is an important step in the synthesis of some non-essential amino acids (amino acids that can be synthesized *de novo* by the organism).

Behavior of fish is commonly monitored when the effects of chemical exposure should be judged. **Choreoathetosis** is the occurrence of involuntary movements in a combination of **chorea** (irregular migrating contractions) and **athetosis** (twisting and writhing).

Infobox: contd....

Infobox: contd....

> **Necropsy** is the examination of a body after death. After death the **gill lamellae** should be investigated. In fish gills, there are two types of lamellae, primary and secondary. The **primary gill lamellae** come out of the interbranchial septum to increase the contact area between the water and the blood capillaries. The **secondary gill lamellae** are small lamellae that come out of the primary ones and are used to further increase of the contact area. Both types of lamellae are used to increase the amount of oxygen intake into the blood. Both types of lamellae contain huge amounts of capillaries and are the sites where the exchange of oxygen from the water and carbon dioxide from the blood occurs. Since these are important functions damage to gill lamellae is often detrimental for fish.

pesticides, including DDT, represent an environmental hazard due to their persistent nature (about 30 years) and potential health effects (Alexander 1980).

The group of organochlorine pesticides includes, e.g., dichlorodiphenyltrichloroethane (DDT), pentachlorophenol, dicofol, heptachlor, endosulphan, chlordane, aldrin, dieldrin, endrin, mirex and kepone. At present, the production and use of organochlorine pesticides are banned for agricultural use worldwide under the Stockholm Convention on Persistent Organic Pollutants. Thus, these pesticides were replaced by organophosphates and pyrethroids.

The high acute toxicity of DDT-based preparations was documented by Mertlik and Svobodova (1973). These pesticides are found to be highly to extremely toxic to fish showing a moderate lethal concentration (lethal concentration that will kill half of the fish during 48 hours—48hLC$_{50}$) of less than 1.0 mg/L (Svobodova et al. 1993). At present, organochlorine pesticides and their residues cause chronic poisonings in aquatic ecosystem (Smith 2000).

Clinical Symptoms

Clinical signs of carp poisoning by organochlorine pesticides on the basis of chlorohydrocarbons include increased activity followed by a long period of reduced activity (Svobodova et al. 1993).

Pathological and Morphological Picture

There is no specific patho-anatomic picture of the intoxication. Dystrophic alterations have been recorded in the liver and kidneys of carp (Svobodova et al. 1993).

In carp living in water contaminated with hexachlorocyclohexane isomers (HCHs) and DDTs, lower values of natural estrogenic steroid, 17β-estradiol (E2) and a reduced gonadosomatic index were found (Singh and Singh 2008). Kozlowski (1983) reported edema of the gills and parenchymal organs and regressive changes in the intestines in common carp exposed to DDT and toxaphene. In carp, aldrin (100 µg/L), DDT (50 µg/L), dieldrin (20 µg/L), endrin (2 µg/L), hexachlorbenzene (100 µg/L) and lindane (100 µl/L) cause a variety of effects such as anemia, increased concentration of glucose and cortisol, lower values of total protein and cholesterol in blood, inhibition of ATPase activity and alterations in nervous function (Gluth and Hanke 1985).

In Indian carp (*Labeo rohita*), chronic exposure to hexachlorocyclohexane (0.35 mg/L and 1.73 mg/L) resulted in swelling of hepatocytes with diffuse necrosis and marked swelling of blood vessels in the liver tissue. Kidney tubules were distended, with tubular cells of posterior kidney exhibiting marked necrotic changes. Gill tissue showed fusion of primary lamellae, congestion of blood vessels and hyperplasia of branchial plates. The pericardial sac was moderately thickened and infiltrated with leukocytes (Das and Mukherjee 2000).

Organophosphate Pesticides

Ethiology, Mechanism of Toxic Action and Toxicity

OP insecticides are derivatives of phosphoric acid (H_3PO_4) or phosphonic acid (H_3PO_3) in which all H atoms are replaced by organic moieties. In 1970s, more than 200 OP insecticides were marketed worldwide (Chambers et al. 2010a). In 1996, OP formed the largest insecticide group with some 38% of worldwide sales (Roberts and Hutson 1999).

OP pesticides are used as insecticides (against mites, cockroaches, ant, flies and fleas) in agriculture, food production, veterinary medicine, hospitals and households. During the last 20 years, they have been replaced by pyrethroid-based pesticides (Chambers et al. 2010a).

In aquaculture, OP insecticides (mostly trichlorphon and diazinon used in commercial formulations, such as Soldep and Diazinon 60 EC) were used for the reduction of excessive growth of daphnian zooplankton population. Soldep was also used for the treatment of parasitic diseases in fish, especially monogeneosis and arthropodosis (Noga 1996, Svobodova et al. 2007). In 2000, trichlorphon production was banned (due to its negative effect on humans and ecosystems) and Soldep was replaced by Diazinon 60 EC. In 2011, diazinon-based preparations were also banned for marketing. Consequently, no effective preparation for the reduction of daphnian zooplankton is available at present.

During OP metabolism, they are subject to phase I reaction (oxidation by cytochromes P450 and flavin monooxygenases, reduction, and hydrolysis) and phase II reaction (conjugation with sulfate, glucuronide, and glutathione). Because of their metabolic and chemical lability, they do not readily remain intact either in the organism or environment (Roberts and Hutson 1999, Tang et al. 2006, Chambers et al. 2010b). OP are rapidly degraded to non-toxic substances which usually do not accumulate in the environment (Cremlyn 1978).

Still OP insecticides are potent inhibitors of serine esterases. The mode of OP action is based on the inhibition of acetylcholinesterase (AchE), a widely distributed enzyme within the vertebrate nervous system, which mediates hydrolysis of the neurotransmitter acetylcholine (Ach) throughout the central and peripheral nervous systems. The phosphorylation of AchE, on the post-synaptic membrane, is relatively persistent, the recovery of the enzyme activity requires hours to days. The inhibition of the enzyme results in a failure to degrade Ach to choline and acetic acid which leads to the accumulation of Ach in cholinergic synapses and neuromuscular/glandular junctions with a subsequent hypercholinergic activity. Ion channels of the nicotinic Ach receptor (i.e., a receptor that forms ligand-gated ion channels in plasma membranes of certain neurons and on postsynaptic side of the neuromuscular junction, directly linked to ion channels and triggered by the binding of the neurotransmitter ACh). Nicotinic receptors are also opened by nicotine (hence the name, nicotinic). They exist permanently in an open conformation and sodium ions pass through, which results in membrane depolarization, continuous neuronal firing and convulsions resulting in death (Roberts and Hutson 1999, Garcia et al. 2006, Chambers et al. 2010b).

Early symptoms of the cholinergic hyperstimulation are the results of parasympathetic stimulation such as bradycardia, miosis, diarrhea, urination, lacrimation and salivation. The stimulation of neuromuscular junctions results in muscle twitching and, at higher doses, paralysis. CNS (i.e., central nervous system) symptoms include tremors, headache and convulsions, with death occurring from depression of respiratory centers in the brain (Garcia et al. 2006). The mode of OP action in mammals and fish is similar (Chambers et al. 2010b). In fish, most of the toxic effects are expressed through the inhibition of AchE (Ray and Ghosh 2006).

OP insecticides are toxic not only to insects (as target organisms), but also to mammals (including humans) and fish. Acute toxicity of organophosphates to fish is variable, values of $96hLC_{50}$ range between 0.1 and 100 mg/L. Salmonid fish are very sensitive to pesticides based on organic compounds of phosphorus (Svobodova et al. 2008). The lethal concentrations (LCs) of selected OP insecticides for common carp are given in Table 12.1. Environmental concentrations of OP insecticides are usually found in units

Table 12.1. Lethal concentrations (LCs) of OP insecticides for common carp.

OP insecticide	Exposure (hr)	LC_{50} (mg/L)	References
profenofos	48	0.081	Ismail et al. 2009
profenofos	96	0.062	Ismail et al. 2009
diazinon	96	10.0–25.0	Machova et al. 2007
diazinon	96	16.0	Svoboda et al. 2001
diazinon(embryos)	48	0.999	Aydin and Köprücü 2005
diazinon (larvae)	48	2.903	Aydin and Köprücü 2005
diazinon (larvae)	96	1.530	Aydin and Köprücü 2005
chlorpyriphos	96	0.160	Halappa and David 2009
quinalphos	96	0.0075	Chebbi and David 2010

of μg/L, in some cases they exceed the LC_{50} values for fish (Smalling et al. 2013) and thus detrimental effects on carp are possible.

Clinical Symptoms

Behavioral responses of common carp exposure to OP are due to the impairment of the nervous system. Common carp fingerlings exposed to profenofos showed loss of balance, moving in a spiral way with sudden jerky movements, lying laterally on their sides and rapid flapping of the operculum with the mouth open (Ismail et al. 2009). In common carp exposed to diazinon, neural paralytic syndrome manifested by strong restlessness, excitation, and cramp movements of fins and mouth were found. Then, loss of movement coordination and orientation began and fish swam in half-circles. Weakening of jerks or arreflexia, paralysis, arrhythmia and block of respiration movements began in the terminal phase of poisoning (Svoboda et al. 2001). Body surface darkening, mainly in the dorsal part, is a typical clinical sign in fish exposed to OP insecticides (Fig. 12.1).

Irregular, erratic, darting swimming movements, hyperexcitability, loss of equilibrium followed by hanging vertically in water and sinking

Figure 12.1. Body surface darkening in carp exposed to 20 mg/L of diazinon for 96 hours, unpublished (A. exposed carp, B. control carp) (Photo: J. Velisek).

to the bottom as well as caudal bending as the main pathomorphological alterations are found in common carp exposed to sublethal concentrations of chlorpyrifos (Halappa and David 2009). Similar clinical signs and alteration of respiratory rates were found in common carp fingerlings exposed to quinalphos (Chebbi and David 2010).

Pathological and Morphological Picture

The autopsy finding is usually characterized by increased mucus production on the body surface and gills, hyperaemia of gills, and small petechias at high OP insecticide concentrations (Svobodova et al. 2008). In hepatocytes of common carp exposed to methidathion, subcellular structure alterations (i.e., contractions of nuclei, slight swelling of mitochondria, and rough endoplasmic reticulum impairment) were found. Large lipid droplets and increased amounts of bile pigments were also detected in hepatocytes (Balint et al. 1995).

In early life stages of common carp, exposure to fenitrothion at a concentration of 10 mg/L for 48 hours resulted in a decrease in hematocrit (Hk) and red blood cell counts as well as in an increase in plasma cortisol level. The increase in cortisol levels can be linked to a response to stress due to a pollutant effect in carp (Sepici-Dincel et al. 2007), which was supported by the study of Carballo et al. (2005). In common carp of 600–800 g body weight exposed to dichlorvosat concentrations of 1 and 5 mg/L for 24 hours, membrane damages measured as a decrease in lipid peroxidation as well as lower concentrations of reduced glutathion in the liver, kidney, gill and muscle were found at both concentrations. The effect on the activity of anti-oxidative enzymes, such as superoxide dismutase (SOD), glutathione peroxidase (GPx), and catalase (CAT) was found to be inconclusive in the tissues (Hai et al. 1997).

Diazinon

The use of the active organophosphate substance diazinon will be reviewed because it was very often used in aquaculture for the suppression of excessive growth of daphnian zooplankton populations. In 2004, U.S. residential use of diazinon was outlawed, except for agricultural purposes and cattle ear tags. Its use in aquaculture in the EU was banned in 2007.

In the study of Machova et al. (2007), 28 days exposure of common carp larvae to 3 mg/L of the preparation Diazinon 60 EC (containing 600 g/L diazinon) caused a significant decrease in total length, a significant increase in Fulton's condition factor as well as a significant retardation of ontogenesis. However, the occurence of morphological anomalies was found to be very low (1.67%).

Diazinon in the preparation Basudin 600 EW (containing 600 g/L diazinon) at a concentration of 32.5 mg/L caused a significant decrease in erythrocyte count, hemoglobin content (Hb), and hematocrit, as well as in leukocyte (white blood cells) count and both, relative and absolute lymphocyte counts, and a significant increase in both, relative and absolute count of developmental forms of neutrophil granulocytes (myelocytes and metamyelocytes) in one-to-two-year-old common carp exposed for 96 hours. These changes may have been due to a disruption in hematopoiesis probably leading to a decrease in the non-specific immunity of the fish (Svoboda et al. 2001).

A significant decrease in the activity of acetylcholine esterase and lactate dehydrogenase, as well as in concentrations of total protein, lactate, calcium and phosphorus were confirmed in blood plasma of one-to-two-year-old common carp exposed to diazinon (Luskova et al. 2002). In the study, plasma glucose, sodium and potassium levels were significantly increased. The results indicate a marked neurotoxic effect of diazinon in fish (Luskova et al. 2002).

Exposure of 9-month-old common carp to sublethal concentrations (0.0036, 0.018, and 0.036 µg/L) of diazinon for 30 days revealed a dose-dependent decrease in AchE activity. Changes in lipid peroxidation and oxidative stress recorded as significant changes in superoxide dismutase, catalase, and glutathione peroxidase were found in carp liver, although common carp was found to possess an efficient antioxidant defense system reducing oxidative damages (Oruc 2011). In one-to-two-year-old common carp exposed to 32.5 mg/L of the preparation Basudin 600 EW (containing 600 g/L of diazinon) for 96 hours, no specific pathological signs were found. However, the body surface was opaque with slightly increased amount of mucus and expressive pigmentation, mainly on the dorsal part. The gills had normal color and were covered with increased amount of mucus. In the body cavity, evident injection of internal organ vessels and hyperaemia of hepatopancreas were found (Svoboda et al. 2001).

Further organophosphate pesticides that are toxic to carp, include alkylphosphates such as the pesticide dimethoate. Dimethoate is a widely used organophosphate insecticide used to kill insects on contact. Common carp fingerlings exposed to a sublethal concentration of dimethoate (2.8 mg/L) for 7 and 14 days showed a decrease in blood sodium and chloride as well as in liver sodium, potassium and calcium levels (Logaswamy et al. 2007). Maintenance of constant internal ion concentrations is essential for active regulation of water influx and ion efflux in aquatic organisms (Mayer et al. 1992). Thus, the exposure of carp fingerlings to dimethoate could lead to an impairment of various physiological activities but further research is needed to verify this.

Carbamate Pesticides

Ethiology, Mechanism of Toxic Action and Toxicity

Carbamates (N-methylcarbamates) are organic compounds derived from carbamic acid ($HOC(O)NH_3$) (Roberts and Hutson 1999). They are mostly used as insecticides, herbicides and fungicides in agriculture and medicine. Active substances, such as aldicarb, carbofuran, fenoxycarb, carbaryl, ethionocarb, and phenobucar belong to the group of carbamate insecticides (Tang et al. 2006, Yu 2008).

Carbamates are moderately to strongly sorbed to soils and fairly rapidly degraded in soil, so they are non-persistent in the environment (Revilla et al. 1996, Roberts and Hutson 1999). Their degradation in sediments is mediated by *Arthrobacter, Pseudomonas, Bacillus*, and *Actinomyces* species (Ambrosoli et al. 1996).

The mode of carbamate action is similar to OP. In fish, carbamates inhibit AchE by carbamylation of the serine hydroxyl group in the active site of the enzyme in the nervous system. Carbamylation of AchE leads to the persistent action of the neurotransmitter, acetylcholine, on cholinergic post-synaptic receptors. Unlike the OP insecticides, the inhibition of AchE is reversible and the onset and recovery of inhibition are rapid (Fukuto 1972, Tang et al. 2006, Zhang et al. 2010a). In phase I of biotransformation, carbamates are hydrolyzed or oxidized (hydroxylated, dealkylated or S-oxidated). Glucuronide, sulfate, and glutathion conjugates are formed in phase II (Tang et al. 2006).

Acute toxicity of carbamates to fish is variable, $96hLC_{50}$ values range from 1 to 1,000 mg/L (Svobodova et al. 2008); for example the $96hLC_{50}$ values of carbosulfan and carbaryl for fish fry (110–340 mg body weight and 20–34 mm total length) of common carp were estimated to be 0.60 and 7.85 mg/L, respectively (DeMel and Pathiratne 2005). Machova et al. (1985) tested the acute toxicity of carbamate pesticides Synbetan P and Betanal (15.1–16.7% of fenmedipham), Synbetan D and Betanal AM (15.1–16.7% of desmedipham), and Synbetan Mix and Betanal AM 11 (7.6–8.4% of fenmedipham and 7.6–8.4% of desmedipham) and found $48hLC_{50}$s in the range of 0.012 to 0.019 ml/L for common carp fry (6.83 ± 1.85 cm and 11.71 ± 9.29 g). The values of $96hLC_{50}$ of diafuran, an insecticide used, e.g., in rice cultivation, in common carp (*C. carpio*), grass carp (*Ctenopharyngodon idella*) and bighead carp (*Hypophthalmichthys nobilis*) fingerlings were found 1.81, 2.71, and 2.37 mg/L, respectively (Golombieski et al. 2008). Thus, no acute hazard from carbamate pesticides is assumed for carp in the environment.

Clinical Symptoms

Clinical signs of fish poisoning with carbamates are non-specific (Svobodova et al. 2008). Increased swimming movements, jumps above the surface, hyperventilation, loss of balance, incoordination, lateral swimming position and death in agony are observed in carp (Hejduk and Svobodova 1980, Dobsikova 2003).

Hyperactivity and erratic movements (especially during the first 48 hours) as well as concentration-dependent inhibition in AchE activity in common carp fry exposed for 14 days to sublethal concentrations of carbaryl (≥ 8 mg/L) and carbosulfan (≥ 0.10 mg/L) were found in the study of DeMel and Pathiratne (2005). Machova et al. (1985) investigated the effects of selected carbamate pesticides containing fenmedipham and desmedipham and found restlessness and accelerated respiration followed by depression, loss of movement, and death in common carp fry tested. In the study of Golombieski et al. (2008), common carp (5.5 ± 0.5 g and 7.7 ± 2.2 cm), grass carp (11.7 ± 3.3 g and 10.4 ± 3.1 cm) and bighead carp (11.3 ± 3.4 g and 10.2 ± 3.0 cm) fingerlings exposed to sublethal concentrations of diafuran (ranging from 0.5 to 4.0 mg/L) for 96 hours were lethargic (at lower concentrations) or immobile. Diafuran inhibited AchE activity in the brain and muscle of all carp species.

Pathological and Morphological Picture

Pathological changes in fish exposed to carbamates are not specific (Svobodova et al. 2008). Increased production of skin mucus, loss of pigmentation, eye hemorrhage, and vascular injection are found in carp (Hejduk and Svobodova 1980, Dobsikova 2003).

Machova et al. (1985) found a slightly increased amount of mucus on the body surface and on the gills, changes of skin color and hyperaemia of internal organs in common carp fry exposed for 48 hours to the preparations containing fenmedipham and desmedipham.

The effect of carbamate pesticide carbofuran in fish are described in detail below.

Carbofuran

Carbofuran (2,2-dimethyl-2,3-dihydro-1-benzofuran-7-yl methylcarbamate) is one of the most toxic carbamate pesticides used in agriculture. Although carbofuran is rapidly degraded by microorganisms in soil and water (Ambrosoli et al. 1996, Yu 2008), its residues may persist, depending on the environmental condition, for several weeks or up to one year (Ray and Ghosh 2006). For example, in water bodies in Bangladesh, carbofuran concentration of 198.7 µg/L was reported in the study of Chowdhury et al. (2012). This level is much higher than LC_{50} for carp (Dembélé et al. 2000) which indicates a risk for carbofuran poisoning in carp at environmental concentrations. In 2008, carbofuran was banned in the EU.

Carbofuran significantly inhibited AchE activity in one-year-old common carp exposed to sub lethal doses (0.05 and 0.10 µg/L) for 96 hours. Carbofuran at higher concentrations (0.22 and 1.0 µg/L) killed all tested carp within 24 hours (Dembélé et al. 2000). The activity of AchE can be recovered in time, indicating reversibility of binding of the pesticide with the enzyme (Dembélé et al. 1999).

Toxic effects of carbofuran on common carp were reported in the study of Pawar (1994), in which egg hatching was decreased and morphological abnormalities were observed, including deformed body curvature, eye pigmentation, enlargement of pericardial sacs, circulatory failure, loss of balance, and abnormal behavior in carp exposed to 2 and 4 mg/L for 96 hours. The 96hLC$_{50}$ for freshwater fishes such as goldfish (*Carassius auratus*), rainbow trout (*Oncorhynchus mykiss*), and brown trout (*Salmo trutta*) were found to be at 1.20, 0.38, and 0.56 mg/L (Johnson and Finley 1980, Munn et al. 2006).

Pyrethrins, Pyrethroids

Ethiology, Mechanism of Toxic Action and Toxicity

More than 1,000 pyrethroids have been synthesized since 1973. Their toxicity to non-target organisms is found in µg/L concentrations (Bradbury and Coats 1989, Haya 1989).

Synthetic pyrethroids are non-systemic insecticides, they remain on the plant surface and kill insects by contact or ingestion of treated foliage. They can be divided, according to chemical structure, into two types (Roberts and Hutson 1999): type I pyrethroids (e.g., bifenthrin, permethrin) do not contain -cyano-3-phenoxybenzyl group, whereas type II pyrethroids (e.g., cypermethrin, deltamethrin) contain the group. Type I pyrethroids block sodium channels in nerve filaments and cause the 'T-syndrome' in mammals (T-tremor). Type II pyrethroids block sodium channels and affect the function of GABA-receptors in nerve filaments.

During the last 15 years, pyrethroids such as deltamethrin and cypermethrin have been used for the control of ectoparasites infesting cattle, sheep, poultry and some companion animals (for example dogs, cats). Recently, these compounds have also been used for the control of ectoparasite infestations (*Lepeophtheirus salmonis* and *Caligus elongatus*) in marine cage culture of Atlantic salmon (*Salmo salar*) (Treasurer and Wadsworth 2004).

Due to their lipophilicity, pyrethroids have a high rate of gill absorption, which in turn would be a contributing factor to the sensitivity of fish to pyrethroid exposures. Fish seem to be deficient in the enzyme system that hydrolyzes pyrethroids. The main reaction involved in the metabolism of pyrethrins and pyrethroids in fish is largely oxidative (Kamalaveni et al. 2003).

Pyrethroids show low toxicity to mammals and birds, but they present a high risk to aquatic organisms. As the 96hLC$_{50}$ for fish is below 10 µg/L, it is assumed that pyrethroids belong to a group of chemicals highly toxic to fish and other aquatic organisms.

The influence and toxicity of three most widely used active substances of pyrethroids, deltamethrin, cypermethrin and bifenthrin, on common carp are presented below.

Deltamethrin

Deltamethrin [(S)-a-cyano-3-phenoxybenzyl (1R,3R)-3-(2,2-dibromvinyl)-2,2-dimethylcyclo-propan-carboxylate] is a widely used pesticide based on pyrethroids. It affects fish by paralyzing the insects' nervous system, thus giving a quick knock-down effect after surface contact or digestion. It is commonly used to control insects on fruits, vegetables, potted plants and ornamentals (Pham et al. 1984, Mueller-Beilschmidt 1990). Deltamethrin is the active ingredient of commercial formulations such as Butoflin, Butoss, Butox, Cislin, Crackdown, Cresus, Decis, Decis-Prime, K-Othrin and K-Otek.

Deltamethrin is very toxic to fish (lethal toxicity data are given in Table 12.2) and its risk to the aquatic environment is very high. For example, in the summers of 1991 and 1995, the pesticide caused massive European eel (*Anguilla anguilla*) kills in Lake Balaton, Hungary, following its application for mosquito control (Balint et al. 1997). In 1995, a presence of deltamethrin was demonstrated in several other fish species and in sediment samples taken from the lake (Balint et al. 1997). Carp are moderately sensitive fish to delthametrin, the mean lethal concentrations are in range of 1.37–3.25 μg/L. Deltamethrin is more toxic to carp at lower temperatures and appears to be more toxic to smaller carp than larger ones (Calta and Ural 2004, Velisek et al. 2006a). Low temperatures entail a slower metabolization of deltamethrin.

During the deltamethrin (3.25 μg/L) poisoning in two-year-old carp the following clinical symptoms of choreoathetosis were observed: loss of movement and coordination, fish laying down on the bottom of the tank and moving on one spot and accelerated respiration (Blahova et al. 2007).

In the study of Velisek et al. 2011, the acute 96 hours exposure of common carp to deltamethrin (3.25 μg/L) did not cause pathological changes in the gills, skin, liver, spleen, cranial and caudal kidney. Nevertheless, in carp blood plasma, a reduction of red blood cell count, hemoglobin concentration, and hematocrit and an increase in ammonia and the activities of aspartate aminotransferase and alanine aminotransferase were found.

Table 12.2. Lethal concentrations (LCs) of deltamethrin for fish.

Fish species	Exposure (h)	LC_{50} (μg/L)	References
Nile tilapia (*Oreochromis niloticus*)	96	14.6	El-Sayed et al. 2007
guppy (*Poecilia reticulata*)	48	5.13	Viran et al. 2003
common carp (*C. carpio*)	96	3.25	Velisek et al. 2006a
common carp (*C. carpio*)	96	1.32–2.07	Calta and Ural 2004
zebrafish (*Danio rerio*)	96	2.0	Lepailleur and Chambon 1984
Atlantic salmon (*S. salar*)	96	1.97	Zitko et al. 1979
fossil catfish (*Heteropneustes fossilis*)	96	1.86	Srivastava et al. 1997
rainbow trout (*O. mykiss*)	96	0.32–1.66	Ural and Saglam 2005
bluegill (*Lepomis macrochirus*)	96	1.2	Buccafusco et al. 1977
mosquitofish (*Gambusia affinis*)	96	1.0	Mulla et al. 1978

Cypermethrin

Cypermethrin   [(RS)-α-cyano-3-phenoxybenzyl   (1RS)-cis,trans-3-(2,2-dichlorovinyl)-2,2-dimethylcyclopropane carboxylate] is a widely used pyrethroid pesticide. The mechanism of its toxicity in fish is the same as that of other type II pyrethroids. Cypermethrin is a synthetic pyrethroid used for the control of ectoparasites infesting cattle, sheep, poultry and some companion animals (Bradbury and Coats 1989). It is the active ingredient of commercial preparations Ammo, Arrivo, Barricade, Basathrin, Cymbush, Cymperator, Cynoff, Cypercopal, Cyperguard, Cyperhard, Cyperkill, Cypermar, Demon, Flectron, Fligene, Kafil, Polytrin, Siperin and Super.

Cypermethrin is very toxic to aquatic organisms; its $96hLC_{50}$ value is generally found within the range of 0.002–5.99 µg/L (acute toxicity data are given in Table 12.3). In general, the fish sensitivity to pyrethroids may be explained by their relatively slow metabolism and elimination of these compounds. The elimination half-lives of several pyrethroids for fish are all longer than 48 hours, while the ones for, e.g., birds and mammals range from 6 to 12 hours (Bradbury and Coats 1989).

Carp are moderately sensitive fish to cypermethrin showing a mean lethal concentration of 2.91 µg/L (Velisek et al. 2006b). In the course of the acute exposure to cypermethrin the following clinical symptoms were observed in carp: increased respiration, loss of coordination, and fish lying on their flank and moving in this orientation. Subsequent short excitation stages with convulsions, jumping above the water surface, and moving in circles alternated with resting. A necropsy performed after the acute toxicity test revealed increased watery mucus on body surfaces. The body cavity contained excess fluid and showed congestion of visceral vessels (Dobsikova et al. 2006).

In one-to-two-year-old common carp, the acute 96 hours exposure to cypermethrin (2.91 µg/L) caused a decrease in total protein, albumin, globulin, ammonia as well as in the activity of lactate dehydrogenase and alkaline phosphatase and an increase in glucose, lactate, creatine kinase as well as in mean volume, mean hemoglobin concentration and red blood cells, also lymphocyte count was increased (Dobsikova et al. 2006). Change in biochemical profile of carp indicates stress and metabolic changes (i.e., glycogen catabolism, glucose shift towards the formation of lactate, primarily in muscle; amplified transamination processes) caused by cypermethrin. Changes of hematological profile of carp indicate a decrease in nonspecific immunity.

Table 12.3. Lethal concentrations (LCs) of cypermethrin for fish.

Fish species	Exposure (h)	LC_{50} (µg/L)	References
guppy (*Poecilia reticulata*)	48	21.4	Polat et al. 2002
Nile tilapia (*O. niloticus*)	96	5.99	Sarikaya 2009
rohu (*Labeo rohita*)	96	4.0	Marigoudar et al. 2009
rainbow trout (*O. mykiss*)	96	3.14	Velisek et al. 2006b
common carp (*C. carpio*)	96	2.91	Dobsikova et al. 2006
rainbow trout (*O. mykiss*)	96	0.72–1.35	Vaishnav and Yurk 1990
fossil catfish (*H. fossilis*)	48	0.67	Saha and Kaviraj 2003

In one-to-two-year-old carp, the acute 96 hours exposure to cypermethrin resulted in hyperaemia and perivascular lymphocyte infiltration in the skin, mild hyperplasia of respiratory epithelium and chloride cells activation in the gills (Fig. 12.2) and vacuolisation of pancreas exocrine cells (Fig. 12.3) (Dobsikova et al. 2006).

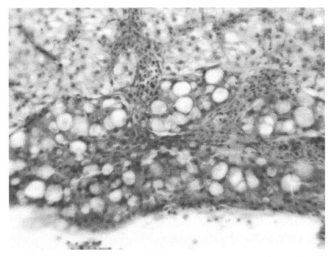

Figure 12.2. Vacuoles in pancreatic exocrine cells of common carp after 96 hours exposure to cypermethrin (2.91 µg/L) (H+E, magnification 100 times) (Photo: L. Novotny).

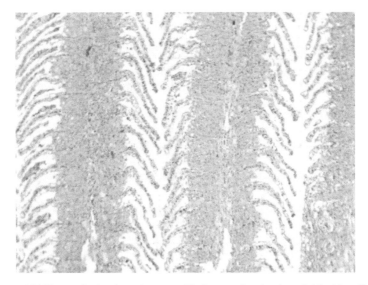

Figure 12.3. Mild hyperplasia of respiratory epithelium and activation of chloride cells in gills of common carp after 96 hours exposure to cypermethrin (2.91 µg/L) (H+E, magnification 100 times) (Photo: L. Novotny).

Bifenthrin

Bifenthrin [2-methylbiphenyl-3-ylmethyl (Z)-(1RS, 3RS)-3-(2-chloro-3,3,3-trifluoroprop-1-enyl)-2,2-dimethylcyclopropane-carboxylate], a newer member of the synthetic pyrethroid family, is an insecticide and acaricide used on a variety of crops, on stored grain, and as a precautionary measure against termites. It was registered in 1985 and is the active ingredient of commercial products such as Talstar, Bifenthrine, Brigade, Capture, Torant and Zipak.

Bifenthrin is a type I pyrethroid (Shan et al. 1997) and has some structural similarities to cypermethrin, tetramethrin and permethrin, but is characterized by greater photostability and insecticidal activity than earlier pyrethroids (Yadav et al. 2003).

Clinical symptoms observed during bifenthrin exposure of common carp correspond to observations reporting the toxicity of other pyrethroid pesticides (Velisek et al. 2009a, 2011). The main hematological and biochemical response of common carp to bifenthrin is the elevation of relative and absolute monocyte counts and levels of plasma glucose, ammonia, and activities of aspartate aminotransferase and creatine kinase. Changes of biochemical and hematological profiles indicate stress and metabolism damage (i.e., glycogen catabolism and amplified transamination processes) and a decrease in nonspecific immunity. Moreover, bifenthrin in carp causes severe telangioectasia in secondary lamellae of the gills, with rupture of pillar cells (Velisek et al. 2009a, 2011).

The acute toxicity of bifenthrin is different for various fish species and ranges from 0.15 to 6.5 µg/L (acute toxicity data are given in Table 12.4).

Carp is not a sensitive fish to bifenthrin, lethal concentration is in the range 2.08–5.75 µg/L. Bifenthrin is more toxic to carp reared at lower temperatures and appears to be more toxic to smaller carp than larger ones (Liu et al. 2005, Velisek et al. 2009a). The low temperature causes lower metabolization of bifenthrin.

Table 12.4. Lethal concentrations (LCs) of bifenthrin for fish.

Fish species	Exposure (h)	LC$_{50}$ (µg/L)	References
zebrafish (*Danio rerio*)	96	3.2–6.5	Zhang et al. 2010b
common carp (*C. carpio*)	96	5.75	Velisek et al. 2009a
common carp (*C. carpio*)	96	2.08	Liu et al. 2005
rainbow trout (*O. mykiss*)	96	1.47	Velisek et al. 2009b
Nile tilapia (*O. niloticus*)	96	0.80	Liu et al. 2005
fathead minnow (*Pimephales promelas*)	96	0.78	Guy 2000
gizzard shad (*Dorosoma cepedianum*)	192	0.51	Drenner et al. 1993
fathead minnow (*P. promelas*)	96	0.21	McAllister 1988
bluegill (*L. macrochirus*)	96	0.15	Horberg 1983a
rainbow trout (*O. mykiss*)	96	0.35	Horberg 1983b

Triazine Pesticides

Ethiology, Mechanism of Toxic Action and Toxicity

Triazines are selective herbicides divided into two groups, asymmetrical (metribuzine) and symmetrical (e.g., simazine, atrazine, ametryne, propazine, terbutryne and prometryne) compounds. In target organisms, triazines are absorbed via plant leaves and roots. They inhibit photosynthesis by blocking electrons within Hill's reaction photosystem II in organisms with oxygen-evolving photosystems (DeLorenzo et al. 2001).

Triazine herbicides are relatively stable in water (half-life 300 days) and in sediments (half-life several years). Recently, most of them have been banned, but triazine residues are still found in surface and ground water worldwide in units to tenths of μg/L (Woudneh et al. 2009, Abrantes et al. 2010, Vryzas et al. 2011, Bottoni et al. 2013). Atrazine, cyanazine, prometryne, propazine, sebuthylzine, simazine, terbutylazine and terbutryne are considered risky to the environment and are listed as a priority group of substances tested in the EU and USA. At present, 14 commercial triazine-based pesticides are still used.

The acute toxicity of triazines is variable, $96hLC_{50}$ values range from units to hundreds of mg/L. Acute poisoning is scarce as the triazine solubility in water is lower than lethal concentrations are.

Clinical Symptoms

Clinical signs are quite non-specific and are characterized by attenuation, i.e., decrease in swimming activity (Svobodova et al. 1987). Incoordination and jerky movements are found in carp, fish swim to the surface and stay in lateral position. Decrease in breathing movements of operculi and death follows (Velisek et al. 2011).

Pathological and Morphological Picture

Transudate in the body cavity and gastrointestinal tract is typical of triazine poisoning in carp. Transudate causes a dilatation of the body cavity (Fig. 12.4). Increased mucus production, darkening of body surface, as well as degeneration of tubular epithelial cells in caudal kidney and hepatocytes and mild proliferation of respiratory epithelium cells of secondary gill lamellae are found in carp. A decrease in hemoglobin and plasma proteins as a result of tubular cells degeneration have been noted (Svobodova et al. 1987, Velisek et al. 2011).

The effect of selected triazine pesticides, i.e., atrazine, metribuzine, simazine, terbutryne and terbutylazine, particularly on carp will be described in detail later, as their residues still remain in water bodies although they have not been used in aquaculture (Pappas and Huang 2008, Mosquin et al. 2012).

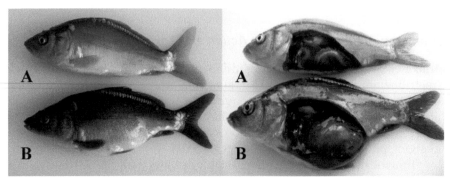

Figure 12.4. Triazine poisoning in carp. Body surface darkening, dilatated body cavity with transudate in one-year-old common carp exposed to 175.15 mg/L of metribuzine for 96 hours. A. Control. B. Experiment (Velisek et al. 2007) (Photo: J. Velisek).

Atrazine

Atrazine is a synthetic herbicide used to protect corn, sugarcane and soya; it is also used in forestry (Graymore et al. 2001). In the EU, it was banned in 2005, but is still used in the USA and other countries. In neutral and alkalic waters, it is less degradable (half-life at least 2 years) than in acid waters (Wackett et al. 2002). Its concentrations in surface waters are found in tenths to tens of µg/L (Pappas and Huang 2008, Mosquin et al. 2012). Atrazine can act as an endocrine disruptor and teratogen, and thus affects a reproduction in fish and amphibians (Storek and Kavlock 2010). Atrazine $48hLC_{50}$ value for common carp is 44.1 mg/L (Neskovic et al. 1993). Exposure of common carp to 100 µg/L of atrazine exhibited an increase in plasma cortisol (Gluth and Hanke 1985). Environmental concentrations of atrazine range usually between tenths to tens of µg/L, so the environmental exposure probably can bring no harm to carp.

Metribuzine

Metribuzine is used as a selective herbicide for the protection of cereals, potatoes and ornamental flowers (Fairchild and Sappington 2002). It degrades via photolysis in water, its half-life is 7 days. Metribuzine is found in tenths to tens of µg/L in surface water (Cerejeira et al. 2000, Haarstad and Ludvigsen 2007).

   In a 30 day toxicity test on early life stages (embryos and larvae) of common carp, metribuzine caused a retardation of growth at concentrations of 0.9, 4, 14, and 32 mg/L as well as a retardation of ontogenetic development and damages of the liver and caudal kidney at 32 mg/L. Exposure to metribuzin at 32 mg/L also led to increased mortality (Stepanova et al. 2012). Negative effect on the hematological and biochemical profile in common carp was found in the studies of Modra et al. (2008) and Velisek et al. (2009c). Modra et al. (2008) found increased hematocrit and red blood cells count in juvenile carp exposed for 28 days to 1.75 mg/L of metribuzin. In the study of Velisek et al. (2009c), exposure of two-year-old common carp to 175.1 mg/L of metribuzin caused a significant decrease in PCV, Hb, MCV, leukocrit value, and WBC as well as an increase in monocytes, band and segmented neutrophil granulocytes, developmental forms myeloid sequence, and basophils and a decrease in the levels of total protein, albumines, globulines, triglycerides, lactate dehydrogenase, lactate and phosphate and increase in glucose, ammonia and calcium in blood plasma.

In general, acute toxicity of triazines ranges from units to hundreds of mg/L. Particularly, the 96hLC$_{50}$ value of metribuzine for common carp is 175.1 mg/L (Velisek et al. 2009c). Consequently, it can be assumed that environmental exposure of carp to metribuzine provides no harm to this fish species.

Simazine

Simazine is used as a selective herbicide to protect potatoes, berries or vineyards, and as algaecide (Arufe et al. 2004). In the EU, it was banned in 2010. In fish, simazine is metabolized rapidly, and its half-life is 7 days (Niimi 1987). In waters, simazine is found in tenths to tens of µg/L (Forget et al. 2003, Woudneh et al. 2009).

Simazine is an endocrine disruptor (Fan et al. 2007). In carp, it has a negative effect on hematological and biochemical profiles (Oropesa et al. 2009, Velisek et al. 2011). Simazine inhibits detoxification enzymes (Uno et al. 2011), causes oxidative stress (Stara et al. 2012) and histopathological changes in the liver and caudal kidney (Oropesa et al. 2009, Velisek et al. 2011). The acute toxicity of simazine for fish ranges from units to hundreds of mg/L. The 48hLC$_{50}$ value for common carp is 40 mg/L (Hashimoto and Nishiuchi 1981), so there is probably no environmental risk for the fish species.

Terbutryne

Terbutryne was used as a selective herbicide for cereals and legumes as well as for the control of submerged and floating plants in water (Nilson and Unz 1977, Tomlin 2003). It was banned in the EU in 2005. Its half-life in water is 180 days (Muir 1980). In the German river Weschmitz, 5.6 µg/L of terbutryne was found (Quednow and Püttmann 2007).

Terbutryne affects hematological and biochemical profiles and detoxification enzymes and causes changes in oxidative stress parameters in carp (Velisek et al. 2011). It furthermore leads to body growth retardation and histopathological alterations of the caudal kidney and liver (Plhalova et al. 2009, Velisek et al. 2012). Acute toxicity (96hLC$_{50}$) for common carp and rainbow trout occurs at 4.0 and 3.0 mg/L, respectively (Bathe et al. 1973,

Bathe et al. 1975, Kidd and James 1991). Environmental concentrations of terbutryne are usually found in units of µg/L, so there is probably no risk for carp.

Terbutylazine

Terbutylazine is used as a broad-spectrum systemic herbicide to protect corn. It degrades via hydrolysis and its half-life exceeds 86 or 200 days in water of pH 5 or pH 7 to 9, respectively (Palm and Zetzsch 1996). In water, the concentration ranges from tenths to units of µg/L (Carafa et al. 2007, Pinto et al. 2012). The acute toxicity (96hLC$_{50}$) for rainbow trout is 3.4 mg/L (Pesticide Ecotoxicity Database 2000). Data on acute toxicity of terbutylazine on carp have not been reported.

Substituted Urea Herbicides (SUHs)

Ethiology, Mechanism of Toxic Action and Toxicity

Urea herbicides form, together with phenoxy derivatives and triazines, the most important group of agricultural pesticides due to their good herbicidal action. The urea-based derivatives are typical pre-emergence herbicides applied usually as aqueous emulsions on the soil surface. Some of them had already been introduced in the 1960s'. Almost all urea compounds with an effective herbicidal action are tri-substituted ureas containing a free iminohydrogen (Lanyi and Dinya 2005).

Most pesticides based on substituted urea are resistant to hydrolysis and oxidation under environmental conditions, but they are only moderately persistent when being degraded by soil microorganisms. Most SUHs have a low volatility. They are mobile in soil and slightly to moderately soluble in water, so they can leach into runoffs and affect wild aquatic plant growth (Wauchope et al. 1992) and further non-target organisms. Many pesticides based on substituted urea are non-toxic to slightly toxic to birds and aquatic insects and vertebrates, although, with massive accidental spills into rivers, fish kills have been reported (Svobodova et al. 1987). Most of the newer SUHs (e.g., sulphonylureas) have a very low toxicity to mammals (Kamrin 1997). Pesticides formulated from substituted ureas are of high to low toxicity to fish (48hLC$_{50}$ values are found in the range of 1 to 1,000 mg/L) (Svobodova et al. 1993).

Clinical Symptoms

The clinical symptoms of the poisoning are not specific and include increased activity, irregular respiration, uncoordinated movement, and a long period of 'distress' (Svobodova et al. 1993).

Pathological and Morphological Picture

The patho-anatomic picture is characterized by an increased amount of mucus on the darkened body surface, hyperaemia of the gills, and the presence of a small amount of exuded fluid in the body cavity of the fish (Svobodova et al. 1993).

A 96hLC$_{50}$ of isoproturon for carp was found to be at 193 mg/L (Tomlin 2000). Lakota (1978) exposed carp larvae to different herbicides (e.g., linuron, diuron, monolinuron and monuron) at the concentrations of 5, 10, 50, and 100 mg/L for 48 hours. All herbicides induced similar changes. The changes mainly concerned branchial damage and manifested in the form of hyperaemia, extravasations, ischaemia and necrotic foci in gills. In addition, significant hyperaemia of branchial leaves, edema of the dorsal laminae with desquamation of epithelial cells and focal regressive lesions of the branchial tissue were also found. This damage to gills restricts breathing capacity of carp and the damaged cell creates the possibility for development of secondary infections.

Pesticides Based on Metal Compounds

Ethiology, Mechanism of Toxic Action and Toxicity

In the past, many pesticides based on metals, such as arsenic (insecticides and rodenticides), phenylmercury (fungicides for seed treatment), tributyltin (fungicides for staining wood) or compounds of thallium (rodenticides) were used, mainly in agriculture and industry (component coating compositions, for the treatment of wood, etc.). At present, copper-based pesticides (copper sulfate and copper oxychloride particularly) are mostly used usually as fungicides, algaecides, and molluscicides.

Pesticides based on copper compounds are toxic to aquatic organisms. Lethal values range from hundreds to tens of mg/L. The toxicity is significantly influenced by physico-chemical properties of the water. Salmonids are very sensitive to pesticides based on copper showing LC$_{50}$ values that are found several orders of magnitude lower than the ones for carp (Svobodova et al. 1987).

In the majority of cases, the toxicity of metal-based pesticides to fish and clinical and patho-anatomic symptoms correspond to the signs found in fish poisoned by the respective metals (Svobodova et al. 1987).

Clinical Symptoms

Clinical symptoms of poisoning by copper-based pesticides are characterized primarily by respiratory changes, e.g., acceleration of respiratory movements, emergency respiration can be also seen in common carp (Svobodova et al. 1987).

Pathological and Morphological Picture

The patho-anatomic picture of metal-based pesticide poisoning is characterized by a strong mucus layer on the skin and gill surface. As for histological examination, damage of gills (vacuolization and decay of respiratory epithelium), liver and kidney (hyperaemia and dystrophic changes) can be found in carp (Svobodova et al. 1987).

In the course of a chronic poisoning with metal-based pesticides, quantitative and qualitative changes in white and red blood cells (reduction of leukocytes and percentage of lymphocytes, developmental disorders, large number of damaged blood cells) can be detected (Svobodova et al. 1985). These changes have resulted in lower production of antibodies and caused lower immunity of carp. For its very frequent use in aquaculture, toxic effects of Kuprikol 50 preparation are described below.

Kuprikol 50

The active substance of Kuprikol 50 is copper oxychloride ($CuCl_2.3Cu(OH)_2$) at the amount of 47.5%. At present, the preparation is used as a fungicide for the control of a wide range of fungal and bacterial diseases of fruit, vegetables and ornamental plants.

This preparation was used in pond aquaculture for the elimination of water cyanobacteria blooms (at the concentration of 0.12 mg/L), vegetation opacity (at the econcentration of 0.2 mg/L), and for antiparasitic baths (elimination of *Cryptobia branchialis*, *Trichodina* spp., *Trichodinella* spp., and *Tripartiella* spp., at the concentration of 30–50 mg/L for 15–30 minutes). The toxicity of Kuprikol 50 is strongly influenced by physico-chemical properties of water. In waters with a high content of organic substances, insoluble matter complexes are formed, which do not readily penetrate into the body and therefore they are less toxic to carp (Svobodova et al. 1985). Table 12.5 provides data on the toxicity of Kuprikol 50 to fish.

Clinical signs in carp exposed to Kuprikol 50 are as follows: initially accelerated and emergency respiration, later the attenuation phase. Typical pathological symptoms are dark skin, increased amounts of watery mucus on the body surface and hemorrhages in the eye (Svobodova et al. 1985).

Table 12.5. Lethal concentrations (LCs) of Kuprikol 50 for fish.

Fish species	48hLC_{50} (mg/L)	References
grass carp (*Ctenopharyngodon idella*)	262.0	Svobodova et al. 1985
guppy (*P. reticulata*)	129.0	Svobodova et al. 1985
common carp (*C.carpio*)	74.0	Svobodova et al. 1985
bighead carp (*Hypophthalmichthys nobilis*)	3.0–4.0	Svobodova et al. 1985
rainbow trout (*O. mykiss*)	0.78	Svobodova et al. 1985

Carp is moderately sensitive to Kuprikol 50; the mean lethal concetration is 74 mg/L (Svobodova et al. 1985). Species-specific differences between carp and other fish species may be due to differences in physiology and metabolism (detoxification).

Conclusion

In aquatic environments, a variety of chemical substances, including drugs, hormones, toxic metals or various groups of pesticides, can be detected. They themselves and their combinations affect all forms of aquatic biota. Carp can be easily affected by chemicals from the surrounding water because their body surface and gills allow the uptake of these foreign substances. They usually act either in an antagonistic or synergistic manner. Their harmful effects on fish, including carp depend, in general, on the presence and the concentration of the substances, their mode of action, and their combinations as well as on the presence of other toxic substances (for example drugs, metals, pesticides) and physico-chemical properties of water. Thus, the final toxicity of these mixtures of chemicals, including pesticides, in the aquatic environment is very unpredictable and is very difficult to evaluate. Usually, combinations of chemicals have long-term, chronic effects on the organisms of aquatic environment.

Diagnostics of Pesticide Poisoning

Fish poisoned by pesticides can be diagnosed either by direct measurement of active substances of pesticides or their metabolites in water, sediment, and tissues of the fish (such as gills, skin, liver, muscle, plasma or in the whole body homogenates in small organisms) or by indirect determination based on the inhibition of, e.g., specific enzymes. For example, fish poisoning by organophosphate and carbamate pesticides can be indicated by the inhibition of acetylcholinesterase (AchE) and butyrylcholinesterase (BchE) activity in brain and blood, respectively (Lotti 1995, Svobodova et al. 2008).

Therapy

Fish poisoned by pesticides usually do not undergo therapy. Fish are removed out of the contaminated water to clean, uncontaminated water. The transfer method is primarily used for the fish bred in tanks in fish hatcheries. Toxic effects of pesticides (e.g., the effects on hematological and biochemical profile) are in most cases reversible, the fish usually finally recover. However, it can be assumed that recovery efficiency depends on the substance and its concentration in water, exposure period, and presence of other contaminants in the aquatic environment as well as on fish health status and quality parameters of water environment. For example the carp brain AchE activity is almost completely recovered within one day after exposure to carbofuran (3 mg/L) and 15 days after exposure to chlorfenvinphos (0.24 mg/L)

(Dembélé et al. 1999). Mikulikova et al. (2013) reported complete recovery of carp within six days after terbuthylazine (3.3 mg/l) exposure. Complete recovery of carp has been reported by Masud and Singh (2013) within seven days after cypermethrin (0.01 and 0.05 µl/L) exposure.

Prevention

The prevention of carp poisoning by pesticides takes place mainly with respect to technological processes (e.g., storage, application, disposal of packaging and remnants of pesticides used by manufacturers) (Svobodova et al. 2008). In aquatic ecosystems, there is a need to reduce and prevent discharge of industrial and municipal wastewater containing pesticides into water bodies.

Acknowledgments

The authors would like to thank to Prof. Zdenka Svobodova for valuable advice and comments and Dr. Veronika Simova for the linguistic revision of the text.

This research was supported by the Ministry of Education, Youth and Sports of the Czech Republic - projects "CENAKVA" (No. CZ.1.05/2.1.00/01.0024) and "CENAKVA II" (No. LO1205 under the NPU I program).

References

Abrantes, N., R. Pereira and F. Gonçalves. 2010. Occurrence of pesticides in water, sediments, and fish tissues in a lake surrounded by agricultural lands: Concerning risks to humans and ecological receptors. Water Air Soil Poll. 212: 77–88.

Alexander, M. 1980. Helpful, harmful and fallible microorganisms: importance in transformation of chemical pollutants. American Society for Microbiology, Washington, pp. 328–332.

Ambrosoli, R., M. Negre and M. Gennar. 1996. Indications of the occurrence of enhanced biodegradation of carbofuran in some Italian soils. Soil Biol. Biochem. 28: 1749–1752.

Arufe, M.I., J. Arellano, M.J. Moreno and C. Sarasquete. 2004. Toxicity of a commercial herbicide containing terbutryn and triasulfuron to seabream (*Sparus aurata* L.) larvae: a comparison with the Microtox test. Ecotox. Environ. Safe. 59: 209–216.

Aydin, R. and K. Köprücü. 2005. Acute toxicity of diazinon on the common carp (*Cyprinus carpio* L.) embryos and larvae. Pest. Biochem. Physiol. 82: 220–225.

Balint, T., J. Ferenczy, F. Katai, I. Kiss, L. Kraczer, O. Kufcsak, G. Lang, C. Polyhos, I. Szabo, T. Szegletes and J. Nemcsok. 1997. Similarities and differences between the massive eel (*Anguilla anguilla* L.) devastations that occurred in Lake Balaton in 1991 and 1995. Ecotox. Environ. Safe. 37: 17–23.

Balint, T., T. Szegletes, Z. Szegletes, K. Halasy and J. Nemcsok. 1995. Biochemical and subcellular changes in carp exposed to the organophosphorus methidathion and the pyrethroid deltamethrin. Aquat. Toxicol. 33: 279–295.

Bathe, R., K. Sachsse, L. Ullmann, W.D. Hormann, F. Zak and R. Hess. 1975. The evaluation of fish toxicity in the laboratory. Proceedings of the European Society of Toxicology 16: 113–124.

Bathe, R., L. Ullmann and K. Sachsse. 1973. Determination of pesticide toxicity to fish. Schriftenr. Ver. Wasser-Boden-Lufthyg. Berlin-Dahlem, 37: 241–256 (in German).

Benitez, F.J., F.J. Real, J.L. Acero and C. Garcia. 2006. Photochemical oxidation processes for the elimination of phenyl-urea herbicides in waters. J. Hazard. Mater. 138: 278–287.

Blahova, J., I. Slatinska, K. Kruzikova, J. Velisek and Z. Svobodova. 2007. The effect of deltamethrin on activity of glutathione S-transferase of common carp (*Cyprinus carpio*). Chem. Listy 101: 168–169.

Bottoni, P., P. Grenni, L. Lucentini and A.B. Caracciolo. 2013. Terbuthylazine and other triazines in Italian water resources. Microchem. J. 107: 136–142.

Bradbury, S.P. and J.R. Coats. 1989. Toxicokinetics and toxicodynamics of pyrethroid insecticides in fish. Environ. Toxicol. Chem. 8: 373–380.

Brundage, K.M. and J.B. Barnett. 2010. Immunotoxicity of pesticides. pp. 483–498. *In*: R. Kriger (ed.). Hayes Handbook of Pesticides Toxicology. Elsevier, Amsterdam.

Buccafusco, R.J., S.J. Ells and G.A. Cary. 1977. Acute toxicity of NRDC 161 to bluegill (*Lepomis macrochirus*) under dynamic test conditions, E.G. & G. Bionomics Aquatic Toxicology Laboratory.

Calta, M. and M.S. Ural. 2004. Acute toxicity of the synthetic pyrethroid deltamethrin to young mirror carp, *Cyprinus carpio*. Fresen. Environ. Bull. 13: 1179–1183.

Carafa, R., J. Wollgast, E. Canuti, J. Ligthart, S. Dueri, G. Hanke, S.J. Eisenreich, P. Viaroli and J.M. Zaldivar. 2007. Seasonal variations of selected herbicides and related metabolites in water, sediment, seaweed and clams in the Sacca di Goro coastal lagoon (Northern Adriatic). Chemosphere 69: 1625–1637.

Carballo, M., J.A. Jimenez, A. Torre, J. Roset and M.J. Munoz. 2005. A survey of potential stressor-induced physiological changes in carp (*Cyprinus carpio*) and barbel (*Barbus bocagei*) along the Tajo River. Environ. Toxicol. 20: 119–125.

Cerejeira, M.J., E. Silva, S. Batista, A. Trancoso, M.S.L. Centeno and A. Silva-Fernandes. 2000. Simazine, metribuzine and nitrates in ground water of agricultural areas of Portugal. Environ. Toxicol. Chem. 75: 245–253.

Chambers, H.W., E.C. Meek and J.E. Chambers. 2010a. Chemistry of organophosphorus insecticides. pp. 1395–1398. *In*: R. Krieger (ed.). Hayes Handbook of Pesticide Toxicology. Elsevier, Amsterdam.

Chambers, J.E., E.C. Meek and H.W. Chambers. 2010b. The metabolism of organophosphorus insecticides. pp. 1399–1407. *In*: R. Krieger (ed.). Hayes Handbook of Pesticide Toxicology. Elsevier, Amsterdam.

Chebbi, S.G. and M. David. 2010. Respiratory responses and behavioural anomalies of the carp *Cyprinus carpio* under quinalphos intoxication in sublethal doses. Scienceasia 36: 12–17.

Chowdhury, A.R., S. Banik, B. Uddin, M. Moniruzzaman, N. Karim and S.H. Gan. 2012. Organophosphorus and carbamate pesticide residues detected in water samples collected from paddy and vegetable fields of the Savar and Dhamrai Upazilas in Bangladesh. Int. J. Environ. Res. Public Health. 9: 3318–3329.

Chowdhury, A.R., A.K. Gautam and H. Venkatakrishna-Bhatt. 1990. DDT (2,2-bis(*p* chlorophenyl) 1,1,1-trichloroethane) induced structural changes in adrenal glands of rats. Bull. Environ. Contam.Toxicol. 45: 193–196.

Cremlyn, R.J.W. 1978. Pesticides: Preparation and Mode of Action. Wiley, Chichester, New York.

Das, B.K. and S.C. Mukherjee. 2000. A histopathological study of carp (*Labeo rohita*) exposed to hexachlorocyclohexane. Vet. Arch. 70: 169–180.

DeLorenzo, M.E., G.I. Scott and P.E. Ross. 2001. Toxicity of pesticides to aquatic microorganisms: A review. Environ. Toxicol. Chem. 20: 84–98.

Dembélé, K., E. Haubruge and C. Gaspar. 1999. Recovery of acetylcholinesterase activity in the common carp (*Cyprinus carpio* L.) after inhibition by organophosphate and carbamate compounds. Bull. Environ. Contam. Toxicol. 62: 731–742.

Dembélé, K., E. Haubruge and C. Gaspar. 2000. Concentration effects of selected insecticides on brain acetylcholinesterase in the common carp (*Cyprinus carpio* L.). Ecotox. Environ. Safe. 45: 49–54.

DeMel, G.W.J.L.M.V.T.M. and A. Pathiratne. 2005. Toxicity assessment of insecticides commonly used in rice pest management to the fry of common carp, *Cyprinus carpio*, a food fish culturable in rice fields. J. Appl. Ichthyol. 21: 146–150.

Dobsikova, R. 2003. Acute toxicity of carbofuran to selected species of aquatic and terrestrial organisms. Plant Prot. Sci. 39: 103–108.

Dobsikova, R., J. Velisek, T. Wlasow, P. Gomulka, Z. Svobodova and L. Novotny. 2006. Effects of cypermethrin on some haematological, biochemical and histopathological parameters of common carp (*Cyprinus carpio* L.). Neuroendocrinol. Lett. 27: 101–105.

Drenner, R.W., K.D. Hoagland, J.D. Smith, W.J. Barcellona, P.C. Johnson, M.A. Palmieri and J.F. Hobson. 1993. Effects of sediment-bound bifenthrin on gizzard shad and plankton in experimental tank mesocosms. Environ. Toxicol. Chem. 12: 1297–1306.

Eichelberger, J.W. and J.J. Lichtenberg. 1971. Persistence of pesticides in river water. Environ. Sci. Technol. 5: 541–544.

El-Sayed, Y.S., T.T. Saad and S.M. El-Bahr. 2007. Acute intoxication of deltamethrin in monosex Nile tilapia, *Oreochromis niloticus* with special reference to the clinical, biochemical and haematological effects. Environ. Toxicol. Pharmacol. 24: 212–217.

Fairchild, J.F. and L.C. Sappington. 2002. Fate and effects of the triazine herbicide metribuzin in experimental pond mesocosms. Arch. Environ. Contam. Toxicol. 43: 198–202.

Falfushynska, H.I. and O.B. Stolyar. 2009. Responses of biochemical markers in carp *Cyprinus carpio* from two field sites in Western Ukraine. Ecotox. Environ. Safe. 72: 729–736.

Fan, W.Q., T. Yanase, H. Morinaga, S. Ondo, T. Okabe, M. Nomura, T. Komatsu, K.I. Morohashi, T.B. Hayes, R. Takayanagi and H. Nawata. 2007. Atrazine-induced aromatase expression is SF-1 dependent: Implications for endocrine disruption in wildlife and reproductive cancers in humans. Environ. Health Perspect. 115: 720–727.

Forget, J., B. Beliaeff and G. Bocquene. 2003. Acetylcholinesterase activity in copepods (*Tigriopus brevicornis*) from the Vilaine River estuary, France, as a biomarker of neurotoxic contaminants. Aquat. Toxicol. 62: 195–204.

Fukuto, T.R. 1972. Metabolism of carbamate insecticides. Drug Metab. Rev. 1: 117–150.

Garcia, S.J., M. Aschner and T. Syversen. 2006. Interspecies variation in toxicity of cholinesterase inhibitors. pp. 145–158. *In*: R.C. Gupta (ed.). Toxicology of Organophosphate and Carbamate Compounds. Elsevier Academic Press, Burlington, Massachusetts.

Gluszczak, L., V.L. Loro, A. Pretto, B.S. Moraes, A. Raabe, M.F. Duarte, M.B. da Fonseca, C.C. de Menezes and D.M. de Sousa Valladão. 2011. Acute exposure to glyphosate herbicide affects oxidative parameters in Piava (*Leporinus obtusidens*). Arch. Environ. Contam. Toxicol. 61: 624–630.

Gluth, G. and W.A. Hanke. 1985. A comparison of physiological changes in carp, *Cyprinus carpio*, induced by several pollutants at sublethal concentrations. I. The dependency on exposure time. Ecotox. Environ. Safe. 9: 179–188.

Golombieski, J.I., E. Marchesan, E.R. Camago, J. Salbego, J.S. Baumart, V.L. Loro, S.L.D. Machado, R. Zanella and B. Baldisserotto. 2008. Acetylcholinesterase enzyme activity in carp brain and muscle after acute exposure to diafuran. Sci. Agric. 65: 340–345.

Graymore, M., F. Stagnitti and G. Allinson. 2001. Impacts of atrazine in aquatic ecosystems. Environ. Int. 26: 483–495.

Guy, D. 2000. Bifenthrin with *Pimephales promelas* in an acute definitive test. Aquatic Toxicology laboratory Report P-2161-2, California Department of Fish and Game, Elk Grove.

Haarstad, K. and G.H. Ludvigsen. 2007. Ten years of pesticide monitoring in Norwegian ground water. Ground Water Monit. Remediat. 27: 75–89.

Hai, D.Q., S.I. Varga and B. Matkovics. 1997. Organophosphate effects on antioxidant system of carp (*Cyprinus carpio*) and catfish (*Ictalurus nebulosus*). Comp. Biochem. Physiol. 117C: 83–88.

Halappa, R. and M. David. 2009. Behavioural responses of the freshwater fish, *Cyprinus carpio* (Linnaeus) following sublethal exposure to chlorpyrifos. Turk. J. Fish. Aquat. Sci. 9: 233–238.

Hashimoto, Y. and Y. Nishiuchi. 1981. Establishment of bioassay methods for the evaluation of acute toxicity of pesticides to aquatic organisms. J. Pestic. Sci. 6: 257–264.

Haya, K. 1989. Toxicity of pyrethroid insecticides to fish. Environ. Toxicol. Chem. 8: 381–391.

Hejduk, I. and Z. Svobodova. 1980. Acute toxicity of carbamate-based pesticides for fish. Acta Vet. Brno 49: 251–257.

Horberg, J.R. 1983a. Acute toxicity of FMC 54800 technical to rainbow trout (*Salmo gairdneri*). FMC Study No. A83/967, Wareham, MA. U.S. Environmental Protection Agency. EPA MRID: 00132539.

Horberg, J.R. 1983b. Acute toxicity of FMC 54800 technical to bluegill (*Lepomis macrochirus*). FMC Study No. A83-987. EG &G, Bionomics study, Environmental Protection Agency. EPA MRID: 00132536.

Huang, D.J., Y.M. Zhang, G. Song, J. Long, J.H. Liu and W.H. Ji. 2007. Contaminants-induced oxidative damage on the carp *Cyprinus carpio* collected from the upper Yellow River, China. Environ. Monit. Assess. 128: 483–488.

Ismail, M., R. Ali, T. Ali, U. Waheed and Q.M. Khan. 2009. Evaluation of the acute toxicity of profenofos and its effects on the behavioural pattern of fingerling common carp (*Cyprinus carpio* L., 1758). Bull. Environ. Contam. Toxicol. 82: 569–573.

Johnson, W.W. and M.T. Finley. 1980. Handbook of acute toxicity of chemicals to fish and aquatic invertebrates. United States Department of the Interior Fish and Wildlife Service, Resource publication 137, Washington, 106 pp.

Kamalaveni, K., V. Gopal, U. Sampson and D. Aruna. 2003. Recycling and utilization of metabolic wastes for energy production is an index of biochemical adaptation of fish under environmental pollution stress. Environ. Monit. Assess. 86: 255–264.

Kamrin, M. 1997. Pesticide Profiles—Toxicity, Environmental Impact, and Fact. CRC-Lewis Publishers, Boca Raton, Florida.

Kelce, W.R., C.R. Stone, S.C. Laws, L.E. Gray, J.A. Kemppainen and E.M. Wilson. 1995. Persistent DDT metabolite p,p'–DDE is a potent androgen receptor antagonist. Nature 375: 581–585.

Kidd, H. and D.R. James. 1991. The Agrochemicals Handbook. Royal Society of Chemistry Information Services, Cambridge, UK.

Kozlowski, F. 1983. Toxic effects of DDT, lindane and toxaphene on the fry of the carp, *Cyprinus carpio*, as revealed by an acute test. Folia Biol. –Krakow 2: 93–100.

Kuivila, K.M. and C.G. Foe. 1995. Concentrations, transport and biological effects of dormant spray pesticides in the San Francisco Estuary, California. Environ. Toxicol. Chem. 14: 1141–1150.

Lakota, S. 1978. Studies on the toxicity of the herbicides diuron, linuron, monolinuron and moniron. Med. Weter. 34: 20–22.

Lanyi, K. and Z. Dinya. 2005. Photodegradation study for assessing the environmental fate of some triazine-, urea- and thiocarbamate-type herbicides. Microchem. J. 80: 79–87.

Lepailleur, H. and A. Chambon. 1984. Etude de la toxicité létale du produit technique deltaméthrine vis à vis du poisson-zebré. Catiche Productions (in French).

Liu, T.L., Y.S. Wang and J.H. Yen. 2005. Separation of bifenthrin enantiomers by chiral HPLC and determination of their toxicity to aquatic organism. J. Food Drug Anal. 12: 357–360.

Logaswamy, S., G. Radha, S. Subhashini and K. Logankumar. 2007. Alterations in the levels of ions in blood and liver of freshwater fish, *Cyprinus carpio* var. communis exposed to dimethoate. Environ. Monit. Assess. 131: 439–444.

Lotti, M. 1995. Cholinesterase inhibition: complexities in interpretation. Clin. Chem. 41: 1814–1818.

Luskova, V., M. Svoboda and J. Kolarova. 2002. The effect of diazinon on blood plasma biochemistry in carp (*Cyprinus carpio* L.). Acta Vet. Brno 71: 117–123.

Machova, J., M. Prokes, Z. Svobodova, V. Zlabek, M. Penaz and V. Barus. 2007. Toxicity of Diazinon 60 EC for *Cyprinus carpio* and *Poecilia reticulata*. Aquacult. Int. 15: 267–276.

Machova, J., Z. Svobodova, R. Faina, J. Kemrova and T. Kuthan. 1985. The acute toxicity of carbamate-based herbicides to aquatic organisms. Prace VURH Vodnany 14: 63–73.

Marigoudar, S.R., R.N. Ahmed and M. David. 2009. Cypermethrin induced respiratory and behavioural responses of the freshwater teleost, *Labeo rohita* (Hamilton). Vet. Arch. 79: 583–590.

Masud, S., I.J. Singh. 2013. Effect of Cypermethrin on some hematological parameters and prediction of their recovery in a freshwater Teleost, *Cyprinus carpio*. African J. Environ. Sci. Technol. 7: 852–856.

Mayer, F.L., D.J. Versteeg, M.J. McKee, L.C. Folmar, R.L. Graney and D.C. McCume. 1992. Physiological and non-specific biomarkers. pp. 5–85. *In*: R.J. Huggett, R.A. Kimerle, P.M. Mehrle Jr. and H.L. Bergman (eds.). Biomarkers: Biochemical, Physiological and Histological Markers of Anthropogenic Stress. Lewis, Boca Raton.

McAllister, W.A. 1988. Full life cycle toxicity of 14C-FMC 54800 to the fathead minnow (*Pimephales promelas*) in a flow-through system. FMC Study No. A86-2100. Analytical Bio-Chemistry Laboratories, Inc. Columbia, Missouri.

Mertlik, J. and Z. Svobodova. 1973. Toxicity of pesticides on the base of chlorinated carbohydrates and phenols and their indication in fish. Bull. VURH Vodnany 9: 3–18. (in Czech).

Mikulikova, I., H. Modra, J. Blahova, K. Kruzikova, P. Marsalek, I. Bedanova and Z. Svobodova. 2013. Recovery ability of common carp (*Cyprinus carpio*) after a short-term exposure to terbuthylazine. Polish J. Vet. Sci. 16: 17–23.

Modra, H., I. Haluzova, J. Blahova, M. Havelkova, K. Kruzikova, P. Mikula and Z. Svobodova. 2008. Effects of subchronic metribuzin exposure on common carp (*Cyprinus carpio*). Neuroendocrinol. Lett. 29: 669–674.

Mosquin, P., R.W. Whitmore and W. Chen. 2012. Estimation of upper centile concentrations using historical atrazine monitoring data from community water systems. J. Environ. Qual. 41: 834–844.

Mueller-Beilschmidt, D. 1990. Toxicology and environmental fate of synthetic pyrethroids. J. Pestic. Reform 10: 32–37.

Mulla, M.S., H.A. Navvab-Gojrati and H.A. Darwazeh. 1978. Toxicity of mosquito larvicidal pyrethroids to four species of freshwater fishes. Environ. Entomol. 7: 428–430.

Muir, D.C. 1980. Determination of terbutryn and its degradation products in water sediments, aquatic plants, and fish. J. Agr. Food Chem. 28: 714–719.

Munn, M.D., R.J. Gilliom, P.W. Moran and L.H. Nowell. 2006. Pesticide toxicity index for freshwater aquatic organisms. Scientific investigations report 2006-5148. U.S. Geological Survey, Reston, Virginia, 81 pp.

Narahashi, T. 2010. Neurophysiological effects of insecticides. pp. 799–818. *In*: R. Kriger (ed.). Hayes Handbook of Pesticides Toxicology. Elsevier, Amsterdam.

Neskovic, N.K., I. Elezovic, V. Karan, V. Poleksic and M. Budimir. 1993. Acute and subacute toxicity of atrazine to carp (*Cyprinus carpio* L.). Ecotox. Environ. Safe. 25: 173–182.

Niimi, A.J. 1987. Biological half-lives of chemicals in fishes. Rev. Environ. Contam. T. 99: 1–46.

Nilson, E.L. and R.F. Unz. 1977. Antialgal substances for iodine-disinfected swimming pools. Appl. Environ. Microbiol. 34: 815–822.

Noga, E.J. 1996. Fish Disease: Diagnosis and Treatment. Mosby, St. Louis, Missouri.

OECD 203. 1992. Fish, Acute Toxicity Test. 9 pp.

OECD 212. 1998. Fish, Short-term Toxicity Test on Embryo and Sac-fry Stages. 20 pp.

Oropesa, A.L., J.P. Garcia-Cambero, L. Gomez, V. Roncero and F. Soler. 2009. Effect of long-term exposure to simazine on histopathology, hematological, and biochemical parameters in *Cyprinus carpio*. Environ. Toxicol. 24: 187–199.

Oruc, E. 2011. Effects of diazinon on antioxidant defense system and lipid peroxidation in the liver of *Cyprinus carpio* (L.). Environ. Toxicol. 26: 571–578.

Palm, W.U. and C. Zetzsch. 1996. Investigation of the photochemistry and quantum yields of triazines using polychromatic irradiation and UV-spectroscopy as analytical tool. Int. J. Environ. Anal. Chem. 65: 313–329.

Pappas, E.A. and C. Huang. 2008. Predicting atrazine levels in water utility intake water for MCL compliance. Environ. Sci. Technol. 42: 7064–7068.

Pawar, K.R. 1994. Toxic and teratogenic effects of fenitrothion, BHC and carbofuran on the embryonic development of *Cyprinus carpio communis*. Environ. Ecol. 12: 284–287.

Peakall, D.B. 1969. Effect of DDT on calcium uptake and vitamin D metabolism in birds. Nature 224: 1219–1220.

Pesticide Ecotoxicity Database. 2000. Office of Pesticide Programs, Environmental Fate and Effects Division. U.S. EPA, Washington, DC.

Pham, H.C., C. Navarro-Dalmasure, P. Clavel, G. VanHaverbeke and S.L. Cheav. 1984. Toxicological studies of deltamethrin. Int. J. Tissue React. 6: 127–133.

Pinto, A.P., C. Serrano, T. Pires, E. Mestrinho, L. Diasa, D.M. Teixeira and A.T. Caldeira. 2012. Degradation of terbuthylazine, difenoconazole and pendimethalin pesticides by selected fungi cultures. Sci. Total Environ. 435-436: 402–410.

Plhalova, L., S. Macova, I. Haluzova, A. Slaninova, P. Dolezelova, P. Marsalek, V. Pistekova, I. Bedanova, E. Voslarova and Z. Svobodova. 2009. Terbutryn toxicity to *Danio rerio*: Effects of subchronic exposure on fish growth. Neuroendocrinol. Lett. 30: 242–247.

Polat, H., F.U. Erkoc, R. Viran and O. Kocak. 2002. Investigation of acute toxicity of beta-cypermethrin on guppies *Poecilia reticulata*. Chemosphere 49: 39–44.

Quednow, K. and W. Püttmann. 2007. Monitoring terbutryn pollution in small rivers of Hesse, Germany. J. Environ. Monitor. 9: 1337–1343.

Ray, A.K. and M.C. Ghosh. 2006. Aquatic toxicity of carbamates and organophosphates. pp. 657–672. *In*: R.C. Gupta (ed.). Toxicology of Organophosphate and Carbamate Compounds. Elsevier Academic Press, Burlington.

Rebich, R.A., R.H. Coupe and E.M. Thurman. 2004. Herbicide concentrations in the Mississippi River Basin—the importance of chloroacetanilide herbicide degradates. Sci. Total Environ. 321: 189–199.

Revilla, E., A. Mora, J. Cornejo and M.C. Hermosin. 1996. Persistence and degradation of carbofuran in Spanish soil suspensions. Chemosphere 32: 1585–1598.

Roberts, T.R. and D.H. Hutson. 1999. Metabolic Pathway of Agrochemicals. Part 2: Insecticides and Fungicides. The Royal Society of Chemistry, Cambridge, UK.

Rossi, S.C., M.D. da Silva, L.D.S. Piancini, C.A.O. Ribeiro, M.M. Cestari and H.C.S. de Assis. 2011. Sublethal effects of waterborne herbicides in tropical freshwater fish. Bull. Environ. Contam. Toxicol. 87: 603–607.

Saha, S. and A. Kaviraj. 2003. Acute toxicity of synthetic pyrethroid cypermethrin to freshwater catfish *Heteropneustes fossilis* (Bloch). Int. J. Toxicol. 22: 325–328.

Sarikaya, R. 2009. Investigation of acute toxicity of alpha-cypermethrin on adult Nile tilapia (*Oreochromis niloticus* L.). Turk. J. Fish. Aquat. Sci. 9: 85–89.

Sepici-Dincel, A., R. Sarikaya, M. Selvi, D. Sahin, C.K. Benli and S. Atalay-Vural. 2007. How sublethal fenitrothion is toxic in carp (*Cyprinus carpio* L.) fingerlings. Toxicol. Mech. Method. 17: 489–495.

Shan, G., R.P. Hammer and J.A. Ottea. 1997. Biological activity of pyrethroid analogs in pyrethroid-susceptible and -resistant tobacco budworms, *Heliothis virescens* (F.). J. Agr. Food Chem. 45: 4466–4473.

Singh, P.B. and V. Singh. 2008. Bioaccumulation of hexachlorocyclohexane, dichlorodiphenyltrichloroethane, and estradiol-17 beta in catfish and carp during the pre-monsoon season in India. Fish Physiol. Biochem. 4: 25–36.

Smalling, K.L., K.M. Kuivila, J.L. Orlando, B.M. Phillips, B.S. Anderson, K. Siegler, J.W. Hunt and M. Hamilton. 2013. Environmental fate of fungicides and other current-use pesticides in a central California estuary. Mar. Pollut. Bull. 73: 144–153.

Smith, A.G. 1991. Chlorinated hydrocarbon insecticides. pp. 731–915. *In*: W.J. Hayes and E.R. Laws (eds.). Handbook of Pesticide Toxicology. Academic Press, San Diego.

Smith, A.G. 2000. How toxic is DDT? Lancet 356: 267–268.

Srivastava, S.K., R. Jaiswal and A.K. Srivastav. 1997. Lethal toxicity of deltamethrin (Decis) to a freshwater fish, *Heteropneustes fossilis*. J. Adv. Zool. 18: 23–26.

Stara, A., J. Machova and J. Velisek. 2012. Effect of chronic exposure to simazine on oxidative stress and antioxidant response in common carp (*Cyprinus carpio* L.). Environ. Toxicol. Pharmacol. 33: 334–343.

Stepanova, S., P. Dolezelova, L. Plhalova, M. Prokes, P. Marsalek, M. Skoric and Z. Svobodova. 2012. The effects of metribuzin on early life stages of common carp (*Cyprinus carpio*). Pest. Biochem. Physiol. 103: 152–158.

Stockholm Convention. 2012. *In*: <http://chm.pops.int/Default.aspx>. Visited 7th December 2012.

Storek, E.T. and R. Kavlock. 2010. Pesticides as endocrine-disrupting chemicals. pp. 551–569. *In*: R. Krieger (ed.). Hayes Handbook of Pesticide Toxicology. Elsevier, Amsterdam.

Svoboda, M., V. Luskova, J. Drastichova and V. Zlabek. 2001. The effect of diazinon on haematological indices of common carp (*Cyprinus carpio* L.). Acta Vet. Brno 70: 457–465.

Svobodova, Z., R. Faina and B. Vykusova. 1985. The use of Kuprikol 50 in pond aquaculture. Methods no. 19, VURH JU, Vodnany (in Czech).

Svobodova, Z., J. Gelnerova, J. Justyn, V. Krauper, J. Machova, L. Simanov, I. Valentova and E. Wohlgemuth. 1987. Toxicology of aquatic organisms. SZN, Prague (in Czech).

Svobodova, Z., R. Lloyd, J. Machova and B. Vykusova. 1993. Water quality and fish health. EIFAC Technical Paper. No. 54. FAO, Rome.

Svobodova, Z., J. Kolarova, S. Navratil, T. Vesely, P. Chloupek, J. Tesarcik and J. Citek. 2007. Deseases of freshwater and aquarium fish. Informatorium Praha, Prague (in Czech).

Svobodova, Z., J. Machova and H. Kroupova. 2008. Fish poisoning. pp. 201–217. *In*: Z. Svobodova (ed.). Veterinary Toxicology in Clinical Practice. Profi Press, Prague (in Czech).

Tang, J., R.L. Rose and J.E. Chambers. 2006. Metabolism of organophosphorus and carbamate pesticides. pp. 127–143. *In*: R.C. Gupta (ed.). Toxicology of Organophosphate and Carbamate Compounds. Elsevier Academic Press, Burlington, Massachusetts.

Tilson, H.A. 1998. Developmental neurotoxicology of endocrine disruptors and pesticides: identification of information gaps and research needs. Environ. Health Persp. 106: 870–811.

Tomlin, C. 2003. The Pesticide Manual. A World Compendium. British Crop Protection Council, Hampshire, 1344 pp.

Tomlin, C.D.S. 2000. The Pesticide Manual. British Crop Protection Council. Farnham Surrey GU9 7PH, UK, 1250 pp.

Treasurer, J.W. and S.L. Wadsworth. 2004. Interspecific comparison of experimental and natural routes of *Lepeophtheirus salmonis* and *Caligus elongatus* challenge and consequences for distribution of chalimus on salmonids and therapeutant screening. Aquacult. Res. 35: 773–783.

Uno, T., S. Kaji, T. Goto, H. Imaishi, M. Nakamura, K. Kanamaru, H. Yamagata, Y. Kaminishi and T. Itakura. 2011. Metabolism of the herbicides chlorotoluron, diuron, linuron, simazine, and atrazine by CYP1A9 and CYP1C1 from Japanese eel (*Anguilla japonica*). Pest. Biochem. Physiol. 101: 93–102.

Ural, M.S. and N. Saglam. 2005. A study on the acute toxicity of pyrethroid deltamethrin on the fry rainbow trout (*Oncorhynchus mykiss* Walbaum, 1792). Pest. Biochem. Physiol. 83: 124–131.

Vaishnav, D.D. and J.J. Yurk. 1990. Cypermethrin (FMC 45806): Acute toxicity to rainbow trout (*Oncorhynchus mykiss*) under flow-through test conditions. FMC Corporation study number A89-3109-01. Laboratory project ID: ESE No. 3903026-0750-3140, Gainesville.

van der Oost, R., J. Beyer and N.P.E. Vermeulen. 2003. Fish bioaccumulation and biomarkers in environmental risk assessment: a review. Environ. Toxicol. Pharmacol. 13: 57–149.

Velisek, J., R. Dobsikova, Z. Svobodova, H. Modra and V. Luskova. 2006a. Effect of deltamethrin on the biochemical profile of common carp (*Cyprinus carpio* L.). Bull. Environ. Contam. Toxicol. 76: 992–998.

Velisek, J., T. Wlasow, P. Gomulka, Z. Svobodova, R. Dobsikova, L. Novotny and M. Dudzik. 2006b. Effects of cypermethrin on rainbow trout (*Oncorhynchus mykiss*). Vet. Med. 51: 469–476.

Velisek, J., G. Polesczuk and Z. Svobodova. 2007. Effect of metribuzine on biochemical profile of common carp (*Cyprinus carpio*) and rainbow trout (*Oncorhynchus mykiss*). Bull. RIFCH Vodnany 48: 126–132. (in Czech).

Velisek, J., Z. Svobodova and J. Machova. 2009a. Effects of bifenthrin on some haematological, biochemical and histopathological parameters of common carp (*Cyprinus carpio* L.). Fish Physiol. Biochem. 35: 583–590.

Velisek, J., Z. Svobodova and V. Piackova. 2009b. Effects of acute exposure to bifenthrin on some haematological, biochemical and histopathological parameters of rainbow trout (*Oncorhynchus mykiss*). Vet. Med. 54: 131–137.

Velisek, J., Z. Svobodova, V. Piackova and E. Sudova. 2009c. Effects of acute exposure to metribuzin on some haematological, biochemical and histopathological parameters of common carp (*Cyprinus carpio* L.). Bull. Environ. Contam. Toxicol. 82: 492–495.

Velisek, J., A. Stara and Z. Svobodova. 2011. The effects of pyrethroid and triazine pesticides on fish physiology. pp. 377–402. *In*: M. Stoytcheva (ed.). Pesticides in the Modern World— Pests Control and Pesticides Exposure and Toxicity Assessment. InTech Open Access Publisher, Rijeka.

Velisek, J., A. Stara, J. Machova, P. Dvorak, E. Zuskova, M. Prokes and Z. Svobodova. 2012. Effect of terbutryn at environmental concentrations on early life stages of common carp (*Cyprinus carpio* L.). Pest. Biochem. Physiol. 102: 102–108.

Viran, R., F.Ü. Erkoc, H. Polat and O. Kocak. 2003. Investigation of acute toxicity of deltamethrin on guppies (*Poecilia reticulata*). Ecotox. Environ. Safe. 55: 82–85.

Vryzas, Z., C. Alexoudis, G. Vassiliou, K. Galanis and E. Papadopoulou-Mourkidou. 2011. Determination and aquatic risk assessment of pesticide residues in riparian drainage canals in northeasternGreece. Ecotox. Environ. Safe. 74: 174–181.

Wackett, L.P., M.J. Sadowsky, B. Martinez and N. Shapir. 2002. Biodegradation of atrazine and related s-triazine compounds: from enzymes to field studies. Appl. Microbiol. Biot. 58: 39–45.

Wauchope, R.D., T.M. Buttler, A.G. Hornsby, P.W.M. Augustijn-Beckers and J.P. Burt. 1992. The SCS/ARS/CES pesticide properties database for environmental decision-making. Rev. Environ. Contam. Toxicol. 123: 1–155.

Whyte, J.J., R.E. Jung, C.J. Schmitt and D.E. Tillitt. 2000. Ethoxyresorufin-O-deethylase (EROD) activity in fish as a biomarker of chemical exposure. Crit. Rev. Toxicol. 30: 347–570.

Woudneh, M.B., Z. Ou, M. Sekela, T. Tuominen and M. Gledhill. 2009. Pesticide multiresidues in waters of the Lower Fraser Valley, British Columbia, Canada. Part II. Groundwater. J. Environ. Qual. 38: 948–954.

Yadav, R.S., H.C. Srivastava, T. Adak, N. Nanda, B.R. Thapar, C.S. Pant, M. Zaim and S.K. Subbarao. 2003. House-scale evaluation of bifenthrin indoor residual spraying for malaria vector control in India. J. Med. Entomol. 40: 58–63.

Yu, S.J. 2008. The Toxicology and Biochemistry of Insecticides. CRC Press, Boca Raton, Florida.

Zhang, X., J.B. Knaak, R. Tornero-Velez, J.N. Blancato and C.C. Dary. 2010a. Application of physiologically based pharmacokinetic/pharmacodynamic modeling in cumulative risk assessment for N-methyl carbamate insecticides. pp. 1591–1605. In: R. Krieger (ed.). Hayes Handbook of Pesticide Toxicology. Elsevier, Amsterdam.

Zhang, Z.Y., X.Y. Yu, D.L. Wang, H.J. Yan and X.J. Liu. 2010b. Acute toxicity to zebrafish of two organophosphates and four pyrethroids and their binary mixtures. Pest Manag. Sci. 66: 84–89.

Zitko, V., D.W. McLeese, C.D. Metcalfe and W.G. Carson. 1979. Toxicity of permethrin, decamethrin, and related pyrethroids to salmon and lobster. Bull. Environ. Contam. Toxicol. 21: 338–343.

13

Impact of Natural Toxins on Common Carp

Constanze Pietsch

Fish are usually exposed to a huge variety of naturally occurring substances, such as secondary metabolites of fungi, toxins produced by algae, toxins and anti-nutritional factors from higher plants and toxins of molluscs. The most important natural compounds displaying negative effects on carp (*Cyprinus carpio* L.) will be reviewed below.

Mycotoxins

More than 400 different substances produced by different fungi have been reported that can be clustered into five major classes: aflatoxins, ochratoxins, fumonisins, zearalenone and trichothecenes (CAST 2003). These groups of toxins are largely prevalent in feedstuffs for animals (Santos et al. 2010). Less frequently, mycotoxins, such as moniliformin, cyclopiazonic acid, citrinin and sterigmatocystin, are detecteted in animal feed (e.g., Balachandran and Parthasarathy 1996, Vesonder et al. 2000, Labuda et al. 2005, Veršilovskis and De Saeger 2010). Mycotoxins, depending on their chemical structure, show diverse effects on fish and their potential impact on carp will be summarized later.

Materials used for production of feeds in aquaculture may contain several mycotoxins (Abdelhamid 1990, Mahmoud 1993, Scudamore et al. 1998, Ellis et al. 2000, Binder et al. 2007, Weidenbörner 2007, Rodrigues and Naehrer 2012, Streit et al. 2012). Accordingly, different mycotoxins have been frequently found in commonly used ingredients (Table 13.1).

Zurich University of Applied Sciences (ZHAW), Institute of Natural Resource Sciences (IUNR), Grüental
P.O. Box, CH-8820 Waedenswil, Switzerland.
Email: constanze.pietsch@zhaw.ch

Table 13.1. Mycotoxins frequently occurring in feed ingredients (according to Abdelhamid 1990, Mahmoud 1993, Scudamore et al. 1998, Ellis et al. 2000, Binder et al. 2007, Weidenbörner 2007, Rodrigues and Naehrer 2010, Streit et al. 2012).

Feed ingredient	Mycotoxins
fishmeal	aflatoxins, ochratoxin-A, DON, ZEN
grains	aflatoxins, citrinin, ochratoxin-A, DON, ZEN
maize and maize flour	aflatoxins, fumonisins, moniliformin, ochratoxin-A, DON, T-2 toxin, ZEN
soybean	aflatoxins, ochratoxin-A, and DON
cottonseed meal or bran	aflatoxin-B1, citrinin, ochratoxin-A, DON and ZEN
peanut meal and sunflower meal	aflatoxins, cyclopiazonic acid, ochratoxin-A, DON and ZEN
rice bran	AFB_1, ochratoxin-A, citrinin, DON, ZEN, cyclopiazonic acid and moniliformin

Figure 13.1. Commercial feeds with different ingredients can be used for carp feeding.

Fusarium toxins (DON, nivalenol, ZEN, T-2, fumonisins, moniliformin, fusaric acid) are formed on the field before harvest of crops while *Penicillium* and *Aspergillus* species generate toxins, such as aflatoxins, ochratoxins, patulin, citrinin, cyclopaizonic acid, sterigmatocystin, and gliotoxin are produced mainly during the storage of feed ingredients and feeds (Frisvad and Samson 1991, Salama 2007, Barbosa et al. 2013).

Products intended for animal feed production including finished fish feeds contain mycotoxins (Måge et al. 2009, Santos et al. 2010, Pietsch et al. 2013) which can endanger fish health or even human health and the environment. Consequently, governmental regulations for products intended for animal feed have been established. Concentrations of mycotoxins in products that are intended for animal feeding are regulated by recommendations of the European Commission (e.g., Commission Recommendation 2006/576/EC of 17th August 2006 on the presence of DON, ZEN, T-2 and HT-2 toxins and fumonisins in products intended for animal feeding) and by separate recommendations in other countries (Van Egmond and Jonker 2004). However, effects on fish health have rarely been considered when recommendations have been released. This is mostly due to the fact that knowledge on effects of many mycotoxins has been gained only recently. The effects of different mycotoxins on carp, or other fish species if necessary information on carp is lacking, will be reviewed here.

Aflatoxins

Aflatoxins are mainly produced by naturally occurring moulds, such as *Aspergillus flavus* and *Aspergillus parasiticus*, which commonly infect many potential feedstuffs such as corn, rice, fishmeal, shrimp and meat meals (Ellis et al. 2000). Four aflatoxins (AFB_1 and AFB_2, AFG_1, and AFG_2) can be found as direct contaminants of grains and finished feeding stuffs. Other common aflatoxins, M_1 and M_2, are in fact metabolites of B_1 and B_2 which were first identified in milk of animals previously exposed to B_1 and B_2 (Stoloff 1976). Limits for aflatoxin B1 in feeds are given, e.g., by the directive 2002/32/EC of the European Parliament and of the Council of 7 May 2002 on undesirable substances in most animal feeds and food, which range from 1–12 $\mu g\ kg^{-1}$ (EU 2002). The US Food and Drug Administration (FDA) has established guidelines on maximum levels of aflatoxins in feedstuff and recommend a concentration of 20 $\mu g\ kg^{-1}$ (Van Egmont and Jonker 2004). Other countries released individual guidelines which show maximum allowable levels for aflatoxins ranging from 2 $\mu g\ kg^{-1}$ to 50 $\mu g\ kg^{-1}$ in fish feeds and possible feed ingredients, while cottonseed and maize and their respective derivatives designated for animal feed production are occasionally regulated differently allowing aflatoxin levels of up to 200 $\mu g\ kg^{-1}$ (Van Egmond and Jonker 2004). Different maximum levels have probably been recommended because regional variations in mycotoxin distribution due to different climatic conditions have been noted (Rodrigues and Naehrer 2012).

The number of studies addressing the effects of aflatoxins in aquatic species is limited but compared to other mycotoxins quite a lot is known on its impact on fish health. Unfortunately, uncertainties in accurately diagnosing aflatoxicosis in fish exist which in part explain the lack of information regarding the incidence of aflatoxicosis in farmed aquatic species. First signs of aflatoxicosis in fish often include pale gills, liver damage, poor growth rates and immunosuppression (Jantrarotai et al. 1990, Sahoo and Mukherjee 2001, Ahn Tuan et al. 2002, Akter et al. 2010).

Carp treated with 2 $\mu g\ kg^{-1}$ AFB_1 did not show any effects on body weight or condition (Svobodova and Piskac 1980). In addition, Svobodova et al. (1982a) reported that AFB_1 at doses of 20 to 200 $\mu g\ kg^{-1}$ feed did not show any effects on feeding and use of protein in carp. In contrast, for carp fingerlings exposed to 100 μg AFB_1 kg^{-1} for 15 and 30 days growth parameters were significantly reduced (Akter et al. 2010). Research on aflatoxins has shown that this group of fungal metabolites often negatively affects growth of other fish species as has been demonstrated in channel catfish (*Ictalurus punctatus* (Rafinesque 1818)) and Nile tilapia (*Oreochromis niloticus* (Linnaeus 1758)) at concentrations ranging from 1.88 to 100 mg kg^{-1} AFB$_1$ (Jantrarotai and Lovell 1990, Chavez-Sanches et al. 1994, Ahn Tuan et al. 2002, Encarnacao et al. 2009). Moreover, haematological parameters such as haematocrit can be influenced by elevated aflatoxin levels (decreased haematocrit at > 250 µg AFB$_1$ kg^{-1} in Nile tilapia and at 80 µg kg^{-1} AFB$_1$ in juvenile sturgeon hybrids (Ahn Tuan et al. 2002, Rajeev Raghavan et al. 2011). These age- and species-dependent differences in the susceptibility of fish towards aflatoxins seem to be based on differences in the hepatic metabolism of AFB$_1$ including formation of AFB$_1$-derived metabolites (Ngethe et al. 1993, Santacroce et al. 2008).

Negative effects of AFB$_1$ on kidney cells and erythrocytes of common carp have been shown indicating a mutagenic potential of this compound (Al-Sabti 1986). But due to the fact that aflatoxin exposure is more commonly known to affect liver functions and histology these have been investigated more frequently. Toxic effects on primary hepatocytes of carp have been identified at AFB$_1$ concentrations of 0.01 µg ml^{-1} (He et al. 2010). Carp treated with 2 µg kg^{-1} AFB$_1$ did not show any liver lesions, but higher doses, such as 20 and 200 µg kg^{-1} feed, caused histo-pathological alterations including dystrophy of liver (Svobodova and Piskac 1980). More prominently severe hepatic necrosis in Nile tilapia occurred at levels of 100 mg kg^{-1} AFB$_1$ (Ahn Tuan et al. 2002) and trout showed increased tumour incidence in the liver after feeding a diet naturally contaminated with aflatoxins (Ashley 1970). Immunosuppressive effects after intraperitoneal injection of Indian major carp (*Labeo rohita*) with 1.25 mg AFB$_1$ per kg body weight have been reported (Sahoo and Mukherjee 2001) but not for other fish species so far.

It has previously been proposed that cold-water species seem to be more susceptible to AFB$_1$. For example, increased mortality was reported in juvenile sturgeon hybrids fed 80 µg kg^{-1} AFB$_1$ after 12 days, which was also observed at 20 and 40 µg kg^{-1} AFB$_1$ after 25 days of feeding (Rajeev Raghavan et al. 2011). Consequently, a value of 10 µg kg^{-1} AFB$_1$ in the diet was established as the maximum allowable level in hybrid sturgeon diet (Rajeev Raghavan et al. 2011). Species like rainbow trout (*Onchorhynchus mykiss* Walbaum) and European seabass (*Dicentrarchus labrax* (L.)) are considered to be more sensitive to aflatoxins than channel catfish, (Manning 2001, Ahn Tuan et al. 2002, El-Sayed and Khalil 2009). However, in Beluga, *Huso huso*, (Linnaeus 1758), no increased mortality was observed after 60 days of feeding although concentration of 75 and 100 µg kg^{-1} AFB$_1$ caused liver damage and influenced weight gain of fish (Sepahdari et al. 2010). Increased mortality was reported in Nile tilapia fed diets with 200 µg kg^{-1} AFB$_1$ (El-Banna et al. 1992). In channel catfish injected inta-peritoneally with AFB$_1$, a lethal dose killing 50% of the fish (LD$_{50}$) of 11.5 mg kg^{-1} body weight was observed after 10 days of exposure (Jantrarotai et al. 1990).

An important issue for fish in aquaculture is the retention of mycotoxins in edible parts of the fish. Svobodova and Piskac (1980) did not observe any accumulation of aflatoxin in carp muscles. However, Akter et al. 2010 concluded that feeds with aflatoxin contents higher that 50 µg kg^{-1} AFB$_1$ should not be used for carp because tissue concentrations in liver and muscle were not appropriate for human consumption. That animal tissues are able to retain relevant concentrations of aflatoxins and might pose a risk if contaminated parts are consumed by humans has also been shown for other fish species. For example, prolonged feeding of European seabass with low levels of AFB$_1$ (1.8 µg kg^{-1} body weight) for 42 days did not only cause serious health problems in exposed fish, but a high risk to consumers through AFB$_1$ residues in fish musculature (with concentrations > 4 µg kg^{-1}) was assumed (El-Sayed and Khalil 2009). Exposure of walleye (*Sander vitreus*) to aflatoxin-contaminated diets (50 and 100 µg kg^{-1}, respectively) for 30 days resulted in residues of aflatoxin B1, G1 and G2 of 5 to 20 µg kg^{-1} in musculature which were, however, no longer detectable after two weeks of mycotoxin withdrawal (Hussain et al. 1993). In line with these results Han et al. (2010) described that gibel carp, *Carassius auratus gibelio* (L.), fed with more than 10 µg AFB$_1$ kg^{-1} diet showed accumulation of AFB$_1$ residues in muscles and ovaries above the safety limitation of European Union of 2 µg kg^{-1} food designated for human consumption (Egmond and Jonker 2004). However, in gibel carp treated with AFB$_1$ concentrations ranging from 3.2–991.5 µg kg^{-1} diet for 16 weeks and in Nile tilapia exposed to 85–1641 µg kg^{-1} AFB$_1$ in the diet for 20 weeks residues exceeding this limit were detected in the liver but not in muscle (Deng et al. 2010, Huang et al. 2011). A cumulative effect on mycotoxin concentrations in flesh depending on the dietary AFB$_1$ levels and duration of exposure was reported in sturgeon (Rajeev Raghavan et al. 2011). These fish showed extremely high levels of 142 µg kg^{-1} AFB$_1$ in liver tissue and approximately four times lower values in muscle samples. Thus, the thread by aflatoxin contamination is present for fish and also in human consumption.

In addition to aflatoxins B$_1$ and B$_2$, *A. flavus* also produces many other mycotoxins including aspertoxin and gliotoxin (Goto et al. 1996). For aspertoxin a LD$_{50}$ of 6.6 mg L^{-1} for zebrafish larvae exposed to this mycotoxin via water for 48 hours was reported (Abedi and Scott 1969). Gliotoxin is another mycotoxin that can be produced by *Aspergillus* species. However, its effects on fish have rarely been investigated. For example, for this mycotoxin a LD$_{50}$ of 280 µg L^{-1} was established for zebrafish larvae exposed for 24 hours (Abedi and Scott 1969). Thus, further research is needed to estimate the toxicological relevance of these unconventional mycotoxins in fish farming.

Sterigmatocystin and Versicolorin A

Sterigmatocystin (STC) and versicolorin A (VA) are precursors in aflatoxin biosynthesis. Similar to aflatoxins STC is not only produced by *Aspergillus* and *Penicillium* species but also by *Bipolaris, Chaetomium, Emiricella* species and naturally contaminates grains (Veršilovskis and De Saeger 2010). Animal feeds have rarely been investigated although concentrations between 92–134 mg kg^{-1} have been found in cattle feed (Vesonder et al. 1985). The occurrence of VA in animal feeds remains unknown so far.

Toxicity symptoms of dietary contamination with STC have been described in common carp exposed to 0 to 1,250 µg kg^{-1} STC for three weeks (Abdelhamid 1988). STC is a hepatocarcinogenic mycotoxin causing decreased growth rates, some pathological findings and increased mortality in carp. Accordingly, a LD$_{50}$ of 211 µg kg^{-1} STC carp diet was estimated. Similarly, for zebrafish larvae exposed to water-borne STC for 24 hours an LD$_{50}$ of 240 µg L^{-1} STC was reported (Abedi and Scott 1969). In addition, reduced muscular protein levels, serum transaminases activity, muscular dry matter and ether extract contents were observed (Abdelhamid 1988). Catfish fed 250 µg kg^{-1} STC for three months showed similar symptoms as observed in carp and to the presence of residual STC in the fish muscles (Abdelhamid 1988). STC was also shown to be carcinogenic in medaka (*Oryzias latipes*) and rainbow trout embryos (Hatanaka et al. 1982, Hendricks et al. 1980). In Nile tilapia repeated oral treatment with STC at 1.6 µg kg^{-1} body weight for four weeks resulted in darker colouration, unbalanced swimming and the death of 25% of the treated fish after three days (Mahrous et al. 2006). Gills of STC-treated animals showed lamellar oedema, hyperplasia and haemorrhages. Haemorrhages and necrosis were also observed in the liver and spleen. In addition, increased chromosome aberrations in kidney tissues of individuals of this fish species treated with the same dose were observed (Abdel-Wahhab et al. 2005).

Up to now, hepatocarcinogenic effects of VA have only been shown in rainbow trout (Hendricks et al. 1980). Therefore, further research is recommended to evaluate its frequency in animal feeds and its effects in fish species that are important for aquaculture.

Citrinin

The nephrotoxic mycotoxin citrinin (CTN) is produced by several species of the genera *Aspergillus, Penicillium* and *Monascus*. It is formed after harvest and occurs in stored grains and therefore also occurs as a contaminant in animal feeds (Abdelhamid 1990).

Effects of CTN in common carp are unkown. However, toxicological studies on other fish species emphasize a toxicological role of CTN in fish feeding in aquaculture. Nephrotoxicity and hepatotoxicity were confirmed by intraperitoneal injection of CTN into fingerlings of Indian major carp (*Labeo rohita*) at 12.5 or 25.0 mg kg^{-1} body weight resulted into damage to the kidney, liver and intestine, depigmentation and congestion of caudal fins and pronounced mortality (Sahoo et al. 1999). Cardiotoxicity of CTN due to modulation of gene expression of relevant genes has been shown in zebrafish (Wu et al. 2013). However, this mycotoxin was not carcinogenic in medaka (Hatanaka et al. 1982). Transfer of CTN from fish products to humans has not yet been investigated. CTN is known to be heat-sensitive which makes a transfer of this mycotoxin to humans unlikely when cooking processes are included. However, heating can lead to the formation of citrinin H1 and citrinin H2, which show higher and weaker cytotoxicity than the parental compound CTN (Trivedi et al. 1993, Hirota et al. 2002). However, the production of fermented salted fish has been shown to lead to CTN contamination of the final product with up to 0.525 mg kg^{-1} CTN, a concentration that also showed high toxicity to brine shrimp (*Artemia salina*) (Youssef et al. 2003).

Cyclopiazonic Acid

Cyclopiazonic acid (CPA) is a mycotoxin produced by several species of *Aspergillus* and *Penicillium* fungi and is therefore frequently found in association with aflatoxins. Feed ingredients which might also be used for fish feed preparation have been shown to contain CPA concentrations of up to 20 mg kg^{-1} (Balachandran and Parthasarathy 1996).

No information on the toxicity of CPA in carp exists. However, Lovell (1992) showed that neurotoxic CPA was even more toxic to channel catfish than aflatoxin. Feeding channel catfish various concentrations of CPA ranging from 0.1–10 mg kg^{-1} reduced growth rates and led to necrosis of the gastric glands and neural damage (Lovell 1992). An LD$_{50}$ of 2.82 mg kg^{-1} body weight was determined by intraperitoneal injection. Another study showed that feeding catfish a diet containing 0.1 mg kg^{-1} CPA in feed for 10 weeks resulted in growth reduction, whereas 10 mg kg^{-1} caused alterations in kidney epithelium and damage to gastric glands (Jantrarotai and Lovell 1990). Unfortunately, no further information on CPA toxicity in fish is currently available.

Ochratoxins

Ochratoxins, which are a group of secondary metabolites consisting of a derivative of an isocoumarin, are produced by fungi belonging to *Aspergillus* and *Penicillium* genera mainly in cereals. Three different ochratoxins can be distinguished: A, B, and C which differ slightly from each other in their chemical structures. These differences, however, have marked effects on their respective toxic potentials. Ochratoxin A (OTA) is the most abundant and hence the most commonly detected member but is also the most toxic of the three (Van der Merwe et al. 1965a,b). A maximum allowable level for OTA in cereals and cereal products designated for animal feeding of 250 µg kg^{-1} has been established by the directive 2006/576/EC of the European Parliament on 17 August 2006. For non-EU countries maximum contamination levels of 0.02 to 2 mg OTA kg^{-1} have been established (Van Egmond and Jonker 2004).

OTA is generally associated with contamination of corn, cereal grains, oilseeds and compound feeds and can affect animal performance through damage to kidney function (CAST 2003, Binder et al. 2007, Kononenko and Burkin 2013). Saad (2002) reported that OTA is immunosuppressive in common carp and Nile tilapia during acute (50 µg/kg fish) and chronic exposure (10 µg/kg fish). Several studies investigated further effects of ochratoxins in other fishes. Manning et al. (2003a) reported a reduction in body weight gain of channel catfish fed diets with 2 mg kg^{-1} of OTA for two weeks and 1 mg kg^{-1} for 8 weeks (Manning et al. 2003a). Reduced feed conversion ratio was also observed in the same species with contamination levels of 4 and 8 mg kg^{-1} in the same study whereby the latter concentration also reduced the haematocrit and the survival (Manning et al. 2003a). In rainbow trout pathological signs due to contamination with OTA included liver necrosis, pale, swollen kidneys and high mortality (Hendricks 1994). An oral LD$_{50}$ of 4.7 mg kg^{-1} OTA has been reported for 6-month-old trout (Lovell 1992) whereas water-borne exposure led to the estimation of an LD$_{50}$ of 1.7 mg L^{-1} in zebrafish larvae exposed for 72 hours (Abedi and Scott 1969). In addition, OTA caused pronounced abnormalities in eggs leading to high mortality in hatchlings of zebrafish (Ali 2007).

Orally applied OTA showed low absorption (1.6%) in common carp (Hagelberg et al. 1989). From these results it may be assumed that the risk with respect to human consumption is low. However, in the study of Farag et al. (2011) OTA concentrations exceeding the permissible limits have been found in 20% of the tested smoked herrings (n = 60) whereby mycotoxin contamination in this case was due to inappropriate storage conditions of the commodities.

Moniliformin

The *Fusarium* toxin moniliformin (MON) is the sodium or potassium salt of 1-hydroxycyclobut-1-ene-3,4-dione produced by *Fusarium* species. Moniliformin in animal feeds has been found at concentrations of up to 1.2 mg kg^{-1} (Labuda et al. 2005).

Yldirim et al. (2000) found that in juvenile channel catfish, diets containing 20 mg kg^{-1} of MON significantly reduced body weight gain after two weeks and led to reduced haematoctrit values at 20 mg kg^{-1} of MON after 10 weeks of feeding. Ahn Tuan et al. (2003) noted that feeding MON at 70 mg kg^{-1} for eight weeks, negatively affected growth performance of Nile tilapia fingerlings despite a lack of mortality or histo-pathological lesions. Compared to channel catfish, Nile tilapia appears to be more resistant to MON in the diet (Ahn Tuan et al. 2003). Up to now, no study has investigated the effects of MON on carp.

Fumonisins

The fumonisins represent a group of mycotoxins produced predominantly by *Fusarium verticilloides* (synonyme: *F. moniliforme*) which display a major problem in maize (Nelson et al. 1993). Thus, fumonisin contamination is of concern to the aquaculture industry. The recommended guidance values by the European Commision for fumonisin B_1 and B_2 in complementary and complete feeding stuffs for fish set maximum levels to 10 mg kg^{-1} (2006/576/EC of the European Parliament on 17 August 2006). Similarly the recommended maximum allowable level in feeding stuff in the USA was established at 10 mg kg^{-1} and for other countries such as China a value of 0.5 mg kg^{-1} has been given although only a few countries have already established particular maximum levels for fumonisins in feeds (Van Egmond and Jonker 2004).

Animal diseases have been linked with this toxin whereby the most important one includes disruption of the sphingolipid metabolism (Wang et al. 1992). The impact of fumonisins on fish remains unclear. The biggest threat that may be posed to fish by this mycotoxin could be its ability to alter the immune system. Levels of 20 mg kg^{-1} have been reported to cause immunological changes (Pepeljnjak et al. 2002). These authors also showed that exposure to 0.5 and 5.0 mg per kg body weight is not lethal to young carp, but can produce adverse physiological effects with kidney and liver being the key target organs for the FB_1 action. Other changes subsequent to fumonisin exposure that have been reported for carp in a study using 10 or 100 mg kg^{-1} FB_1 including symptoms such as scattered lesions in the exocrine and endocrine pancreas, and interrenal tissue, probably due to ischemia and/or increased endothelial permeability and damage to blood vessels, heart and brain tissue (Petrinec et al. 2004). In contrast, few adverse effects have been reported in channel catfish fed *F. moniliforme* culture material containing 313 mg kg^{-1} FB_1 for five weeks (Brown et al. 1994). However, for the same fish species, dietary levels of FB_1 of 20 mg kg^{-1} or above have been shown to result in lower weight gain and significant decrease in haematocrit and red and white blood cells compared to fish fed lower doses (Lumlertdacha et al. 1995). Likewise, Yldirim et al. (2000) found that in channel catfish, diets containing 20 mg kg^{-1} FB_1 significantly reduced body weight gain after two weeks. Increased mortality was observed in channel catfish when diets contained 240 mg kg^{-1} FB_1 (Li et al. 1994). Another study by Ahn Tuan et al. (2003) noted that feeding FB_1 at 40 mg kg^{-1} negatively affected growth performance of Nile tilapia fingerlings despite no effect on mortality or histopathological lesions. According to the same authors, FB_1 is more toxic than MON to channel catfish and Nile tilapia. Compared to channel catfish, Nile tilapia appears to be more resistant to FB_1 in the diet (Ahn Tuan et al. 2003). The reasons for species-dependent differences in fumonisin toxicity should be clarified by further research.

Zearalenone

Zearalenone (ZEN), a resorcylic acid lactone, is an important mycotoxin produced by fungi of the genus *Fusarium*. This *Fusarium*-derived mycotoxin has already been identified as an aquatic micropollutant in aquatic ecosystems (Bucheli et al. 2008, Hoerger et al. 2009). In feed materials, limits for ZEN in cereals and cereal products have been set at 2 mg kg^{-1} in the EU while maize by-products should not exceed levels of 3 mg kg^{-1}. In other countries individual guidelines have been established recommending maximal ZEN levels of 0.02 to 1 mg kg^{-1} (Van Egmond and Jonker 2004).

It has been reported that mycotoxins also contribute to the estrogenic environmental exposure in aquatic systems (Hartmann et al. 2008), and especially ZEN has been shown to act as a typical estrogenic compound also in fish (Bucheli et al. 2005). A study assessing the toxic effects of ZEN in fish cell lines indicated that it is unlikely that the toxicity of this substance is solely due to its endocrine potential. Therefore, other mechanisms, such as oxidative stress or toxicity to lysosomes, could be involved (Pietsch et al. 2014a).

Woźny et al. (2013) investigated the ZEN content in trout in fish farms in Poland. The authors reported ZEN contamination of less than 2 µg kg^{-1} tissue in the liver and intestine but concentrations of up to 7.1 µg kg^{-1} in ovaries. No obvious health risk could be assumed for humans since ZEN was not detected in trout muscle, although exposure of carp to ZEN concentrations which were in the range found in commercial feed in aquaculture led to detection of ZEN and traces of its metabolite a-ZEL in carp fillet after four weeks of experimental feeding (unpubl. results, C. Pietsch). Due to its high estrogenic potency effects of ZEN on reproduction in fish have been observed. Carp fed diets contaminated with zearalenone showed a reduced number and quality of sperm (Sándor and Ványi 1990). Developmental problems in zebrafish (*Danio rerio*) and early life stages of fathead minnow (*Pimephales promelas*) due to ZEN exposure have also been described (Johns et al. 2009, Schwartz et al. 2010). In addition, negative effects on development and iron metabolism were observed in rainbow trout (Woźny et al. 2012, Bakos et al. 2013). Effects on white blood cell counts of common carp, leading to pronounced modulations of immune parameters have been noted in carp fed ZEN contaminated diets (unpubl. results, C. Pietsch). Furthermore, the micronuclei formation in carp blood *in vivo* was paralleled by genotoxic effects in an *in vitro* assay using the salmonid fish cell line RTL W-1 (Pietsch et al. 2014a). The latter results indicate that not only reproduction is targeted by ZEN. Thus, further research is needed to establish the roles of possible modes of action of ZEN in fish *in vivo* independent of its estrogenic potency.

Patulin

Patulin is a cyclic γ-lactone mycotoxin which is also produced by *Penicillium* and *Aspergillus* species. The extent of patulin contamination in animal feeds and its effect on fish are unknown so far. However, patulin exposure via water led to a LD$_{50}$ of 18 mg L^{-1} for zebrafish larvae exposed for an hour (Abedi and Scott 1969). Thus, further research is essential to judge the importance of patulin intoxication in farming of fish species such as carp.

Trichothecenes

Many trichothecenes are also produced by fungi of the genus *Fusarium*. This group of mycotoxins is sub-divided into two groups, type A and type B trichothecenes which will be regarded separately in the chapter.

Type A Trichothecenes

This group of toxins comprises toxicological important mycotoxins such as T-2 and HT-2 toxin, neosolaniol and diacetoxyscirpenol (DAS). The European Commission established maximum allowable levels of 250 mg kg^{-1} for T-2 toxin in cereal products in compound feeds (Commission Recommendation 2013/165/EU of 27th March 2013 on the presence of T-2 and HT-2 toxin in cereals and cereal products). Some countries also released individual recommendations (maximum levels of 80 to 100 mg kg^{-1}) on the occurrence of T-2 toxin in complete feed and all grains (Van Egmond and Jonker 2004).

The effects of sublethal concentrations of T-2 (2.45 mg kg^{-1} feed) and HT-2 (0.52 mg kg^{-1} feed) toxin produced by *Fusarium trichioides* for four weeks were investigated in common carp (Balogh et al. 2009) which revealed fast reduction of feed consumption, weight gain and effects on the antioxidant system. The latter had already been observed previously in carp treated with 0.46 mg kg^{-1} body weight whereby also elevation of lysosomal enzymes and alkaline phosphatase were observed (Kravchenko et al. 1989). T-2 toxin produced by *Fusarium tricintum* reduced feed consumption, growth, lowered haematocrit, and lowered blood haemoglobin in rainbow trout at levels higher than 2.5 mg kg^{-1} (Poston 1983). Also channel catfish fed diets with levels of T-2 toxin ranging from 0.625 to 5.0 mg kg^{-1} showed reduced growth rate and increased mortality above 2.5 mg kg^{-1} (Manning et al. 2003b). Additionally disease resistance and survival of channel catfish challenged with *Edwardsiella ictaluri* was reduced when fish were fed with contaminated feed (Manning et al. 2005) which may have a great implication for susceptibility of fish to diseases in aquaculture. Land animals are able to hydrolyze T-2 toxin to HT-2 toxin but this metabolization pathway is not as important in carp (Wu et al. 2011a) which may cause differences in sensitivity to these mycotoxins in fish compared to terrestrial animals.

The toxicity of DAS was rarely investigated in fish. A LD_{50} of 4.8 mg L^{-1} DAS in water was established for zebrafish larvae exposed for 24 hours (Abedi and Scott 1969).

Type B Trichothecenes

Among others this group of trichothecenes comprises the important mycotoxins deoxynivalenol (DON) and nivalenol. DON is an important contaminant of wheat and corn. A guidance value in feeding stuff 5 mg kg^{-1} with a moisture content of 12% in complementary and complete feeding stuffs was established by the EU (2006/576/EC of the European Parliament on 17 August 2006). In non-EU countries recommended maximum values range from 0.2 to 5 mg kg^{-1} feed (Van Egmond and Jonker 2004).

Effects of nivalenol on fish remain unknown so far. Nevertheless, toxicity of DON on fish cell lines which is comparable to mammalian cell lines has been shown including effects on cell viability and production of oxidative stress (Pietsch et al. 2011). Toxic effects on primary hepatocytes of common carp have been identified at concentrations of 0.5 µg ml^{-1} DON (He et al. 2010). Effects of DON feeding on juvenile carp (Pietsch et al. 2014b,c) showed no influence on weight gain after four weeks of feeding concentrations of up to 953 µg kg^{-1}. In contrast to carp, rainbow trout showed reduction of feed intake, reduced growth and feed efficiency at dietary concentrations of DON of 0.5 to 13 mg kg^{-1} of diet (Woodward et al. 1983, Hooft et al. 2010). Significant histopathological changes in liver with increasing dietary levels (> 1.4 mg kg^{-1}) of DON were also reported (Hooft et al. 2010) which was also observed in common carp treated with DON concentration of 352 µg kg^{-1} to 953 µg kg^{-1} (Pietsch et al. 2014c). Similarly, in Atlantic salmon (*Salmo salar*) diets with 3.7 mg kg^{-1} of DON resulted in reduction in feed intake, increase in feed conversion, and reduction of specific growth rates (Döll et al. 2010).

In carp, suppression of proinflammatory immune responses occurred (Pietsch et al. 2014b, Pietsch et al. 2015). Furthermore nutritional status of carp was influenced by DON feeding (Pietsch et al. 2014c). Sanden et al. (2012) observed effects of DON on fecundity in zebrafish. Although this shows that the effects of DON have at least been investigated in several studies investigating different endpoints, further research is needed to estimate its actual impact on fish in aquaculture.

Ergot Alkaloids

Carp may also be intoxicated by alkaloids produced by ergot (*Claviceps purpurea*), living on cereals. Ergotamine as one of the toxins produced by ergot has already been shown as a contaminant in ingredients for fish feeds (van Leeuwen et al. 2008). Only few governmental regulations for ergot alkaloids regulated in animal feed exist which allow maximum levels of 450 µg kg^{-1} (Van Egmont and Jonker 2004). Research showed that carp are able to avoid uptake of ergot-contaminated grains to some extent (Oberle 2000, Oberle 2002a,b). Despite its quite low toxicity, presence of ergotamine in feed leads to histopathological alterations in carp (Svobodová et al. 1982b). However, detailed investigations on the effects of these toxins are lacking in fish and mostly also in other farm animals.

Mixtures of Mycotoxins

It is necessary and environmentally increasingly important to investigate the effects of mycotoxin mixtures on fish since various mycotoxins often occur together in the same feed ingredient and/or finished feed. Recent research identified synergism and additivity that exist between several mycotoxins. This means that mycotoxins alone can occur at unproblematic levels but together they can be harmful due to synergistic or additive effects. Similar to other vertebrates, fish are more sensitive to the effects of mycotoxins when multiple forms are present within the feed. Toxic effects of mycotoxin combinations on fish have been shown *in vitro* and *in vivo*. For example, different mixtures of AFB_1 and DON have been tested on primary hepatocytes of common carp (He et al. 2010). The authors concluded that both mycotoxins in combination led to an additive response in these cells.

Combined effect of aflatoxin B1 and T-2 toxin have been demonstrated by McKean et al. (2006) in mosquito fish (*Gambusia affinis*). The study revealed that the LD_{50} for aflatoxin was 681 $\mu g\ L^{-1}$ and the LD_{50} for T-2 was 147 $\mu g\ L^{-1}$. For a combination of the two mycotoxins a LD_{50} of 414 $\mu g\ L^{-1}$ was expected but in reality it was found to be much lower (at 234 $\mu g\ L^{-1}$).

Prevention of Toxic Effects by Mycotoxins

Commercial probiotics and adsorbents are available that counteract the negative effects of AFB_1 and OTA on carp fingerlings (Agouz and Anwer 2011). In other fishes also clay minerals as adsorbents could prevent toxicity of sterigmatocystin as has been shown for example in Nile tilapia (Abdel-Wahhab et al. 2005) but clay components may also have disadvantages, e.g., by binding vitamins in feeds. Plant-derived materials, such as pepper (*Piper nigrum* L.) and coriandrum (*Coriandrum sativum*) meal proved to be valuable as detoxifying agents of aflatoxin (Salem et al. 2010). In addition, antioxidants such as vitamin C have been used as antidotes in ochratoxin-contaminated feeds for tilapia (Shalaby 2009).

Antibiotics and Antifungal Compounds in *Aspergillus* Species

Aspergillus species are known to secrete antibiotics, antifungals, and anti-tumour drugs (Keller et al. 2005, Hoffmeister and Keller 2007). The antibiotic penicillic acid had no lethal effect on zebrafish larvae exposed to water contaminated with up to 5 mg L^{-1} of this mycotoxin (Abedi and Scott 1969). In the same study, stemphone showed an LD_{50} of 1.2 mg L^{-1} in zebrafish treated similarly for 24 hours. For anti-fungals such as griseofulvin no LD_{50} value could be established for concentrations of up to 10 mg L^{-1} (Abedi and Scott 1969). No reports on effects of these substances on carp could be found in the literature.

Cyanobacterial Toxins

Cyanobacteria, which are also called 'blue-green algae', are highly prevalent photosynthetic prokaryotes. Several characteristics promote the success of many planktonic forms including the higher temperature optimum compared with many eukaryotic algae (Castenholz and Waterbury 1989) and their ability to efficiently utilize light for photosynthesis even at low light intensities (van Liere and Walsby 1982). In addition, a competitive advantage can also be gained due to the ability of many species to fix atmospheric nitrogen where environmental nitrogen concentrations may be low.

In freshwater surface waters cyanobacterial blooms usually occur worldwide. Despite their detrimental effects on light penetration and oxygen availability in lakes, ponds, rivers, streams and canals, toxic metabolites known as cyanotoxins occur. Mostly eutrophic freshwater ecosystems are affected by planktonic cyanobacteria but toxic effects of benthic cyanobacteria have also been reported and cyanobacteria can appear in oligotrophic lakes as well (Owen 1984, Edwards et al. 1992, Gunn et al. 1992, Mez et al. 1997). Nutrient load of surface water seems to be important since it has been found that nitrogen shortage favours the progress of nitrogen-fixing cyanobacteria (Fig. 13.2) due to their competitive advantage (Howarth et al. 1988, Kardinaal and Visser 2005). Interestingly, lakes that especially showed winterkills of carp exhibited higher cyanobacteria densities which are presumably due to release of nutrients from the decaying fish carcasses (Schoenebeck et al. 2012). Thus, the reasons for cyanobacterial blooms are quite complex, but influences of climate characteristics including warmer temperatures, earlier ice-outs, more intense rainfall and greater run-off, strong anthropogenic influences, e.g., via agricultural industry often leading to high nutrient loads, as well as certain hydrological characteristics such as water mixing frequency and water retention often contribute to the occurrence of cyanobacteria blooms (Journey et al. 2013). In addition, certain environments, particularly streams, undergo frequent shifts in conditions and some cyanobacteria species appear to be especially adapted to deal with such shifts. This is notably true for genera which can form heterocysts (Whitton et al. 1998).

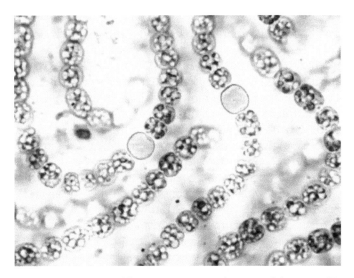

Figure 13.2. Microscopic picture of heterocysts in cyanobacteria of the genus *Nostoc*; picture provided by the California State Water Resources Control Board.

In fact, not only toxigenic cyanobacteria affect fish health but also factors such as high pH values caused by photosynthetic activity contribute to the detrimental effects of toxic algal blooms. Low oxygen concentrations may occur since cyanobacterial blooms often cover the entire water surface and cut off light drastically. This results in critical reduction in primary productivity of the pond. Finally, oxygen problems are also caused by decomposition of cyanobacterial and algal blooms.

Mortalities of freshwater fish including carp due to the occurrence of cyanobacterial blooms have been reported (Mackenthun et al. 1948, Rodger et al. 1994). Cyanobacterial toxins are commonly listed according to their effects in organisms. This will also be applied in this chapter, since these characteristics also determine their toxicological importance.

Hepatotoxins

Toxic effects on liver tissue by cyanobacterial toxins are often caused by toxins that belong to the groups of microcystins and nodularins. These cyclic peptides are produced by different genera, such as *Microcystis, Planktothrix, Anabaena, Nodularia, Nostoc* and *Umezakia* (Carmichael 1997, Codd 1995, Sivonen 1996). Both groups of toxins differ in their amino acid content. Especially the content of the hydrophobic amino acid 2S,3S,8S,9S-3-amino-9-methoxy-2,6,8-trimethyl-10-phenyl-deca-4,6-dienoic acid (ADDA) determines the biological activity of these molecules (Rinehart et al. 1988, Stotts et al. 1993).

Microcystins

Microcystins have been identified for the first time in *Microcystis aeruginosa*. Members of this group of toxins consist of seven amino acids. More than 70 different microcystin variants are known (Sivonen 1996, Chorus and Bartram 1999, Carmichael 2001) that differ from each other by the L amino acid at two non-conserved positions in the molecule (Botes et al. 1982, Duy et al. 2000). The most toxic microcystin variant microcystin-LR contains the common L-amino acids, leucine (L) and arginine (R) at this position, whereas the also frequently occurring microcystin-YR contains leucine (L) and tyrosine (Y), and microcystin-RR contains two arginines (RR) (Duy et al. 2000). The WHO provisional limit for microcystin in drinking water was set to $1 \, \mu g \, L^{-1}$ (WHO 1998). In natural blooms microcystin concentrations between 5 and 7,500 $\mu g \, L^{-1}$ can occur (Williams et al. 2007) which may persist for several weeks. The toxins can, furthermore, accumulate and can be found at low concentrations in viscera and muscle of carp (Vasconcelos 1999). Other reports showed that purified microcystin shows lower toxicity than cyanobacterial crude extracts in fish which might be related to the fatty acid content in the cyanobacterial cells that are able to interact with epithelial membranes especially in gills (Bury et al. 1996, Oberemm et al. 1997, Bury et al. 1998, Oberemm et al. 1999).

Moreover, fish eggs are thought to be protected against toxic effects of pure microcystins (Wang et al. 2005).

It is known that intraperitoneal injection of microcystin, oral uptake and water-borne exposure has different effects on fish (Tencalla et al. 1994). Particularly, the detrimental effects of microcystins on epithelial cells of carp gills have even been observed after intraperitoneal injection of the toxin (Carbis et al. 1996). Compared to other fish species carp are more sensitive to cyanobacteria after oral uptake because of the rather long digestion process which probably leads to more absorption of microcystins by gut epithelial cells (Fischer and Dietrich 2000, Fischer et al. 2000). Interestingly, silver carp (*Hypophthalmichthys molitrix*) and tilapia (*Oreochromis niloticus*) rather chose to feed on a non-toxic strain of *Microcystis* than on a toxic strain (Beveridge et al. 1993, Keshavanath et al. 1994). Similar avoidance of contact to cyanobacteria in common carp is so far unknown.

A crude extract of cyanobacteria consisting of microcystins LR, YR and RR (130 µg L^{-1}) increased embryonic mortality, prolonged hatching, and caused malformed and dead larvae when administered to carp eggs and larvae for eight and 30 days (Palíková et al. 2003). In addition, yolk sac edema were observed. Exposure to 13 µg L^{-1} increased dead larvae after eight days and caused significantly more malformed and dead larvae after 30 days of exposure. Carp larvae exposed to microcystin-containing *Nostoc muscurum* and *Anabaena variabilis* for 12 days showed decreased weight gain, length and survival (Al-Sultan 2011). However, exposure to 1.3 µg L^{-1} microcystin resulted in dead larvae only after the long-term exposure (Palíková et al. 2003).

In general, it can be concluded that extracts with high concentration result in acute toxicity to carp. Acute toxicity of a cyanobacterial crude extract mainly containing microcystin-LR was found at 12 µg L^{-1} in juvenile carp after exposure for 72 hours (Wu et al. 2011b). Similar to mammals, the main targets of microcystins in carp are the liver and the kidney which show dysfunction due to loss of cell architecture and necrosis (Råberg et al. 1991, Carbis et al. 1996). Gill impairment including specific inhibition of ion pumps has also been reported (Gaete et al. 1994, Carbis et al. 1996, Zambrano and Canelo 1996).

The influence of extracts with sub-lethal concentrations manifests after long-term exposure. Adult carp exposed for an entire season to 0.3–9.5 µg L^{-1} microcystins showed changes of plasma biochemical parameters (Kopp et al. 2009). Microcystins are known to specifically inhibit serine/threonine phosphatases which lead to ROS production and DNA damage (Runnegar et al. 1995, Ding et al. 1998). Different fractions of cyanobacterial extracts have also been shown to induce oxidative stress in carp embryos after 120 hours of exposure (Palíková et al. 2007) although hypoxia during the incubations complicates the direct attribution of effects to the presence or absence of microcystins in the exposure media. Oxidative stress often affects DNA integrity. Genotoxic signs (micronuclei) in red blood cells and morphological changes of cells in liver and gills of carp were observed (Wu et al. 2011b). In addition, effects of different cyanobacterial biomasses and fractions of these on eye pigmentation after 48 hours of exposure and swim bladder filling after 120 hours of exposure have been reported (Palíková et al. 2007). Besides different physiological and toxicological effects, abnormal behaviour of larvae exposed to microcystins LR, YR and RR (130 µg L^{-1}) was reported (Palíková et al. 2003).

Carp larvae accumulated half of the applied microcystin concentration after being fed contaminated *Artemia* nauplii for 12 days (El Ghazali et al. 2010). Although accumulation of microcystins in carp could not be shown in a study on carp from a hypertrophic pond in the Czech Republic (Kopp et al. 2009), their accumulation in the liver, kidney and muscle of carp has been shown in fish from a naturally contaminated lake in Greece (Mitsoura et al. 2013) whereby the concentration in muscle (mean value 114 ng g^{-1}) was lower than in the other organs (mean values of 732 and 362 ng g^{-1}, respectively). Under controlled conditions oral uptake of 50 µg microcystins per kg of body weight for 28 days in a laboratory study also led to retarded growth, influence on serum parameter, and to detection of microcystin levels above maximum allowable limits in the muscle of carp (Li et al. 2004). The recommended Tolerable Daily Intake (TDI) for microcystin-LR in human consumption is 40 ng per day (WHO 1998). Despite the fact that exposure to microcystin-containing blooms may lead to undesired accumulation of toxins, Mares et al. (2009) showed that fatty acid contents in common carp are affected thus decreasing the nutritional value of edible parts in this and other cyprinids.

Nodularins

Nodularins consist of five amino acids (Rinehart et al. 1988, Duy et al. 2000). At least seven variants of nodularin from the filamentous cyanobacterium *Nodularia spumigina* have been isolated (Lahti 1997, Chorus and Bartram 1999, Falconer 1999, Codd 2000). In natural blooms nodularin concentrations between 0.5 to 2.6 µg L^{-1} water may occur. Blooms are common within estuaries, coastal lagoons and lakes worldwide including the German North Sea and Baltic Sea, Australia, New Zealand and North America (Carmichael et al. 1988, Galat et al. 1990, Heresztyn et al. 1997, Kankaanpää et al. 2001).

Nodularins show similar biological activity to microcystins in spite of their different chemical structures and their slightly different mode of action on serine/threonine phosphatases (Honkanen et al. 1991). Fish kills due to occurrence of nodularins which affected, e.g., Three-spined sticklebacks have been reported (Kankaanpää et al. 2001). However, the effects of nodularins on carp remain unkown.

Cytotoxins

A prominent cytotoxic compound of cyanobacteria is cylindrospermopsin (Froscio et al. 2001). It is a tricyclic alkaloid that can be found less frequently in natural blooms than microcystins reaching concentrations of up to 1.6 μg L^{-1} (Williams et al. 2007). Although this toxin is heat-stable it is degraded by sunlight in natural algal extracts (Chiswell et al. 1999). Cylindrospermopsin is biosynthesized mainly by *Cylindrospermopsis raciborskii* and *Umezakia natans* as well as by *Aphanizomenon ovalisporum, Anabaena bergii*, and *Raphidiopsis curvata* (Ohtani et al. 1992, Chorus and Bartram 1999, Duy et al. 2000, Li et al. 2001, Schrembi et al. 2001) although non-toxic strains also exist. In mammals it is known that this cytotoxin generally blocks protein synthesis (Hawkins et al. 1985). In mammals, early clinical symptoms of cylindrospermopsin poisoning are kidney and liver failure and damages to the spleen, intestine, heart and thymus (Codd 2000). Cylindrospermopsin poisoning is considered to be especially dangerous because clinical symptoms may become obvious only several days after exposure, so that toxic effects can be difficult to correlate. Up to now, effects of cylindrospermopsin in fish remain unknown.

Neurotoxins

Neurotoxins are very important toxic compounds produced by cyanobacteria. But cyanobacteria producing neurotoxins are often less prevalent, so that these toxins have generally a narrower environmental impact than hepatotoxins. Neurotoxins are produced by certain strains of species such as *Anabaena* (Mahmood and Carmichael 1987), *Aphanizomenon* (Mahmood and Carmichael 1986), *Nostoc* (Davidson 1959), *Oscillatoria* (Mutawie 2012), *Lyngbya* and *Cylindrospermopsis* (Humpage et al. 1994). Up to now four groups of neurotoxins have been well described: anatoxin-a, anatoxin-a(s), saxitoxin, and neosaxitoxin (Ressom et al. 1994, Chorus and Bartram 1999).

Anatoxin-a and homoanatoxin-a are alkaloids described as secondary amines whereas anatoxin-a(s) is a phosphate ester of a cyclic N-hydroxyguanidine structure (Devlin et al. 1977, Matsunaga et al. 1989). Anatoxin-a(s) is thus structurally unrelated to anatoxin-a, but their effects are similar. Although anatoxin-a(s) occurs at relevant concentrations in environmental samples (Devic et al. 2002), no effects on fish have been investigated so far. Other neurotoxic cyanobacterial toxins comprise neosaxitoxins and saxitoxins which are unique tricyclic molecules with hydropurine rings known as potent Paralytic Shellfish Poisons (PSP) (Ressom et al. 1994). They are produced by the genera *Anabaena* and *Aphanizomenon, Lyngbya* and *Cylindrospermopsis* (Humpage et al. 1994). Effects of saxitoxins have only been described in marine fishes although freshwater cyanobacteria, such as *Anabaena circinalis*, also produce this toxin (Llewellyn et al. 2001). Nevertheless, the effects of saxitoxins in carp are up to now unknown.

In natural blooms anatoxin-a can be found less frequently than microcystins reaching concentrations of up to 7 μg L^{-1} (Williams et al. 2007), although concentrations of anatoxin-a-producing cyanobacteria may reach densities of 10^5 to 10^7 cells/ml in natural blooms (Pereira et al. 2000, Briand et al. 2002). Anatoxin-a interferes with the neurotransmitter acetylcholine often leading to overstimulated muscle cells (Spivak et al. 1980, Ressom et al. 1994,

Dow and Swoboda 2000). Pure Anatoxin-a showed less pronounced adverse effects on early life stages of carp than cyanobacterial extracts containing similar toxin concentrations (Osswald et al. 2009) which was attributed to the presence of other cyanobacterial toxic substances or metabolites or putative bacterial contamination. High mortality of carp eggs occurred at an exposure concentration of 666 μg L^{-1} of the cyanobacterial extract after four days of exposure and 333 μg L^{-1} of the cyanobacterial extract after seven days which resulted in no hatching and 10% hatching, respectively. Length of carp larvae was decreased by high concentration of the pure toxin (640 μg L^{-1}) and 333 μg L^{-1} of the cyanobacterial extract. In this study skeletal malformations such as bent tails and bent spines of newly hatched larvae were also observed. Juvenile carp (three months old) exposed to an anatoxin-a-producing strain of *Anabaena* at environmental relevant concentrations showed abnormal swimming behaviour and increased opercular movement (Osswald et al. 2007). At a cyanobacterial concentration of 10^7 cells per ml the fish died after exposure for 24 to 29.5 hours. A concentration of 73 ng toxin g^{-1} fish (fresh weight) has been noted. An *in vitro* study using lymphocytes of carp showed rapid toxicity of anatoxin-a concentrations ranging from 0.1 to 10 μg ml^{-1} and a reduced proliferative response of these cells to mitogens (Bownik et al. 2012). The effects on lymphocytes were confirmed by Rymuszka (2012) who, furthermore, showed that the anti-oxidative potential of phagocytes is affected by this toxin. Since in the natural environment mixtures of cyanobacteria toxins also occur, Rymuszka and Sieroslawska (2013) investigated the effects of microcystin and anatoxin-a together on carp immune responses *in vitro* and found evidence for synergistic effects of these toxins.

Dermatotoxins

So called dermatotoxins are mainly produced by tropical and subtropical marine benthic cyanobacteria such as *Oscillatoria, Lyngbya* and *Schizothrix* (Chorus and Bartram 1999). They include aplysiatoxins, debromoaplysiatoxins and lyngbyatoxins. Since their occurrence is restricted to marine environments, effects in freshwater fish have not yet been described.

Irritant toxins—Lipopolysaccharides (LPS)

Generally, lipopolysaccharides (LPS) are integral components of the cell wall of all gram-negative bacteria, including cyanobacteria. LPS consist of lipid A, core polysaccharides and an outer polysaccharide chain. LPS are apparently ichthyotoxic compounds. For example, LPS from *Escherichia coli* was toxic for carp after intaperitoneal injection of more than 200 mg per kg body weight but does not cause an endotoxic shock (Berczi et al. 1966). Immersion of halibut larvae (*Hippoglossus hippoglossus* L.) in LPS from *Aeromonas* bacteria proved that LPS can be incorporated via the integument and/or the gut in fish (Dalmo et al. 2000). LPS as well as other foreign components such as polysaccharides, peptidoglycan, bacterial DNA and viral RNA are recognized by the immune system of carp via receptor proteins and as a result immune reactions are elicited. The reactions and functions of the immune system are further explained in Chapter 8 of this book.

LPS from cyanobacteria show a greater variety of long chain fatty acids and hydroxyl fatty acids and lack phosphate. In addition, it was suggested that the backbone sugar is glucosamine connected to various amounts of 2-keto 3-deoxyoctonate, galactoses and heptoses (Martin et al. 1989). In mammals it is known that these LPS can cause gastroenteritis and inflammation, but they have been reported to be less toxic than LPS from pathogenic gram-negative bacteria such as *Salmonella* (Bell and Codd 1996). In contrast, the comparison of cyanobacterial LPS to LPS of enteric bacteria showed that the former was more potent in zebrafish (Best et al. 2002). Effects of the cellular pellet after fractioning of cyanobacterial biomasses on glutathione S-transferase (GST) activity were observed which was attributed to the occurrence of LPS in carp as well as in zebrafish (Best et al. 2002, Palíková et al. 2007). Similarly, downregulation of biotransformation enzymes such as cytochrome P450 enzymes and reduced induction of GST activity by its model inductor 3-methylcholanthrene was observed in carp which may indicate reduced biotransformational ability (Marionnet et al. 1998). Still, the toxicity mechanisms of LPS endotoxins produced by cyanobacteria remain largely unknown.

Toxins of Algae

Toxins of algal origin are more important in marine ecosystems with 300 species of marine phytoplankton that can occur in 'red tides' (Hallegraeff 1993). However, even in freshwater systems problems with the appearance of algae including the occurrence of algal toxins may play a role.

Aquaculture ponds are often prepared for stocking which is especially important for stocking with fry or fingerlings since these early life stages are generally more sensitive to environmental conditions than others. This pre-stocking management includes the removal of aquatic weeds and algae since these may cause poor survival or low fish growth. Proper control of nutrients in the water often prevents algal blooms which not only may lead to toxin contamination but also cause depletion of dissolved oxygen and wide diurnal fluctuations of dissolved oxygen values. The decomposition of resulting dead organic material further sharpens the oxygen problem which has already been discussed in the section on cyanobacteria. Blooms of microalgae are not a new phenomenon occurring in fresh, brackish and marine waters (Burkholder 1998, Barkoh and Fries 2010). Algal blooms are easily recognizable because the proliferation of a particular species of alga to high densities often leads to visible discolouration of the water. Interestingly, the occurrence of harmful algal blooms has increased in numbers during the past decades (Hallegraeff 1993). This might be due to the facts that 1) there is an increased scientific awareness of harmful algal blooms, 2) waters are increasingly used for aquaculture, 3) harmful algal blooms have been stimulated by cultural eutrophication, 4) harmful algae species have been transported to other environments in ballast water of ships, and 5) water temperature has increased due to climate change (Smayda 1990). Common aquatic algae causing problems in fish culture ponds that lead to oxygen problems due to creation of algal mats are often filamentous algae belonging to the genera *Pithophora*, *Spiroqyra* or *Cladophora*.

Harmful unicellular algae that are closely related to the green alga *Chlorella* (Chlorophyta), are *Prototheca* species. These ubiquitous algae are spherical to oval in shape and their size ranges from 3 to 30 µm in diameter. Two of these species, *P. zopfii* and *P. wickerhamii*, have been reported to cause infections in humans and animals (Pore 1998). Loupal et al. (1992) reported inflammatory signs in the swimbladder of carp due to the non-photosynthetic, obligate heterotrophic alga *Prototheca*, but further investigations are lacking.

Ichthyotoxic algae are mostly restricted to marine ecosystems that comprise at least 60 phytoplankton species including raphidophytes, dinoflagellates, dictyochophytes and a few haptophytes have been identified (Aure et al. 2001, Backe-Hansen et al. 2001, Landsberg 2002). However, harmful algal blooms which may affect freshwater ecosystems also include planktonic species belonging to the groups of dinoflagellates, raphidiophyta, prymnesionmonads, silicoflagellates and diatoms (Rensel and Whyte 2004). Some species belonging to these groups that may also be important for carp will be reviewed later.

Effects of Dinoflagellates

Most dinoflagellates are marine species (genus: *Amphidinium*, *Gambierdiscus*, *Gymnodinium*, *Gyrodinium*, *Noctiluca*, *Peridinium*, *Pyrodinium*, *Ptychodiscus*, *Protogonyaulax*, etc.) (Burkholder 1998), but Hashimoto et al. (1968) reported mass mortality of carp during extensive blooms of *Peridinium polonicum* (= *Glenodima* sp.) in a lake near Tokyo (Japan).

Effects of Raphidiophyta

Raphidiophyta which are often also called chloromonads are a small group of photosynthetic motile unicellular organisms. They can become locally abundant in the marine plankton and often cause large-scale fish kills in near shore and brackish water environments. Thus, effects of toxic raphidiophyta on freshwater fish species are rare, but may occur in estuaries. Single-cell raphidophyte alga *Chattonella* cf. *verruculosa* was suggested to produce neurotoxic brevetoxins which have up to now only been reported in dinoflagellates such as *Karenia brevis* (Bourdelais et al. 2002). A lysis assay carried out with red blood cells of carp used the raphidiophyta *Heterosigma akashiwo* and *Chattonella* cf. *subsalsa* which showed low haemolytic effects. This led to the assumption that the toxicity of these species to carp is insignificant since similar cell abundances can never be detected in the environment for these species (Pistocchi et al. 2012). Nevertheless, neurotoxins prepared from

Chattonella marina and *Gymnodinium* ssp. led to depressed heart rate and reduced blood circulation in carp (Endo et al. 1992).

Effects of Prymnesiomonads

The most prominent member of prymnesiomonads in freshwater ecosystems is the alga *Prymnesium parvum* which is also called the 'golden alga' (Fig. 13.3) due to the yellow colour of the resulting blooms (Barkoh and Fries 2010). *P. parvum* is a flagellated haptophyte that is normally found suspended in the water column. *P. parvum* first appeared in 1947 in fish ponds in Israel causing extensive mortality of farmed carp and other fish species (Reich and Aschner 1947). Large-scale fish kills have also frequently occurred in Texas (USA) for least 30 years (Roelke et al. 2011) but have also occasionally been reported in Northern Germany (Hickel 1976, Dietrich and Hesse 1990). Cell densities of 150 million cells per litre were recorded during the Pecos River fish kill in 1986 (James and De La Cruz 1989). All species of fish including common carp can be affected in fish kill areas (James and De La Cruz 1989). Early investigations revealed that the toxin of *P. parvum* consists of several substances and not a single compound (Shilo and Sarig 1989) showing proteinaceous, acid-labile, thermostable and non-dialyzable characteristics (Prescott 1968). The haemolytic component of the *P. parvum* toxin mixture was reported to be a lipopolysaccharide (Padilla 1970). The *P. parvum* toxin, prymnesin, shows ichthyotoxicity, haemolytic activity and several chemical properties which are similar to the properties of saponins (Yariv and Hestrin 1961). Igarashi et al. (1999) reported that *P. parvum* produces two glycosidic toxins, prymnesin-1 and prymnesin-2, of almost similar biological activity including ichthyotoxicity. Prymnesin attaches to gill cell membranes and the attachment imposes a rearrangement on the membrane making it more

Figure 13.3. Microscopic picture of *Prymnesium parvum*, reprinted with permission from Manning and La Claire (2010). Marine Drugs 8(3): 678–704.

permeable. The ichthyotoxin leads to increased permeability of the gill and therefore affects all gill-breathing species (Yariv and Hestrin 1961, Paster 1973). The observed toxicity is characterized by a reversible damage to gill tissues that depends on availability of potassium and calcium, and suitable pH (Ulitzer and Shilo 1966, Glass et al. 1991). This is followed by the mortality stage where the target organisms are affected by the toxin present in micelles in the water.

Increased permeability of gill membranes imposed by prymnesin also makes fish more susceptible to other foreign compounds in water (Yariv and Hestrin 1961) and may even include increased susceptibility to the cytotoxic and haemolytic activity of the *P. parvum* toxin (Ulitzer and Shilo 1966). It was also observed that intraperitoneal injection of the toxin in mosquitofish (*Gambusia* sp.) leads to less toxicity than toxin immersion. This led to the assumption that the toxin may be altered (inactivated) in the gastrointestinal tract and liver (Spiegelstein et al. 1969). Consequently, this may explain why the toxin is not toxic to non-gill breathers, but toxic to gill breathers.

Padan et al. (1967) showed that the ichthyotoxin and hemolysin are both sensitive to inactivation by light. In addition, they concluded that light is needed for the appearance of extracellular hemolysin. In contrast, Spiegelstein et al. (1969) showed that light was not necessary for ichthyotoxin production. Larsen and Bryant (1998) assumed that growth phase and nutrient status of the algae probably have a greater impact on toxicity than variable salinity, light and temperature. Past studies suggested that the *P. parvum* toxin appears when the conditions for growth are limited and growth is disturbed, while with substantial concentrations of nitrogen and phosphorous, *P. parvum* will not produce or release toxins (Holdway et al. 1978). Dafni et al. (1972) found that decreased phosphate concentrations caused an increase in toxicity possibly due to increased membrane leakiness and the toxin escaping which was confirmed by Kaartvedt et al. (1991) and Larsen et al. (1993). Later on it was also observed that nitrogen limitation causes increased toxicity which led to the hypothesis that the ratio of nitrogen to phosphorus could be the governing factor of toxicity by *P. parvum* (Johansson and Graneli 1999). The authors speculated that the toxin might function as a suppressor of competing organisms during nutrient limitation.

When intoxication of fish occurred several strategies have been suggested to lower toxicity including transferring fish to uncontaminated water, increasing dissolved oxygen levels by blowing bubbles into the water, increasing the sodium chloride concentration in water or using absorbents such as charcoal or kaolin (Ulitzer and Shilo 1966, Glass et al. 1991).

Although the occurrence of *P. parvum* is often restricted to certain water bodies, the economic importance can be overwhelming since fish are killed across the globe. Thus, the presence of *P. parvum* is still a cause for concern most importantly because this alga is ever present once it first appears.

Toxicity and Anti-nutritive Effects of Compounds Derived from Higher Plants

Plant-derived ingredients are also often considered for replacement of fish meal in fish feeds because the increased demand for protein sources for feed production for aquaculture cannot be satisfied by the globally available fishmeal. Especially, oilseeds and pulses are important crops and their worldwide demand has increased considerably in the recent years (Tacon et al. 2011). Deficiencies in essential amino acids of the most plant materials can be counteracted by dietary supplementations with the required free amino acids or by using complementary protein sources (Tacon and Jackson 1985). However, various anti-nutrients occur in oilseeds, pulses and other plant materials that are considered for feed production (Tacon 1995, Gatlin et al. 2007). These anti-nutrients can (I) affect protein utilization or digestions, e.g., protease inhibitors; (II) affect utilization of minerals, e.g., glucosinolates; or comprise (III) antivitamins; and (IV) detrimental compounds such as phytoestrogens and saponins.

The most important anti-nutrients of group I are protease inhibitors since they can often occur due to ingredients in fish feed such as legumes (Norton 1991). Despite the presence of several other anti-nutrients, soybeans (*Glycine max*) contain different types of protease inhibitors with different sensitivity to heat and acid which also show differences in inhibiting trypsin (Tacon 1995). Commercial soybean products usually show trypsin inhibitors ranging from 2.6 to 8.5 mg g^{-1} (Mustakas et al. 1981). Carp are often able to maintain their growth rates when trypsin inhibitors are present in their diet. However, hydrothermally- and intensely thermally-treated soybean meal included at 50% in the diet led to better growth performance of carp than untreated soybean meal (Abel et al. 1984). Soybean products often also contain high concentrations of isoflavones which have high estrogenic potentials. Accordingly, up to 273 µg g^{-1} genistein and 176 µg g^{-1} daidzein and high estrogenic potential have been found in experimental feed for carp (Miyahara et al. 2003). These phytoestrogens have also been identified in wastewater and surface waters (Liu et al. 2010). However their effects on carp remain unknown so far.

Rapeseed meal (from *Brassica campestris napus*) often contains protease inhibitors and other anti-nutritional factors (Tacon 1995). When given to carp via the diet it affected the levels of free aminoacids in the muscle and hepatopancreas and influenced the carotenoid content in fish (Dabrowski et al. 1981, Czeczuga and Dabrowski 1983). Similarly, crucian carp showed lower growth performance and higher feed conversion ratios after feeding rapeseed meal at high inclusion rates in the diet for eight weeks (Cai et al. 2013). Even the application of processing techniques such as dehulling, use of high temperature and organic solvents during oil extraction, and sieving of the obtained meal did not sufficiently reduce the content of anti-nutritional factors in rapeseed meal (Anderson-Haferman et al. 1993, Mawson et al. 1993, 1994a,b, 1995, Leming et al. 2004). Although protein extraction from rapeseed meal by methanol-ammonia treatment or ethanol treatment increased the protein content and removed glucosinolates and phenolic compounds, the amount of non-digestible fibre increased (Naczk and Shahidi 1990, McCurdy and March 1992, Mwachireya et al. 1999, Chabanon et al. 2007). In a study using juvenile carp, the cold-pressed rapeseed was hexane treated to further reduce the oil content, which was followed by liquid water extraction, diafiltration and ultrafiltration to reduce the content of undesired anti-nutritive compounds (Slawski et al. 2010). However, an inclusion of this rapeseed protein concentrate into the carp diet as a replacement of more than one third of the fishmeal component in the diet was not advisable since the growth performance of carp was reduced. It was assumed that the bitter taste of glucosinolates led to a reduced acceptance of the diet although the glucosinolate content in the diet in which all fishmeal had been replaced by rapeseed protein concentrate was as low as 0.05 $\mu M/g$. These results indicate that carp are quite sensitive to glucosinolates as compared to rainbow trout and turbot which reject diets containing glucosinolate concentrations of 7.3 and 18.7 $\mu M/g$ respectively (Burel et al. 2000a,b).

Indian mustard, *Brassica juncea*, is known to contain trypsin inhibitors, glucosinolate and anti-thiamine factors which might affect fish growth performance (Tacon 1995). Carp fry fed a mustard cake-containing diet (at 25 and 50% of the total protein) for five weeks showed reduced growth and increased feed conversion ratios (Hasan et al. 1997). In particular, liver histology was affected showing intracellular lipid deposition. Moreover, inclusion of 50% mustard cake led to more than 50% mortality. Effects of Indian mustard on common carp were also investigated by Hossain and Jauncey (1988, 1989a,b). These studies revealed that different glucosinolate concentrations originating from mustard oilcake significantly reduced weight gain, specific growth rates, protein efficiency ratios and feed conversion ratios in fish fed with the experimental diets for 8 weeks. Although, carp appear to be more resistant to glucosinolates from rapeseed than salmonids (Yurkowski et al. 1978, Hilton and Slinger 1986), mustard-based diets are not appropriate for carp feeding.

Carp fry fed a diet containing linseed at 25% of the total protein for five weeks showed growth and feed conversion similar to control fish (Hasan et al. 1997). A diet with 50% linseed showed poorer acceptability. In addition, sesame inclusion in the diet for carp fry at 25% of the total protein reduced the weight gain of carp fry (Hasan et al. 1997). At 75% sesame in the diet some fish showed malformations and 26.7% mortality occurred. Linseed meal and sesame meal were also included in the diet of carp at 30 and 27%, respectively (Hossain and Jauncey 1990). Linseed is known to contain several anti-nutritional factors, such as the cyanogen linamarin, phytic acid, estrogenic factors and anti-vitamins whereas in sesame phytic acid is often present (Montgomery 1980, Tacon 1995). Sodium phytate fed separately to carp led to decreased growth, food utilization and protein digestibility (Hossain and Jauncey 1993). The study of Hossain and Jauncey (1990) showed that the nutritional value of linseed and sesame could be increased by aqueous extraction or autoclaving steps. However, the growth performance could still not compete with control fish fed a fishmeal-based diet.

Inclusion of copra at 25% of the total protein for feeding carp fry led to reduced weight gain and increased the feed conversion ratio after five weeks of feeding (Hasan et al. 1997) which can be attributed to the presence of anti-nutritional factors in copra. Peanut (*Arachis hypogaea*) oil cakes were also tested in diets for carp fry at two concentrations (at 25 and 75% of total protein) for five weeks (Hasan et al. 1997). Peanut material is known to contain anti-nutritional factors, such as protease inhibitors, phytohaemagglutinins, phytic acid and saponins (Tacon 1995), and also aflatoxins (Ashley 1970). Carp fry showed sufficient growth when fed with 25% peanut oil cakes. However, inclusion of 75% peanut oil cake in the total protein of the diet for carp fry considerably reduced weight gain, led to 20% mortality and induced severe deformations of fish after five weeks of feeding (Hasan et al. 1997).

Seeds of *Jatropha curcas* (which is a plant belonging to the family of *Euphorbiaceae*) show high oil contents and are used mostly for production of bio-diesel fuel. Most varieties of this species are inedible and even non-toxic strains are still known to contain trypsin inhibitors. Carp fed a diet containing 23% *J. curcas* seed meal showed lower growth performance than control fish fed a fishmeal-based diet (Makkar and Becker 1999). Since *J. curcas* seed meal also often contains phorbol esters as anti-nutritional factors, these compounds have been investigated separately in the study of Becker and Makkar (1998) who reported feed rejection, intestinal mucus production and decreased growth at concentrations of 31 mg kg^{-1} phorbol esters or higher in feed for carps.

Sunflower seed meal is known to contain tannins and has been shown to affect the anti-oxidative status of liver in carp after 12 weeks of feeding (Meriç 2013). Tannic acid which belongs to the hydrolysable tannins completely suppressed feeding of carp after 28 days although condensed tannins did not affect the performance of fish (Becker and Makkar 1999). Cassava (*Manihot esculenta*) or rice meal at 45% in the diet showed high weight increase and allowed protein sparing (Ufodike and Matty 1983) and therefore might be more suitable plant-derived ingredients in fish feeds.

Bitter lupin (*Lupinus luteus*), a member of the legumes family *Fabaceae*, is unsuitable for feed production due to its high content in alkaloids. However, sweet lupin (*Lupinus angustifolius*) seed meal can be included in fish feeds due to its low alkaloid contents. At levels of 30 and 45% in carp diet higher or similar growth performance as compared to the control group was observed (Viola et al. 1988). In the study of Oberle et al. (1997) even 50% lupin seed in the diet for carp showed similar growth performance as cereal-based diets in a feeding trial for 105 days.

Sesbania is a genus that also belongs to the group of peas, *Fabaceae*. Their value as a protein source in carp nutrition has been investigated (Hossain et al. 2001a,b,c). Juvenile carp fed with diets which contained 10–40%

Sesbania seed meal in the dietary crude protein for seven weeks showed at 10% inclusion of *Sesbania* seed meal similar growth performance compared to fish fed a meal-based control diet whereas higher inclusion ratios clearly had detrimental effects on growth performance (Hossain et al. 2001a). The negative effects of *Sesbania* were particularly pronounced when endosperm was used for feed preparation (Hossain et al. 2001b). It was noticeable, that the inclusion of endosperm led to a high viscosity of intestinal contents within the eight-weeks feeding trial using common carp probably caused by non-starch polysaccharides such as galactomannan. Nevertheless, *Sesbania* species have been shown to contain different anti-nutritional factors, such as phenols, tannins, condensed tannins, phytic acid, saponins, trypsin inhibitors and lectins at occasionally substantial concentrations (Hossain and Becker 2001). Soaking the *Sesbania aculeata* seeds in water for 24 hours as well as an additional autoclaving step significantly increased growth performance of juvenile carp fed diets prepared from this for eight weeks at inclusion levels of 20 and 30% of the total dietary protein in the diet. However, the performance of fishmeal-based control feeds could not be reached which was attributed to the presence of thermo-stable anti-nutrients and low solubility of proteins of *Sesbania aculeata* (Hossain et al. 2001c).

Another member of the family *Fabaceae*, *Mucuna pruriens*, was used in a feeding trial with juvenile carp (Siddhuraju and Becker 2001). In this study fish were fed diets in which 10–40% of the total dietary protein was replaced by *Mucuna* seed meal whereby concentrations above 10% *Mucuna* significantly reduced growth which was not reversed by autoclaving the seed meal.

Leucaena leucocephala of the group of *Mimosaceae* also belongs to the family of *Fabaceae*. This plant shows a high content of anti-nutritional mimosine. Its leaf meal is used for animal feeding in East Asia, but their use for feeding carp fry was not recommended by Hasan et al. (1997) since inclusion at 25% of total protein already led to high mortality and very low growth compared to a fishmeal-based control group after five weeks of feeding. Mimosine has also been shown to directly induce cell death in primary cultured carp hepatocytes which further emphasizes its unsuitability in carp nutrition (Vogt et al. 1994).

The use of red clover (*Trifolium pratense*) as a feed additive at 100 mg/kg for carp promoted growth, and increased protein and fat content (Turan et al. 2007) possibly due to its estrogenic potential. In contrast to other fishes this plant also increased survival of carp (Turan and Akyurt 2005, Turan 2006, Turan et al. 2007).

Clove oil distilled from the clove tree (*Eugenia aromatica* or *E. caryophyllata* (Isaacs 1983, Soto and Burhanuddin 1995, Keene et al. 1998) is usually used for anaesthesia of fish. The active compound is called eugenol (4-allyl-2-methoxyphenol). For this purpose 25 to 100 mg L^{-1} was recommended for carp by Hikasa et al. (1986). Acute toxicity of clove oil was investigated in juvenile carp and the 96 hours LC_{50} was reported to be at 18.1 mg L^{-1} (Velisek et al. 2005). Dead fish showed increased mucus excretion by body surfaces and darkened gills. The LC_{50} for short exposure duration (10 minutes) was identified at 74.3 mg L^{-1}. Changes of blood biochemistry (glucose levels) have already occurred at lower concentrations (30 mg L^{-1}) in fish exposed for 10 minutes. These effects were abrogated within 24 hours after clove oil application.

Toxins of Mollusks

Mollusks can be part of the diet of fish which also includes carp (García-Berthou 2001). Toxin accumulation occurs via uptake of toxigenic microalgae in mollusks rather than by toxins produced directly by molluscs in freshwater ecosystems. However, marine gastropods of the family *Buccinidae* (whelk, sea snails) are know to produce the neurotoxin tetramine as the major component of their salivary poison. For example, the salivary gland of *Neptunea arthritica* contains 7 to 9 mg tetramine g^{-1} of gland (Asano and Itoh 1960). The Scandinavian species *Neptunea antiqua* contains as much as 20 to 30 mg tetramine g^{-1} of gland (Anthoni et al. 1989). Thus, contamination with toxins from these gastropods may occur in brackish water.

Toxicity tests of salivary poison and of authentic tetramine using fish (*Cyprinus carpio*) were carried out by submuscular injection, or by per os administration using polyethlene catheters. After either injection or administration, the following symptoms have been observed: gradual loss of balance, inversion of the abdomen, delayed respiration rate, clinical convulsion and, finally death (Asano and Itoh 1960). Further effects of toxins from brackish water on carp have not yet been investigated.

Conclusions

Carp in natural habitats and aquaculture are exposed to a multitude of natural toxins and detrimental factors which are mostly not fully understood so far. More research is needed to further explore the importance of natural toxins for carp health.

References

Abedi, Z.H. and P.M. Scott. 1969. Detection of toxicity of aflatoxins, sterigmatocystin, and other fungal toxins by lethal action on zebra fish larvae. J. Assoc. Off. Anal. Chem. 52: 962–969.

Abdelhamid, A.M. 1988. Effect of sterigmatocystin contaminated diets on fish performance. Arch. Anim. Nutr. Berlin 38: 833–846.

Abdelhamid, A.M. 1990. Occurrence of some mycotoxins (aflatoxin, ochratoxin-A, citrinin, zearalenone and vomitoxin) in various Egyptian feeds. Arch. Anim. Nutr. 40: 647–664.

Abdel-Wahhab, M.A., A.M. Hasan, S.E. Aly and K.F. Mahrous. 2005. Adsorption of sterigmatocystin by montmorillonite and inhibition of its genotoxicity in the Nile tilapia fish (*Oreochromis niloticus*). Mut. Res. 582: 20–27.

Abel, H.J., K. Becker, C.H.R. Meske and W. Friedrich. 1984. Possibilities of using heat-treated full-fat soybeans in carp feeding. Aquacult. 42: 97–108.

Agouz, H.M. and W. Anwer. 2011. Effect of Biogen® and Myco-Ad® on the growth performance of common carp (*Cyprinus carpio*) fed a mycotoxin contaminated aquafeed. J. Fish. Aquatic Sci. 6(3): 334–345.

Ahn Tuan, N.A., J.M. Grizzle, R.T. Lovell, B.B. Manning and G.E. Rottinghaus. 2002. Growth and hepatic lesions of Nile tilapia (*Oreochromis niloticus*) fed diets containing aflotoxin B1. Aquacult. 212: 311–319.

Ahn Tuan, N.A., B.B. Manning, R.T. Lovell and G.E. Rottinghaus. 2003. Responses of Nile tilapia (*Oreochromis niloticus*) fed diets containing different concentrations of moniliformin of fumonisin B1. Aquacult. 217: 515–528.

Akter, A., M. Rahman and M. Hasan. 2010. Effects of aflatoxin B$_1$ on growth and bioaccumulation in common carp fingerling in Bangladesh. Asia-Pacif. J. Rural Dev. 20(2): 1–13.

Ali, N. 2007. Teratology in zebrafish embryo: A tool for risk assessment. Report—Master of Science Programme in Veterinary Medicine for International Students Faculty of Veterinary Medicine and Animal Science, Swedish University of Agricultural Sciences, Report no. 65, ISSN 1403-2201.

Al-Sabti, K. 1986. Clastogenic effects of five carcinogenicmutagenic chemicals on the cells of the common carp, *Cyprinus carpio* L. Comp. Biochem. Physiol. Part C: Comp. Pharmacol. 85: 5–9.

Al-Sultan, E.Y.A. 2011. The isolation, the purification and the identification of Hepatotoxin Microcystin-LR from two cyanobacterial species and studying biological activity on some aquatic orgamisms. J. Basrah Res. (Sci.) 37(1): 39–57.

Anderson-Hafermann, J.C., Y. Zhang and C.M. Parsons. 1993. Effects of processing on the nutritional quality of canola meal. Poult. Sci. 72: 326–333.

Anthoni, U., L. Bohlin, C. Larsen, P. Nielsen, N.H. Nielsen and C. Christophersen. 1989. The toxin tetramine from the "edible" whelk *Neptunea antiqua*. Toxicon 27: 717–723.

Asano, M. and M. Itoh. 1960. Salivary poison of a marine gastropod, *Neptunea arthritica* Bernardi, and the seasonal variation of its toxicity. Ann. New York Acad. Sci. 90: 674–688.

Ashley, L.M. 1970. Pathology of fish fed aflatoxins and other antimetabolites *In*: S.F. Sniesko (ed.). Proceedings of the Symposium on Diseases of Fishes and Shellfishes. Am. Fish. Soc. Spec. Publ. 5: 366–379.

Aure, J., D.S. Danielssen, M. Skogen, E. Svendsen, E. Dahl, H. Søiland and L. Petterson. 2001. Environmental conditions during the *Chattonella* bloom in the North Sea and Skagerrak in May 1998. *In*: G.M. Hallegraeff, C.J.S. Bolch, I. Blackburn and R. Lewis (eds). Harmful algal blooms 2000. Intergovernmental Oceanographic Commission of UNESCO, Paris, pp. 82–85.

Backe-Hansen, P., E. Dahl and D.S. Danielssen. 2001. On the bloom of Chattonella in the North-Sea/Skagerrak in April–May 1998. *In*: G.M. Hallegraeff, C.J.S. Bolch, I. Blackburn and R. Lewis (eds.). Harmful algal blooms 2000. Intergovernmental Oceanographic Commission of UNESCO, Paris, pp. 78–81.

Bakos, K., R. Kovács, Á. Staszny, D.K. Sipos, B. Urbányi, F. Müller, Z.Csenki and B. Kovács. 2013. Developmental toxicity and estrogenic potency of zearalenone in zebrafish (*Danio rerio*). Aquatic Toxicol. 136-137: 13–21.

Balachandran, C. and K.R. Parthasarathy. 1996. Occurrence of cyclopiazonic acid in feeds and feedstuffs in Tamil Nadu, India. Mycopathol. 133(3): 159–162.

Balogh, K., M. Heincinger, J. Fodor and M. Mézes. 2009. Effect of long term feeding of T-2 and HT-2 toxin contaminated diet on the glutathione redox status and lipid peroxidation processes in common carp (*Cyprinus carpio* L.). Acta Biol. Szeged 53: Suppl.1.

Barbosa, T.S., C.M. Pereyra, C.A. Soleiro, E.O. Dias, A.A. Oliveira, K.M. Keller, P.P.O. Silva, L.R. Cavaglieri and C.A.R. Rosa. 2013. Mycobiota and mycotoxins present in finished fish feeds from farms in the Rio de Janeiro State, Brazil. Int, Aquatic Res. 5: 3.

Barkoh, A. and L.T. Fries. 2010. Aspects of the origins, ecology, and control of golden alga *Prymnesium parvum*: Introduction to the featured collection. J. Am. Water Resour. Ass. 46(1): 1–5.

Becker, K. and H.P.S. Makkar. 1998. Effect of phorbol esters in carp *Cyprinus carpio* L. Vet. Hum. Toxicol. 40(2): 82–86.

Becker, K. and H.P.S. Makkar. 1999. Effects of dietary tannic acid and quebracho tannin on growth performance and metabolic rates of common carp *Cyprinus carpio* L. Aquacult. 175: 327–335.

Bell, S.G. and G.A. Cood. 1996. Detection, analysis and risk assessment of cyanobacterial toxins. *In*: Hester, R.E. and Harrison R.M. (eds.). Agricultural chemicals and the environment. Royal Society of Chemistry, Cambridge, UK. 5: 109–122.

Berczi, I., L. Bertók and T. Bereznai. 1966. Comparative studies on the toxicity of *Escherichia coli* lipopolysaccharide edotoxin in various animal species. Can. J. Microbiol. 12: 1070–1071.

Best, J.H., S. Pflugmacher, C. Wiegand, F.B. Eddy, J.S. Metcalf and G.A. Codd. 2002. Effects of enteric bacterial and cyanobacterial *lipopolysaccharides*, and of microcystin-LR, on gluthaione *S*-transferase activities in zebra fish (*Danio rerio*). Aquatic Toxicol. 60: 223–231.

Beveridge, M.C.M., D.J. Baird, S.M. Rahmatulla, L.A. Lawton, K.A. Beattie and G.A. Codd. 1993. Grazing rates on toxic and non-toxic strains of cyanobacteria by *Hypophthalamichtys molitrix* and *Oreochromis niloticus*. J. Fish Biol. 43: 901–907.

Binder, E.M., L.M. Tan, L.J. Chin, J. Handl and J. Richard. 2007. Worldwide occurrence of mycotoxins in commodities, feeds and feed ingredients. Anim. Feed Sci. Technol. 137: 265–282.

Botes, D.P., H. Kruger and C.C. Viljoen. 1982. Isolation and characterization of four toxins from the blue-green algae, *Microcystis aeruginosa*. Toxicon 20: 945–954.

Bourdelais, A.J., C.R. Tomas, J. Naar, J. Kubanek and D.G. Baden. 2002. New fish-killing alga in coastal Delaware produces neurotoxins. Environ. Health Perspect. 110(5): 465–470.

Bownik, A., A. Rymuszka, A. Sierosławska and T. Skowroński. 2012. Anatoxin-a induces apoptosis of leukocytes and decreases the proliferative ability of lymphocytes of common carp (*Cyprinus carpio* L.) *in vitro*. Pol. J. Vet. Sci. 15(3): 531–535.

Briand, J.F., C. Robillot, C. Quiblier-Lloberas, J.F. Humbert, A. Coute and C. Bernard. 2002. Environmental context of *Cylindrospermopsis raciborskii* (Cyanobacteria) blooms in a shallow pond in France. Water Res. 36(13): 3183–3192.

Brown, D.W., C.P. McCoy and G.E. Rottinghaus. 1994. Experimental feeding of *Fusarium moniliforme* culture material containing fumonisin B1 to channel catfish (*Ictalurus punctatus*). J. Vet. Diagn. Invest. 6(1): 123–124.

Bucheli, T.D., M. Erbs, N. Hartmann, S. Vogelsang, F.E. Wettstein and H.R. Forrer. 2005. Estrogenic mycotoxins in the environment. Mitt. Lebensm. Hyg. 96: 386–403.

Bucheli, T.D., F.E. Wettstein, N. Hartmann, M. Erbs, S. Vogelsang, H.-R. Forrer and R.P. Schwarzenbach. 2008. *Fusarium* mycotoxins: Overlooked aquatic micropollutants. J. Agricult. Food Chem. 56: 1029–1034.

Burkholder, J.M. 1998. Implications of harmful microalgae and heterotrophic dinoflagellates in management of sustainable marine fishes. Ecol. Appl. 8(1): S37–S62.

Burel, C., T. Boujard, S.J. Kaushik, G. Boeuf, S. van der Geyten, K.A. Mol, E.R. Kuhn, A. Quinsac, M. Krouti and D. Ribaillier. 2000a. Potential of plant-protein sources as fishmeal substitutes in diets for turbot (*Psetta maxima*): growth, nutrient utilisation and thyroid status. Aquacult. 188: 363–382.

Burel, C., T. Boujard, A.M. Escaffre, S.J. Kaushik, G. Boeuf, K. Mol, S. van der Geyten and E.R. Kühn. 2000b. Dietary low glucosinolate rapeseed meal affects thyroid status and nutrient utilization in rainbow trout (*Oncorhynchus mykiss*). Brit. J. Nutr. 83: 653–664.

Bury, N.R., F.B. Eddy and G.A. Codd. 1996. The effects of cyanobacterium and the cyanobacterial toxin microcystin-LR on Ca^{2+} transport and Na^+/K^+-ATPase in tilapia gills. J. Exp. Biol. 199: 1319–1326.

Bury, N.R., G.A. Codd, S.E. Wendelaar Bonga and G. Flik. 1998. Fatty acids from the cyanobacterium *Microcystis aeruginosa* with potent inhibitory effects of fish gill Na^+/K^+-ATPase activity. J. Exp. Biol. 201: 81–89.

Cai, C., L. Song, Y. Wang, P. Wu, Y. Ye, Z. Zhang and C. Yang. 2013. Assessment of the feasibility of including high levels of rapeseed meal and peanut meal in diets of juvenile crucian carp (*Carassius auratus gibelio* ♀ × *Cyprinus carpio* ♂): Growth, immunity, intestinal morphology, and microflora. Aquacult. 410–411: 203–215.

Carbis, C.R., G.F. Mitchel, J.W. Anderson and I. McCauley. 1996. The effects of microcystins on the serum biochemistry of carp, *Cyprinus carpio* L., when the toxins are administered by gavage, immersion and intraperitoneal routes. J. Fish Dis. 19: 151–159.

Carmichael, W.W., J.T. Eschedor, G.M. Patterson and R.E. Moore. 1988. Toxicity and partial structure of a hepatotoxic peptide produced by the cyanobacterium *Nodularia spumigena* Mertens emend. L575 from New Zealand. Appl. Environ. Microbiol. 54: 2257–2263.

Carmichael, W.W. 1997. The cyanotoxins. pp. 211–256. *In*: J.A. Callow (ed.), Advances in Botanical Research,. Academic Press, London, UK.

Carmichael, W.W. 2001. Health effect of toxin-producing Cyanobacteria: "The CyanoHABs". Hum. Ecol. Risk Ass. 7: 1393–1407.

Castenholz, R.W. and J.B. Waterbury. 1989. Cyanobacteria. pp. 1710–1727. *In*: J.T. Staley, M.P. Bryant, N. Pfennig and J.G. Holt (eds.). Bergey's Manual of Systematic Bacteriology, Volume 3. Williams & Wilkins, Baltimore.

CAST Report. 2003. Mycotoxins: risks in plant, animal, and human systems. *In*: J.L. Richard and G.A. Payne (eds.). Council for Agricultural Science and Technology Task Force Report No. 139, Ames, Iowa, USA.

Chabanon, G., I. Chevalot, X. Framboisier, S. Chenu and I. Marc. 2007. Hydrolysis of rapeseed protein isolates: kinetics, characterization and functional properties of hydrolysates. Process Biochem. 42: 1419–1428.

Chavez-Sanches, M.C., C.A. Martinez and I.O. Moreno. 1994. Pathological effects of feeding young *Oreochromis niloticus* diets supplemented with different levels of aflatoxin B1. Aquacult. 127: 49–60.

Chorus, I. and J. Bartram. 1999. Toxic Cyanobacteria in Water. A Guide to their Public Health Consequences, Monitoring and Management. E&FN Spon, London.

Chiswell, R.K., G.R. Shaw, G. Eaglesham, M.J. Smith, R.L. Norris, A.A. Seawright and M.R. Moore. 1999. Stability of cylindrospermopsin, the toxin from the cyanobacterium, *Cylindrospermopsis raciborskii*: effect of pH, temperature, and sunlight on decomposition. Environ. Toxicol. 14: 155–161.

Codd, G.A. 1995. Cyanobacterial toxins: occurrence, properties and biological significance. Wat. Sci. Technol. 32: 149–156.

Codd, G.A. 2000. Cyanobacterial toxins, the perception of water quality, and the prioritisation of eutrophication control. Ecol. Engin. 16: 51–60.

Czeczuga, B. and K. Dabrowski. 1983. Rapeseed meal in the diet of common carp reared in heated waters. Zeitschr. Tierphysiol. Tierern. Futtermittelk. 50: 52–61.

Dabrowski, K., H. Dabrowska and H. Kozlowska. 1981. Rapeseed meal in the diet of common carp reared in heated waters. Zeitschr. Tierphysiol. Tierern. Futtermittelk. 45: 66–76.

Dafni, Z., S. Ulitzur and M. Shilo. 1972. Influence of light and phosphate on toxin production and growth of *Prymnesium parvum*. J. Gen. Microbiol. 70: 199–207.

Dalmo, R.A., A.A. Kjerstad, S.M. Arnesen, P.S. Tobias and J. Bogwald. 2000. Bath exposure of Atlantic halibut (*Hippoglossus hippoglossus* L.) yolk sac larvae to bacterial lipopolysaccharide (LPS): Absorption and distribution of the LPS and effect on fish survival. Fish Shellfish Immunol. 10: 107–128.

Davidson, F.F. 1959. Poisoning of wild and domestic animals by a toxic waterbloom of *Nostoc rivulare* Küertz. J. Am. Water Works Assoc. 51: 1277–1287.

Deng, S.X., L.X. Tian, F.J. Liu, S.J. Jin, G.Y. Liang, H.J. Yang, Z.Y. Du and Y.J. Liu. 2010. Toxic effects and residue of aflatoxin B. Aquacult. 307: 233–240.

Devic, E., D. Li, A. Dauta, P. Henriksen, G.A. Codd, J.-L. Marty and D. Fournier. 2002. Detection of anatoxin-a(s) in environmental samples of cyanobacteria by using a biosensor with engineered acetylcholinesterases. Appl. Environ. Microbiol. 68: 4102–4106.

Devlin, J.P., O.E. Edwards, P.R. Gorham, M.R. Hunter, R.K. Pike and B. Stavric. 1977. Anatoxin-a, a toxic alkaloid from *Anabaena flosaquae* NCR-44h. Can. J. Chem. 55: 1367–1371.

Dietrich, W. and K.J. Hesse. 1990. Local fish kill in a pond at the German North Sea coast associated with a mass development of *Prymnesium* sp. Meeresforschung/Report. Mar. Res. 33(1): 104–106.

Ding, W.X., H.M. Shen, Y. Shen, H.G. Zhu and C.N. Ong. 1998. Microcystic cyanobacteria causes mitochondrial membrane potential alteration and reactive oxygen species formation in primary cultured rat hepatocytes. Environ. Health Perspect. 106: 409–413.

Döll, S., G. Baardsen, P. Möller, W. Koppe, I. Stubhaug and S. Dänicke. 2010. Effects of increasing concentrations of the mycotoxins deoxynivalenol, zearalenone or ochratoxin A in diets for Atlantic salmon (Salmo salar) on growth performance and health. Book of abstracts, International Symposium of Fish Nutrition and Feeding, Qingdao, China. 120.

Dow, C.S. and U.K. Swoboda. 2000. Cyanotoxins. pp. 614–632. *In*: B.A. Wihtton and M. Potts (eds.). The Ecology of Cyanobacteria. Kluwer Academic Publishers, Dordrecht.

Duy, T.N., P.K.S. Lam, G.R. Shaw and D.W. Connell. 2000. Toxicology and risk assessment of freshwater cyanobacterial (blue-green algal) toxins in water. Rev. Environ. Contam. Toxicol. 163: 113–186.

Edwards, C., K.A., Beattie, C.M., Scrimgeour and G.A. Codd. 1992. Identification of anatoxin-a in benthic cyanobacteria (blue-green-algae) and in associated dog poisonings at Loch Insh, Scotland. Toxicon 30: 1165–1175.

El-Banna, R., H.M. Teleb and F.M. Fakhry. 1992. Performance and tissue residues of tilapias fed dietary aflatoxin. Vet. Med. J. 40: 17–23.

Ellis, R.W., M. Clements, A. Tibbetts and R. Winfree. 2000. Reduction of the bioavailability of 20 g/kg aflotoxin in trout feed containing clay. Aquacult. 183: 179–188.

El Ghazali, I., S. Saqrane, A.P. Carvalho, Y. Ouahid, F.F. Del Campo, B. Oudra and V. Vasconcelos. 2010. Effect of different microcystin profiles on toxin bioaccumulation in common carp (*Cyprinus carpio*) larvae via *Artemia* nauplii, Ecotoxicol. Environ. Saf. 73(5): 762–770.

El-Sayed, Y.S. and R.H. Khalil. 2009. Toxicity, biochemical effects and residue of aflatoxin B1 in marine water-reared sea bass (*Dicentrarchus labrax* L.). Food Chem. Toxicol. 47: 1606–1609.

Encarnacao, P., B. Srikhum, I. Rodrigues and U. Hofstetter. 2009. Growth performance of red tilapia (*O. niloticus* x *O. mossambicus*) fed diets contaminated with aflatoxin b1 and the use of a commercial product to suppress negative effects. Book of abstracts, World Aquaculture 2009, Veracruz, Mexico. September 2009.

Endo, M., Y. Onoue and A. Kuroki. 1992. Neurotoxin-induced cardiac disorder and its role in the death of fish exposed to *Chattorzella marina*. Mar. Biol. 112: 372–376.

European Commission, Commission Recommendation (EC) 576 of 17 August 2006 on the presence of deoxynivalenol, zearalenone, ochratoxin A, T-2 and HT-2 and fumonisins in products intended for animal feeding. Off. J. Eur. Un. 2006, L229/7-L229/9.

European Union: Directive 2002/32/EC of the European Parliament and of the Council on undesirable substances in animal feed. Official Journal L 140, 30 May 2002, pp. 10–22 07 May 2002.

European Commission, Commission Recommendation (EC) 165 of 27 March 2013 on the presence of T-2 and HT-2 in cereals and cereal products. Off. J. Eur. Un. 2006, L91/12.

Falconer, I.R. 1999. An overview of problems caused by toxic blue-green algae (cyanobacteria) in drinking and recreational water. Environ. Toxicol. 14: 5–12.

Farag, H.E.M., A.A. El-Tabiy and H.M. Hassan. 2011. Assessment of ochratoxin A and aflatoxin B1 levels in the smoked fish with special reference to the moisture and sodium chloride content. Res. J. Microbiol. 6(12): 813–825.

Fischer, W.J. and D.R. Dietrich. 2000. Pathological and biochemical characterization of microcystin-induced hepatopancreas and kidney damage in carp (*Cyprinus carpio*). Toxicol. Appl. Pharmacol. 164: 73–81.

Fischer, W.J., B.C. Hitzfeld, F. Tencalla, J.E. Eriksson, A. Mikhailov and D.R. Dietrich. 2000. Microcystin-LR toxicodynamics, induced pathology, and immunohistochemical localization in livers of blue-green algae exposed rainbow trout (*Oncorhynchus mykiss*). Tox. Sci. 54: 365–373.

Frisvad, J.C. and R.A. Samson. 1991. Mycotoxins produced by species of *Penicillium* and *Aspergillus* occurring in cereals. pp. 441–476. *In*: J. Chełkowski (ed.). Cereal Grain: Mycotoxins, Fungi and Quality in Drying and Storage. Elsevier Science Publishers B.V., Amsterdam, Netherlands.

Froscio, S.M., S. Fanok and A.R. Humpage. 2009. Cytotoxicity screening for the cyanobacterial toxin cylindrospermopsin. J. Toxicol. Environ. Health Part A 72: 345–349.

Gaete, V., E. Canelo, N. Lagos and F. Zambrano. 1994. Inhibitory effects of Microcystis aeruginosa toxin on ion pumps of the gill of freshwater fish. Toxicon. 32(1): 121–127.

Galat, D.L., J.P. Verdin and L.L. Sims. 1990. Large-scale patterns of *Nodularia spumigena* blooms in Pyramid Lake, Nevada, determined from Landsat imagery: 1972–1986. Hydrobiol. 197: 147–164.

García-Berthou, E. 2001. Size- and depth-dependent variation in habitat and diet of the common carp (*Cyprinus carpio*). Aquatic Sci. 63: 466–476.

Gatlin, D.M., F.T. Barrows, P. Brown, K. Dabrowski, T.G. Gaylord, R.W. Hardy, E. Herman, G. Hu, Å. Krogdahl, R. Nelson, K. Overturf, M. Rust, W. Sealey, D.J. Skonberg, E. Souza,

D. Stone, R. Wilson and E. Wurtele. 2007. Expanding the utilization of sustainable plant products in aquafeeds: a review. Aquacult. Res. 38: 551–579.

Glass, J., G. Linam and J. Ralph. 1991. The association of *Prymnesium parvum* with fish kills in Texas. Texas Parks and Wildlife Document. 8 pp.

Goto, T., D.T. Wicklow and Y. Ito. 1996. Aflatoxin and cyclopiazonic acid production by a sclerotium-producing *Aspergillus tamarii* strain. Appl. Environ. Microbiol. 62: 4036–4038.

Gunn, G.J., A.G. Rafferty, G.C. Rafferty, N. Cockburn, C. Edwards, K.A. Beattie and G.A. Codd. 1992, Fatal canine neurotoxicosis attributed to blue-green algae (cyanobacteria). Vet. Rec. 4: 301–302.

Hagelberg, S., K. Hult and R. Fuchs. 1989. Toxicokinetics of ochratoxin A in several species and its plasma-binding properties. J. Appl. Toxicol. 9(2): 91–96.

Hallegraeff, G.M. 1993. A review of harmful algal blooms and their apparent global increase. Phycol. 32: 79–99.

Han, D., S. Xie, X. Zhu, Y. Yang and Z. Guo. 2010. Growth and hepatopancreas performances of gibel carp fed diets containing low levels of aflatoxin B1. Aquacult. Nutr. 16(4): 335–342.

Hartmann, N., M. Erbs, H.-R. Forrer, S. Vogelsang, F.E. Wettstein, R.P. Schwarzenbach and T.D. Bucheli. 2008. Occurrence of zearalenone on *Fusarium graminearum* infected wheat and maize field in crop organs, soil, and drainage water. Environ. Sci. Technol. 42: 5455–5460.

Hasan, M.R., D.J. Macintosh and K. Jauncey. 1997. Evaluation of some plant ingredients as dietary protein sources for common carp (*Cyprinus carpio* L.) fry. Aquacult. 151: 55–70.

Hashimoto, Y., T. Okaichi, L.D. Dang and T. Noguchi. 1968. Glenodinine, an ichthyotoxic substance produced by a dinoflagellate, *Peridinium polonicum.* Bull. Jap. Soc. Sci. Fish. 34: 528–533.

Hatanaka, J., N. Doke, T. Harada, T. Aikawa and M. Enomoto. 1982. Usefulness and rapidity of screening for the toxicity and carcinogenicity of chemicals in medaka, *Oryzias latipes*. Jap. J. Exp. Med. 52(5): 243–53.

Hawkins, P.R., M.T.C. Runnegar, A.R.B. Jackson and I.R. Falconer. 1985. Severe hepatotoxicity caused by the tropical cyanobacterium (bluegreen alga) *Cylindrospermopsis raciborskii* (Woloszynska) Seenaya and Subba Raju isolated form a domestic supply reservoir. Appl. Environ. Microbiol. 50: 1292–1295.

He, C.-H., Y.-H. Fan, Y. Wang, C.-Y. Huang, X.-C. Wang and H.-B. Zhang. 2010. The Individual and combined effects of deoxynivalenol and aflatoxin B1 on primary hepatocytes of *Cyprinus carpio*. Int. J. Mol. Sci. 11: 3760–3768.

Hendricks, J.D., R.O. Sinnhuber, J.H. Wales, M.E. Stack and D.P. Hsieh. 1980. Hepatocarcinogenicity of sterigmatocystin and versicolorin A to rainbow trout (*Salmo gairdneri*) embryos. J. Natl. Cancer Inst. 64: 1503–1509.

Hendricks, J.D. 1994. Carcinogenecity of aflatoxins in nonmammalian organisms. *In*: D.L. Eaton and J.D. Groopman (eds.). Toxicology of Aflatoxins: Human Health, veterinary, and Agricultural Significance. Academic Press, San Diego. pp. 103–136.

Heresztyn, T. and B.C. Nicholson. 1997. Nodularin concentrations in Lakes Alexandrina and Albert, South Australia, during a bloom of the cyanobacterium (blue-green alga) *Nodularia spumigena* and degradation of the toxin. Environ. Toxicol. Wat. Qual. 12: 273–282.

Hickel, B. 1976. Fischsterben in einem Karpfenteich bei einer Massenentwicklung des toxischen Phytoflagellaten *Prymnesium parvum* Carter (Haptophyceae). Arch. Fischereiwiss. 27: 143–148 (in German).

Hikasa, Y., K. Takase, T. Ogasawara and S. Ogasawara. 1986. Anaesthesia and recovery with tricaine methanesulphonate, eugenol and thiopental sodium in the carp (*Cyprinus carpio*). Jap. J. Vet. Sci. 48: 341–351.

Hilton, J.W. and S.J. Slinger. 1986. Digestibility and utilization of canola meal in practical-type diets for rainbow trout (*Salmo gairdneri*). Can. J. Fish. Aquat. Sci. 43: 1149–1155.

Hirota, M., A. Menta, K. Yoneyama and N. Kitabatake. 2002. A major decomposition product, citrinin H2, from citrinin on heating with moisture. Biosci. Biotechnol. Biochem. 66: 206–210.

Hoerger, C., J. Schenzel, B. Strobel and T. Bucheli. 2009. Analysis of selected phytotoxins and mycotoxins in environmental samples. Anal. Bioanal. Chem. 395: 1261–1289.

Hoffmeister, D. and N.P. Keller. 2007. Natural products of filamentous fungi: enzymes, genes, and their regulatio. Nat. Prod. Rep. 24: 393–416.

Holdway, P.A., R.A. Watson and B. Moss. 1978. Aspects of the ecology of *Prymnesium parvum* (Haptophyta) and water chemistry in the Norfolk Broads, England. Freshw. Biol. 8(4): 295–311.

Hooft, J.M., H. Elmor, P. Encarnação and D.P. Bureau. 2010. Effects of low levels of naturally occurring fusarium mycotoxins on the performance and health of rainbow trout (*Oncorhynchus mykiss*). Book of Abstracts of the World Aquaculture 2010, San Diego, USA.

Honkanen, R.E., M. Dukelow, J. Zwiller, R.E. Moore, B.S. Khatra and A.L. Boynton. 1991. Cyanobacterial nodularin is a potent inhibitor of type 1 and type 2A protein phosphatases. Mol. Pharmacol. 40: 577–583.

Hossain, M.A. and K. Jauncey. 1988. Toxic effects of glucosinolate (allyl isothiocyanate) (synthetic and from mustard oilcake) on growth and food utilization in common carp. Indian J. Fish. 35(3): 186–196.

Hossain, M.A. and K. Jauncey. 1989a. Studies on the protein, energy and amino acid digestibility of fishmeal, mustard oilcake, linseed and sesame meal for common carp *Cyprinus carpio* L. Aquacult. 83: 59–72.

Hossain, M.A. and K. Jauncey. 1989b. Nutritional evaluation of some Bangladesh oilseed meals as partial substitutes for fishmeal in the diet of common carp, *Cyprinus carpio* L. Aquacult. Fish. Manage. 20: 255–268.

Hossain, M.A. and K. Jauncey. 1990. Detoxification of linseed and sesame meal and evaluation of their nutritive value in the diet of carp *Cyprinus carpio* L. Asian Fish. Sci. 3: 169–183.

Hossain, M.A. and K. Jauncey. 1993. The effects of varying dietary phytic acid, calcium and magnesium levels on the nutrition of common carp, *Cyprinus carpio*. Fish Nutrition in Practice. 4th International Symposium on Fish Nutrition and Feeding, Biarritz, France, 705–715 P. June 24–27, INRA, Paris, France.

Hossain, M.A. and K. Becker. 2001. Nutritivve value and antinutritional factors in different varieties of *Sesbania* seeds and their morphological fractions. Food Chem. 73: 421–431.

Hossain, M.A., U. Focken and K. Becker. 2001a. Evaluation of an unconventional legume seed, *Sesbania aculeata*, as a dietary protein source for common carp, *Cyprinus carpio* L. Aquacult. 198: 129–140.

Hossain, M.A., U. Focken and K. Becker. 2001b. Galactomannan-rich endosperm of Sesbania (Sesbania aculeate) seeds responsible for retardation of growth and feed utilization in common carp, *Cyprinus carpio* L. Aquacult. 203: 121–132.

Hossain, M.A., U. Focken and K. Becker. 2001c. Effect of soaking and soaking followed by autoclaving of Sesbania seeds on growth and feed utilization in common carp, *Cyprinus carpio* L. Aquacult. 203: 133–148.

Howarth, R.W., R. Marino, J. Lane and J.J. Cole 1988. Nitrogen fixation in freshwater, estuarine, and marine ecosystems. 1. Rates and importance. Limnol. Oceanogr. 33: 669–687.

Huang, Y., D. Han, X. Zhu, Y. Yang, J. Jin, Y. Chen and S. Xie. 2011. Response and recovery of gibel carp from subchronic oral administration of aflatoxin B1. Aquacult. 319: 89–97.

Humpage, A.R., J. Rositano, A.H. Breitag, R. Brown, P.D. Baler, W.C. Nicholson and A.D. Steffensen. 1994. Paralytic shellfish poisons from Australian cyanobacterial blooms. Austr. J. Mar. Freshw. Res. 45: 761–777.

Hussain, M., M.A. Gabal, T. Wilson and R.C. Sumerfelt. 1993. Effect of aflatoxin-contaminated feed on morbidity and residues in walleye fish. Vet. Hum. Toxicol. 35(5): 396–398.

Igarashi, T., M. Satake and T. Yasumoto. 1999. Structures and partial stereochemical assignments for prymnesin-1 and prymnesin-2: potent hemolytic and ichthyotoxic glycosides isolated from the red tide alga *Prymnesium parvum*. J. Am. Chem. Soc. 121(37): 8499–8511.

Isaacs, G. 1983. Permanent local anaesthesia and anhydrosis after clove oil spillage. Lancet 1: 882–883.

James, T. and A. De La Cruz. 1989. *Prymnesium parvum* Carter (Chrysophyceae) as a suspect of mass mortalities of fish and shellfish communities in western Texas. Texas J. Sci. 41(4): 429–430.

Jantrarotai, W. and R.T. Lovell. 1990. Subchronic toxicity of dietary Aflatoxin B1 to channel catfish. J. Aquat. Anim. Health 2: 248–254.

Jantrarotai, W., R.T. Lovell and J.M. Grizzle. 1990. Acute toxicity of Aflatoxin B1 to channel catfish. J. Aquat. Anim. Health 2: 237–247.

Johansson, N. and E. Graneli. 1999. Influence of different nutrient conditions on cell density, chemical composition and toxicity of *Prymnesium parvum* (Haptophyta) in semi-continuous cultures. J. Exp. Mar. Biol. Ecol. 239(2): 243–258.

Johns, S.M., N.D. Denslow, M.D. Kane, K.H. Watanabe, E.F. Orlando and M.S. Sepulveda. 2009. Effects of estrogens and antiestrogens on gene expression of fath,ead minnow (*Pimephales promelas*) early life stages. Environ. Toxicol. 26(2): 195–206.

Journey, C.A., K.M. Beaulieu and P.M. Bradley. 2013. Environmental factors that influence cyanobacteria and geosmin occurrence in reservoirs. *In*: Paul Bradley (ed.). Current Perspectives in Contaminant Hydrology and Water Resources Sustainability. InTech, DOI: 10.5772/54807. Available from: http://www.intechopen.com/books/current-perspectives-in-contaminant-hydrology-and-water-resources-sustainability/environmental-factors-that-influence-cyanobacteria-and-geosmin-occurrence-in-reservoirs.

Kaartvedt, S., T.M. Johnsen, D.L. Aksnes, U. Lie and H. Svendsen. 1991. Occurrence of the toxic phytoflagellate *Prymnesium parvum* and associated fish mortality in a Norwegian fjord system. Can. J. Fish. Aquat. Sci. 48(12): 2316–2323.

Kankaanpää, H.T., V.O. Sipiä, J.S. Kuparinen, J.L. Ott and W.W. Carmichael. 2001. Nodularin analyses and toxicity of a *Nodularia spumigena* (Nostocales, Cyanobacteria) water-bloom in the western Gulf of Finland, Baltic Sea, in August 1999. Phycol. 40(3): 268–274.

Kardinaal, W.E.A. and P.M. Visser. 2005. Dynamics of cyanobacterial toxins. *In*: J. Huisman, C.P. Matthijs and P.M. Visser (eds.). Harmful Cyanobacteria. Springer Dordrecht 241: 41–63.

Keene, J.L., D.L.G. Noakes, R.D. Moccia and C.G. Soto. 1998. The efficacy of clove oil as an anaesthetic for rainbow trout, *Oncorhynchus mykiss* (Walbaum). Aquacult. Res. 29: 89–101.

Keller, N.P., G. Turner and J.W. Bennett. 2005. Fungal secondary metabolism—from biochemistry to genomics. Nature Reviews Microbiology 3: 937–947.

Keshavanath, P., M.C.M. Beveridge, D.J. Baird, L.A. Lawton, A. Nimmo and G.A Codd. 1994. The functional grazing response of a phytoplanktivorous fish *Oreochromis niloticus* to mixtures of toxic and non-toxic strains of the cyanobacterium *Microcystis aeruginosa*. J. Fish Biol. 45: 123–129.

Kononenko, G.P. and A.A. Burkin. 2013. Peculiarities of feed contamination with citrinin and ochratoxin A. Agricult. Sci. 4(1): 34–38.

Kopp, R., J. Mareš, M. Palíková, S. Navrátil, Z. Kubíček, A. Ziková, J. Hlávková and L. Bláha. 2009. Biochemical parameters of blood plasma and content of microcystins in tissues of common carp (*Cyprinus carpio* L.) from a hypertrophic pond with cyanobacterial water bloom. Aquacult. Res. 40: 1683–1693.

Kravchenko, L.V., V.T. Galash, L.T. Avreneva, A.E. Kranauskas. 1989. On the sensitivity of carp, *Cyprinus carpio*, to mycotoxins T-2. J. Ichthyol. 29: 156–160.

Labuda, R., A. Parich, E. Vekiru and D. Tancinová. 2005. Incidence of fumonisins, moniliformin and Fusarium species in poultry feed mixtures from Slovakia. Ann. Agricult. Environ. Med. 12(1): 81–86.

Lahti, K. 1997. Cyanobacterial hepatotoxins and drinking water supplies—aspect of monitoring and potential health risks. *In*: Monographs of boreal environment research, no. 4. Finnish Environ. Inst., Finland.

Landsberg, J.H. 2002. The effects of harmful algal blooms on aquatic organisms. Rev. Fish. Sci. 10(2): 113–390.

Larsen, A., W. Eikrem and E. Paasche. 1993. Growth and toxicity in *Prymnesium patelliferum* (Prymnesiophycea) isolated from Norwegian waters. Can. J. Bot. 71: 1357–1362.

Larsen, A. and S. Bryant. 1998. Growth rate and toxicity of *Prymnesium parvum* and *Prymnesium patelliferum* (Haptophyta) in response to changes in salinity, light and temperature. Sarsia 83(5): 409–418.

Leming, R., A. Lember and T. Kukk. 2004. The content of individual glucosinolates in rapeseed and rapeseed cake produced in Estonia. Agraaeteadus 15: 21–27.

Llewellyn, L.E., A.P. Negri, J. Doyle, P.D. Baker, E.C. Beltran and B.A. Neilan. 2001. Radioreceptor assays for sensitive detection and quantitation of saxitoxin and its analogues from strains of the freshwater cyanobacterium, *Anabaena circinalis*. Environ. Sci. Technol. 35: 1445–1451.

Li, M.H., S.A. Raverty and E.H. Robinson. 1994. Effects of dietary mycotoxins produced by the mold *Fusarium moniliforme* on channel catfish (*Ictalurus punctatus*) J. World Aquacult. Soc. 25(4): 512–516.

Li, R.H., W.W. Carmichael, S. Brittain, G.K. Eaglesham, G.R. Shaw, Y.D. Liu and M.M. Watanabe. 2001. First report of the cyanotoxins cylindrospermopsin and deoxycylindrospermopsin from *Raphidiopsis curvata* (Cyanobacteria). J. Phycol. 37: 1121–1126.

Li, X.-Y., I.-K. Chung, J.-I. Kin and J.-A. Lee. 2004. Subchronic oral toxicity of microcystin in common carp (*Cyprinus carpio* L.) exposed to Microcystis under laboratory conditions. Toxicon 44(8): 821–827.

Liu, Z.-H., Y. Kanjo and S. Mizutani. 2010. A review of phytoestrogens: their occurrence and fate in the environment. Wat. Res. 44: 567–577.

Loupal, G., E.S. Kuttin and O. Kölbl. 1992. Prototheca as the cause of swim bladder inflamation in carp (*Cyprinus carpio*). Tierarztl. Umschau 47(11): 850–854.

Lovell, R.T. 1992. Mycotoxins: hazardous to farmed fish. Feed International 13: 24–28.

Lumlertdacha, S., R.T. Lovell, R.A. Shelby, S.D. Lenz, B.W. Kemppainen. 1995. Growth, hematology, and histopathology of channel catfish (*Ictalurus punctatus*), fed toxins from *Fusarium moniliforme*. Aquacult. 130: 201–218.

Mackenthun, K.M., E.F. Herman and A.F. Bartsch. 1948. A heavy mortality of fishes resulting from the decomposition of algae in the Yahara river, Wisconsin. Transact. Am. Fish. Soc. 75: 175–180.

Måge, A., K. Julshamn and B.T. Lunestad. 2009. Overvåkningsprogram for fôrvarer til fisk og andre akvatiske dyr - Årsrapport 2008 og 2009. NIFES, Bergen (in Norwegian only).

Mahmood, N.A. and W.W. Carmichael. 1986. Paralytic selfish poisons produced by the freshwater cyanobacterium *Aphanizomenon flos-aque* NH-5. Toxicon 24: 175–186.

Mahmood, N.A. and W.W. Carmichael. 1987. Anatoxin-a(s), an anticholinesterase from the cyanobacterium *Anabaena flos-aquae* NRC-525-17. Toxicon 25: 1221–1227.

Mahmoud, A.L.E. 1993. Toxigenic fungi and mycotoxin content in poultry feedstuff ingredients. J. Basic Microbiol. 33: 101–104.

Mahrous, K.F., W.K.B. Khalil and M.A. Mahmoud. 2006. Assessment of toxicity and clastogenicity of sterigmatocystin in Egyptian Nile tilapia. Afr. J. Biotechnol. 5: 1180–1189.

Makkar, H.P.S. and K. Becker. 1999. Nutritional studies on rats and fish (carp, *Cyprinus carpio*) fed diets containing unheated and heated *Jatropha curcas* meal of a non-toxic provenance. Plant Food. Hum. Nutr. 53: 183–192.

Manning, B.B. 2001. Mycotoxins in fish feeds. pp. 267–287. *In*: C. Lim and C.D. Webster (eds.). Nutrition and Fish Health. Food Products Press, New York.

Manning, B.B., R.M. Ulloa, M.H. Li, E.H. Robinson and G.E. Rottinghaus. 2003a. Ochratoxin A fed to channel catfish (*Ictalurus punctatus*) causes reduced growth and lesions of hepatopancreatic tissue. Aquacult. 219: 739–750.

Manning, B.B., M.H. Li, E.H. Robinson, P.S. Gaunt, A.C. Camus and G.E. Rottinghaus. 2003b. Response of catfish to diets containing T-2 toxin. J. Aquatic Anim. Health 15(3): 229–238.

Manning, B.B., J.S. Terhune, M.H. Li, E.H. Robinson, D.J. Wise and G.E. Rottinghaus. 2005. Exposure to feedborne mycotoxins T-2 toxin or ochratoxin A causes increased mortality of channel catfish challenged with *Edwardsiella ictaluri*. J. Aquatic Anim. Health 17(2): 147–152.

Manning, S.R. and J.W. II. La Claire. 2010. Prymnesins: Toxic Metabolites of the Golden Alga, *Prymnesium parvum* Carter (Haptophyta). Mar. Drugs 8: 678–704.

Marionnet, D., C. Chambras, L. Taysse, C. Bosgireaud and P. Deschaux. 1998. Modulation of drug-metabolizing systems by bacterial endotoxin in carp liver and immune organs. Ecotoxicol. Environ. Saf. 41: 189–194.

Mares, J., M. Palikova, R. Kopp, S. Navratil and J. Pikula. 2009. Changes in the nutritional parameters of muscles of the common carp (*Cyprinus carpio*) and the silver carp (*Hypophthalmichthys molitrix*) following environmental exposure to cyanobacterial water bloom. Aquacult. Res. 40: 148–156.

Martin, C., G.A. Codd, H.W. Siegelman and J. Weckesser. 1989. Lipopolysaccharides and polysaccharides of the cell envelope of toxic *Microcystis aeruginosa* strains. Arch. Microbiol. 152: 90–94.

Matsunaga, S., R.E. Moore, W.P. Niemczura and W.W. Carmichael. 1989. Anatoxin-a(s), a potent anticholinesterase from *Anabaena flos-aquae*. J. Am. Chem. Soc. 111: 8021–8023.

Mawson, R., R.K. Heaney, Z. Zdunczyk and H. Kozlowska. 1993. Rapeseed meal— glucosinolates and their antinutritional effects. Part 1. Rapeseed production and chemistry of glucosinolates. Nahr. 37: 131–140.

Mawson, R., R.K. Heaney, Z. Zdunczyk and H. Kozlowska. 1994a. Rapeseed meal— glucosinolates and their antinutritional effects. Part 3. Animal growth and performance. Nahr. 38: 167–177.

Mawson, R., R.K. Heaney, Z. Zdunczyk and H. Kozlowska. 1994b. Rapeseed meal— glucosinolates and their antinutritional effects. Part 4. Goitrogenicity and internal organs abnormalities in animals. Nahr. 38: 178–191.

Mawson, R., R.K. Heaney, Z. Zdunczyk and H. Kozlowska. 1995. Rapeseed meal—glucosinolates and their antinutritional effects. Part 7. Processing. Nahr. 39: 32–41.

McCurdy, C.M. and B.E. March. 1992. Processing of canola meal for incorporation in trout and salmon diets. J. Am. Oil Chem. Soc. 69: 213–220.

McKean, C., L. Tang, M. Tang, M. Billam, Z. Wang, C.W. Theodorakis, R.J. Kendall and J.-S. Wang. 2006. Comparative acute and combinative toxicity of aflatoxin B1 and fumonisin B1 in animals and human cells. Food Chem. Toxicol. 44(6): 868–876.

Meriç, I. 2013. Evaluation of sunflower seed meal in feeds for carp: Antinutritional effects on antioxidant defense system. Food Agricult. Environ. 11(2): 1128–1132.

Mez, K., K.A. Beattie, G.A. Codd, K. Hanselmann, B. Hauser, H. Naegeli and H.R. Preisig. 1997. Identification of a microcystin in benthic cyanobacteria linked to cattle deaths on alpine pastures in Switzerland. Europ. J. Phycol. 32: 111–117.

Mitsoura, A., I. Kagalou, N. Papaioannou, P. Berillis, E. Mente and T. Papadimitriou. 2013. The presence of microcystins in fish *Cyprinus carpio* tissues: a histophathological study. Int. Aquatic Res. 5: 8–24.

Miyahara, M., H. Ishibashi, M. Inudo, H. Nishijima, T. Iguchi, L.J. Jr., Guilette and K. Arizono. 2003. Estrogenic activity of a diet to estrogen receptors-α and -β in an experimental animal. J. Health Sci. 49(6): 481–491.

Montgomery, R.D. 1980. Cyanogens. pp. 143–160. *In*: I.E. Liener (ed.). Toxic Constituents of Plant Foodstuff. Academic Press, New York.

Mustakas, G.C., K.J. Moulton, E.C. Baker and W.F. Kwolek. 1981. Critical processing factors in desolventizing-toasting soybean meal for feed. J. Am. Oil Chem. Soc. 58: 300–305.

Mutawie, H.H. 2012. Assesment of *hepatotoxins* and *neurotoxins* from five *Oscillatoria* species isolated from Makkah area, KSA using HPLC. International Research Journal of Agricultural Science and Soil Science 2(10): 440–444.

Mwachireya, S.A., R.M. Beames, D.A. Higgs and B.S. Dosanjh. 1999. Digestibility of canola protein products derived from the physical, enzymatic and chemical processing of commercial canola meal in rainbow trout, *Oncorhynchus mykiss* (Walbaum) held in fresh water. Aquacult. Nutr. 5: 73–82.

Naczk, M. and F. Shahidi. 1990. Carbohydrates of canola and rapeseed. pp. 211–220. *In*: F. Shahidi (ed.). Canola, Rapeseed: Production, Chemistry, Nutrition & Processing Technology. Van Nostrand Reinhold, New York.

Nelson, P.E., A.E. Desjardins and R.D. Plattner. 1993. Fumonisins, mycotoxins produced by Fusarium species: biology, chemistry and significance. Ann. Rev. Phytopathol. 31: 233–252.

Ngethe, S., T.E. Horsberg, E. Mitema and K. Ingebrigtsen. 1993. Species differences in hepatic concentration of orally administered ^{3}H-AFB1 between rainbow trout (*Oncorhynchus mykiss*) and tilapia (*Oreochromis niloticus*). Aquacult. 114: 355–358.

Norton, G. 1991. Proteinase inhibitors. *In*: F.J.P. D'Mello, C.M. Duffus and J.H. Duffus (eds.). Toxic Substances in Crop Plants, pp. 68–106. The Royal Society of Chemistry, Thomas Graham House, Science Park, Cambridge CB4 4WF, Cambridge.

Oberemm, A., J. Fastner and C.E.W. Steinberg. 1997. Effects of microcystin-LR and cyanobacterial crude extracts on embryo-larval development of zebrafish (*Danio rerio*). Wat. Res. 31(11): 2918–2921.

Oberemm, A., J. Becker, G.A. Codd and C.E.W. Steinberg. 1999. Effects of cyanobacterial toxins and aqueous crude extrakt of cyanobacteria on the development of fish and amphibians. Environ. Toxicol. 14: 77–88.

Oberle, M., F.J. Schwarz and M. Kirchgessner. 1997. Growth and carcass quality of carp (*Cyprinus carpio* L.) fed different cereals, lupin seed or zooplankton. Arch. Tierern. 50(1): 75–86.

Oberle, M. 2000 Einfluss der Fütterung von mutterkornhaltigem Getreide auf Karpfen. Presentation at the Tagung für Fischhaltung und Fischzucht, Landesanstalt für Fischerei, Starnberg, Deutschland, 11.1.-12.1.2000.

Oberle, M. 2002. Verfütterung von mutterkornhaltigem Getreide an Karpfen. Fischer & Teichwirt 53: 97 (in German).

Oberle, M. 2002. Verfütterung von mutterkornhaltigem Getreide an Karpfen. Europäisches Netzwerk zur Verbreitung und Informationen über Aquakulturforschung, Aqua-Flow-Ref.: TLDE2001-002 (in German).

Ohtani, I., R.E. Moore and M.T.C. Runnegar. 1992. Cylindrospermopsin: a potent hepatotoxin from the blue-green alga *Cylindrospermopsis raciborskii*. J. Am. Chem. Soc. 114: 7941–7942.

Osswald, J., S. Rellan, A.P. Carvalho, A. Gago and V. Vasconcelos. 2007. Acute effects of an anatoxin-a producing cyanobacterium on juvenile fish—*Cyprinus carpio* L. Toxicon 49: 693–698.

Osswald, J., A.P. Carvalho, J. Claro, V. Vasconcelos. 2009. Effects of cyanobacterial extracts containing anatoxin-a and of pure anatoxin-a on early developmental stages of carp. Ecotoxicol. Environ. Saf. 72: 473–478.

Owen, R.P. 1994. Biological and economic significance of benthic cyanobacteria in two Scottish Highland Lochs. pp. 145–148. *In*: G.A. Codd, T.M. Jefferies, C.W. Keevil and E. Potter (eds.). Detection Methods for Cyanobacterial Toxins. The Royal Society for Chemistry. Cambridge.

Padan, E., D. Ginzburg, M. Shilo. 1967. Growth and colony formation of the phytoflagellate *Prymnesium parvum* Carter on solid media. J. Protozool. 14(3): 477–480.

Padilla, G.M. 1970. Growth and toxigenesis of the chrysomonad *Prymnesium parvum* as a function of salinity. J. Protozool. 17: 456–462.

Palíková, M., S. Navrátil, B. Maršálek and L. Bláha. 2003. Toxicity of crude extract of Cyanobacteria for embryos and larvae of carp (*Cyprinus carpio* L.). Acta Vet. Brno 72: 437–443.

Palíková, M., R. Krejčí, K. Hilscherová, P. Babica, S. Navrátil, R. Kopp and L. Bláha. 2007. Effect of different cyanobacterial biomasses and their fractions with variable microcystin content on embryonal development of carp (*Cyprinus carpio* L.). Aquat. Toxicol. 81: 312–318.

Pepeljnjak, S., Z. Petrinec, S. Kovacic and M. Segvic. 2002. Screening toxicity study in young carp (*Cyprinus carpio*) on feed amended with fumonisin B1. Mycopathol. 156: 139–145.

Pereira, P., H. Onodera, D. Andrinolo, S. Franca, P. Araujo, N. Lagos, Y. Oshima. 2000. Paralytic shellfish poisoning toxins in the freshwater cyanobacterium *Aphanizomenon flosaquae* isolated from Montargil reservoir, Portugal. Toxicon 38: 1689–1702.

Paster, Z. 1973. Pharmacology and mode of action of prymnesin. *In*: D.F. Martin and G.M. Padilla (eds.). Marine Pharmacognosy: Action of Marine Biotoxins at the Cellular Level. Academic Press; New York, NY, USA, pp. 241–263.

Petrinec, Z., S. Pepeljnjak, S. Kovacic and A. Krznaric. 2004. Fumonisin B1 causes multiple lesions in common carp (*Cyprinus carpio*). Deut. Tierärztl. Wochenschr. 111(9): 358–363.

Pietsch, C., T. Bucheli, F.E. Wettstein and P. Burkhardt-Holm. 2011. Frequent biphasic cellular responses of permanent fish cell cultures to deoxynivalenol (DON). Toxicol. Appl. Pharmacol. 256: 24–34.

Pietsch, C., S. Kersten, P. Burkhardt-Holm, H. Valenta and S. Dänicke. 2013. Occurrence of deoxynivalenol and zearalenone in commercial fish feed. An initial study. Toxins 5: 184–192.

Pietsch, C., J. Noser, F.E. Wettstein and P. Burkhardt-Holm. 2014a. Unravelling mechanisms involved in zearalenone-mediated toxicity in permanent fish cell cultures. Toxicon 88: 44–61.

Pietsch, C., S. Kersten, H. Valenta, S. Dänicke, C. Schulz, W. Kloas and P. Burkhardt-Holm. 2014b. *In vivo* effects of deoxynivalenol (DON) on innate immune responses of carp (*Cyprinus carpio* L.). Food Chem. Toxicol. 68: 44–52.

Pietsch, C., B.A. Katzenback, E. Garcia-Garcia, C. Schulz, M. Belosevic and P. Burkhardt-Holm. 2015. Acute and subchronic effects on immune responses of carp (*Cyprinus carpio* L.) after exposure to deoxynivalenol (DON) in feed. Mycotox. Res., accepted.

Pietsch, C., C. Schulz, P. Rovira, W. Kloas and P. Burkhardt-Holm. 2014c. Organ damage and altered nutritional condition in carp (*Cyprinus carpio* L.) after food-borne exposure to the mycotoxin deoxynivalenol (DON). Toxins 6: 756–778.

Pistocchi, R., F. Guerrini, L. Pezzolesi, M. Riccardi, S. Vanucci, P. Ciminiello, C. Dell'Aversano, M. Forino, E. Fattorusso, L. Tartaglione, A. Milandri, M. Pompei, M. Cangini, S.Pigozzi and E. Riccardi. 2012. Toxin levels and profiles in microalgae from the North-Western Adriatic Sea—15 years of studies on cultured species. Mar. Drugs 10: 140–162.

Pore, R.S. 1998. Prototheca and Chlorella. pp. 631–643. *In*: L. Ajello and R. J. Hay (eds.). Topley & Wilson's Microbiology and Microbial Infections. Vol. 4, 9th ed. Arnold Publications, London, UK.

Poston, H.A. 1983. Biological effects of dietary T2 toxins on rainbow trout. Aquat. Toxicol. 2: 79–88.

Prescott, G.W. 1968. The Algae: A Review. Houghton Mifflin Co., Boston. 436 pp.

Råberg, C.M.I., G. Bylund and J.E. Eriksson. 1991. Histopathological effects of microcystin-LR, a cyclic peptide toxin from the cyanobacterium (blue-green alga) *Microcystis aeruginosa* on common carp (*Cyprinus carpio*). Aquat. Toxicol. 20: 131–146.

Rajeev Raghavan, P., X. Zhu, W. Lei, D. Han, Y. Yang and S. Xie. 2011. Low levels of Aflatoxin B1 could cause mortalities in juvenile hybrid sturgeon, *Acipenser ruthenus* ♂× *A. baeri* ♀. Aquacult. Nutr. 17: e39–e47.

Reich, K. and M. Aschner. 1947. Mass development and control of the phytoflagellate *Prymnesium parvum* in fish ponds in Palestine. Palestine Journal of Botany, Jerusalem 4: 14–23.

Rensel, J.E. and J.N.C. Whyte. 2004. Finfish mariculture and harmful algal blooms. pp. 693–722. *In*: G.M. Hallegraeff, D.M. Anderson and A.D. Cembella (eds.). Manual on Harmful Marine Microalgae. Unesco Publishing, Paris.

Ressom, R., F. San Soong, J. Fitzgerald, L. Turczynowicz, O. El Saadi, D. Roder, T. Maynard and I. Falconer. 1994. Health effects of toxic Cyanobacteria (Blue-Green Algae). pp. 27–69. Australian Government Publishing Service, Canberra.

Rinehart, K.L., K. Harada, M. Namikoshi, C. Chen, C.A. Harvis, M.H.G. Munro, J.W. Blunt, P.E. Mulligan, V.R. Beasley, A.M. Dahlem and W.W. Carmicheal. 1988. Nodularin, microcystin, and the configuration of Adda. J. Am. Chem. Soc. 110: 8557–8558.

Roelke, D.L., J.P. Grover, B.W. Brooks, J. Glass, D. Buzan, G.M. Southard, L. Fries, G.M. Gable, L. Schwierzke-Wade, M. Byrd and J. Nelson. 2011. A decade of fish-killing *Prymnesium parvum* blooms in Texas: roles of inflow and salinity. J. Plankton Res. 33(2): 243–253.

Rodger, H.D., T. Turnbull, C. Edwards and G.A. Codd. 1994. Cyanobacterial (blue-green alga) bloom associated pathology on brown trout, *Salmo trutta* L., in Loch Leven, Scottland. J. Fish Dis. 17: 177–181.

Rodrigues, I. and K. Naehrer. 2012. Prevalence of mycotoxins in feedstuffs and feed surveyed worldwide in 2009 and 2010. Phytopathol. Mediterr. 51: 175–192.

Runnegar, M., N. Berndt, S.M. Kong, E.Y.C. Lee and L.F. Zhang. 1995. *In vivo* and *in vitro* binding of microcystin to protein phosphatase 1 and 2A. Biochem. Biophys. Res. Commun. 216: 162–169.

Rymuszka, A. 2012. Cytotoxic activity oft he neurotoxin anatoxin-a on fish leukocytes *in vitro* and *in vivo* studies. Acta Vet. Brno 81: 175–182.

Rymuszka, A. and A. Sierosławska. 2013. Cytotoxic and immunotoxic effects of the mixture containing cyanotoxins on carp cells following *in vitro* exposure. Centr. Europ. J. Immunol. 38(2): 159–163.

Saad, T.T. 2002. Some studies on the effects of ochratoxin-A on cultured *Oreochromis niloticus* and carp species. M.V.SC. Faculty of Veterinary Medicine. Alexandria University.

Sahoo, PK., S.C. Mukherjee, S. Mohanty, S. Dey and S.K. Nayak. 1999. A preliminary study of acute citrinin toxicity in rohu (*Labeo rohita*) fingerlings. Indian J. Microbiol. Immunol. Infect. Dis. 20(1): 62–64.

Sahoo, P.K. and S.C. Mukherjee. 2001. Immunosuppressive effects of aflotoxin B1 in Indian major carp (*Labeo rohita*). Comp. Immunol. Microbiol. Infect. Dis. 24: 143–149.

Salama, A.J. 2007. Preliminary studies on the occurrence of mycotic and mycotoxin contamination in aquaculture feeds used in Saudi Arabia. Egypt. J. Aquat. Biol. Fish. 11(2): 23–42.

Salem, M.F.I., M.T. Shehab El-Din, M.M.M.A. Khalafallah, S.H. Sayed and S.H. Amal. 2010. Nutritional attempts to detoxify aflatoxic effects in diets if tilapia fish (*Oreochromis niloticus*). J. Arab. Aquacult. Soc. 5(2): 195–206.

Sanden, M., S. Jørgensen, G.-I. Hemre, R. Ørnsrud and N.H. Sissener. 2012. Zebrafish (*Danio rerio*) as a model for investigating dietary toxic effects of deoxynivalenol contamination in aquaculture feeds, Food Chem. Toxicol. 50(12): 4441–4448.

Sándor, G. and A. Ványi. 1990. Mycotoxin research in the Hungarian Central Veterinary Institute. Acta Vet. Hung. 38(1-2): 61–68.

Santacroce, M.P., M.C. Conversano, E. Casalino, O. Lai, C. Zizzadoro, G. Centoducati and G. Crescenzo. 2008. Aflatoxins in aquatic species: metabolism, toxicity and perspectives. Reviews of Fish Biology and Fisheries 18: 99–130.

Santos, G.A., I. Rodrigues, K. Naehrer and P. Encarnacao. 2010. Mycotoxins in aquaculture: Occurrence in feed components and impact on animal performance. Aquacult. Europ. 35: 6–10.

Schembri, M.A., B.A. Neilan and C.P. Saint. 2001. Identification of genes implicated in toxin production in the cyanobacterium *Cylindrospermopsis raciborskii*. Environ. Toxicol. 16: 413–421.

Schoenebeck, C.W., M.L. Brown, S.R. Chipps and D.R. German. 2012. Nutrient and algal responses to winterkilled fish-derived nutrient subsidies in eutrophic lakes. Lake Reserv. Manage. 28: 189–199.

Schwartz, P., K.L. Thorpe, T.D. Bucheli, F.E. Wettstein and P. Burkhardt-Holm. 2010. Short-term exposure to the environmentally relevant estrogenic mycotoxin zearalenone impairs reproduction in fish. Sci. Tot. Environ. 409: 326–333.

Scudamore, K.A., S. Nawaz and M.T. Hetmanski. 1998. Mycotoxins in ingredients of animal feeding stuffs: II. determination of mycotoxins in maize and maize products. Food Add. Contam. 15: 30–55.

Sepahdari, A., H.A. Ebrahimzadeh Mosavi, I. Sharifpour, A. Khosravi, A.A. Motallebi, M. Mohseni, S. Kakoolaki, H.R. Pourali and A. Hallajian. 2010. Effects of different dietary levels of AFB_1 on survival rate and growth factors of Beluga (*Huso huso*). Iran. J. Fish. Sci. 9(1): 141–150.

Shalaby, A.M.E. 2009. The opposing effects of ascorbic acid (Vitamin C) on ochratoxin toxicity in Nile tilapia (*Oreochromis niloticus*). http://www.ag.arizona.edu/ista/ista6web/pdf/209.pdf. Retrieved: 18-04-14.

Shilo, M. and S. Sarig. 1989. Fish Culture in Warm Water Systems: Problems and Trends. Franklin Book Co., Inc., Elkins Park, Pennsylvania, USA.

Siddhuraju, P. and K. Becker. 2001. Preliminary nutritional evaluation of Mucuna seed meal (*Mucuna pruriens* var. *utilis*) in common carp (*Cyprimus carpio* L.): an assessment by growth performance and feed utilization. Aquacult. 196: 105–123.

Sivonen, K. 1996. Cyanobacterial toxins and toxin production. Phycol. 35: 12–24.

Slawski, H., H. Adem, R.-P. Tressel, K. Wysujack, U. Koops and C. Schulz. 2010. Replacement of fishmeal by rapeseed protein concentrate in diets for common carp (*Cyprinus carpio* L.). Israel. J. Aquacult. – Bamidgeh, IIC: 63.2011.605, 6 pages.

Smayda, T.J. 1990. Novel and Nuisance Phytoplankton Blooms in the Sea: Evidence for a Global Epidemic. Elsevier Science Publishing Co. Inc., New York, Amsterdam & London.

Soto, C.G. and S. Burhanuddin. 1995. Clove oil as a fish anaesthetic for measuring length and weight of rabbitfish (*Siganus lineatus*). Aquacult. 136: 149–152.

Spiegelstein, M., K. Reich and F. Bergmann. 1969. The toxic principles of *Ochromonas* and related chrysomonadina. Verh. Int. Ver. Limnol. 17: 778–783.

Spivak, C.E., B. Witkop and E.X. Albuquerque. 1980. Anatoxin-a: a novel, potent agonist at the nicotinic receptor. Mol. Pharmacol. 18: 384–394.

Stoloff, L. 1976. Occurrence of mycotoxins in foods and feeds. pp. 23–51. *In*: J.V. Rodricks (ed.). Mycotoxins and Other Fungal Related Food Problems, Advances in Chemistry Series No.149. American Chemical Society, Washington, DC.

Stotts, R.R., M. Namikoshi, W.M. Haschek, K.L. Rinehart, W.W. Carmichael, A.M. Dahlem and V. Beasley. 1993. Structural modifications imparting reduced toxicity in microcystins from *Microcystis* spp. Toxicon 31: 783–789.

Streit, E., G. Schatzmayr, P. Tassis, E. Tzika, D. Marin, I. Taranu, C. Tabuc, A. Nicolau, I. Aprodu, O. Puel and I.P. Oswald. 2012. Current Situation of Mycotoxin Contamination and Co-occurrence in Animal Feed-Focus on Europe. Toxins 4: 788–809.

Svobodova, Z. and A. Piskac. 1980. Effect of feeds with a low content of aflatoxin B1 on the health of carp Cyprinus carpio. Zivocisna Vyroba–UVTIZ, 25(11): 809–814.

Svobodova, Z., A. Piskac, J. Havlikova and L. Groch. 1982a. Influence of feed with different contents of B_1 aflatoxin on the carp health condition. Zivocisna Vyroba–UVTIZ, 27(11): 811–820.

Svobodova, Z., R. Faina, A. Piskac and L. Groch. 1982b. The effect of ergot (*Claviceps purpurea*) in feed on the health of carp. Zivocisna Vyroba–UVTIZ 26(11): 837–844.

Tacon, A.G.J. and A.J. Jackson. 1985. Utilization of conventional and unconventional protein sources in practical fish feeds—a review. pp. 119–145. *In*: C.B. Cowey, A.M. Mackie and J.G. Bell (eds.). Nutrition and Feeding Fish. Academic Press, London and New York.

Tacon, A.G.J. 1995. Fishmeal replacers: Review of antinutrients within oilseeds and pulses—A limiting factor for the aquafeed *Green Revolution*. Presented at Feed Ingredients Asia 1995, 19–20 September 1995, Singapore.

Tacon, A.G.J., M.R. Hasan and M. Metian. 2011. Demand and supply of feed ingredients for farmed fish and crustaceans: trends and prospects. FAO Fisheries and Aquaculture Technical Paper No. 564. FAO, 2011. 87 pp.

Tencalla, F.G., D.R. Dietrich and C. Schlatter. 1994. Toxicity of *Microcystis aeruginosa* peptide toxin to yearling rainbow trout (*Oncorhynchus mykiss*). Aquat. Toxicol. 30(3): 215–224.

Trivedi, A.B., M. Hirota, E. Doi and N. Kitabatake. 1993. Formation of a new toxic compound, citrinin H1, from citrinin on mild heating in water. J. Chem. Soc. Perkin Transact. 1: 2167–2171.

Turan, F. and I. Akyurt. 2005. Effects of red clover extract on growth performance and body composition of African catfish *Clarias gariepinus* (Burchell, 1822). Fish. Sci. 71: 618–620.

Turan, F. 2006. Improvement of growth performance in tilapia (*Oreochromis aureus* Linnaeus) by supplementation or red clover (*Trifolium pratense*) in diets. Israel. J. Aquacult.—Bamidgeh 58(1): 34–38.

Turan, F., M. Gürlek and D. Yaglioglu. 2007. Dietary red clover (*Trifolium pratense*) on growth performance of common carp (*Cyprinus carpio*). J. Anim. Vet. Adv. 6(12): 1429–1433.

Ufodike, E.B.C. and A.J. Matty. 1983. Growth responses and nutrient digestibility in mirror carp *Cyprinus carpio* fed different levels of cassava and rice. Aquacult. 31: 41–50.

Ulitzer, S. and M. Shilo. 1964. A sensitive assay system for determination of the ichthyotoxicity of *Prymnesium parvum*. J. Gen. Microbiol. 36(2): 161–169.

Van der Merwe, K.J., P.S. Steyn, L. Fourie, D.B. Scott and J.J. Theron. 1965a. Ochratoxin A, a toxic metabolite produced by *Aspergillus ochraceus* Wilh. Nature 205: 1112–1113.

Van der Merwe, K.J., P.S. Steyn and L. Fourie. 1965b. Mycotoxins. Part II. The constitution of ochratoxin A, B, C, metabolites of *Aspergillus ochraceus* Wilh. J. Chem. Soc. 7083–7088.

Van Egmond, H.P. and M.A. Jonker. 2004. Worldwide regulations for mycotoxins in food and feed in 2003. Food and Agriculture Organization (FAO) Food and Nutrition Paper No. 81. Rome, Italy.

van Leeuwen, S., M. van Velzen, K. Swart, I. van der Veen, W. Traag, M. Spanjer, J. Scholten, H. van Rhijn and J. de Boer. 2008. Contaminants in popular farmed fish consumed in The Netherlands and their levels in fish feed. Report R-08/03 of the Dutch Food and Consumer Product Safety Authority, September 17, 2008, pp. 61.

Van Liere, L. and A.E. Walsby. 1982. Interactions of cyanobacteria with light. *In*: N.G. Carr and B.A. Whitton (eds.). The Biology of Cyanobacteria. Blackwell, Oxford, and Univ. California Press, Berkeley 655: 9–45.

Vasconcelos, V. 1999. Cyanobacterial toxins in Portugal: effects on aquatic animals and risk for human health. Brazil. J. Med. Biol. Res. 32: 249–254.

Velisek, J., Z. Svobodova, V. Piackova, L. Groch and L. Nepejchalova. 2005. Effects of clove oil anaesthesia on common carp (*Cyprinus carpio* L.). Vet. Med. –Czech 50(6): 269–275.

Veršilovskis, A. and S. De Saeger. 2010. Sterigmatocystin: occurrence in foodstuffs and analytical methods—an overview. Mol. Nutr. Food Res. 54: 136–147.

Vesonder, R.F. and B. Horn. 1985. Sterigmatocystin in dairy cattle feed contaminated with *Aspergillus versicolor*. Appl. Environ. Microbiol. 49: 234–235.

Vesonder, R.F., W. Wu, D. Weisleder, S.H. Gordon, T. Krick, W. Xie, H.K. Abbas and C.E. McAlpin. 2000. Toxigenic strains of *Fusarium moniliforme* and *Fusarium proliferatum* isolated from dairy cattle feed produce fumonisins, moniliformin and a new C21H38N2O6 metabolite phytotoxic to *Lemna minor* L. J. Nat. Toxins 9(2): 103–12.

Viola, S., Y. Arieli and G. Zohar. 1988. Unusual feedstuffs (tapioca and lupin) as ingredients for carp and tilapia feeds in intensive culture—Bamidgeh 40: 29–34.

Vogt, G., R. Bohm and H. Segner. 1994. Mimosine-induced cell-death and reöated chromatin changes. J. Submicroscop. Cytol. Pathol. 26(3): 319–330.

Wang, E., P.F. Ross, T.M. Wilson, R.T. Riley, A.H. Jr. Merril. 1992. Increase in serum sphingosine and sphinganine and decreases in complex shpingolipids in ponies given feed containing fumonisins, mycotoxins produced by *Fusarium moniliforme*. J. Nutr. 122: 1706–1716.

Wang, P.J., M.S. Chien, F.J. Wu, H.N. Chou and S.J. Lee. 2005. Inhibition of embryonic development by microcystin-LR in zebrafish *Danio rerio*. Toxicon 45: 303–308.

Weidenbörner, M. 2007. Mycotoxins in Feedstuffs. Springer-Verlag, New York, USA.

Williams, C.D., M.T. Aubel, A.D. Chapman and P.E. D`Aiuto. 2007. Identification of cyanobacterial toxins in Florida`s freshwater systems. Lake Reserv. Manage. 23: 144–152.

WHO. 1998. Guidelines for Drinking Water Quality. World Health Organization, Geneva.

Whitton, B.A., J.M. Yelloly, M. Christmas and I. Hernandez. 1998. Surface phosphatase activity of benthic algal communities in a stream with highly variable ambient phosphate concentrations. Verh. Int. Ver. Limnol. 26: 967–972.

Woodward, B., L.G. Young and A.K. Lun. 1983. Vomitoxin in diets of rainbow trout (*Salmo gairdneri*). Aquacult. 35: 93–101.

Woźny, M., K. Obremski, E. Jakimiuk, M. Gusiatin and P. Brzuzan. 2013. Zearalenone contamination in rainbow trout farms in north-eastern Poland. Aquacult. 416-417: 209–211.

Woźny, M., P. Brzuzan, M. Gusiatin, E. Jakimiuk, S. Dobosz and H. Kuźmiński. 2012. Influence of zearalenone on selected biochemical parameters in juvenile rainbow trout (*Oncorhynchus mykiss*). Polish J. Vet. Sci. 15(2): 221–225.

Wu, Q., L. Huang, Z. Liu, M. Yao, Y. Wang, M. Dai and Z. Yuan. 2011a. A comparison of hepatic in vitro metabolism of T-2 toxin in rats, pigs, chickens, and carp. Xenobiot. 41(10): 863–73.

Wu, Q., M. Li, X. Gao, J.P. Giesy, Y. Cui, L. Yang and Z. Kong. 2011b. Genotoxicity of crude extracts of cyanobacteria from Taihu Lake on carp (*Cyprinus carpio*). Ecotoxicol. 20: 1010–1017.

Wu, T.-S., J.-J. Yang, F.-Y. Yu and B.-H. Liu. 2013. Cardiotoxicity of mycotoxin citrinin and involvement of microRNA-138 in zebrafish embryos. Tox. Sci. 136(2): 402–412.

Yariv, J. and S. Hestrin. 1961. Toxicity of the extracellular phase of *Prymnesium parvum* cultures. J. Gen. Microbiol. 24(2): 165–175.

Yldirim, M., B.B. Manning, R.T. Lovell, J.M. Grizzle and G.E. Rottinghaus. 2000. Toxicity of moniliformin and fumonisin B1 fed singly and in combination in diets for young channel catfish (*Ictalurus punctatus*). J. World Aquacult. Soc. 31(4): 599–608.

Youssef, M.S., N.F. Abo-Dahab and R.M. Farghaly. 2003. Studies on mycological status of salted fish "Moloha" in Upper Egypt. Mycobiol. 31(3): 166–172.

Yurkowski, M., J.K. Bailey, R.E. Evans, J.A.L. Tabachek, G.B. Ayles and J.G. Eales. 1978. Acceptability of rapeseed proteins in diets of rainbow trout (*Salmo gairdneri*). J. Fish. Res. Board Can. 35: 951–962.

Zambrano, F. and E. Canelo. 1996. Effects of microcystin-LR on the partial reactions of the Na$^+$–K$^+$ pump of the gill of carp (*Cyprinus carpio* Linnaeus). Toxicon 34: 451–458.

Index